Function Modules in ABAP®

SAP PRESS

SAP PRESS is a joint initiative of SAP and Galileo Press. The know-how offered by SAP specialists combined with the expertise of the Galileo Press publishing house offers the reader expert books in the field. SAP PRESS features first-hand information and expert advice, and provides useful skills for professional decision-making.

SAP PRESS offers a variety of books on technical and business-related topics for the SAP user. For further information, please visit our website: *www.sap-press.com*.

Thorsten Schneider, Eric Westenberger, and Hermann Gahm
ABAP Development for SAP HANA
2014, approx. 530 pp., hardcover
ISBN 978-1-59229-859-4

Tanmaya Gupta
ABAP Data Dictionary
2011, 403 pp., hardcover
ISBN 978-1-59229-379-7

Rich Heilman and Thomas Jung
Next Generation ABAP Development (2nd Edition)
2011, 735 pp., hardcover
ISBN 978-1-59229-352-0

James Wood
Object-Oriented Programming with ABAP Objects
2009, 357 pp., hardcover
ISBN 978-1-59229-235-6

Tanmaya Gupta

Function Modules in ABAP®

A Quick Reference Guide

Galileo Press

Bonn • Boston

Galileo Press is named after the Italian physicist, mathematician, and philosopher Galileo Galilei (1564—1642). He is known as one of the founders of modern science and an advocate of our contemporary, heliocentric worldview. His words *Eppur si muove* (And yet it moves) have become legendary. The Galileo Press logo depicts Jupiter orbited by the four Galilean moons, which were discovered by Galileo in 1610.

Editor Laura Korslund
Acquisitions Editor Kelly Grace Weaver
Copyeditor Julie McNamee
Cover Design Graham Geary
Photo Credit iStockphoto.com/17691336/© 3dmentat
Layout Design Graham Geary, Kelly O'Callaghan
Production Graham Geary
Typesetting SatzPro, Krefeld (Germany)
Printed and bound in the United States of America, on paper from sustainable sources

ISBN 978-1-59229-850-1

1st edition 2014

Library of Congress Cataloging-in-Publication Data
Gupta, Tanmaya.
Function modules in ABAP : a quick reference guide / Tanmaya Gupta. -- 1st edition.
pages cm
ISBN 978-1-59229-850-1 -- ISBN 1-59229-850-8 -- ISBN 978-1-59229-852-5 (e-book) --
ISBN 978-1-59229-851-8 (e-book)
1. ABAP/4 (Computer program language) 2. Modular programming. 3. Functions--Computer programs. I. Title.
QA76.73.A12G873 2014
005.1'12--dc23
2014033883

Contents at a Glance

Dear Reader,

A dictionary is a masterful piece of work. Think about it—tens of thousands of entries are all in alphabetical order, the formatting is the same, and you're provided with reliable, consistent information. But have you ever taken the time to wonder who writes and compiles dictionaries? Who has that sort of time, dedication, and knowledge?

If you assume that dictionaries are put together by a very knowledgable team, you're probably right. However, the "dictionary" of function modules you now read was compiled by a single person, who has some of the best knowledge, patience, and dedication of any author I've worked with to date. Because this book was such a large—even Herculean—task we could find no one to consider this feat until we approached Tanmaya. With his dedication to researching thousands of function modules to pick the 700+ covered in this resource, you can stop relying on desktop post-it notes and repeated Internet searches. Finally, be confident when approaching your ABAP tasks.

As always, we appreciate your business and welcome your feedback—and want to know which other function modules are vital to your work as an ABAP professional, super user, or consultant. Your comments and suggestions are the most useful tools to help us improve our books for you, the reader. We encourage you to visit our website at *www.sap-press.com* and share your feedback about *Function Modules in ABAP: A Quick Reference Guide*.

Laura Korslund
Editor, SAP PRESS

Galileo Press
Boston, MA

laura.korslund@galileo-press.com
www.sap-press.com

Contents

Acknowledgments

Apart from the efforts of oneself, the success of any project depends largely on the encouragement and guidelines of many others. I take this opportunity to express my gratitude to the people who have been instrumental in the successful completion of this book. I express my gratitude to all the people who have given their overwhelming full support in making this compilation a magnificent experience.

To almighty God the father of all, I am thankful for the strength that keeps me standing and for the hope that kept me believing that this book would be possible and more interesting. I am forever indebted to my beloved parents and my family for their blessings; they have always inspired, encouraged, fully supported, and boosted my self-esteem in every trouble that came my way.

I take immense pleasure in thanking all my friends who have always encouraged me. Special gratitude goes to my best friend, Ritika Varshney, who has always motivated, supported, and scolded me to write and finish this book well and in time. I wish to express my deep sense of gratitude to all the readers of *ABAP Data Dictionary* who liked and appreciated my work and encouraged me to write another book for them. Valuable comments that I have received from readers always inspired and motivated me to complete this enormous and challenging book on time.

Words are inadequate in offering my thanks to Kelly Weaver, who believed in me and gave me an opportunity to work with Galileo Press again. I am grateful for her constant support and help, without which this book would not have materialized.

I would like to show my appreciation to the copyeditor, typesetter, designer, and the entire team at Galileo Press for their encouragement and cooperation in carrying out this book.

I would like to thank my tireless editor, Laura Korslund, who was constantly engaged with this work and provided amazing support and encouragement during the creation of this book. Her thoroughness and attention to detail make the difference between a book that works and one that bewilders.

Last but not least, I would like to thank everyone who is related to this book for being a source of inspiration and constant support and guidance, without which I would not have been able to bring this book to a successful completion.

Before we begin, I want to add few lines of my own:

When you want to do something, go for it,
Never sit back and think you can't do it,
Motivate, inspire, and encourage yourself to achieve it,
Your passion, desire, willingness, and dedication will accomplish it.

Tanmaya Gupta

Introduction

Function modules (or FMs, as we'll often refer to them throughout the book) are templates that can be used and re-used within ABAP code in order to perform specific tasks. Instead of writing code to perform a certain functionality every time that functionality is needed, programmers can simply call a function module template within the logic. Then the programmer needs only to fill out the template with the details that are specific to his particular program, and the functionality is executed.

For example, a common functionality required in an ABAP program is the need to convert a value from an internal format (how the backend SAP system stores the value) to an external format (how the user should view the value). If this functionality is required, there is no need for programmers to write a separate program every time they need it; rather, they can simply call a function module and supply the necessary details that control how the value should be converted.

The most efficient programmers are those who have a commanding knowledge of function modules, which can save the time of writing unnecessary custom code. However, with about 7,000 function modules provided and released by SAP, understanding all of these FMs isn't an easy task.

This book is your guide to the most commonly used FMs in direct programming activities. We've also included common FMs that are related to SAP functionality but are used in the background. For example, a user may receive an error, and the error root cause is the FM executed in the background. Additionally, SAP uses FMs in its standard delivered transactions. A user might want to redesign and customize those functionalities with some additional options and might need to use an FM.

During our research for the book, we created, analyzed, and worked on a list of thousands of released FMs, as well as some very common but not released FMs. Based on our research, this book names and describes the most commonly used FMs from that list of thousands, which we identified by searching where-used lists for each FM, as well as extensive help and Google searches. Our goal was to provide an invaluable reference for all those who work in ABAP development.

> **Note**
>
> Although you can use unreleased FMs in your programming, SAP offers no guarantee that it will support these FMs in the future or that these FMs are correct. You can go to a FM's ATTRIBUTE tab via Transaction SE37 to see if the FM is released.

Structure of the Book

Based on the functionality of the FMs we've included, we've structured this book to make it very easy to understand. You can search by FM name in the index or table of contents, as well as by key word for what the FM accomplishes or works with. This book contains 15 parts, each of which are created and based on different functionalities and different areas in the SAP system. You may find that some FMs are part of a larger sequence or process and must be used chronologically in order to complete a task. In these cases, the first FM used in the process will contain a list of or mention the other FMs that are used sequentially.

This book has been structured to be as user- and reader-friendly as possible. Each FM is first explained in general terms and then defined with an example to help you understand when, where, and why you can use that FM. Some of the FMs will also contain figures to show how you can use the FM and what results you can expect from the FM call.

We'll now discuss the different parts of this book.

Part I

Part I focuses on the FMs associated with systems. "System" here refers to SAP ERP or SAP NetWeaver systems. This part deals with more of the Basis-related operations such as calling a new transaction, calling SAP NetWeaver from legacy systems, such as Java or Visual Basic (VB), or calling from an FTP server or a remote location. This part also provides you with the FMs related to IP addresses, user authorizations, login and logoff of the user in the system, calling a transaction in another window, and so on.

Part II

This part focuses on the ABAP Data Dictionary. This is the central repository of ABAP, which describes and manages all of the data definitions used in the ABAP system and the database. ABAP Data Dictionary supports the definition of user-defined types, such as data elements, structures, and table types. These types are used in the ABAP processors and programs. It also defines the structure of database objects, such as tables, views, and indexes. These database objects are automatically created in the underlying database with the definition of the ABAP Data Dictionary when the objects are activated. It also provides editing tools such as search help and lock objects.

Part III

This part includes FMs that are related to conversions of values from one format to another, such as from character to float, from one SI unit to another, from spool to PDF, and so on. FMs related to date, time, unit of measure, round off, and so on are also covered in this part. This part also includes CONVERSION_EXIT FMs, which are used to convert the SAP internal format to the user output format and vice versa.

Part IV

This part includes FMs related to Batch Data Communication (BDC), ABAP List Viewer (ALV), report programming, and screen programming. Report and screen (dialog or module pool) programming are the main crux of the ABAP language. Reports are the executable programs that mainly deal with reporting; that is, fetching the data from the database, processing it, and displaying the results. Results are displayed by ALV. This part defines various FMs such as those for calling an ALV, creating a field catalog for ALV, calling the BDC, reading and displaying the screen values, and so on.

Part V

This part focuses on FMs related to form printing. SAP's first form technology was SAPscript. In release 4.6, SAP introduced Smart Forms, which eased the creation of form output by allowing modeling of logic and output using graphical tools. In release 6.40, SAP introduced SAP Interactive Forms by Adobe, where the graphical editor was much easier to use, provided many more functionalities, and rendered output exactly as designed in the editor (PDF). This part contains the FMs used to call the Smart Forms or Adobe Interactive Forms; reading, saving, and updating the SAPscript texts; and so on.

Part VI

This part contains various functions such as archives, number ranges, texts, and strings that you'll use while developing your programs. You'll use archives to store the old data and information (which isn't in use) in your secondary storage so that you free the current memory for usage while still being able to refer to the information whenever required.

Database records of the commercial objects such as purchase documents, sales documents, addresses, and so on are uniquely identified by a number or an ID. Number ranges assign the next available numbers to these commercial object database records and help provide unique identification. This part also contains FMs for text and string operations such as splitting the text based on some separator, concatenating strings, and so on.

Part VII

This part contains the FMs related to frontend services and Microsoft Office integration. Frontend services contain the FMs related to SAP GUI frontends such as GUI_DOWNLOAD, which downloads the FMs to your computer, and GUI_UPLOAD, which uploads the FMs.

Office integration deals with the FMs related to Microsoft Office features such as opening the file in Word and Excel.

Part VIII

An Intermediate Document (IDoc) is the data container used to transfer data between two systems. Application Link Enabling (ALE) and Electronic Data Interchange (EDI) are used for exchanging business data between different systems. Data is distributed through ALE when you want to communicate from one SAP system to one or more other (mostly SAP) systems. ALE transfers the data in IDoc format and uses the tRFC (transactional RFC) method for data transfer. EDI, on the other hand, is generally used if you want to transfer business application document with the external partner system like vendor or customer. This part focuses on the ALE, EDI, and IDoc FMs; that is, FMs used to send IDoc data from one system to another.

Part IX

This part contains the FMs related to SAP users and their authorizations to execute transactions. Users are maintained in Transaction SU01 (User Master), and data are saved in user-specific tables, whose names generally start with "USR" such as

USR01, USR02, USR03, and so on. This part contains the FMs that are used to set user details such as an address, print the address in Smart Forms, and so on. This part also contains the FM to check whether the user has relevant authorizations to execute any transaction, to execute any ABAP Data Dictionary table or view, or to access any file.

Part X

This part contains the FMs related to spools and jobs. A spool request is the document for which a print function has been selected and not yet been printed. The spool system uses the spool request to store the print data temporarily and access it. An output request is created for the spool request document that outputs the print data of the spool request to a particular output device (e.g., printer).

SAP background processing automates routine tasks and helps you optimize your organization's SAP computing resources. Background processing tells the SAP system to run programs for you. It also lets you move long-running or resource-intensive program runs to times when the system load is low. You can delegate the task of running reports or programs to the system. Background processing isn't tied up with the user interaction and dialog screens. This part covers FMs used for creating a spool request, finding the spool request, opening the job, submitting the job for background processing, closing the job, and so on.

Part XI

This part deals with communication with the user via messages or popups. Message texts (standard or custom) are generally used to provide information, warning, success, or error details to the user.

This part contains the FMs for creating a message, sending and collecting the messages for the message handler, and taking user confirmation in popup screens. Popup screens are used to show the time and date, to display help, and so on.

Part XII

The Transport Management System plays an important role in the development and customizing environment. All of the development is locked under the task of the transport in the development system, which is moved to the quality system and finally to the production system. The Transport Management System administers and controls the new development, and records the version of the changes made for a particular program. This part focuses on FMs used in the Transport Management System.

Part XIII

Application logs are used for logging application events. This process is similar to using a system log where system event information is recorded. The advantage of an application log is that it's a system-wide standardized process and is convenient to analyze. Several application logs can be written at the same time by an application. This part defines the FMs related to application logs.

Part XIV

Business Application Programming Interfaces (BAPIs) define the API methods for SAP business objects. These business object types and their BAPIs are described and stored in the Business Object Repository (BOR). BAPIs are always remote function call (RFC) enabled, so they can also be called from external application systems and other programs. BAPIs are also related to SAP components such as Sales and Distribution (SD), Materials Management (MM), Production Planning (PP), Quality Management (QM), Warehouse Management (WM), and so on. This part contains all of the commonly used BAPIs.

Part XV

This part contains miscellaneous FMs that are important and commonly used but can't be categorized into any of the preceding parts. Many component-specific FMs that aren't BAPIs are covered in this part.

Structure of Each Function Module Entry

Each FM can be defined by its parameters (import, export, changing, tables, and exceptions). This book describes all of these parameters and uses them in the FM call. All of these parameters, excluding exceptions, are associated with a data type. In each FM, we've identified each type of parameter with an icon and heading for easier identification.

[↓] Import

Import parameters are the parameters or values that you pass on to the FM to execute it correctly when you're in an ABAP program. Some of the import parameters are mandatory, and the rest are optional. Some import parameters are defined with the default values; that is, if you don't pass any value, then the default values are automatically passed.

[↑] Export

Export parameters contain the results returned from the FM execution.

[Δ] Changing

Changing parameters indicate those parameters whose value is changed in the FM. These parameters can be considered as import and export parameters; for example, an input value is passed in the FM. The value is used, and a new value after execution of the FM is returned in the same parameter.

⊞ Tables

Table parameters contain the internal table that you want to pass in the FM. Table parameters are also like changing parameters, in that some of the values can be changed after the execution of the FM.

[!] Exceptions

Exceptions are the errors or some exemptions that might have occurred during FM execution. They generally return the messages that define what problem has occurred. Exceptions should be handled in the program that is calling the FM, and the message explaining the exception should be displayed to the user.

Part I
System

In this part of the book, we refer to *system* as the SAP ERP or SAP NetWeaver system. This part deals with more of the Basis-related operations that can include calling a new transaction, calling SAP NetWeaver from legacy systems such as Java or Visual Basic, or from an FTP server or a remote location.

Function modules (FMs) are important when working with the system because it helps get the system information—such as the IP address, user authorizations, user login and logoff—by calling a transaction in another window without interrupting the execution of the current window. In this part, we've included the important FMs that are commonly used and interact in some way with SAP systems and Basis operations.

1 ABAP4_CALL_TRANSACTION

This FM is used to call a transaction code in the same task or a separate task (i.e., in the same session or window). If you want to call the transaction in a new task, you need to call the FM with `Starting New Task`. This is an RFC-enabled FM, so it can be called from non-SAP systems as well. Authorization rules of the calling transaction apply.

Example: You have an ALV output with the list of materials. Clicking on the material, you want to open the details of that material (i.e., opening Transaction MM03) in a separate window.

[↓] Import

TCODE

Specifies the transaction code you want to call. This is a mandatory parameter. Type: SY-TCODE.

SKIP_SCREEN

If this parameter is set as X, it skips the initial screen of the transaction. The default value is [Space]. Type: SY-FTYPE.

MODE_VAL

Defines how the transaction should be displayed. It can have the following values:

- A(default): Display the screen.
- E: Display only if an error occurs.
- N: Do not display, for background mode.
- Type: SY-FTYPE.

UPDATE_VAL

Defines how the update will occur. It can have the following values.

- A: Asynchronous update.
- S: Synchronous update.
- L: Local updates.
- Type: SY-FTYPE.

[↑] Export

SUBRC

This is a SY-SUBRC check. It returns the value as 0 if the FM is executed successfully. Type: SY-SUBRC.

Table

USING_TAB

Contains the BDC data for the transaction; that is, if some BDC recording is sent to call the transaction. Type: BDCDATA.

SPAGPA_TAB

Holds SPA\GPA, that is, SAP memory parameters to fill input fields. Type: RFC_SPAGPA.

MESS_TAB

Contains any error messages from the transaction. Type: BDCMSGCOLL.

[!] Exceptions

CALL_TRANSACTION_DENIED

No authorization to call the transaction.

TCODE_INVALID

Transaction code entered is invalid.

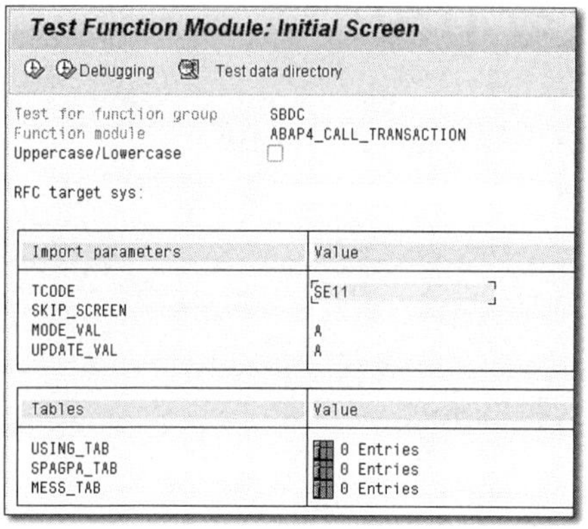

Calling Transaction SE11 from Function Module

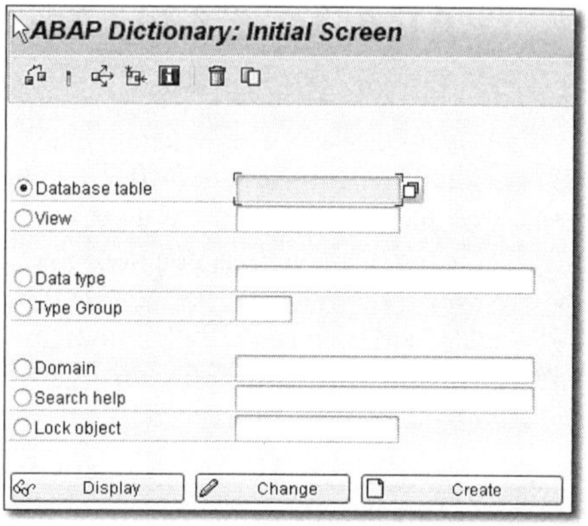

Output of the Function Module Call When Transaction SE11 Is Called

---

2 ABAP_DOCU_DOWNLOAD

This FM downloads the ABAP keyword documentation on the frontend (SAP Help) and generates it into a tree structure. The tree is produced in two variants; that is, JavaScript for Internet Explorer and JavaScript for Netscape navigator.

Example: You want to download all of the SAP Help files available in your local presentation server so that you can refer to them anytime. Pass the parameters in this FM and call this FM, and all of the files are downloaded to the folder specified.

[↓] Import

TARGET

Target directory. Type: Default: C:\ABAP_DOCU_HTML\.

LANGU

Language. Default: SY-LANGU. Type: SY-LANGU.

[!] Exceptions

NO_DOCU

No documentation found.

ERROR_DURING_DOWNLOAD

Error during download.

---

3 ARFC_GET _TID

When you're looking for the IP address of the system you're working on, execute this FM. The IP address is returned in hexadecimal format, so you'll need to convert it into the normal dotted notation to read it.

Example: When the user executes the FM, the following IP address of the user system is returned as shown in the figure. The HOST ID remains the same but the PROCESS ID, TIME STAMP, and TRANSACTION ID can differ.

[↑] Export

TID
Returns the IP address of the user computer that runs this FM. Transactional ID type ARFCTID is the combination of host ID, process ID, timestamp, and transaction ID. HOST ID is the IP address of the system, PROCESS ID is the ID of the LUW. TIME STAMP is the time at which the FM is executed, and TRANSACTION ID is the sequential four-digit hexadecimal number. Type: ARFCTID.

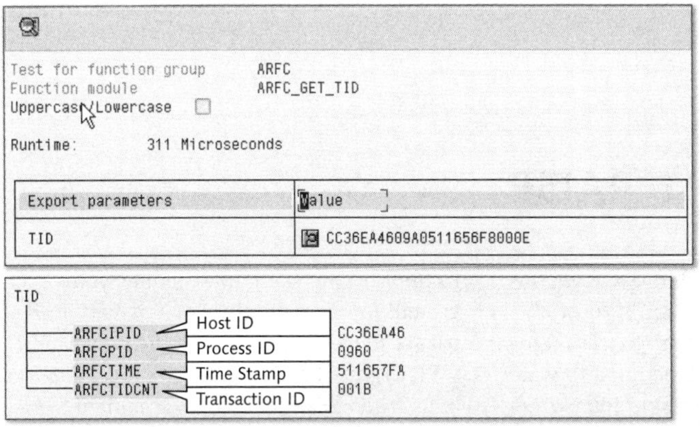

IP Address Returned When This FM Is Executed

---

4 AUTHORITY_CHECK_RFC

This FM is used to check the authorization of a user who is trying to access RFC FMs.

This FM should be executed before the RFC FM is executed. An authorization check is performed by authorization object S_RFC.

Example: A user wants to execute FM BP_EVENT_RAISE (FM 5), which is in a function group, say FGBP. When you see this occur in the system, you'll execute FM AUTHORITY_CHECK_RFC and give the details of FGBP in the parameters in the code. If the user is authorized, he can execute BP_EVENT_RAISE; if he isn't, he'll get the exception RFC_NO_AUTHORITY.

[↓] Import

USERID

Name of user who will execute the RFC function and whose authorization needs to be checked. Default/Type: SY-UNAME.

FUNCTIONGROUP

Name of the function group that is to be executed. Default: *. Type: AUTHB-RFC_NAME.

PROFILECHECK

Check the profile parameter if required. Type: SY-BATCH.

[!] Exception

USER_DONT_EXIST

User does not exist.

RFC_NO_AUTHORITY

No authorization to execute the RFC function.

---

5 BP_EVENT_RAISE

This FM triggers a background event in the background processing system. It tells the background processing system to start any background jobs that were scheduled to wait for the event (i.e., the background job is triggered when the event is triggered, and the event is triggered with this FM). You can define the event ID in Transaction SM62. Events have meaning only in the background processing system and can be used only to start background jobs. Triggering an event notifies the background processing system that a particular condition has been met. The background processing system reacts by starting any jobs that were waiting for that event. Events aren't saved. After an event occurs and triggers any jobs that were waiting for that event, the event is discarded.

Example: You want to schedule your background job every night. For this you need to trigger the SAP system event SAP_OPMODE_SWITCH in the background processing system. When this event has an argument `NIGHT`, your job will start when this FM is triggered.

[↓] Import

EVENTID

Name of the event to be triggered. This is a mandatory parameter. Type: BTCEVENTID.

EVENTPARM

Specifies an event parameter that can be passed to the background processing system. Default: [Space]. Type: BTCEVTPARM.

TARGET_INSTANCE

Defines the instance on which the event is processed. Normally, this parameter is left blank. Default: [Space]. Type: MSXXLIST-NAME.

[!] Exceptions

BAD_EVENTID

Invalid event ID.

EVENTID_DOES_NOT_EXIST

Unknown event.

EVENTID_MISSING

Event ID missing.

RAISE_FAILED

Error when triggering the event.

---

6 CAT_CHECK_RFC_DESTINATION

This FM checks the RFC destination and connections to make sure that they're working properly. RFC destinations are defined and can be checked from Transaction SM59.

Example: There are two SAP landscapes in your organization whose configuration boxes are C01 and C02. The RFC connection is maintained from C01 to C02 through RFC destination RFC_C02. Let's say that you're working on C01 and want to call the RFC FM defined in C02. First, you need to check whether the RFC RFC_C02 is working properly. To check this, use this FM. If the RFC destination is working fine, you'll get the RFC_SUBRC = 0; otherwise, you'll get RFC_SUBRC not equal to 0 with an error message in the MSGV1 and MSGV2 parameters.

[↓] Import

RFCDESTINATION

Name of the RFC destination that is to be tested. This is a mandatory parameter. Type: RSCAT-RFCDEST.

[↑] Export

MSGV1

First part of error message. Type: SY-MSGV1.

MSGV2

Second part of error message. Type: SY-MSGV2.

RFC_SUBRC

This parameter returns the code as 0 if successful, 1 if no destination is found, 2 for communication error, and 3 for system error. Type: SY-SUBRC.

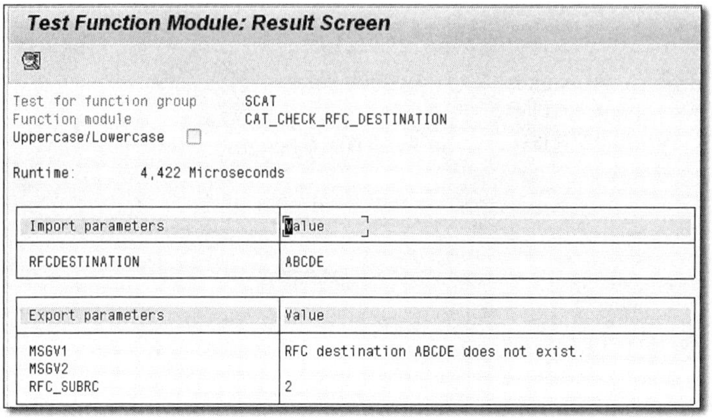

Output of the Function Module When RFC Destination Does Not Exist

---

7 DB_COMMIT

When a user makes any change in any table (i.e., creates, deletes, or modifies a record), the changes must get committed in the system. Users use this FM to commit changes to the database by native SQL statement. Rollback is also not possible after commit.

Example: Press the New entries button from the program in the table ZTABLE1. The created new entry will be stable only when the user commits those entries by entering the name of this FM. If you press Save, the data is tentative until you commit.

---

8 FTP_COMMAND

This FM is used to pass the FTP command to the FTP server and execute it. FTP is a standard network protocol used to transfer the files from one host to another using a TCP-based network.

To use FTP_COMMAND, the FTP connection must be established by FTP_CONNECT.

Example: When you're connected to the FTP server, execute the FTP command PUT through FM FTP_COMMAND. For an example, refer to FM FTP_CONNECT (FM 9).

[↓] Import

HANDLE

Unique ID by which the FTP connection is established. Type: I.

COMMAND

FTP command that is to be executed. This is a mandatory parameter. Type: C.

COMPRESS

Defines whether the FTP command needs to be compressed. Type: C.

RFC_DESTINATION

Defines the logical destination. Type: RFCDES-RFCDEST.

VERIFY

Verifies the FTP. Type: C.

[↑] Export

FILESIZE

Size of the file. Type: I.

FILEDATE

Date of the file. Type: C.

FILETIME

Time of the file. Type: C.

Table

DATA

Returns the data of the file.

[!] Exception

TCPIP_ERROR

Error while connecting to TCP/IP.

COMMAND_ERROR

Error in execution of command.

DATA_ERROR

Error in data.

---

9 FTP_CONNECT

You can use this FM to open and establish the connection to the FTP server. FTP is a standard network protocol used to transfer the files from one host to another using on a TCP-based network. For FTP to connect, a user ID and password must be sent, and it returns the unique ID (handle), which can be used for further FTP commands execution. The password should be encrypted, which can be done by using FM HTTP_SCRAMBLE (FM 11).

Refer to SAP standard report RSFTP002 for more about FTP connections and the sequence of the FTP FMs. Other FMs that are used in this process include the following:

- FTP_COMMAND (FM 8)
- HTTP_SCRAMBLE (FM 11)
- FTP_DISCONNECT (FM 10)

Example: Open the FTP connection through this FM. The password to be passed in this FM should be scrambled, which is received from FM HTTP_SCRAMBLE (FM 11).

[↓] Import

USER

User name to the FTP server. This is a mandatory parameter. Type: C.

PASSWORD

Password of the FTP server. This is a mandatory parameter. Type: C.

ACCOUNT

Account details for the connection. Type: C.

HOST

FTP server host name. This is a mandatory parameter. Type: C.

RFC_DESTINATION

RFC destination name that is configured in SAP. This is a mandatory parameter. Type: RFCDES-RFCDEST.

GATEWAY_USER

User name of the gateway. Type: C.

GATEWAY_PASSWORD

Password of the gateway. Type: C.

GATEWAY_HOST

Name of the gateway. Type: C.

[↑] Export

HANDLE

Unique ID by which the FTP connection is established. Type: I.

[!] Exceptions

NOT_CONNECTED

FTP destination not connected.

10 FTP_DISCONNECT

Use this FM to disconnect the connection with the FTP server. FTP is a standard network protocol used to transfer the files from one host to another using a TCP-based network. After the FTP commands are executed, use this FM to close the FTP connection. Then use RFC_CONNECTION_CLOSE to close the RFC connection after the FTP disconnects.

Example: After you've executed the FTP command, you'll close the FTP connection through this FM. After the FTP is disconnected, you should close the RFC connection by using FM RFC_CONNECTION_CLOSE (FM 14). Refer to FM FTP_CONNECT (FM 9).

[↓] Import

HANDLE

Unique ID by which the FTP connection is established. This is the mandatory parameter. Type: I.

11 HTTP_SCRAMBLE

This FM scrambles the source parameter. This is used to send the encrypted password that is required to be passed for the FTP connection FM FTP_CONNECT.

Example: You want to connect to the FTP server. The password you will enter for the connection should be scrambled. You need to use this FM to return the scrambled password. See also FM FTP_CONNECT (FM 9).

[↓] Import

SOURCE

String or password that is to be scrambled. This is a mandatory parameter. Type: C.

SOURCELEN

Length of the password source. This is a mandatory parameter. Type: I.

KEY

Key value. For FTP connections, it's 26101957, and if it's to be used before FM HTTP_GET, it should be 1303621. This is a mandatory parameter. Type: I.

[↑] Export

DESTINATION

Returned password string. This is passed to the FM FTP_CONNECT (FM 9). Type: C.

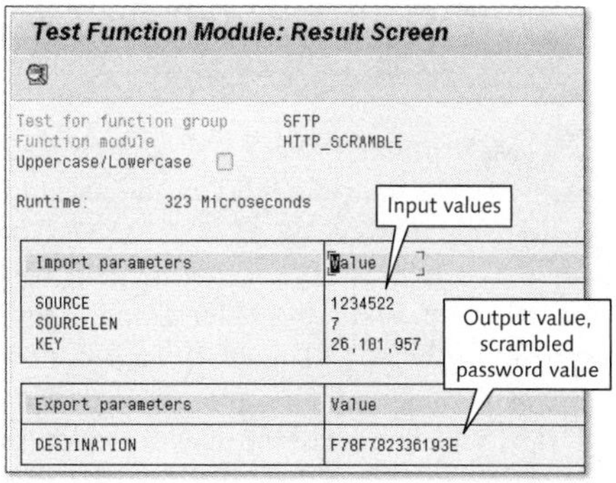

Output of Function Module

12 OWN_LOGICAL_SYSTEM_GET

You use this FM to specify the name of the logical system of the current logon system). The FM accesses table T000 to find specific details of the system and the logical system.

You use the logical system information when you work with ALE, IDoc, or RFC. The logical system is used to recognize the client in the server. You can use this FM to get the logical system name and then use it.

Example: You want to connect to the browser link *http://test123.xyz.com/SAPAction.do?USER-ID=UNAME&SSID=SYSID&CLIENT=MANDT*. When on the browser page, execute this FM and replace the variables SYSID and MANDT in this URL link with the value you get in the export parameter OWN_LOGICAL_SYSTEM after executing this FM.

[↑] Export

OWN_LOGICAL_SYSTEM

Returns the name of the logical system. Type: TBDLS-LOGSYS.

[!] Exception

OWN_LOGICAL_SYSTEM_NOT_DEFINED

Logical system is not defined in the table T000.

13 PRETTY_PRINTER

This FM rearranges the ABAP source code (i.e., improves the readability of the program source code passed in table OTEXT). The source code is structured according to its control statements and logical blocks and then is returned in the NTEXT table. If the first noncommented line in the source code is indented by *n* characters, all of the lines are indented by at least *n* characters. This FM won't change or check the syntax or semantics of the program.

Example: You want to perform the same operations of pretty printer (via the PRETTY PRINTER button), which is found in the ABAP Workbench or Editor. However, you're on a screen where the PRETTY PRINT button isn't available, so you use this FM instead. You can also use this if you don't want to use the Pretty Printer in the entire code, so you switch off the Pretty Printer functionality and use this FM in some part of your code.

[↓] Import

INCTOO

If the parameter is set as X, include programs declared in the main program are also processed and written back to the database. Mandatory. Type: Boolean.

[↑] Export

INDENTATION_MAYBE_WRONG

If there are syntax errors in the source text OTEXT, which results in the indentation error, this parameter gets the line number from which indentation may be wrong. Type: I.

Tables

NTEXT

Table of formatted source code that is returned after this FM is executed. Type: TY_RSWMAXSOURCE.

OTEXT

Table with source code to be processed. Type: TY_RSWMAXSOURCE.

[!] Exceptions

ENQUEUE_TABLE_FULL

Error in lock management.

INCLUDE_ENQUEUED

An include program is locked.

INCLUDE_READERROR

Error reading an include program.

INCLUDE_WRITEERROR

Include program is no longer used.

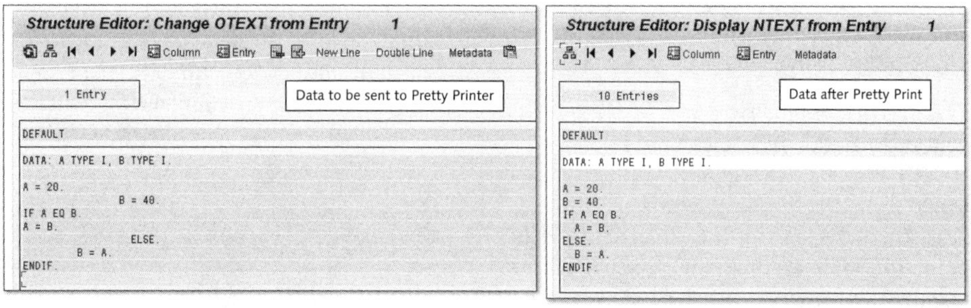

Data Before and After Pretty Print

---

14 RFC_CONNECTION_CLOSE

This FM is used to close the open RFC connection. The RFC context and the connection are retained until it's explicitly closed or the calling program is finished. This FM is used to close the RFC connection explicitly.

Example: You have opened an RFC connection between a non-SAP system and an SAP system. Once you have performed all the RFC operations, you need to close the RFC connection. Pass the parameters in this FM to close the RFC connection.

[↓] Import

DESTINATION

Name of the RFC destination you want to close. Type: C.

TASKNAME

Name of the task. Type: C.

[!] Exception

DESTINATION_NOT_OPEN

Specified destination is not open.

---

15 RFC_READ_TABLE

This FM is used to read the table defined in SAP from any other server. This FM is important when communication is required from a non-SAP system or any other SAP client to the SAP system.

Example: There are two SAP landscapes in your organization whose configuration boxes are C01 and C02. The RFC connection is maintained from C01 to C02 through RFC destination RFC_C02. You want to use the data present in table ZTABLE, which is defined in C02 from C01. You can use this FM to read the table ZTABLE defined in C02.

[↓] Import

QUERY_TABLE

Name of the table to be read. This is a mandatory parameter. Type: DD02L-TABNAME.

DELIMITER

Sign for indicating field limits in a DATA output table. The default value is [Space], which means data is continuous, and no delimiter is used. The delimiter differentiates the contents of two fields. Type: SONV-FLAG.

NO_DATA

If this parameter isn't a blank space, no data comes to the DATA table. Only the table definition (i.e., the structure of the table) is returned in the FIELDS table. Type: SONV-FLAG.

ROWSKIPS

Skips the data of the row. Default: 0. Type: SOID-ACCNT.

ROWCOUNT

Number of rows to be returned. Type: SOID-ACCNT.

Tables

OPTIONS

Passes the selection criteria to the DATA table to be passed for reading entries. These conditions are executed in the `Where` clause of the `SELECT` statement. Type: RFC_DB_OPT.

FIELDS

Inputs the list of fields to be read and outputs the details of the inputted fields, its position in the table, name, type, and description. Type: RFC_DB_FLD.

DATA

Returns the data that is read if the NO_DATA option is left blank. Type: TAB512.

[!] Exception

TABLE_NOT_AVAILABLE

QUERY_TABLE, that is, the table to be read is not active in the dictionary.

TABLE_WITHOUT_DATA

Data is not found in the QUERY_TABLE.

OPTION_NOT_VALID

Selection entries entered in the OPTIONs are incorrect.

FIELD_NOT_VALID

Field to be read not in table.

NOT_AUTHORIZED

User not authorized to access QUERY_TABLE.

DATA_BUFFER_EXCEEDED

Selected fields and the output data do not fit into structure DATA.

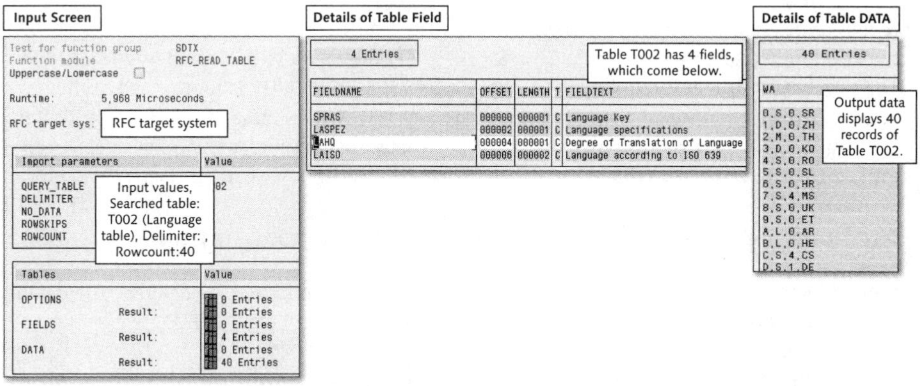

Execution of Functional Module to Call Language Table T002

16 RS_DELETE_PROGRAM

This FM is used to delete a program and its objects in the development system. When the program or include is deleted, all of its subobjects such as text elements, screens, documentations in all languages, menus, variants, and indexes in the ABAP Workbench are also deleted.

You need to be extremely careful before using this FM because it will delete the entire program.

Example: You've created a program ZREP_TEST, along with subobjects such as text, screen, and documentation. Now, the business doesn't want to use this program, and you need to delete it and its subobjects. Instead of deleting everything one by one, you can delete the program and its subobjects by filling the parameters of this FM and executing it.

[↓] Import

CORRNUMBER

This is the transport request number that is required for the deletion of the program if the development class isn't equal to T or $. The correction number is required in this parameter if the development class (Package) isn't equal to $XXX or TXXX to delete the program. Type: E071-TRKORR.

PROGRAM

Program name or include name to be deleted. Only the first 30 bytes are allowed for external use. This is a mandatory parameter. Type: SY-REPID.

SUPPRESS_CHECKS

Suppress authorization checks, if any. The default value is [Space]. Type: SY-INPUT.

SUPPRESS_COMMIT

This is for internal SAP use. The default value is [Space]. Type: SY-INPUT.

SUPPRESS_POPUP

Use this to suppress dialogs. If this parameter is set, no screen is executed, but the correct program name needs to be given in the PROGRAM parameter, and the correction number is required. Also, the development class (that is, package), doesn't equal $ or T. The default value is [Space]. Type: SY-INPUT.

MASS_DELETE_CALL

Deletion takes place en mass. The default value is [Space]. Type: SY-INPUT.

WITH_CUA

Set the menu to be copied by default. If SUPRESS_POPUP is set to X, all of the WITH_XXX parameters get initialized. If it's set to a blank space, no popup screen or message appears. Default: 0. Type: SY-INPUT.

WITH_DOCUMENTATION

Documentation will be deleted. The default value is X. Type: SY-INPUT.

WITH_DYNPRO

Screens will be deleted. The default value is X to delete the screen. Type: SY-INPUT.

WITH_INCLUDES

Standard includes will be deleted. The default value is [Space]. Type: SY-INPUT.

WITH_TEXTPOOL

Text pool will be deleted. The default value is X. Type: SY-INPUT.

WITH_VARIANTS

Variants will be deleted. The default value is X. Type: SY-INPUT.

TADIR_DEVCLASS

Development class or package in which the program is present. Type: TADIR-DEVCLASS.

SKIP_PROGRESS_IND

Suppress the Progress indicator. The default value is [Space]. Type: SY-INPUT.

FORCE_DELETE_USED_INCLUDES

Deletion of used includes, that is, includes that are also used in other programs are deleted forcefully. The default value is [Space]. Type: SY-INPUT.

SUPPRESS_DELETE_ENH

Suppress deletion of ENHO, that is, enhancement framework implementations. The default value is [Space]. Type: SY-INPUT.

[↑] Export

CORRNUMBER

This is the transport request number that is required for the deletion of the program if the development class isn't equal to T or $. Type: E071-TRKORR.

PROGRAM

Program Name that is deleted. Type: SY-REPID.

[!] Exception

ENQUEUE_LOCK

Deletion aborted since SAP lock exists.

OBJECT_NOT_FOUND

Object not found.

PERMISSION_FAILURE

Deletion unsuccessful.

REJECT_DELETION

User did not want to delete.

17 RSLG_WRITE_SYSLOG_ENTRY

This FM writes an entry in the SAP system log. The SAP system logs all system errors, warnings, user locks due to failed logon attempts from known users, and process messages in the system log. There are to two different types of logs created by the system log:

- Local logs
- Central logs

Use Transaction SM21 to access the system log output screen. With this transaction, you can read any of the messages that are contained in the system logs. A system log entry contains four relatively independent parts:

- **Detail information**
  This information includes time, user process, transaction, and so on, and is added automatically.
- **Message ID**
  This is used to identify the message text and documentation. The message text may use placeholders.
- **Variable part of the message**
  This part has 32 bytes available.
- **Predefined parameters**
  These are used by hidden system log entries.

This FM uses the message texts that are created in Transaction SE92 to create the system log message. You can analyze messages from Transaction SM21.

Example: You can use this FM if you want to add your message to the system log file. For instance, you can add the message "User & unlocked," so the user knows that he can use this FM for the same.

[↓] Import

DATA_AREA

32-character variable parts for the message.

DATA_STRUCTURE

Code letter for structure in DATA_AREA. The letter is defined in the C file *rslg.h* as well in the table TSL3T. If you're unsure what to use, insert a [Space] character.

DATA_WORD1

Variable word for &A in the message. Variable words inserted in the message at &A, &B, and so on. Words do not have a structure; do not fill DATA_STRUCTURE. CAUTION: This FM removes any [Space] characters from the words.

This FM does support the mixing of word and structure variable. Therefore, use DATA_AREA or DATA_WORD1 through DATA_WORD5.

DATA_WORD2

Variable word for &B in the message.

DATA_WORD3

Variable word for &C in the message.

DATA_WORD4

Variable word for &D in the message.

DATA_WORD5

Variable word for &E in the message.

SL_MESSAGE_AREA

First and second character of the message. ID. The message ID in the format 2 + 1. Messages are maintained using Transaction SE92. You can add variable characters and words to the message text. These are mandatory parameters. Type: TSL1D-AREA.

SL_MESSAGE_SUBID

Third character of the message ID. Manatory. Type: TSL1D-SUBID.

PRE_PARAM_A

Variable word for &a in the message.

PRE_PARAM_B

Variable word for &b in the message.

PRE_PARAM_C

Variable word for &c in the message.

PRE_PARAM_D

Variable word for &d in the message.

PRE_PARAM_E

Variable word for &e in the message.

PRE_PARAM_F

Variable word for &f in the message

PRE_PARAM_G

Variable word, for &g in the message.

PRE_PARAM_H

Variable word for &h in the message.

PRE_PARAM_I

Variable word for &i in the message.

PRE_PARAM_LONG

Long predefined parameters inserted in the message at &a&b&c&d or similar places. Individual predefined parameters can't be mixed with long predefined parameters. Therefore, use PRE_PARAM_LONG or PRE_PARAM_A through PRE_PARAM_I, or none at all.

[!] Exception

DATA_MISSING

DATA_AREA not filled in spite of DATA_STRUCTURE.

DATA_WORDS_PROBLEM

If both DATA_WORDi and DATA_AREA are added simultaneously, this exception is raised.

OTHER_PROBLEM

Problem within the FM.

PRE_PARAMS_PROBLEM

If both PRE_PARAM_i and PRE_PARAM_LONG are added simultaneously, this exception is raised.

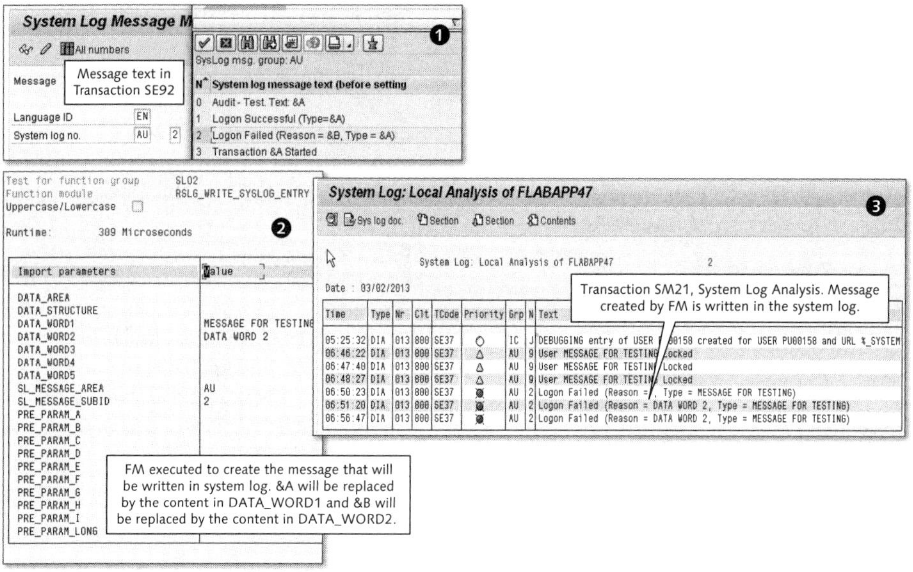

System Log in Detail

---

18 SXPG_CALL_SYSTEM

This FM is used to execute an external operating system command. This FM first checks the user authorization to run the command in the target system and then executes the command. This FM should be used carefully, as you have access to multiple systems.

Example: You want to copy a file from one directory to another in the SAP system. To achieve this, you can use this FM to execute the external UNIX command `MV` defined in Transaction SM69, which helps you copy the file from one directory to another.

[↓] Import

COMMANDNAME

Name of external command as specified in the maintenance function Transaction SM69. This is a mandatory parameter. Type: SXPGCOLIST-NAME.

ADDITIONAL_PARAMETERS

Arguments for the external command as specified by the definition in the SAP system and the calling program. The default value is [Space]. These arguments are checked for impermissible characters such as ";" in UNIX. Any problems are registered with the exception SECURITY_RISK. Type: SXPGCOLIST-PARAMETERS.

TRACE

If this parameter is set to X, the external program for the command call is traced Type: EXTCMDEXIM-TRACE.

[↑] Export

STATUS

Status of external command that is returned after the execution of the external command. Value O specifies that the external command was executed successfully while value E specifies that an error occurred while executing the external command. Type: EXTCMDEXEX-STATUS.

EXITCODE

Exits code of an external program. Type: EXTCMDEXEX-EXITCODE.

Tables

EXEC_PROTOCOL

Output of the external command. This contains the output data that is received from the target host system. Type: BTCXPM.

[!] Exception

NO_PERMISSION

No authorization to execute the external command. Authorization object S_LOG_COM fails because the user isn't authorized to execute the command on the target system.

COMMAND_NOT_FOUND

Command name isn't defined in the maintenance function Transaction SM69.

PARAMETERS_TOO_LONG

Parameter longer than 128 characters, that is, the parameter defined in the ADDITIONAL_PARAMETERS exceeds the maximum length of 128 characters.

SECURITY_RISK

Execution rejected for security reasons. Some impermissible characters may be passed, or it may have been rejected for some other security reason by the target system.

WRONG_CHECK_CALL_INTERFACE

Check module does not exist. The command defines that an extra check FM should be executed that does not exist, or the interface defined to execute the FM doesn't match the SAP standard FM SXPG_DUMMY_COMMAND_CHECK (FM 22).

PROGRAM_START_ERROR

Error while starting the external command. The system field SY-MSGV1 contains additional information about the problem.

PROGRAM_TERMINATION_ERROR

Return code of the external program contains errors. The system field SY-MSGV1 contains additional information of the problem.

X_ERROR

Reserved for internal use only.

PARAMETER_EXPECTED

No additional arguments specified. If the command definition contains a placeholder and isn't specified as an argument, then this exception is raised.

TOO_MANY_PARAMETERS

Impermissible parameter. More parameters are passed than actually required.

ILLEGAL_COMMAND

Illegal command. The external command definition was modified illegally, that is, not modified with maintenance function Transaction SM69. The modified command is registered in the system log in the substituted form.

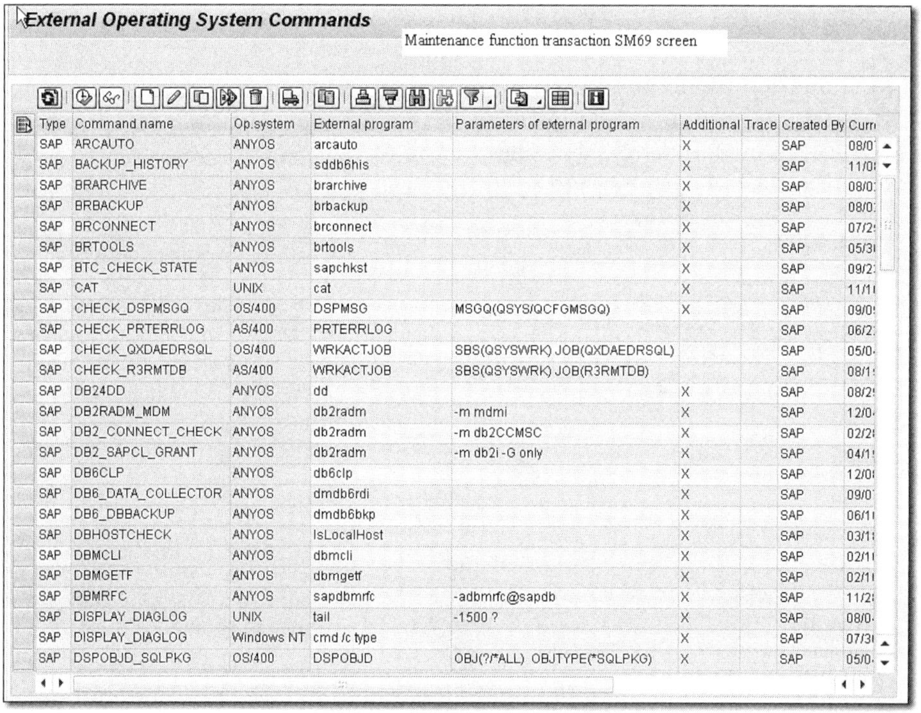

Type	Command name	Op.system	External program	Parameters of external program	Additional	Trace	Created By	Curr
SAP	ARCAUTO	ANYOS	arcauto		X		SAP	08/0
SAP	BACKUP_HISTORY	ANYOS	sddb6his		X		SAP	11/0
SAP	BRARCHIVE	ANYOS	brarchive		X		SAP	08/0
SAP	BRBACKUP	ANYOS	brbackup		X		SAP	08/0
SAP	BRCONNECT	ANYOS	brconnect		X		SAP	07/2
SAP	BRTOOLS	ANYOS	brtools		X		SAP	05/3
SAP	BTC_CHECK_STATE	ANYOS	sapchkst		X		SAP	09/2
SAP	CAT	UNIX	cat		X		SAP	11/1
SAP	CHECK_DSPMSGQ	OS/400	DSPMSG	MSGQ(QSYS/QCFGMSGQ)	X		SAP	09/0
SAP	CHECK_PRTERRLOG	AS/400	PRTERRLOG				SAP	06/2
SAP	CHECK_QXDAEDRSQL	OS/400	WRKACTJOB	SBS(QSYSWRK) JOB(QXDAEDRSQL)			SAP	05/0
SAP	CHECK_R3RMTDB	AS/400	WRKACTJOB	SBS(QSYSWRK) JOB(R3RMTDB)			SAP	08/1
SAP	DB24DD	ANYOS	dd		X		SAP	08/2
SAP	DB2RADM_MDM	ANYOS	db2radm	-m mdmi	X		SAP	12/0
SAP	DB2_CONNECT_CHECK	ANYOS	db2radm	-m db2CCMSC	X		SAP	02/2
SAP	DB2_SAPCL_GRANT	ANYOS	db2radm	-m db2i -G only	X		SAP	04/1
SAP	DB6CLP	ANYOS	db6clp		X		SAP	12/0
SAP	DB6_DATA_COLLECTOR	ANYOS	dmdb6rdi		X		SAP	09/0
SAP	DB6_DBBACKUP	ANYOS	dmdb6bkp		X		SAP	06/1
SAP	DBHOSTCHECK	ANYOS	IsLocalHost		X		SAP	03/1
SAP	DBMCLI	ANYOS	dbmcli		X		SAP	02/1
SAP	DBMGETF	ANYOS	dbmgetf		X		SAP	02/1
SAP	DBMRFC	ANYOS	sapdbmrfc	-adbmrfc@sapdb	X		SAP	11/2
SAP	DISPLAY_DIAGLOG	UNIX	tail	-1500 ?	X		SAP	08/0
SAP	DISPLAY_DIAGLOG	Windows NT	cmd /c type		X		SAP	07/3
SAP	DSPOBJD_SQLPKG	OS/400	DSPOBJD	OBJ(?/*ALL) OBJTYPE(*SQLPKG)	X		SAP	05/0

Transaction SM69 Screen

19 SXPG_COMMAND_CHECK

This FM checks the user's authorization to execute an external command on the host system. Users can execute the command only if they pass the authorization check. This is a RFC-enabled FM. This FM checks if the user is authorized to execute the command with the arguments specified in ADDITIONAL_PARAMTERS on the target host system as specified in the TARGETSYSTEM and OPERATINGSYSTEM.

Example: You want to copy a file from one directory to another in the SAP system. To achieve this, you can use the external UNIX command `MV` defined in Transaction SM69. Before calling the FM to execute the command, first you should check if the user is authorized to execute the command. You can use this FM to check the authorization. If the user isn't authorized, then this FM raises an exception.

[↓] Import

COMMANDNAME

Name of external command as specified in the maintenance function Transaction SM69. This is a mandatory parameter. Type: SXPGCOLIST-NAME.

OPERATINGSYSTEM

Operating system or host system in which the command is to be executed. The default value is SY-OPSYS, that is, the operating system of the application server. The operating system of the command is defined in Transaction SM69. Type: SXPGCOLIST-OPSYSTEM.

TARGETSYSTEM

Target system where the command is to be executed. It defines the RFC destination of the host system. Type: RFCDISPLAY-RFCHOST.

ADDITIONAL_PARAMETERS

Arguments for the external command as specified by the definition in the SAP system and the calling program. The default value is [Space]. These arguments are checked for impermissible characters such as semicolons (;) in UNIX. Any problems are registered with the exception SECURITY_RISK. Type: SXPGCOLIST-PARAMETERS.

DESTINATION

Defines the logical RFC destination of the target system. Type: RFCDES-RFCDEST.

LONG_PARAMS

Specifies the long parameters arguments for an external command. The long parameter maximum is up to character 1,024. Type: CHAR1024.

EXT_USER

User name of the person whose authorization is to be checked. Type: SY-UNAME.

[↑] Export

PROGRAMNAME

Full name of the command to be executed. Type: SXPGCOLIST-OPCOMMAND.

DEFINED_PARAMETERS

Returns the predefined arguments back from the command definition in the SAP system, such as some variables that need to be returned from the target system. Type: SXPGCOLIST-PARAMETERS.

ALL_PARAMETERS

Returns all of the arguments for the command. This parameter includes all of the parameters defined in ADDITIONAL_PARAMETERS and DEFINED_PARAMETERS. This parameter contains the parameters that are sent to the target system and are returned back from it. Type: SXPCOLIST-PARAMETERS.

[!] Exception

NO_PERMISSION

No authorization to execute the external command. Authorization object S_LOG_COM fails because the user isn't authorized to execute the command on the target system.

COMMAND_NOT_FOUND

Command name isn't defined in maintenance function Transaction SM69.

PARAMETERS_TOO_LONG

Parameter longer than 128 characters; that is, the parameter defined in the ADDITIONAL_PARAMETERS exceeds the maximum length of 128 characters.

SECURITY_RISK

Execution rejected for security reasons. Some impermissible characters may have been passed, or it was rejected for some other security reason by the target system.

WRONG_CHECK_CALL_INTERFACE

Check module does not exist. The command defines that an extra check FM should be executed that does not exist, or the interface defined to execute the FM does not match the SAP standard FM SXPG_DUMMY_COMMAND_CHECK (FM 22).

X_ERROR

Reserved for internal SAP use.

PARAMETER_EXPECTED

No additional arguments specified. If the command definition contains a placeholder (?) and isn't specified as an argument, then this exception is raised.

TOO_MANY_PARAMETERS

Impermissible parameter. More parameters are passed than actually required.

ILLEGAL_COMMAND

Illegal command. The external command definition was modified illegally, that is, not modified with maintenance function Transaction SM69. The modified command is registered in the system log in the substituted form.

COMMUNICATION_FAILURE

Error occurred while communicating between the systems.

SYSTEM_FAILURE

System failure occurred.

20 SXPG_COMMAND_EXECUTE

This RFC-enabled FM checks the user's authorization to execute the external command on the host system and then executes the command if the user is authorized. This FM checks whether the user is authorized to execute the command with the arguments specified in ADDITIONAL_PARAMETERS on the target host system as specified in the TARGETSYSTEM and OPERATINGSYSTEM. IF the SAP profile parameter is present in the command stored in the database, then the value of the parameter is substituted into the command. If an SAP application server is active on the target system (TARGETSYSTEM), then the profile parameter values are read from the profile in effect on that system. No parameter substitution is made in ADDITIONAL_PARAMETERS.

Example: You want to copy a file from one directory to another in the SAP system. To achieve this, you can use the external UNIX command `MV` defined in Transaction SM69. You can fill in the desired parameters and execute this FM, which helps you copy the file from one directory to another.

[↓] Import

COMMANDNAME

Name of external command as specified in the maintenance function Transaction SM69. This is a mandatory parameter. Type: SXPGCOLIST-NAME.

ADDITIONAL_PARAMETERS

Arguments for the external command as specified by the definition in the SAP system and the calling program. The default value is [Space]. These arguments are checked for impermissible characters such as semicolons (;) in UNIX. Any problems are registered with the exception SECURITY_RISK. Type: SXPGCOLIST-PARAMETERS.

OPERATINGSYSTEM

Operating system or host system in which the command is to be executed. The default value is SY-OPSYS, that is, the operating system of the application server. The operating system of the command is defined in Transaction SM69. Type: SXPGCOLIST-OPSYSTEM.

TARGETSYSTEM

Target system where the command is to be executed. It defines the RFC destination of the host system. The default value is SY-HOST. Type: RFCDISPLAY-RFCHOST.

DESTINATION

Defines the logical RFC destination of the target system. Type: RFCDES-RFCDEST.

STDOUT

If this parameter is set, the output is logged in the table EXEC_PROTOCOL. Logging is only possible if parameter TERMINATIONWAIT is also not [Space]. The default value is X. Type: EXTCMDEXIM-STDOUT.

STDERR

If this parameter is set to X, the error is logged in the table EXEC_PROTOCOL. Logging is only possible if parameter TERMINATIONWAIT is also not [Space]. The default value is X. Type: EXTCMDEXIM-STDERR.

TERMINATIONWAIT

Wait for termination of external command. This parameter defines the logging of output and errors in the table EXEC_PROTOCOL if this parameter isn't [Space] and is synchronous; that is, the FM waits for the command to complete. However, if this is [Space], then the command starts asynchronously, and the FM won't wait for the completion of the command. Neither the output nor errors are logged. The default value is X. Type: EXTCMDEXIM-TERMWAIT.

TRACE

Perform trace. This is done only for testing. If this parameter isn't [Space], tracing is active; otherwise, it is not. Type: EXTCMDEXIM-TRACE.

DIALOG

If this parameter is [Space], then this FM is called in background processing. Type: BTCH0000-CHAR1.

[↑] Export

STATUS

Status of external command that is returned after the execution of the external command. If the include RSXPGDEF is called in your program, there are the following possible values:

- O: The external command was started and successfully run.
- F: Error; the external command was not executed.
- E: Error, the external command was terminated with an error.
- S: The external command was terminated due to an operating system message.
- C: The external command was terminated in an unknown way.

If you do not use RSXPGDEF, only the following values can be delivered back:

- O: The external command was started and successfully run.
- E: Error, the external command wasn't executed.

Type: EXTCMDEXEX-STATUS.

EXITCODE

Exit code of an external program. Type: EXTCMDEXEX-EXITCODE.

Tables

EXEC_PROTOCOL

If parameter TERMINATIONWAIT is active, this table contains the STDOUT and STDERR data for the external command and the target host system. Type: BTCXPM.

[!] Exceptions

NO_PERMISSION

No authorization to execute the external command. Authorization object S_LOG_COM fails because the user isn't authorized to execute the command on the target system.

COMMAND_NOT_FOUND

Command name is not defined in the maintenance function Transaction SM69.

PARAMETERS_TOO_LONG

Parameter longer than 128 characters; that is, the parameter defined in the ADDITIONAL_

PARAMETERS exceeds the maximum length of 128 characters.

SECURITY_RISK

Execution rejected for security reasons. Some impermissible characters may have been passed, or it was rejected for some other security reason by the target system.

WRONG_CHECK_CALL_INTERFACE

Check module does not exist. Command defines that an extra check FM should be executed that does not exist, or the interface defined to execute the FM does not match the SAP standard FM SXPG_DUMMY_COMMAND_CHECK (FM 22).

PROGRAM_START_ERROR

Error while starting the external command. The system field SY-MSGV1 contains additional information for the problem.

PROGRAM_TERMINATION_ERROR

Return code of the external program contains errors. The system field SY-MSGV1 contains additional information for the problem.

X_ERROR

Reserved.

PARAMETER_EXPECTED

No additional arguments specified. If the command definition contains a placeholder (?) and is not specified as an argument, then this exception is raised.

TOO_MANY_PARAMETERS

Impermissible parameter. More parameters are passed than actually required.

ILLEGAL_COMMAND

Illegal command. The external command definition was modified illegally, that is, not modified with maintenance function Transaction SM69. The modified command is registered in the system log in the substituted form.

WRONG_ASYNCHRONOUS_PARAMETERS

Reserved by SAP.

JOBCOUNT_GENERATION_ERROR

Reserved by SAP.

CANT_ENQ_TBTCO_ENTRY

Reserved by SAP.

21 SXPG_COMMAND_LIST_GET

This FM reads a list of the external commands that have been defined in the SAP system and imports them into an internal table. You can then loop over the table to select a command or offer the list to your user for selection. You can then pass the selection on to SXPG_COMMAND_EXECUTE for an authorization check and execution of the command.

Example: You want to copy a file from one directory to another in the SAP system. However, you don't know the command that you need to use to perform the operation. You need to access the list of all of the external commands so that you can identify the command you need. This FM is used to get the list of all of the external commands defined in the system. You can then use FM SXPG_COMMAND_EXECUTE (FM 20) to execute the chosen command.

[↓] Import

COMMANDNAME

Name of external command as specified in maintenance function Transaction SM69. The default value is the wildcard character (*), You can also represents the wildcard (*) anywhere in the value to represent any number of characters. Type: SXPGCOLIST-NAME.

OPERATINGSYSTEM

Operating system or host system in which the command is to be executed. The operating system of the command is defined in Transaction SM69. The default value is *; that is all of the operating systems can be searched and defined in Transaction SM69. You can also represent * anywhere in the value for any number of characters. An operating system value of ANYOS means it's a SAP command that can be executed on any operating system. Type: SXPGCOLIST-OPSYSTEM.

TARGETSYSTEM

Target system where the command is to be executed. It defines the RFC destination of the host system. This parameter is used to check the user's authorization to run the command in the target system. If the user is authorized, table field COMMAND_LIST-PERMISSION is set to X. Default: *. Type: RFCDISPLAY-RFCHOST.

Tables

COMMAND_LIST

This table contains a list of the selected commands in the format shown in Transaction SM49 or Transaction SM69. The user can select a command from the list to run FM SXPG_COMMAND_EXECUTE (FM 20).

The table is empty if no command is found that meets the selection criteria. No exception is raised in this case.

Notes on fields:

- COMMAND_LIST-PERMISSION shows whether the calling user is authorized to run a command on the system that you name in TARGETSYSTEM (optional). X means that the user is authorized; [Space] means that the user does not have authorization to run commands on TARGETSYSTEM.
- COMMAND_LIST-TRACEON and COMMAND_LIST-CHECKALL are set to X if the user has the authorization to maintain external commands.
- COMMAND_LIST-SAPCOMMAND is set to X if the command in that row is an SAP standard command. These commands may not be modified in customer systems.

Type: SXPGCOLIST.

22 SXPG_DUMMY_COMMAND_CHECK

When you define an external command, you can specify that an additional FM be called to check a user's authorization to run a command. These FMs must have the interface defined in the FM SXPG_DUMMY_COMMAND_CHECK. To ensure that such a user exit FM can't be misused, the system checks that the interface of the authorization-checking FM matches the interface of SXPG_DUMMY_COMMAND_CHECK. It's therefore essential that SXPG_DUMMY_COMMAND_CHECK never be changed.

Example: You want to copy a file from one location to another, and you don't find the appropriate external command in Transaction SM69. You have the option to create your own FM and define the logic per your requirement. The new FM must have the interface defined in this FM.

[↓] Import

PROGRAMNAME

The fully substituted path name of the external command that is to be run. Mandatory. Type: SXPGCOLIST-OPCOMMAND.

PARAMETERS

Arguments for the external command as specified by the definition in the SAP system and by the calling program or user.

These arguments are checked for impermissible characters, such as the semicolon (;) under UNIX. Problems are registered with the SECURITY_RISK exception. Mandatory. Type: SXPGCOLIST-PARAMETERS.

[!] Exception

NO_PERMISSION

No authorization to execute the external command. Authorization object S_LOG_COM fails because user is not authorized to execute the command on the target system. No permission triggers the SECURITY_RISK exception in the calling FM.

23 TH_DELETE_USER

The system administrator can use this FM to log off the user. It's recommended to use this FM only when necessary because the user's unsaved work can be lost.

Example: Your friend forgot to log off the SAP session and has left work for the day. You want to log him off from the system. You can fill in the details in this FM to log him off the system.

[↓] Import

USER

Name of the user to be logged off. This is a mandatory parameter. Type: SY-UNAME.

CLIENT

Client name for which the session is to be deleted. This is a mandatory parameter. Type: SY-MANDT.

ONLY_POOLED_USER

Deletes the pool session of the user. The default value is [Space]. Type: SM04DIC-FLAG.

TID

Terminal ID. The default value is –1. Type: SY-INDEX.

[!] Exception

AUTHORITY_ERROR

No authorization to execute the FM. Authorization object S_ADMI_FCD is checked for the authorization.

---

24 TH_ENVIRONMENT

This FM gets the value in the SAP environment variables. Environment variables define values used by the Oracle database and BR* tools. These environment variables contain the important details of the system, such as operating system details, number of processors, DBMS type, name of the computer, location of the program files, and so on.

Example: You want to know system details such as the path of the program files. After you execute this FM, the system returns the path in the program files environment variable.

Tables

ENVIRONMENT

This table returns all of the SAP environment variables. Type: THENV.

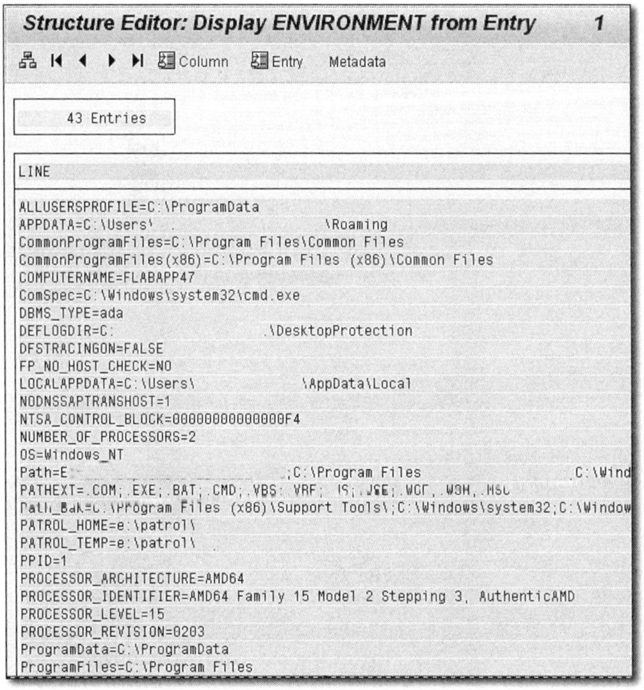

Environment Variables Returned from This FM

25 TH_GET_OWN_WP_NO

This FM gets the work process details over which the particular session is running. You can kill or delete the work process if you pass the WP_PID details to another FM TH_STOP_WP (FM 27).

Example: You executed one transition, but it became hung up, and your system became very slow. To stop that transaction, you need the work process details. You can get these details by calling this FM. After you get the work process details, you can use those details in FM TH_STOP_WP (FM 27) to stop that work process and terminate the transaction as well.

[↑] Export

SUBRC

Return code of the FM, that is, whether FM execution is a success or failure. Type: SY-SUBRC.

WP_NO

Work process number. Type: WPINFO-WP_NO.

WP_PID

Work process ID number. Type: WPINFO-WP_PID.

WP_INDEX

Index of the work process. Type: WPINFO-WP_INDEX.

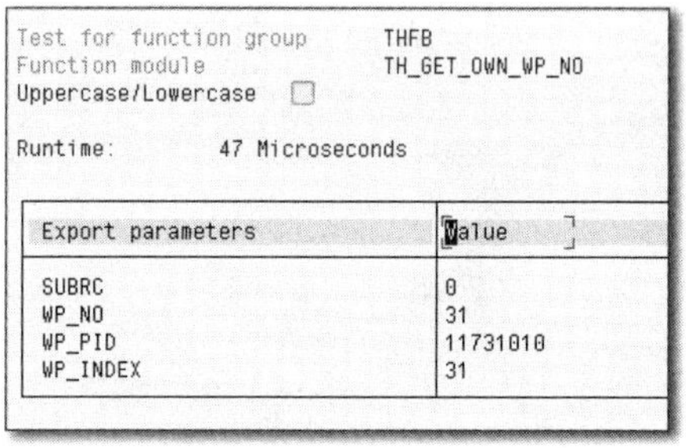

Work Process Details

---

26 TH_POPUP

This FM sends a popup to the user. The user receives the message only when he logs on to the system. This FM is useful to the system administrator who wants to send a message to any particular user. This customized message appears in the popup screen of the user after he logs on to the system. Transaction SM02 can be used to send the message to all users. This FM can also be used to send any type of general information message to the user. This FM should be used judiciously and should not spam other users.

Example: You need to access a specific transaction, say Transaction MM02, but find that a different user has locked that transaction. You can therefore call this FM, fill in the parameter details, and send the message "Please exit Transaction MM02."

[↓] Import

CLIENT

Client name for which the popup needs to be send. This is a mandatory parameter. Type: SY-MANDT.

USER

User name to whom the popup needs to be sent. This is a mandatory parameter. Type: SY-UNAME.

MESSAGE

This indicates which message is to be sent. This is a mandatory parameter. Type: SM04DIC-POPUPMSG.

MESSAGE_LEN

Length of the message that will be displayed. The default value is zero. Zero displays the entire message; however, if any length is set, then the message up to that length is displayed. Type: SY-INDEX.

CUT_BLANKS

Language key. The default value is a [Space]. Type: SY-LANGU.

[!] Exception

USER_NOT_FOUND

The user entered in the USER import parameter is not found.

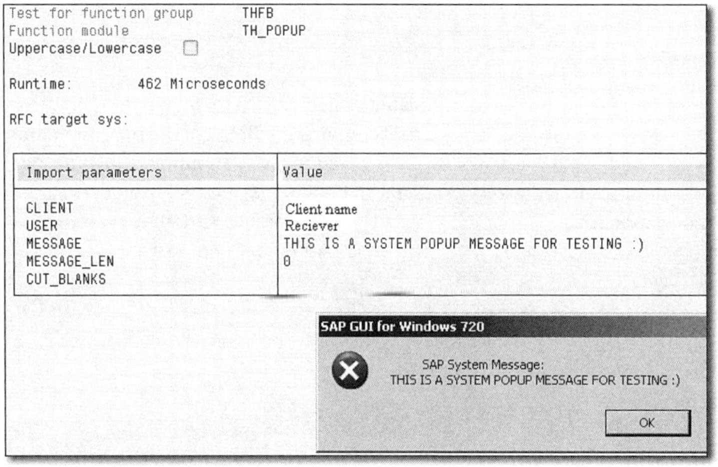

Popup Message after Execution of This FM

27 TH_REMOTE_TRANSACTION

This FM is used to execute the transaction remotely (i.e., from a remote server). This transaction can be executed with BDC by filling in the BDCTAB table parameter. Value for the DEST parameter can be received from the table parameter LIST-NAME of FM TH_SERVER_LIST (FM 29).

Example: Working in Java, you can change the description of the material M123 by using this FM remotely from the Java program, and you can give TCODE parameter as MM02 to change the material description.

[↓] Import

TCODE

Transaction code you want to call. This is a mandatory parameter. Type: SM04DIC-TCODE1.

DEST

Application server destination name. This is a mandatory parameter. Type: MSXXLIST-NAME.

USE_BDC

Determines whether you want to use BDC to call the transaction or not. The default value is 0. If this value isn't 0, you should pass BDC data to table BDCTAB. Type: SM04DIC-VBGETINFO.

[↑] Export

COMM_MESSAGE

Indicates whether a communication failure occurs while calling the transaction.

SYST_MESSAGE

Indicates whether a system failure occurs while calling the transaction.

Table

BDCTAB

Table structure to pass BDC data. This structure contains the detail of the program name to be recorded, screen number, beginning screen, field name, and field value. If parameter USE_BDC is checked, you need to pass the BDCTAB data. Type: BDCDATA.

28 TH_SERVER_LIST

This FM provides a list of registered application servers to the caller. The list can be limited by specifying work process types and system services.

Example: You want the list of application server and its details, including host name, ID, service number, and so on, which you want to pass as import parameters to some other FM. You can use this FM to get those details.

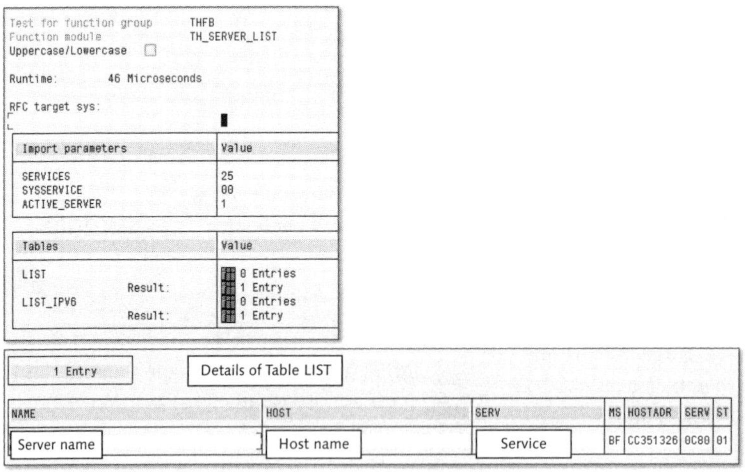

Output Screen after Executing the Function Module

[↓] Import

SERVICES

BIT vector of the services. Default: 255. Type: MSXXLIST-MSGTYPES.

SYSSERVICE

Services of the application server. Default: 0. Type: MSSYSSERVICE.

ACTIVE_SERVER

Display active servers only. Type: I.

Tables

LIST

Internal table for storing the List. This structure includes the application server name, host name, host IP address, services details, services port number, server status, and so on. Type: MSXXLIST.

LIST_IPV6

Internal table for storing the list with IPv6 addresses. This structure includes the details of table LIST along with the host IPv4 address, host IPv6 address, and system services registered on the message server. Type: MSXXLIST_V6.

[!] Exception

NO_SERVER_LIST

No server list received.

---

29 TH_STOP_WP

This FM stops or deletes the work process session. You can receive the WP_PID details from the FM TH_GET_OWN_WP_NO (FM 25).

Example: You executed one transition, but it became hung up and your system became very slow. To stop and terminate the transaction, call this FM.

[↓] Import

WP_PID

Process ID number of the work process. This is a mandatory parameter. Type: WPINFO-WP_PID.

[!] Exception

NO_AUTHORITY

No authorization to execute the FM. Authorization object S_ADMI_FCD is checked for the authorization.

---

30 TH_USER_INFO

This FM provides a user's (the person who executes the FM) details. Users can use this FM to see the technical specifications required for information purposes, to execute a different FM, or to use the details in the program itself.

Example: User1 is logged on to the system and executes this FM, so he sees his details such as the host address, terminal name, active session, maximum allowed sessions, current session, task status, terminal ID of user, IP address of the frontend computer, and return code.

[↓] Import

CLIENT

SAP system client number. Type: SY-MANDT.

USER

SAP system user name whose details are required. Type: SY-UNAME.

CHECK_GUI

Number of loop iterations to check. Type: SY-INDEX.

[↑] Export

HOSTADDR

IP address of the frontend computer. Type: MSXXLIST-HOSTADR.

TERMINAL

Terminal of the user.

ACT_SESSIONS

Total number of active sessions. Type: SM04DIC-COUNTER.

MAX_SESSIONS

Maximum number of sessions allowed. Type: SM04DIC-COUNTER.

MY_SESSION

External session in which the user finds the user details. Type: SM04DIC-COUNTER.

MY_INTERNAL_SESSION

Total number of active internal sessions. Type: SM04DIC-COUNTER.

TASK_STATE

Task state number. Type: SM04DIC-COUNTER.

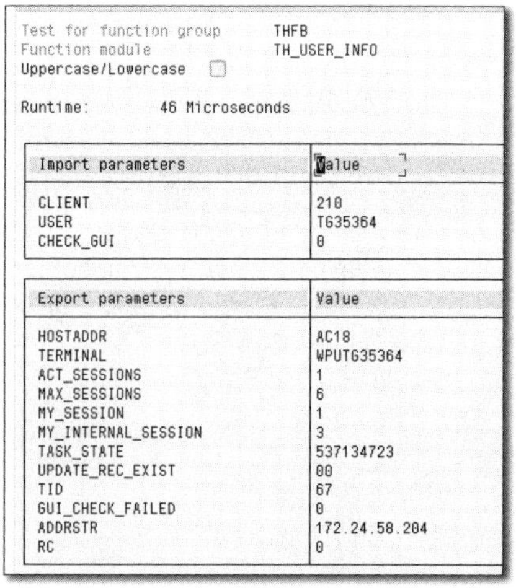

User Details Received after Execution of the Function Module

UPDATE_REC_EXIST

X will be returned if REC instructions are updated Type: THFB_BOOL.

TID

Terminal ID of the user. Type: SY-INDEX.

GUI_CHECK_FAILED

Number of failed iterations in the loop. Type: SY-INDEX.

ADDRSTR

IP address of the frontend computer as a string and IPv6 compatible. Type: NI_NODEADDR.

RC

Return code received after the execution of the FM, that is, success or failure. Type: I.

31 TH_USER_LIST

This FM returns the list of all users who are logged in to the system. The list includes the terminal ID, client, user name, transaction code, terminal ID, dialog time in Transaction SM04, host IP address, user trace, number of external sessions, number of internal sessions, type, status, GUI version of the server, and RFC type. Either table parameter LIST or USRLIST is returned with the list of users who are logged in to the system.

Example: The administrator can execute this FM to get all of the details of all users logged on to the system and their detail for monitoring or reporting purposes.

Tables

LIST

Internal table to store the list. Type: UINFO.

USRLIST

Internal table to store the new user list. Type: USRINFO.

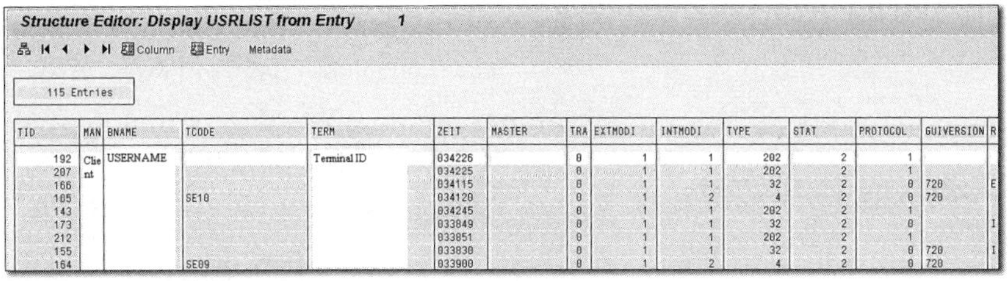
Structure Editor: Display USRLIST from Entry 1

Column Entry Metadata

115 Entries

TID	MAN	BNAME	TCODE	TERM	ZEIT	MASTER	TRA	EXTMODI	INTMODI	TYPE	STAT	PROTOCOL	GUIVERSION	R
192	Client	USERNAME		Terminal ID	034226		0	1	1	202	2	1		
207					034225		0	1	1	202	2	1		
166					034115		0	1	1	32	2	0	720	E
105			SE10		034120		0	1	2	4	2	0	720	
143					034245		0	1	1	202	2	1		
173					033849		0	1	1	32	2	0		I
212					033851		0	1	1	202	2	1		
155					033830		0	1	1	32	2	0	720	I
164			SE09		033900		0	1	2	4	2	0	720	

User List Returned after Execution of TH_USER_LIST FM

32 TRFC_RECEIVER_INFO

This FM returns the information about the transactional RFC (tRFC) at the receiver's side. This FM is used to tell the receiver side about the current transaction ID while the FM is still running at the server side.

Example: There are two SAP landscapes in your organization whose configuration boxes are C01 and C02. An RFC connection is maintained from C01 to C02 through RFC destination RFC_C02. Now, you want the tRFC of the C02 system at the C01 system so that you can use the information at the receiver side. You can execute this FM and all of the details are returned in the export parameters.

[↑] Export

TRFCTYPE

Type of the RFC transaction: queued, outbound, or inbound. Type: SY-INPUT.

MODE

Mode in which RFC is running. It can have value as L, which is the local execution of the current queued RFC in a LUW. Type: SY-INPUT.

TID

Transaction ID on the receiver side. Type: ARFCTID.

FNAME

Function name. Type: ARFCRSTATE-ARFCFNAM.

FNUM

Function number within the LUW. Type: ARFCRSTATE-ARFCLUWCNT.

FTOTAL

Number of functions in the LUW. Type: ARFCRSTATE-ARFCLUWCNT.

S_DATE

Sender date. Type: SY-DATUM.

S_TIME

Sender time. Type: SY-UZEIT.

S_CLIENT

Sender client. Type: SY-MANDT.

S_USER

ABAP system, user logon name. Type: SY-UNAME.

S_HOST

Sender host name. Type: ARFCRSTATE-ARFCRHOST.

S_PROG

Sender program. Type: SY-CPROG.

S_TCODE

Sender transaction code. Type: SY-TCODE.

S_ID

Sender ID (hostname_sysid_sysnr). Type: RFCDES-RFCDEST.

33 USER_EXISTS

This FM checks for the existence of a specific user.

Note that an older version of this FM had one parameter LOCKED, which does not indicate the lock reason. If you read SAP Note 1317803, a new lock is introduced called the Global Lock. This is helpful when you implement the Central User Administration (CUA). The user is locked in all of the CUA systems. This lock is ignored in the parameter LOCKED. Refer to SAP Note 1317803 for more information.

Example: The administrator can check whether a specific user ID exists in the SAP system or whether it's locked by entering a wrong password multiple times by using this FM.

[↓] Import

BNAME

SAP user ID to be checked. This is a mandatory parameter. Type: USR02-BNAME.

CLIENT

Client to be checked. If [Space] is given, then the current client is taken; if * is given, all clients are considered. The default value is [Space]. Type: SY-MANDT.

NEW_EXPORT_STYLE

This parameter describes the locking state of the user. If you enter the value X, the new implementation is executed. In the new implementation, export parameter LOCKSTATE is returned instead of parameter LOCKED. If you enter a [Space], the old behavior is executed. Type: BOOLE_D.

[↑] Export

LOCKED

If the value is X, it means the user exists but is locked.

LOCKSTATE

This parameter contains one field according to the type of lock. The CUA-related lock is returned in this field. This contains one field for each lock reason and can have values U – (unlocked) or L (locked). Type: USLOCK.

[!] Exceptions

USER_DONT_EXIST

The user does not exist in the system and client.

USER_EXISTS

The user exists.

Part II
ABAP Data Dictionary

The ABAP Data Dictionary is the SAP ABAP central repository, which describes and manages all of the data definitions used in the ABAP system and the database. The ABAP Data Dictionary supports the definition of user-defined types, such as data elements, structures, and table types. These types are used in the ABAP processors and programs. It also defines the structure of database objects, such as tables, views, and indexes. These database objects are automatically created in the underlying database with the definition of the ABAP Data Dictionary when the objects are activated. It also provides editing tools such as search help and locking tools. Thus, objects present in the ABAP Data Dictionary are tables, views, types (data elements, structures, and table types), domains, search helps, and lock objects.

This part includes function modules (FMs) that are related to working with the ABAP Data Dictionary and its subobjects.

34 DDIF_DOMA_GET

This FM is used to read the domain from the ABAP Data Dictionary. The name of the domain to be read is passed in the NAME parameter.

Example: To read the fixed value information of the domain to process it in your program, you can use this FM.

[↓] Import

NAME

Name of the domain to be read. This is a mandatory parameter ILLEGAL_INPUT is raised. Type: DDOBJNAME.

STATE

Read status of the domain. Parameter STATE defines how the domain source should be read. STATE = A defines to read an active version and STATE = M defines to read a latest version. The default value is A. Type: DDOBJSTATE.

LANGU

Language in which texts are read. The default value is [Space]. Type: SY-LANGU.

[↑] Export

GOTSTATE

Status in which reading took place. If STATE = A, GOSTATE can have values A (object has an active version) or [Space] (no active version of the object exists). If STATE = M, the parameter GOSTATE can have values [Space] (object doesn't exist in the ABAP Data Dictionary), N (object is in status New in the ABAP Data Dictionary), A (object is in status Active in the ABAP Data Dictionary), or M (object is in status Revised in the ABAP Data Dictionary). Type: DDGOTSTATE.

DD01V_WA

Header of the domain. Type: DD01V.

Tables

DD07V_TAB

Domain information, including fixed domain values, is returned in this parameter when requested by the caller. If not required, fixed values should not be defined in the parameter DD07V_TAB. Type: DD07V.

[!] Exceptions

ILLEGAL_INPUT

Value not allowed for parameter.

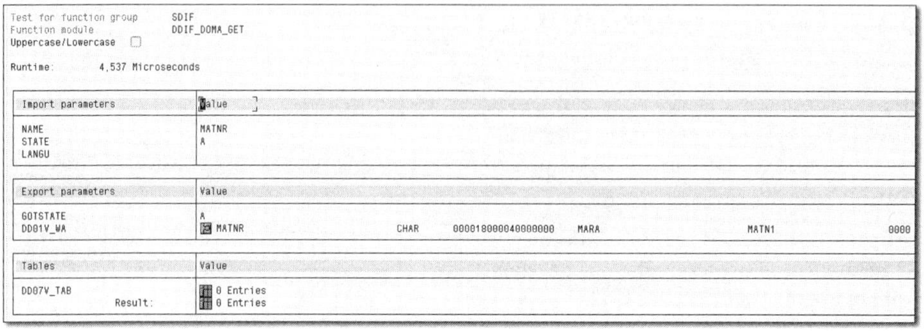

Domain Information Returned after Execution of DDIF_DOMA_GET

35 DDIF_DTEL_GET

This FM is used to read the data element from the ABAP Data Dictionary. The name of the data element to be read is passed in the NAME parameter. If you want to read the runtime information of the data element, use the more effective FM DDIF_NAMETAB_GET (FM 37; runtime information without text) or DDIF_FIELDINFO_GET (FM 36; runtime information with text).

Example: Use this FM if you need the data element definition in your program, or you need information for the data elements that isn't contained in the runtime objects. You may need this information to get other details in your program.

[↓] Import

NAME

Name of the data element to be read. This is a mandatory parameter. If the name is invalid, exception ILLEGAL_INPUT is raised. Type: DDOBJNAME.

STATE

Read status of the data element. Parameter STATE defines how the data element source should be read. STATE = A defines to read an active version, and STATE = M defines to read the latest version. The default value is A. Type: DDOBJSTATE.

LANGU

Language in which texts are read. The default value is [Space]. Type: SY-LANGU.

[↑] Export

GOTSTATE

Status in which the object was read. If STATE = A, GOTSTATE can have values A (object has an active version) or [Space] (no active version of the object exists). If STATE = M, GOTSTATE can have values [Space] (object doesn't exist in the ABAP Data Dictionary), N (object is in status New in the ABAP Data

Dictionary), A (object is in status Active in the ABAP Data Dictionary), or M (object is in status Revised in the ABAP Data Dictionary). Type: DDGOTSTATE.

DD04V_WA

Header of the data element. Type: DD04V.

TPARA_WA

This parameter returns the information about the parameter ID if assigned to the data element. This parameter should be defined only if the corresponding information is required. Type: TPARA.

[!] Exceptions

ILLEGAL_INPUT

Value not allowed for parameter.

36 DDIF_FIELDINFO_GET

This FM is used to read the text on tables or types; for example, this FM reads the runtime information of a table or any *types* of the ABAP Data Dictionary. This FM also returns the text information of types. If text information isn't needed, you should use FM DDIF_NAMETAB_GET (FM 37).

Example: To get the runtime information of the fields of the table to use in other FM or program, use this FM. For instance, you might want to know the name of and call the search help used in the field FIELD1 of a table.

[↓] Import

TABNAME

Name of the table (or the type) for which information is required. This is a mandatory parameter. If the name is invalid, exception NOT_FOUND_INPUT is raised. Type: DDOBJNAME.

FIELDNAME

FIELDNAME whose details are required. If this parameter doesn't have an initial value, only fields whose names are given in this parameter are returned. If no such field is found, exception NOT_FOUND is triggered. Type: DFIES-FIELDNAME.

LANGU

Language of the texts. The default value is SY-LANGU, that is, the logon language. Type: SY-LANGU.

LFIELDNAME

If filled, only fields whose long name is the same as the content of the parameter are taken into consideration. Because there can be at most one such field, the information is then returned in the export parameter DFIES_WA, and table DFIES_TAB is undefined. If there is no such type, exception NOT_FOUND is triggered. Type: DFIES-LFIELDNAME.

ALL_TYPES

Take all types into consideration. If this parameter is [Space], only flat structures are read, and exception NOT_FOUND is triggered even if name specified in TABNAME is a data element, deep structure or table types. However, if this parameter is set to X, exception NOT_FOUND is triggered only if there is no active runtime object for the type with the name specified by TABNAME. By default value is [Space] for this parameter. Type: DDBOOL_D.

GROUP_NAMES

Take named includes into consideration. If the value is:

- [Space]: Don't take named includes into consideration.
- X: Take named includes into consideration.

Type: DDBOOL_D.

UCLEN

Unicode length with which the runtime object was generated. Type: UNICODELG.

[↑] Export

X030L_WA

Name of the tab header of the table (or the Type). Type: X030L.

DDOBJTYPE

This parameter returns the type category. It can have one of the following values: TRANSP (transparent table), POOL (logical pool table), CLUSTER (logical cluster table), VIEW (database or projection view), INTTAB (structure view, help view, maintenance view, and other structures that can't be accessed by Open SQL such as deep types), TPOOL (physical pooled table), TCLUSTER (physical cluster table), DTEL (data element), and TTYP (table type). The values DTEL and TTYP are only allowed if parameter ALL_TYPES is set. Type: DD02V-TABCLASS.

DFIES_WA

Information in work area is returned. If the type is a data element or table type and the ALL_TYPES parameter is X, the description is returned in this parameter. Type: DFIES.

LINES_DESCR

Information about other referenced types.

If the type of a field of a structured type or the row of a table type is defined by a data element, the information about this data element is directly copied to the corresponding row of the field list or to the description of the table type. These data elements are therefore not contained in parameter LINES_DESCR.

If the type of a field of a structured type is defined in turn by a structured type, the fields of the referenced type are expanded into the field list of the referencing type. Such a reference therefore only causes the referenced type in parameter LINES_DESCR to be executed if IMPORT parameter GROUP_NAMES is set.

If the type of a field of a structured type is defined by a table type or if the type of the row of a table type is defined by a structured type or a table type, the description of the referenced type of the field list or of the description of the table type can't be taken directly. In these cases, the referenced type is copied to parameter LINES_DESCR. If the referenced type is a table type, component FIELDS contains a row describing the table type. If the referenced type is a structured type, component FIELDS contains the field list of this structured type.

If this FM is called with the name of a structured type also specifying a (long) field name, referenced table types and structured types are copied to LINES_DESCR.

If GROUP_NAMES parameter is set, named includes are also listed for all of the structured types listed. In this case, the structures referenced as named includes are also included in LINES_DESCR.

The set of types described in LINES_DESCR is just the transitive cover of the types to be described according to the preceding rules. Because the definition of this cover can be quite complicated for deeply nested types, this parameter should only be defined if it's really necessary.

Table LINES_DESCR is sorted according to component TYPENAME.

Type: DDTYPELIST.

Tables

DFIES_TAB

Returns the field list details. If the TABNAME is a structure type, its field list is returned in this table. Type: DFIES.

FIXED_VALUES

Description of domain fixed values. If the TABNAME, in particular the data element, is associated with the domain that has fixed values defined, that detail will come in this table. If there are no field values, this table is blank. Type: DDFIXVALUES.

[!] Exceptions

NOT_FOUND

This exception is triggered if there is no active version found as specified in parameter TABNAME.

INTERNAL_ERROR

Internal error occurred.

37 DDIF_NAMETAB_GET

Similar to FM DDIF_FIELDINFO_GET (FM 36), this FM is also used to read the runtime information of a table or any *types* of the ABAP Data Dictionary. Unlike DDIF_FIELDINFO_GET, this FM doesn't return the text information of types.

Example: Use this FM when you want the actual runtime information of the table or other ABAP Data Dictionary type in your program. For example, you can get the complete details of sales order number, item, and material from table VBAP for further dynamic programming.

[↓] Import

TABNAME

Name of the table (or the type) for which information is required. This is a mandatory parameter. If the name is invalid, exception NOT_FOUND_INPUT is raised. Type: DDOBJNAME.

ALL_TYPES

Take all types into consideration. If this parameter is [Space], only flat structures are read, and exception NOT_FOUND is triggered even if the name specified in TABNAME is a data element, deep structure, or table type. However, if this parameter is set to X, exception NOT_FOUND is triggered only if there is no active runtime object for the type with the name specified by TABNAME. By default, the value is [Space] for this parameter. Type: DDBOOL_D.

LFIELDNAME

If filled, only fields whose long name is the same as the content of the parameter are taken into consideration. Because there can be at most one such field, the information is then returned in the export parameter DFIES_WA, and table DFIES_TAB is undefined. If there is no such type, exception NOT_FOUND is triggered. Type: DFIES-LFIELDNAME.

GROUP_NAMES

Take named includes into consideration. If the value is:

- [Space]: Don't take named include into consideration.
- X: Take named includes into consideration.

Type: DDBOOL_D.

UCLEN

Unicode length with which the runtime object was generated. Type: UNICODELG.

[↑] Export

X030L_WA

Name of the tab header of the table (or the type). Type: X030L.

DTELINFO_WA

Decoded information about a data element. This parameter is defined only if the type specified by the TABNAME parameter is a data element and ALL_TYPES is set to X. Type: DTELINFO.

TTYPINFO_WA

Encoded information about a table type. This parameter is defined only if the type specified by the TABNAME parameter is a table type and ALL_TYPES is set to X. Type: TTYPINFO.

DDOBJTYPE

This parameter returns the type category. It can have one of the following values: TRANSP (transparent table), POOL (logical pool table), CLUSTER (logical cluster table), VIEW (database or projection view), INTTAB (structure view, help view, maintenance view, and other structures that can't be accessed by Open SQL such as deep types), TPOOL (physical pooled table), TCLUSTER (physical cluster table), DTEL (data element), and TTYP (table type). The values DTEL and TTYP are only allowed if parameter ALL_TYPES is set. Type: DD02V-TABCLASS.

DFIES_WA

Information in work area is returned. If the type is a data element or table type, and the ALL_TYPES parameter is X, the description is returned in this parameter. Type: DFIES.

LINES_DESCR

Information about other referenced types.

If the type of a field of a structured type or the row of a table type is defined by a data element, the information about this data element is directly copied to the corresponding row of the field list or to the description of the table type. These data elements are therefore not contained in this parameter.

If the type of a field of a structured type is defined in turn by a structured type, the fields of the referenced type are expanded into the field list of the referencing type. Such a reference therefore only causes the referenced type in parameter LINES_DESCR to be executed if the IMPORT parameter GROUP_NAMES is set.

If the type of a field of a structured type is defined by a table type or if the type of the row of a table type is defined by a structured type or a table type, the description of the referenced type of the field list or of the description of the table type can't be taken directly. In these cases, the referenced type is copied to this parameter. If the referenced type is a table type, component FIELDS contains a row describing the table type. If the referenced type is a structured type, component FIELDS contains the field list of this structured type.

If this FM is called with the name of a structured type also specifying a (long) field name, referenced table types and structured types are copied to LINES_DESCR.

If parameter GROUP_NAMES is set, named includes are also listed for all of the structured types listed. In this case, the structures referenced as named includes are also included in LINES_DESCR.

The set of types described in LINES_DESCR is just the transitive cover of the types to be described according to the preceding rules. Because the definition of this cover can be quite complicated for deeply nested types, this parameter should only be defined if it's really necessary.

Table LINES_DESCR is sorted according to component TYPENAME.

Type: DDTYPELIST.

Tables

X031L_TAB

Field list of runtime object (technical view). Type: X031L.

DFIES_TAB

Returns the field list details. If the TABNAME is a structure type, its field list is returned in this table. Type: DFIES.

[!] Exceptions

NOT_FOUND

Nothing found.

38 DDIF_TABL_ACTIVATE

This FM creates and activates tables in the ABAP Data Dictionary.

Example: Use this FM if you have a requirement to create a table from a program dynamically and then activate the table.

[↓] Import

NAME

Name of the table to be activated. This is a mandatory parameter. Type: DDOBJNAME.

AUTH_CHK

If this parameter is set with value X, user authorization is checked for activation of the table. Default: X. Type: DDBOOL_D.

PRID

Defines an ID that should be written in the ABAP Data Dictionary log writer. This FM always writes a log with the name ACTTABL<Name> in database table DDPRH/DDPRS (Log Header and Log Lines). Default: –1. Type: SY-TABIX.

EXCOMMIT

If this parameter is set to X, database commit is performed on the operation before this function call. Default: X. Type: DCTABLACT-EXCOMMIT.

[↑] Export

RC

Result of activation. It can have the following values:

- RC = 0 "The table is activated successfully."
- RC = 4 "The table is activated with warnings."
- RC > 4 "The table could not be activated."

Type: SY-SUBRC.

[!] Exceptions

NOT_FOUND

The table isn't found or can't be read.

PUT_FAILURE

Table isn't successfully written.

---

39 DDIF_TABL_GET

This FM is used to read the table from the ABAP Data Dictionary. This FM should be used when you need the table definition in your program or information that isn't contained in the runtime objects. If you want to read the runtime information of the table, use the more effective FM DDIF_NAMETAB_GET (FM 37) or DDIF_FIELDINFO_GET (FM 36).

Example: If you want the table definition information in your program for dynamic programming, use this FM to get that table information.

[↓] Import

NAME

Name of the table to be read. This is a mandatory parameter. If the name is invalid, exception ILLEGAL_INPUT is raised. Type: DDOBJNAME.

STATE

Read status of the table. Parameter STATE defines how the table source should be read. STATE = A defines to read an active version, and STATE = M defines to read the latest version. The default value is A. Type: DDOBJSTATE.

LANGU

Language in which texts are read. The default value is [Space]. Type: SY-LANGU.

[↑] Export

GOTSTATE

Status in which the object was read. If STATE = A, GOTSTATE can have values A (object has an active version) or [Space] (no active version of the object exists). If STATE = M, GOTSTATE can have values [Space] (object doesn't exist in the ABAP Data Dictionary), N (object is in status New in the ABAP Data Dictionary), A (object is in status Active in the ABAP Data Dictionary), or M (object is in status Revised in the ABAP Data Dictionary).

If this parameter has a value of A, an expanded version of the table is returned; that is, all of the included structures as well as their foreign keys and search help assignments are expanded.

If this parameter has a value of M or N, the table is returned in compressed form; that is, the included structures, foreign keys, and search help won't appear in the field list. Type: DDGOTSTATE.

DD02V_WA

Header of the table. Type: DD02V.

DD09L_WA

Returns the technical settings of the table. Type: DD09V.

Tables

DD03P_TAB

Table fields sorted by position. Type: DD03P.

DD05M_TAB

Foreign key fields of the table sorted by FIELDNAME PRIMPOS. Type: DD05M.

DD08V_TAB

Foreign keys of the table sorted by FIELDNAME. Type: DD08V.

DD12V_TAB

Table indexes. Type: DD12V.

DD17V_TAB

Index fields of the table. Type: DD17V.

DD35V_TAB

Header of the search help assignments of the table sorted by FIELDNAME. Type: DD35V.

DD36M_TAB

Allocations of the search help assignments of the table sorted by FIELDNAME SHLPFIELD. Type: DD36M.

[!] Exceptions

ILLEGAL_INPUT

Value not allowed for parameter.

40 DDIF_TABL_PUT

This FM writes the table information in the ABAP Data Dictionary for a new or revised version of the table. If there is an active version of the table, it isn't affected. This FM doesn't perform any authorization checks and doesn't write any correction entries. This FM doesn't provide `COMMIT` or `ROLLBACK`, so you should perform `COMMIT` or `ROLLBACK` to save your changes in the ABAP Data Dictionary.

Example: Use this FM when you need to create a table dynamically through a program, or you want to add some fields in the ABAP Data Dictionary table programmatically. You first create this table, then put the new table in the ABAP Data Dictionary from this FM, and finally activate the new table by using FM DDIF_TABL_ACTIVATE (FM 38).

[↓] Import

NAME

Name of the domain to be read. This is a mandatory parameter. Type: DDOBJNAME.

DD02V_WA

Header of the table. Type: DD02V.

DD09L_WA

Definitions of the technical settings of the table. Type: DD09V.

Tables

DD03P_TAB

Table fields sorted by position. Type: DD03P.

DD05M_TAB

Foreign key fields of the table sorted by FIELDNAME PRIMPOS. Type: DD05M.

DD08V_TAB

Foreign keys of the table sorted by FIELDNAME. Type: DD08V.

DD35V_TAB

Header of the search help assignments of the table sorted by FIELDNAME. Type: DD35V.

DD36M_TAB

Allocations of the search help assignments of the table sorted by FIELDNAME SHLPFIELD. Type: DD36M.

[!] Exceptions

TABL_NOT_FOUND

Table header could not be found. If parameter DD02V_WA isn't defined, and the table isn't in the ABAP Data Dictionary, this exception is raised.

NAME_INCONSISTENT

If the parameter defined has a value other than NAME, this exception is triggered.

TABL_INCONSISTENT

Inconsistent sources. If the position entry in table DD03P_TAB occurs more than once, this exception is raised. Table DD08V_TAB is the foreign key header, and DD05M_TAB is the foreign key field. If only one of these tables are defined, this exception is raised. The same occurs with DD35V_TAB and DD36V_TAB, which are used for search help. If only one of these tables is defined, this exception is raised.

PUT_FAILURE

Write error (`ROLLBACK` recommended). If the write operation fails, this exception is raised. However, part of the source might already be written, so `ROLLBACK` should be performed to reverse the changes.

PUT_REFUSED

The exception occurred while writing. For example, the name of the object to be written duplicates with the already existing object in the ABAP Data Dictionary.

---

41 DDIF_VIEW_GET

This FM is used to read the view from the ABAP Data Dictionary. The name of the view to be read is passed in the NAME parameter.

Example: Use this FM when you need the view definition in your dynamic program or need to understand how the different views join.

[↓] Import

NAME

Name of the view to be read. This is a mandatory parameter. If the name is invalid, exception ILLEGAL_INPUT is raised. Type: DDOBJNAME.

STATE

Read status of the view. Parameter STATE defines how the view source should be read. STATE = A defines to read an active version, and STATE = M defines to read the latest version. The default value is A. Type: DDOBJSTATE.

LANGU

Language in which texts are read. The default value is [Space]. Type: SY-LANGU.

[↑] Export

GOTSTATE

Status in which the object was read. If STATE = A, GOTSTATE can have values A (object has an active version) or [Space] (no active version of the object exists). If STATE = M, GOTSTATE can have values [Space] (object doesn't exist in the ABAP Data Dictionary), N (object is in status New in the ABAP Data Dictionary), A (object is in status Active in the ABAP Data Dictionary), or M (object is in status Revised in the ABAP Data Dictionary). Type: DDGOTSTATE.

DD025V_WA

Header of the view. Type: DD025V.

DD09L_WA

Returns the technical settings of the view. Type: DD09V.

Tables

DD26V_TAB

Basis tables of the view. Type: DD26V.

DD27P_TAB

View fields. Type: DD27P.

DD28J_TAB

Join conditions of the view. Type: DD28J.

DD28V_TAB

Selection conditions of the view. Type: DD28V.

[!] Exceptions

ILLEGAL_INPUT

Value not allowed for parameter.

---

42 DEQUEUE_ALL

This FM releases all of the locks of an LUW. Locks are assigned to the objects (tables, records, etc.) to lock them for changes; that is, if one user is changing a record, then another user can't change the same record until the first user completes and commits his changes. When the lock objects (`lobj`) are created in the ABAP Data Dictionary, two FMs are generated: ENQUEUE_LOBJ and DEQUEUE_LOBJ. ENQUEUE_LOBJ grants the lock, and

DEQUEUE_LOBJ releases the lock. Hence, DEQUEUE_ALL releases all of the locks that are assigned.

Example: Use this FM to release all of the locks after changes are complete and committed. If locks aren't released, then they will be released automatically at the end of the transaction. For an example, see FM ENQUEUE_E_TABLE (FM 44).

[↓] Import

_SYNCHRON

If this parameter is set, the lock is released only when the entries are deleted from lock table. Type: DDENQ_LIKE-SYNCHRON.

43 DEQUEUE_E_TABLE

This FM releases the lock of an LUW for the table that is passed in the parameter TABNAME. Tables locked with FM ENQUEUE_E_TABLE (FM 44) can be unlocked with this FM.

Example: Use this FM to release the lock after changes are complete and committed. If locks aren't released, then they will be released automatically at the end of the transaction. See FM ENQUEUE_E_TABLE (FM 44).

[↓] Import

MODE_RSTABLE

This parameter defines the lock mode. Lock mode can be E (write lock), S (read lock), or X (exclusive, noncumulative). The default value is E. Type: DD26E-ENQMODE.

TABNAME

Name of the ABAP Data Dictionary table to be locked. Type: RSTABLE-TABNAME.

VARKEY

This parameter defines the lock key that is to be locked. This includes the key fields of the record that the user wants to update. Type: RSTABLE-VARKEY.

X_TABNAME

X_FIELD defines the lock behavior when the initial value is passed in the lock field (i.e., TABNAME in this case). If the initial value is assigned to <field> and X_<field>, then a generic lock is initialized with respect to <field>. If <field> is assigned the initial value and X_<field> is defined as X, the lock is set with exactly the initial value of <field>. The default value is [Space].

X_VARKEY

X_field defines the lock behavior when the initial value is passed in the lock field (i.e., TABNAME in this case). If the initial value is assigned to <field> and X_<field>, then a generic lock is initialized with respect to <field>. If <field> is assigned the initial value, and X_<field> is defined as X, the lock is set with exactly the initial value of <field>. The default value is [Space].

_SCOPE

This defines the lock behavior when the update task is called.

- _Scope = 1: The lock or lock release isn't passed to the update program. The lock is removed when the transaction is ended.
- _Scope = 2: The lock or lock release is passed to the update program. The update program is responsible for removing the lock. The interactive program with which the lock was requested no longer has an influence on the lock behavior. This is the standard setting for the ENQUEUE FM.
- _SCOPE = 3: The lock or lock release is also passed to the update program. The lock must be removed in both the interactive program and in the update program. This is the standard setting for the DEQUEUE FM.

The default value for this parameter is 3. Type: DDENQ_LIKE-SCOPE.

_SYNCHRON

If X is passed, the DEQUEUE function waits until the entry has been removed from the lock table. Otherwise, it's deleted asynchronously; that is, if the lock table of the system is read directly after the lock is removed, the entry in the lock table may still exist. The default value is [Space]. Type: DDENQ_LIKE-SYNCHRON.

_COLLECT

This parameter controls whether the lock request or lock release should be performed directly or whether it should first be written to the local lock container.

- [Space]: The lock request or lock release is sent directly to the lock server.
- X: The lock request or lock release is placed in the local lock container. The lock requests and lock releases collected in this lock container can then be sent to the lock server at a later time as a group by calling the FM FLUSH_ENQUEUE (FM 59).

The default value is [Space]. Type: DDENQ_LIKE-COLLECT.

---

44 ENQUEUE_E_TABLE

This FM assigns the lock to an LUW for the table, which is passed in the parameter TABNAME. Locked tables can be unlocked by DEQUEUE_E_TABLE or DEQUEUE_ALL. Locks can also be released automatically when the transaction ends. Refer to the other FMs that are used in this process:

- DEQUEUE_E_TABLE (FM 43)
- DEQUEUE_ALL (FM 42)

This FM increases data consistency and integrity by locking the entry a user is modifying. Changes used by one user won't be lost if another user tries to change the same object because he can't change it until the lock is released by the first user. Created locks can be viewed in the lock entry list Transaction SM12.

Example: To change the material detail from your customized program, lock the record you want to modify of table MARA from this FM to ensure that no other user can make any modification while you are modifying it.

[↓] Import

MODE_RSTABLE

This parameter defines the lock mode. Lock mode can be E (write lock), S (read lock), or X (exclusive, noncumulative). The default value is E. Type: DD26E-ENQMODE.

TABNAME

Name of the ABAP Data Dictionary table to be locked. Type: RSTABLE-TABNAME.

VARKEY

This parameter defines the lock key that is to be locked. This includes the key fields of the record that the user wants to update. Type: RSTABLE-VARKEY.

X_TABNAME

X_field defines the lock behavior when the initial value is passed in the lock field (i.e., TABNAME in this case). If the initial value is assigned to <field> and X_<field>, then a generic lock is initialized with respect to <field>. If <field> is assigned the initial value, and X_<field> is defined as X, the lock is set with exactly the initial value of <field>. The default value is [Space].

X_VARKEY

X_field defines the lock behavior when the initial value is passed in the lock field (i.e., TABNAME in this case). If the initial value is assigned to <field> and X_<field>, then a generic lock is initialized with respect to <field>. If <field> is assigned the initial value, and X_<field> is defined as X, the lock is set with exactly the initial value of <field>. The default value is [Space].

_SCOPE

This defines the lock behavior when the update task is called.

- _Scope = 1: The lock or lock release isn't passed to the update program. The lock is removed when the transaction is ended.
- _Scope = 2: The lock or lock release is passed to the update program. The update program is responsible for removing the lock. The interactive program with which the lock was requested no longer has an influence on the lock behavior. This is the standard setting for the ENQUEUE FM.
- _SCOPE = 3: The lock or lock release is also passed to the update program. The lock must be removed in both the interactive program and in the update program. This is the standard setting for the DEQUEUE FM.

The default value for this parameter is 2. Type: DDENQ_LIKE-SCOPE.

_WAIT

This parameter determines the lock behavior when there is a lock conflict.

- [Space]: If a lock attempt fails because there is a competing lock, the exception FOREIGN_LOCK is triggered.
- X: If a lock attempt fails because there is a competing lock, the lock attempt is repeated after waiting for a certain time. The exception FOREIGN_LOCK is triggered only if a certain time limit has elapsed since the first lock attempt. The waiting time and the time limit are defined by profile parameters.

The default value is [Space]. Type: DDENQ_LIKE-WAITFLAG.

_COLLECT

This parameter controls whether the lock request or lock release should be performed directly or whether it should first be written to the local lock container.

- [Space]: The lock request or lock release is sent directly to the lock server.
- X: The lock request or lock release is placed in the local lock container. The lock requests and lock releases collected in this lock container can then be sent to the lock server at a later time as a group by calling the FM FLUSH_ENQUEUE (FM 59).

The default value is [Space]. Type: DDENQ_LIKE-COLLECT.

[!] Exceptions

FOREIGN_LOCK

If a lock attempt fails because there is a competing lock, the exception FOREIGN_LOCK is triggered. See parameter _WAIT for more details.

SYSTEM_FAILURE

Internal error from ENQUEUE server.

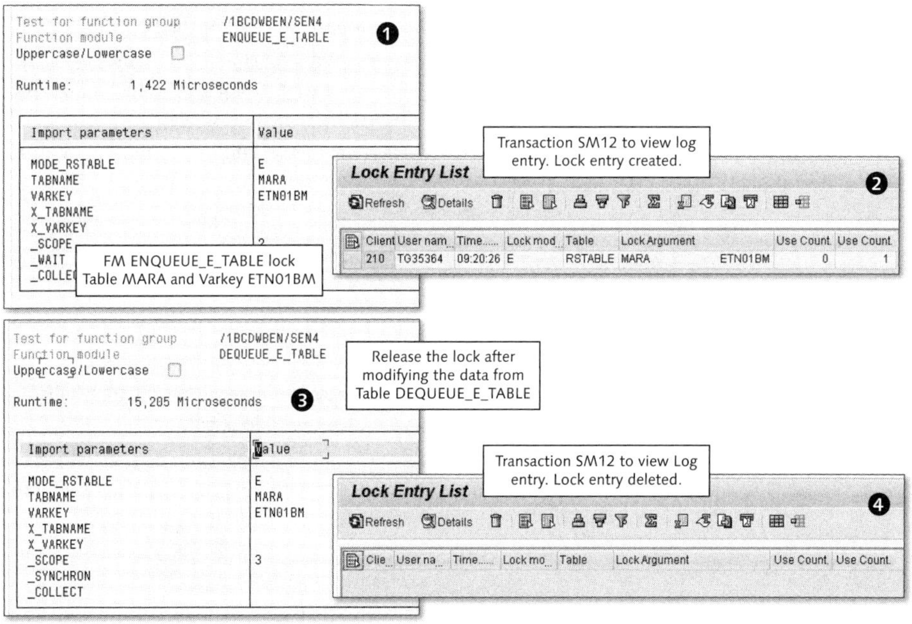

ENQUEUE_E_TABLE and DEQUEUE_E_TABLE Execution

45 F4_DATE

This FM displays the factory calendar or the Gregorian calendar in which you can select a specific day.

Example: You want to run the report on all of the sales orders received on a particular date, such as today's date. You have a DATE parameter in the selection screen. On clicking DATE, you want the date calendar to pop up so that you can select a date and your date is returned to the screen. You can use this FM to pass the relevant parameters to make this happen.

[↓] Import

DATE_FOR_FIRST_MONTH

Determines the first month to be displayed. If the date is passed in this parameter, the month containing that date is displayed. If this parameter isn't passed, the first month is determined from parameter PROGNAME_FOR_FIRST_MONTH (if passed); otherwise, the date is determined with the default value, which is SY-DATUM. Type: WORKFLDS-GKDAY.

DISPLAY

If this parameter is [Space] (default), the date can be selected and returned. If this parameter is X, the date is displayed only and not selected. Type: WORKFLDS-DISPL.

FACTORY_CALENDAR_ID

This parameter displays the factory calendar. If the defined factory calendar ID isn't found, exception FACTORY_CALENDAR_NOT_FOUND is raised. The default value is [Space]. Type: WORKFLDS-CALID.

GREGORIAN_CALENDAR_FLAG

This parameter displays the Gregorian calendar. The default value is [Space]. This is the default calendar that is displayed if no calendar was set explicitly. If this parameter is passed with X, the Gregorian calendar is displayed irrespective of FACTORY_CALENDAR_ID or HOLIDAY_CALENDAR_ID.

HOLIDAY_CALENDAR_ID

This parameter displays the public holiday calendar. If the defined factory calendar ID isn't found, exception FACTORY_CALENDAR_NOT_FOUND is raised. The default value is [Space]. Type: WORKFLDS-HIDENT.

PROGNAME_FOR_FIRST_MONTH

To determine the date at the cursor position. If no DATE_FOR_FIRST_MONTH was passed, but a program name was passed with parameter PROGNAME_FOR_FIRST_MONTH, the FM determines the month to be displayed using the contents of the field to which the cursor is pointing. The default value is [Space]. Type: SY-REPID.

[↑] Export

SELECT_DATE

If the parameter display is [Space], the date selected by the user is returned in this parameter. Type: WORKFLDS-GKDAY.

SELECT_WEEK

If the week is selected, then the week number is returned here. Type: SCAL-WEEK.

SELECT_WEEK_BEGIN

When week is selected, the begin date of the week is returned here. Type: SY-DATUM.

SELECT_WEEK_END

When week is selected, the end date of the week is returned here. Type: SY-DATUM.

[!] Exceptions

CALENDAR_BUFFER_NOT_LOADABLE

Factory calendar read error.

DATE_AFTER_RANGE

Date is later than factory calendar.

DATE_BEFORE_RANGE

Date is earlier than factory calendar.

DATE_INVALID

Date in invalid format.

FACTORY_CALENDAR_NOT_FOUND

Factory calendar isn't available.

HOLIDAY_CALENDAR_NOT_FOUND

Holiday calendar isn't available.

PARAMETER_CONFLICT

Pass factory calendar and holiday calendar.

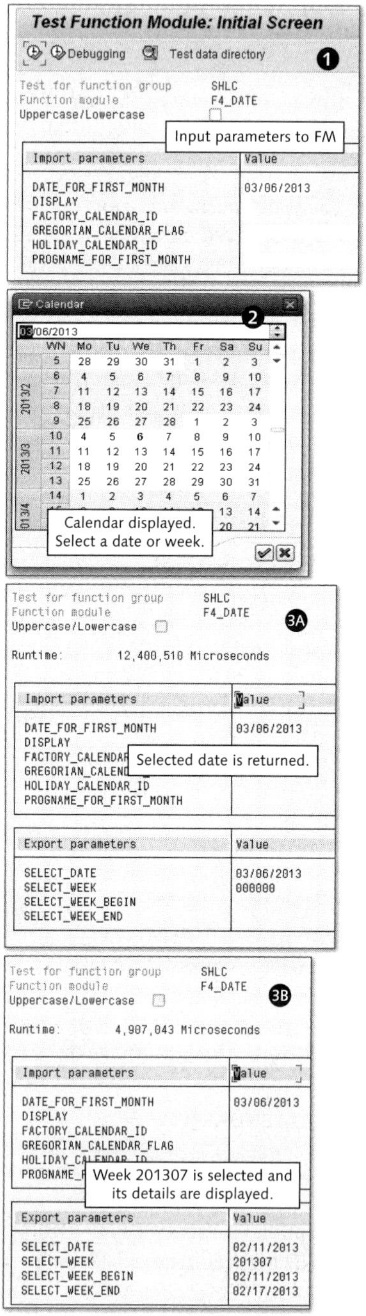

Execution of F4_DATE

46 F4IF_DETERMINE_SEARCHHELP

This FM determines the search help that is assigned to the field, according to the ABAP Data Dictionary. It returns exactly the search help that is effective on a screen when the field has been taken from the ABAP Data Dictionary on the screen.

Example: On a selection screen that contains a parameter of storage location MARC-LGPRO, you can first use this FM to determine the search help to be used from the ABAP Data Dictionary and then use FM F4IF_START_VALUE_REQUEST (FM 52) to start that returned search help.

[↓] Import

TABNAME

Table/structure name from the ABAP Data Dictionary. This is a mandatory parameter. Type: DFIES-TABNAME.

FIELDNAME

Field name from the ABAP Data Dictionary for which search help is required. This is a mandatory parameter but can be [Space]. Type: DFIES-FIELDNAME.

SELECTION_SCREEN

Behavior of the selection screen. The value table of the domains for the F4 help is also used in the selection screens of ABAP programs if the reference field comes from the value table itself.

In all other cases, the value table of a domain has no effect on the F4 help.

- [Space]: Value table isn't taken into consideration.
- X: Value table is taken into consideration.

Type: DDBOOL_D.

[↑] Export

SHLP

Search help details are returned in this FM. Type: SHLP_DESCR.

[!] Exceptions

FIELD_NOT_FOUND

Field does not exist in the dictionary.

NO_HELP_FOR_FIELD

No F4 help is defined for the field.

INCONSISTENT_HELP

F4 help for the field is inconsistent.

47 F4IF_FIELD_VALUE_REQUEST

This FM provides the F4 help for the screen fields whose references to the ABAP Data Dictionary are only known at runtime. If the FM is called at the event `Process On Value Request`, it behaves like standard help. The reference to the ABAP Data Dictionary is given in the form of a table with a field name as input. The FM then determines the search help defined directly or indirectly in the ABAP Data Dictionary such as the domain fixed value or data element attached to the search help. If many parameters of the search help are linked with the ABAP Data Dictionary table or structure, the additional fields are searched for on the screen as in standard help, and the parameters are given their contents.

If the screen information DYNPPROG, DYNPNR, DYNPROFIELD, and possibly STEPL are also specified, the selected value is automatically stored in the screen field. Further export parameters are also stored there if they are found as input fields on the screen.

Example: The screen has a P_MATNR parameter of type `char18`. You can use this FM to attach the explicit search help `MAT1` to the parameter.

[↓] Import

TABNAME

Table/structure name from the ABAP Data Dictionary. This is a mandatory parameter. Type: DFIES-TABNAME.

FIELDNAME

Field name from the ABAP Data Dictionary for which search help is required. This is a mandatory parameter but can be [Space]. Type: DFIES-FIELDNAME.

SEARCHHELP

This parameter defines the search help defined in the ABAP Data Dictionary. The values of TABNAME and FIELDNAME aren't considered important if this parameter is provided. Type: SHLPNAME.

SHLPPARAM

If the parameter SEARCHHELP is defined with the search help name, this parameter can be used to attach the contents to the screen field parameter. If no value is defined, the first parameter is attached to the screen field. This corresponds to the behavior when the search help is directly attached to the field for the field attributes of the screen painter. Type: SHLPFIELD.

DYNPPROG

Current program name. Type: SY-REPID.

DYNPNR

Screen number. Type: SY-DYNNR.

DYNPROFIELD

Name of the screen field for value return. Type: HELP_INFO-DYNPROFLD.

STEPL

Step loop line of the screen field for value return. Default: 0. Type: SY-STEPL.

VALUE

Screen field contents are passed to the F4 call. The contents are passed only if they contain a wildcard character. If the parameter has been assigned the default value Space, and the screen information (DYNPPROG, DYNPNR, STEPL) is available, the FM itself obtains the current contents. Type: HELP_INFO-FLDVALUE.

MULTIPLE_CHOICE

Multiple selection is active if this parameter has the value X. You can also select multiple lines in the hit list. Type: DDBOOL_D.

DISPLAY

This parameter defines if the F4 help should behave like an input field or a pure display field. The F4 help for a field that isn't an input field doesn't permit you to select values from the hit list.

- Space: Selected value from the F4 hit list comes back to the screen field if the screen field is an input field.
- X: F4 help is called in display mode. It isn't possible to select the value from the hit list.
- F (force input): Selected value from the F4 hit list comes back to the screen field even if the screen field is a display only field.

The default value is Space. Type: DDBOOL_D.

SUPPRESS_RECORDLIST

Skips display of the hit list. If this parameter is set to X, the hit list is selected but not displayed. The hits are placed in table RETURN_TAB. The maximum number of hits for the selection isn't limited. Type: DDSHF4CTRL-HIDE_LIST.

CALLBACK_PROGRAM

Program for callback before F4 start. Type: SY-REPID.

CALLBACK_FORM

Name of the form to be called immediately before calling the F4 processor.

The form must have the following interface:

```
FORM <NAME> TABLES RECORD_TAB
  STRUCTURE SEAHLPRES
    CHANGING SHLP TYPE SHLP_DESCR
    CALLCONTROL LIKE DDSHF4CTRL.
```

The F4 help works internally with a complex description of the search help passed in parameter SHLP. It can be a search help defined in the ABAP Data Dictionary. The description of a search help can also be built internally temporarily without the search help actually existing in the ABAP Data Dictionary. Type: SY-XFORM.

SELECTION_SCREEN

Behavior for the selection screen. The value table of the domains for the F4 help is also used in the selection screens of ABAP programs if the reference field comes from the value table itself.

In all other cases, the value table of a domain has no effect on the F4 help.

- Space: Value table isn't taken into consideration.
- X: Value table is taken into consideration.

Type: DDBOOL_D.

[↑] Export

USER_RESET

Parameter is returned with X if the user has chosen Deselect on the hit list with multiple selection, or if the user has deselected the selected values individually. Type: C.

Tables

RETURN_TAB

The selected value is returned in table RETURN_TAB if it wasn't copied to screen fields in the FM itself. In most cases, table RETURN_TAB has exactly one line containing the contents of the return column from the selected line of the hit list in field FIELDVAL.

If there are multiple return columns, SHLPNAME and FIELDNAME are the names of the search help parameters to which the value in FIELDVAL belongs.

In a multiple selection (several lines of the hit list are selected), field RECORDPOS is used for numbering the lines; for example, if three lines were selected, all of the columns of the first selected line are numbered with RECORDPOS = 1, and all of the columns of the second line with RECORDPOS = 2, and so on.

Field RETFIELD specifies the screen field belonging to the column if it's known. If there is only one return column, RETFIELD contains the field name passed in DYNPROFIELD. Type: DDSHRETVAL.

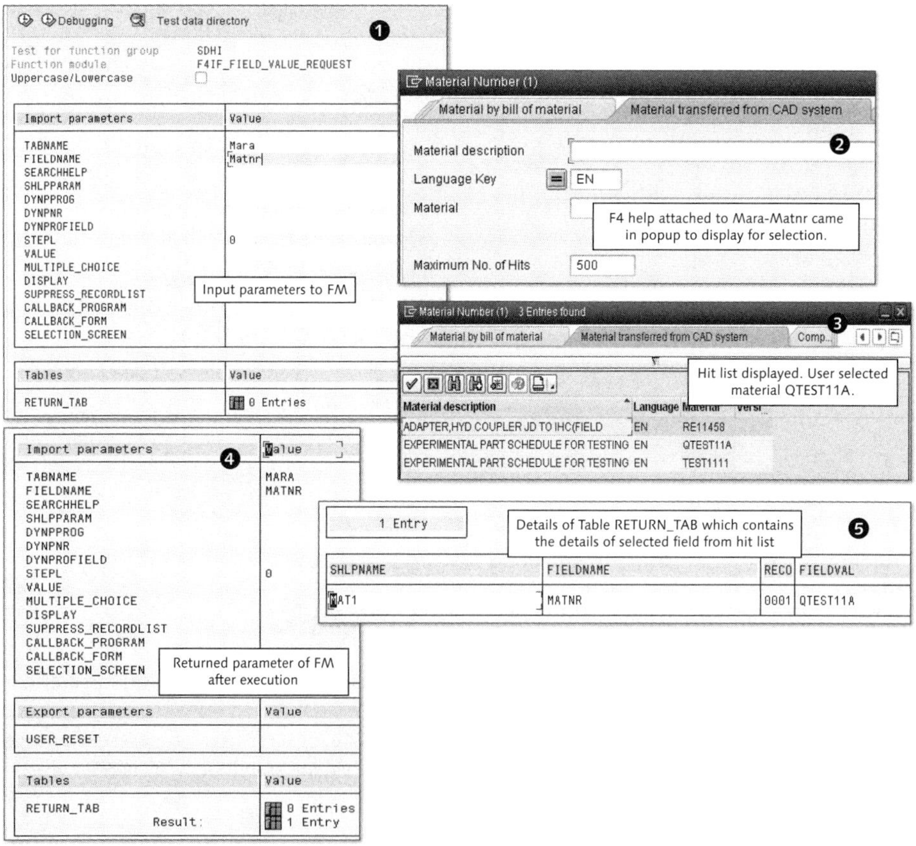

Function Module in Action

[!] Exceptions

FIELD_NOT_FOUND

Field does not exist in the dictionary.

NO_HELP_FOR_FIELD

No F4 help is defined for the field.

INCONSISTENT_HELP

F4 help for the field is inconsistent.

NO_VALUES_FOUND

No values found.

48 F4IF_GET_SHLP_DESCR

The FM returns the complete description of the search help. The function reads the description of the search help that is passed in SHLPNAME, and the search help can thus be passed to the block F4IF_START_VALUE_REQUEST (FM 52).

Example: You want to start the search help for a work center by clicking a button (instead of using F4 onscreen) in the output screen, and you've designed an elementary search help for the same. Also, you only want those work centers to appear that are assigned to that plant. In this case, you can get the structure of the search help from this FM. After you get the structure, you can pass the plant details in the SHLP_DESCR structure and call the Search help by calling the FM F4IF_START_VALUE_REQUEST (FM 52).

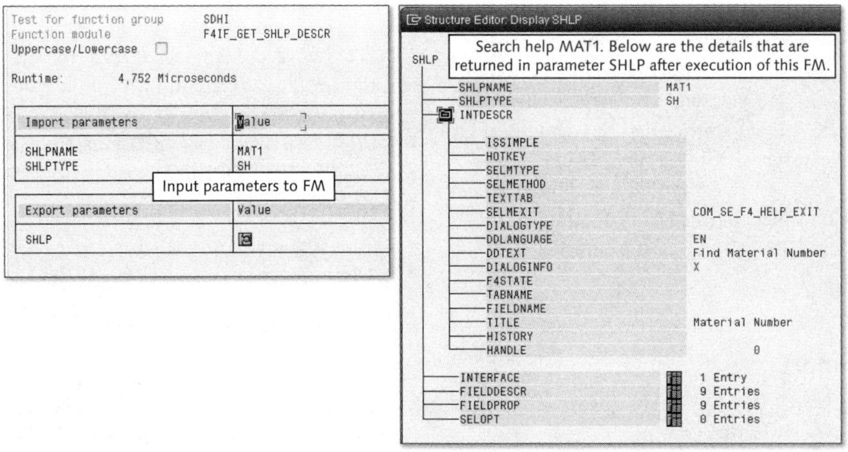

Search Help Structure Returned

[↓] Import

SHLPNAME

Search help name. Mandatory. Type: SHLPNAME.

SHLPTYPE

Type of search help. The default value is SH. Type: DDSHLPTYP.

[↑] Export

SHLP

Search help details are returned in this FM. Type: SHLP_DESCR.

---

49 F4IF_INT_TABLE_VALUE_REQUEST

The function displays the F4 help from the field of internal table. This FM implements the standard help at POV (Process on Value Request) while passing the values to be displayed in a table. The entire dialog behaves as a standard F4 help. This FM also supports input help control.

Example: In most scenarios, you want to display the F4 help from the values you get from an internal table. You can do so by using this FM. Suppose you have a work center field in your screen, and you want to display the F4 help for it. You have processed the content of F4 in the internal table I_WC_TABLE. You pass this table, work center field, and other required parameters in this FM to get the F4 displayed with the content of the internal table you selected.

[↓] Import

DDIC_STRUCTURE

Structure of VALUE_TAB. This is helpful when VALUE_ORG = S. If the table passed as VALUE_TAB has a structure defined in the ABAP Data Dictionary, the name of the structure can be specified here. In this case, parameter FIELD_TAB is ignored. Type: TABNAME.

RETFIELD

Name of return field in FIELD_TAB. Table VALUE_TAB consists of fields KEY and TEXT. When a line is selected from the hit list, the contents of column KEY should be copied to the screen field or to table RETURN_TAB. Field name KEY must then be passed as RETFIELD. This is a mandatory parameter. Type: DFIES-FIELDNAME.

PVALKEY

Key for personal help. If you've personalized any search help with some value, it's displayed instead of the value passed in VALUE_TAB. Leave this parameter if you don't want to display the personal help. Type: DDSHPVKEY.

DYNPPROG

Current program name from where the search help is triggered. Type: SY-REPID.

DYNPNR

Current screen number from where the search help is triggered. Type: SY-DYNNR.

DYNPROFIELD

Name of screen field for value return. Default: 0. Type: HELP_INFO-DYNPROFLD.

STEPL

Step loop line of the screen field. Type: SY-STEPL.

WINDOW_TITLE

Title for the hit list. Type: C.

VALUE

Screen field contents are passed to the F4 call. The contents are passed only if they contain a wildcard character. If the parameter has been assigned the default value [Space], and the screen information (DYNPPROG, DYNPNR, STEPL) is available, the FM itself-obtains the current contents. Type: HELP_INFO-FLDVALUE.

VALUE_ORG

This defines how the table VALUE_TAB should be organized. This can have value S (structure) or C (column). The default value is C. Type: DDBOOL_D.

MULTIPLE_CHOICE

Multiple selection is active if this parameter has the value X. You can also select multiple lines in the hit list. Default: C. Type: DDBOOL_D.

DISPLAY

This parameter defines whether the F4 help should behave like an input field or a pure display field. The F4 help for a field that isn't an input field doesn't permit you to select values from the hit list.

- [Space]: Selected value from the F4 hit list comes back to the screen field if the screen field is the input field.
- X: F4 help is called in display mode. It isn't possible to select the value from the hit list.
- F (force input): Selected value from the F4 hit list comes back to the screen field even if the screen field is a display-only field.

The default value is [Space]. Type: DDBOOL_D.

CALLBACK_PROGRAM

Program for callback before F4 start. Type: SY-REPID.

CALLBACK_FORM

Name of the form to be called immediately before calling the F4 processor.

The form must have the following interface:

```
FORM <NAME> TABLES RECORD_TAB
STRUCTURE SEAHLPRES
                  CHANGING SHLP TYPE SHLP_
DESCR
                               CALLCONTROL
LIKE DDSHF4CTRL.
```

The F4 help works internally with a complex description of the search help passed in parameter SHLP. It can be a search help defined in the ABAP Data Dictionary. The description of a search help can also be built internally temporarily without the search help actually existing in the ABAP Data Dictionary. Type: SY-XFORM.

CALLBACK_METHOD

This is an alternative of CALLBACK_FORM. This method must exist in the program CALLBACK_PROGRAM. Type: IF_F4CALLBACK_VALUE_REQUEST.

MARK_TAB

This defaults for selected lines when multiple selection is switched on. Type: DDSHMARKS.

[↑] Export

USER_RESET

This parameter is returned with X if the user has chosen DESELECT on the hit list with multiple selection, or if the user has deselected the selected values individually. Type: C.

Tables

VALUE_TAB

Table of values: entries cell by cell. Here you pass the internal table that you want to display in the F4 help.

FIELD_TAB

Fields of the hit list. Type: DFIES.

RETURN_TAB

The selected value is returned in table RETURN_TAB if it was not copied to the screen fields in the FM itself. In most cases, table RETURN_TAB has exactly one line containing the contents of the return column from the selected line of the hit list in field FIELDVAL.

If there are multiple return columns, SHLPNAME and FIELDNAME are the names of the search help parameters to which the value in FIELDVAL belongs.

In a multiple selection (several lines of the hit list are selected) field, RECORDPOS is used for numbering the lines; for example, if three lines are selected, all of the columns of the first selected line are numbered with RECORDPOS = 1, all columns of the second line with RECORDPOS = 2, and so on.

Field RETFIELD specifies the screen field belonging to the column if it's known. If there is only one return column, RETFIELD contains the field name passed in DYNPROFIELD. Type: DDSHRETVAL.

DYNPFLD_MAPPING

Assignment of the screen fields to the internal table. This internal table is used when you want to return other columns of the hit list in addition to the field RETFIELD to the screen. This table creates the link from the table fields to the screen fields. Components of this table are:

- FLDNAME: Field name from table FIELD_TAB or from the structure DDIC_STRUCTURE.
- DYNFLDNAME: Name of the screen field where the selected value should be returned. If the DDIC_STRUCTURE or FIELD_TAB isn't defined, the field names for the column of VALUE_TAB are assigned internally as F0001, F0002, and so on.

Type: DSELC.

[!] Exceptions

PARAMETER_ERROR

Incorrect parameter.

NO_VALUES_FOUND

No values found.

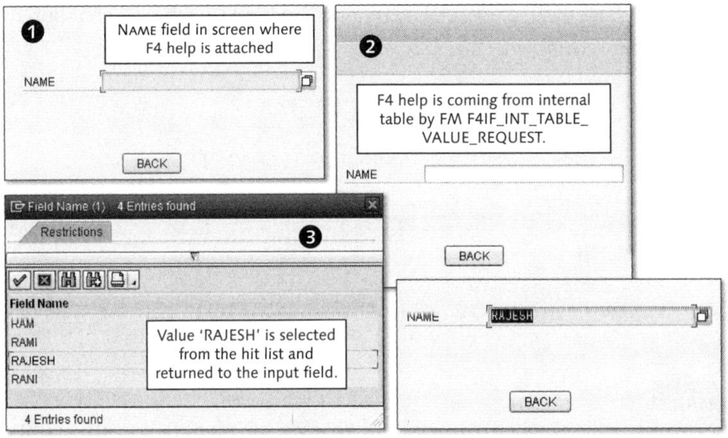

F4 Help Coming from Internal Table

---

50 F4IF_SELECT_VALUES

The FM selects the value of a search help. This FM calls the search help, and the contents returned from the hit list are collected in an internal table. You do this after using FM F4IF_GET_SHLP_DESCR (FM 48).

Example: You have a requirement to call the material search help MAT1, select some records from the hit list, and use those records for further processing. After you return the structure of the search help, execute this FM by passing the search help structure. You get the RESULT_TAB, which has all of the selected entries. You can use that table for further processing.

[↓] Import

SHLP

Search help details are returned in this FM. This is a mandatory parameter. Type: SHLP_DESCR.

MAXROWS

This defines the maximum number of rows that are returned from the selection into the internal table. With MAXROWS <> 0, the number of results are limited. Default: 0. Type: I.

SORT

When SORT = X, the selection is done directly on the database with ORDER BY PRIMARY KEY. This may mean that, despite a reduction of the number of hits with MAXROWS, selection takes a long time. Setting this parameter is therefore recommended only for small numbers of hits. Type: DDBOOL_D.

CALL_SHLP_EXIT

Search help exit is called as well if this parameter is set to X. Type: DDBOOL_D.

[↑] Export

MAXROWS_EXCEEDED

If there are more than MAXROWS hit, MAXROWS_EXCEEDED is returned as X. Type: DDBOOL_D.

Tables

RECORD_TAB

Do not use. The use of the parameters RECORD_TAB and RECDESCR_TAB isn't recommended because the interpretation of the data in Unicode systems is complicated.

RECDESCR_TAB

Do not use. Type: DFIES.

RETURN_TAB

This returns the selected value. The results of the selection cells are included. The number of entries in RETURN_TAB thus results from the product of the number of selected records with the number of export parameters of the search help. Type: DDSHRETVAL.

---

51 F4IF_SHLP_EXIT_EXAMPLE

This FM acts as a template and example for the interface and design of the search help exits in search help. All of the interface parameters defined here are mandatory for a FM to be used as a search help exit because the calling program doesn't know which parameters are actually used internally.

Example: You've defined search help ZSH_MATNR to select material. Now, you have to add some additional functionality in this search help, and you're required to create search help exit ZSH_MATNR_EXIT (which is a FM) for the search help. ZSH_MATNRZSH_MATNR_EXIT should have the same parameters that are present in this FM. It isn't recommended to add any other parameters in your new FM or to modify anything in this FM.

[Δ] Changing

SHLP

Elementary or current search help (if collective) is defined in this parameter. This defines the structure of the search help. Type: SHLP_DESCR.

CALLCONTROL

This parameter defines the control structure of the F4 process. This structure has fields such as display only, max records, multiselect, no max display, hide hit list, and so on. Type: DDSHF4CTRL.

Tables

SHLP_TAB

Table of elementary search helps. If you are in collective search help, this parameter defines the elementary search help for those collective search helps. Type: SHLP_DESCT.

RECORD_TAB

This table defines the hit list and its details. Type: SEAHLPRES.

---

52 F4IF_START_VALUE_REQUEST

The FM starts the F4 dialog screen with the known search help. The import parameter SHLP of this FM should have been filled with the export parameter SHLP returned from FM call F4IF_DETERMINE_SEARCHHELP (FM 46) or F4IF_GET_SHLP_DESCR (FM 48).

Example: You want to start the search help for a work center by clicking a button in the output screen. You've designed an elementary search for this. Also, only those work centers assigned to that plant should be displayed. In this case, you can call the search help by calling this FM. See FM F4IF_GET_SHLP_DESCR (FM 48).

[↓] Import

SHLP

Elementary or current search help (if collective) is defined in this parameter. This defines the structure of the search help. Mandatory. Type: SHLP_DESCR.

DISPONLY

This parameter defines whether the F4 help should behave like an input field or a pure display field. The F4 help for a field that isn't an input field doesn't permit you to select values from the hit list.

- [Space]: Selected value from the F4 hit list comes back to the screen field if the screen field is the input field.
- X: F4 help is called in display mode. It isn't possible to select the value from the hit list.

The default value is [Space]. Type: DDSHF4CTRL-DISPONLY.

MAXRECORDS

Maximum number of hits that can be displayed in the hit list table. The default value is 500. Type: DDSHF4CTRL-MAXRECORDS.

MULTISEL

IF this parameter is X, multiple selections are allowed. The default value is [Space]. Type: DDSHF4CTRL-MULTISEL.

CUCOL

Relative column position of the first popup. The default value is SY-CUCOL. Type: SY-CUCOL.

CUROW

Relative row position of the first popup. The default value is SY-CUROW. Type: SY-CUROW.

[↑] Export

RC

Return code. If the search help is displayed, the return code is 0. If some error occurred, the return code isn't equal to 0. Type: SY-SUBRC.

Tables

RETURN_VALUES

This table returns the selected value with the search help parameter. Type: DDSHRETVAL.

53 F4UT_OPTIMIZE_COLWIDTH

The FM is used to adjust the column width in the dialog box for displaying the hit list of an elementary search help.

Example: Your search help's hit list contains a NAME column whose width is very large but the actual values have lengths that are small. Because of this, you aren't able to see the other columns in the hit list. To shorten the width of the NAME column, you can fill in parameters and call this FM in the search help exit.

[↓] Import

PARAMETER_OPTIMIZE

List of parameters to be optimized if necessary. If you only want to optimize the selected columns of the hit list, you must pass the list of corresponding parameters this parameter. If you have more than one parameter, you must separate them with a comma (,) without a blank. Type: C.

[Δ] Changing

SHLP

Elementary or current search help (if collective) is defined in this parameter. This defines the structure of the search help. Type: SHLP_DESCR_T.

CALLCONTROL

This parameter defines the control structure of the F4 process. This structure has fields such as display only, max records, multiselect, no max display, hide hit list, and so on. Type: DDSHF4CTRL.

Tables

SHLP_TAB

Table of elementary search helps. If you are in collective search help, this parameter defines the elementary search help for those collective search helps. Type: SHLP_DESCR_TAB_T.

RECORD_TAB

This table defines the hit list and its details. Type: SEAHLPRES.

54 F4UT_PARAMETER_ALLOCATE

This FM reserves space for the parameter in RECORD_TAB. In some cases, the search help exit also needs the contents of additional fields of the selection method. Calling this FM ensures that space is reserved for the field in RECORD_TAB and its contents are also selected. In this case, the search help must have the parameter with the same name as the corresponding field. This parameter isn't required to be an export parameter or to appear in the hit list. The following other FMs are also used in this process:

- F4UT_PARAMETER_VALUE_GET (FM 57)
- F4UT_PARAMETER_RESULT_PUT (FM 55)

Example: Consider the standard search help SFLIGHT. In this search help, search help parameters SEATSMAX and SEATSOCC aren't displayed on the list of possible entries. To compute the number of free seats, you are required to get these field details. Fill in the parameter details and call this FM to ensure that space is reserved for the field in RECORD_TAB and its contents are also selected.

[↓] Import

PARAMETER

Name of the search help parameter. This is a mandatory parameter. Type: SHLPFIELD.

DEALLOCATE

If this parameter is set to X, the parameter won't be administered in RECORD_TAB. Type: DDBOOL_D.

[Δ] Changing

SHLP

Elementary or current search help (if collective) is defined in this parameter. This defines the structure of the search help. Type: SHLP_DESCR.

CALLCONTROL

This parameter defines the control structure of the F4 process. This structure has fields such as display only, max records, multiselect, no max display, hide hit list, and so on. Type: DDSHF4CTRL.

Tables

SHLP_TAB

Table of elementary search helps. If you are in collective search help, this parameter defines the elementary search help for those collective search helps. Type: SHLP_DESCT.

RECORD_TAB

This table defines the hit list and its details. Type: SEAHLPRES.

[!] Exceptions

PARAMETER_UNKNOWN

If the search help doesn't have any parameter with the given name, this exception is raised.

NO_SPACE

No more space in RECORD_TAB.

---

55 F4UT_PARAMETER_RESULTS_PUT

This FM can be used to change the result of the search help parameter, if this FM is called in the search help exit.

Example: In continuation from the example of FM F4UT_PARAMETER_VALUE_GET (FM 57), after the parameters for SEATSMAX and SEATSOCC are filled in, compute the free seats. The changed or new values of SEATSMAX and SEATSOCC should be saved into the search help parameter by calling this FM. Refer to FM F4UT_PARAMETER_ALLOCATE (FM 54).

[↓] Import

PARAMETER

Name of the search help parameter. This is a mandatory parameter. Type: SHLPFIELD.

OFF_SOURCE

Do not use due to problem with Unicode. Default: 0. Type: SY-FDPOS.

LEN_SOURCE

Do not use due to problem with Unicode. Type: SY-FDPOS.

VALUE

Results of parameter if only a header line.

FIELDNAME

Name of the source field in SOURCE_TAB. Type: DFIES-LFIELDNAME.

[Δ] Changing

SHLP

Elementary or current search help (if collective) is defined in this parameter. This defines the structure of the search help. Type: SHLP_DESCR_T.

CALLCONTROL

This parameter defines the control structure of the F4 process. This structure has fields such as display only, max records, multiselect, no max display, hide hit list, and so on. Type: DDSHF4CTRL.

Tables

SHLP_TAB

Table of elementary search helps. If you are in collective search help, this parameter defines the elementary search help for those collective search helps. Type: SHLP_DESCR_TAB_T.

RECORD_TAB

The results of the input help process are managed by the F4 processor in this table. This FM changes the values for a parameter in RECORD_TAB. You can decide if you only want to change the header of RECORD_TAB or if you want to change all of the rows of RECORD_TAB. Type: SEAHLPRES.

SOURCE_TAB

Results of parameter if a hit list. You can put the results in this table. Pass the name of the field of SOURCE_TAB containing the contents of the parameter to the FM in the parameter FIELDNAME. If the row type of SOURCE_TAB is elementary, pass the value * in FIELDNAME. The results are copied to the RECORD_TAB row by row. First rows are deleted at the end, or the initial row is added to RECORD_TAB to ensure that RECORD_TAB and SOURCE_TAB have the same number of rows.

[!] Exceptions

PARAMETER_UNKNOWN

If the search help doesn't have any parameter with the given name, this exception is raised.

---

56 F4UT_PARAMETER_SORT

The FM can be used to change how the hit list of an input help is sorted. Normally, the hit list is sorted according to the first 10 columns displayed. This FM can be used to sort the hit list according to another characteristic (which might not be in the hit list at all). The input help must be defined with an elementary search help. If the hit list is sorted according to more than one parameter, their names must be defined in PARAMETER_SORT separated with commas, and the commas may not be followed by a space. The parameters must be defined in decreasing order of their sort relevance.

Example: The hit list of your search help contains four columns: col1, col2, col3, and col4. You want the hit list to be displayed as col3 ascending and col4 ascending. You can pass col3 and col4 in the import parameter PARAMETER_SORT and execute this FM in the search help exit in the DISP time point to sort the hit list by col3 and col4.

[↓] Import

PARAMETER_SORT

Name of parameter to be used for sorting. Type: C.

LEADING_COLUMNS

Number of lead columns to be taken into consideration. Default: 0. Type: DDSHLCOL.

DESCENDING

If this parameter is set, then sort occurs in descending order. Type: DDBOOL_D.

[Δ] Changing

SHLP

Elementary or current search help (if collective) is defined in this parameter. This defines the structure of the search help. Type: SHLP_DESCR_T.

CALLCONTROL

This parameter defines the control structure of the F4 process. This structure has fields such as display only, max records, multiselect, no max display, hide hit list, and so on. Type: DDSHF4CTRL.

Tables

SHLP_TAB

Table of elementary search helps. If you are in collective search help, this parameter defines the elementary search help for those collective search helps. Type: SHLP_DESCR_TAB_T.

RECORD_TAB

This table defines the hit list and its details. Type: SEAHLPRES.

---

57 F4UT_PARAMETER_VALUE_GET

By calling this FM in the search help exit, you can find the current contents of a search help parameter. You must distinguish between the parameter contents that are used in the selection process of the input help process and the parameter contents that are determined by the selection process and the user selection.

This FM enables you to transfer the information about the specified parameters from RECORD_TAB to RESULTS_TAB in edited form.

Example: To fill the contents of the parameter in RECORD_TAB, you must get the current value of the search help parameter by calling this FM. When you get this value, you can modify it and save it.

[↓] Import

PARAMETER

Search help parameter whose contents are required. Mandatory. Type: SHLPFIELD.

OFF_RESULT

Do not use due to problems with Unicode. Type: SY-FDPOS.

LEN_RESULT

Do not use due to problem with Unicode. Type: SY-FDPOS.

FIELDNAME

Name of the target field in RESULTS_TAB. Type: DFIES-LFIELDNAME.

[↑] Export

VALUE

Contents of the parameter if a single value is returned. If VALUE has the value [Space], this means that either there are complex restrictions for the parameter or the unique contents are [Space]. SELOPT_TAB must be evaluated to determine which of these is true. Type: DDSHIFACE-VALUE.

[Δ] Changing

SHLP

Elementary or current search help (if collective) is defined in this parameter. This defines the structure of the search help. Type: SHLP_DESCR_T.

CALLCONTROL

This parameter defines the control structure of the F4 process. This structure has fields such as display only, max records, multiselect, no max display, hide hit list, and so on. Type: DDSHF4CTRL.

Tables

SHLP_TAB

Table of elementary search helps. If you are in collective search help, this parameter defines the elementary search help for those collective search helps. Type: SHLP_DESCR_TAB_T.

RECORD_TAB

This table defines the hit list and its details. Type: SEAHLPRES.

SELOPT_TAB

Possibly parameter assignments. The contents of the specified parameter used in the selection process are stored as complex restrictions in this parameter. Type: DDSHSELOPT.

RESULTS_TAB

Parameter information from the hit list.

[!] Exceptions

PARAMETER_UNKNOWN

If the search help doesn't have any parameters with the given name, this exception is raised.

58 F4UT_RESULTS_MAP

This FM is used to fill the hit list or to transfer the hit list from your own data structure into the data structure managed by the F4 processor. This FM is typically used whenever a search help exit itself-selects the values to be displayed at the SELECT point and thus fills an internal table with its own structure. This table must be passed to this FM in the parameter SOURCE_TAB.

This FM is usually called at SELECT time after the search help exit has executed the data selection itself. Afterward, CALLCONTROL-STEP should be changed to DISP because the F4 processor will otherwise again delete the values entered in RECORD_TAB at SELECT time.

Example: You want an F4 help on screen field P_SCR1, and you've created a search help but want the data for this search help to come from a different table, TAB1. You've created the search help exit, written your selection logic in the exit, and filled the internal table, ITAB1. Now, you want this ITAB1 to be displayed in the F4 help. For this, you need to use this FM to move the content of ITAB1 to RECORD_TAB. After you do this, the content of ITAB1 is displayed in the hit list.

[↓] Import

SOURCE_STRUCTURE

Name of the DDIC structure that SOURCE_TAB describes. If this parameter isn't supplied with data, the structure used to define SOURCE_TAB is taken from the program. Type: TABNAME.

APPLY_RESTRICTIONS

Only those lines are passed from SOURCE_TAB to RECORD_TAB that meet these selection criteria. If you observe these limitations yourself-already when setting up the SOURCE_TAB, you can leave this parameter at the initial value. Type: DDBOOL_D.

[Δ] Changing

SHLP

Elementary or current search help (if collective) is defined in this parameter. This defines the structure of the search help. Type: SHLP_DESCR_T.

CALLCONTROL

This parameter defines the control structure of the F4 process. This structure has fields such as display only, max records, multiselect, no max display, hide hit list, and so on. Type: DDSHF4CTRL.

Tables

SHLP_TAB

Table of elementary search helps. If you are in collective search help, this parameter defines the elementary search help for those collective search helps. Type: SHLP_DESCR_TAB_T.

RECORD_TAB

The results of the input help process are managed by the F4 processor in this table. This FM changes the values for a parameter in RECORD_TAB. You can decide if you only want to change the header of RECORD_TAB or if you want to change all of the rows of RECORD_TAB. Type: SEAHLPRES.

SOURCE_TAB

A user-defined internal table must be passed to this table parameter. The results are copied to RECORD_TAB row by row.

[!] Exceptions

ILLEGAL_STRUCTURE

The line type for SOURCE_TAB must have a flat structure; that is, a structure whose components all have an elementary type. Otherwise, this exception may trigger.

59 FLUSH_ENQUEUE

The FM sends the locks collected in the local lock container to the lock server. The lock request or lock release is placed in the local lock container. The lock requests and lock releases collected in this lock container can then be sent to the lock server at a later time as a group by calling this FM. The _COLLECT parameter of the FMs ENQUEUE and DEQUEUE of the lock determine if the lock should be collected or not. (Refer to parameter _COLLECT of FM DEQUEUE_E_TABLE [FM 43]).

Example: You want to lock four tables before modifying them. You call FM ENQUEUE_E_TABLE (FM 43) where _COLLECT = X; that is, the lock request or lock release is placed in the local lock container. You then use FM FLUSH_ENQUEUE to send the locks to the lock server.

[!] Exceptions

FOREIGN_LOCK

Locks aren't granted.

SYSTEM_FAILURE

System error in the lock management.

60 HELP_DOCULINES_SHOW

The FM is helpful to display the help text. Text can be passed from the table parameter HELPLINES.

Example: You want to display the help programmatically for the parameter field P_PARA1, which is to be displayed on event `AT SELECTION SCREEN ON HELP REQUEST`. You want that help text to be passed from the program. You can fill in the parameter details and call this FM, which displays the F1 help.

[↓] Import

CUCOL

This parameter displays the start column. Default: 10. Type: SY-CUCOL.

CUROW

This parameter displays the start row. Default: 3. Type: SY-CUROW.

HELP_INFOS

This structure transfers values to the help processor. This structure contains the details such as help processor call type, screen object ID, called program name, called screen number, and so on. This is a mandatory parameter. Type: HELP_INFO.

LINENR

Not used in the FM. Default: 1. Type: SY-LINCT.

NOT_HELP

If this parameter is set to X, and CLASSIC_SAPSCRIPT is set to X, additional buttons to start the workflow and maintain entries are displayed in the popup.

OVERLAY_HEADER

Long text saved in SAPscript. Type: THEAD.

SUPPRESS_VIEWER

Not used in the FM. Type: SY-BATCH.

CLASSIC_SAPSCRIPT

If this parameter is checked, the Performance Assistant screen isn't called, but help is displayed in a popup window. Type: CHAR1.

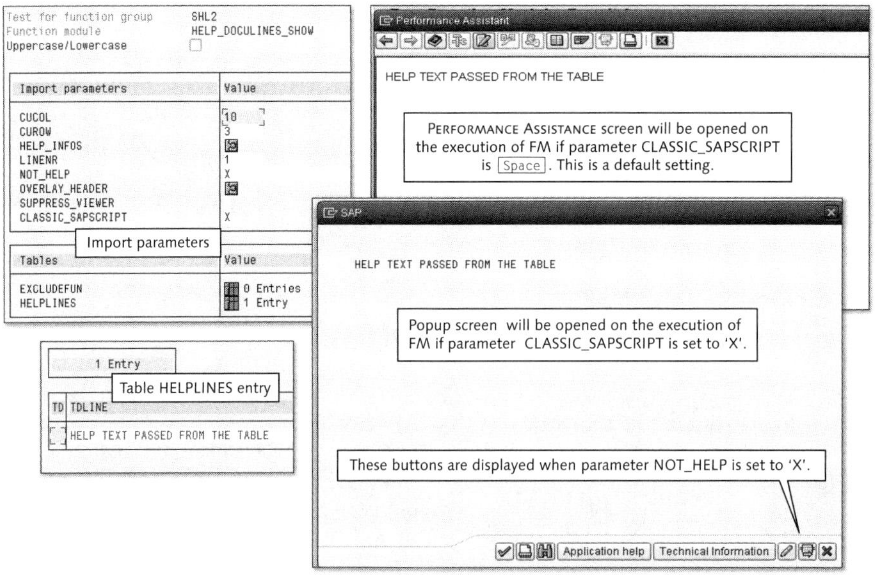

F1 Help Displayed from the Table Parameter

[↑] Export

HELP_POPUP_CANCEL

If you select the parameter CLASSIC_SAPSCRIPT, help is displayed in a popup. If you cancel the popup, this parameter is returned with a value of X.

Tables

EXCLUDEFUN

This table returns the list of function codes excluded from the help display.

HELPLINES

This table should include the message you want to display in F1 help. Type: TLINE.

61 HELP_OBJECT_SHOW

The FM shows the document in a popup. This self-custom help is available only if the standard F1 help isn't used. The document created from Transaction SE61 can be shown in F1. Parameters DOKCLASS and DOKNAME can be referred from table DOKIL.

Example: You want to display the help for the parameter field P_PARA1, which is to be displayed on event `AT SELECTION SCREEN ON HELP REQUEST`. The help should be the same one that you saved for the data element from Transaction SM61. You can fill in the parameter details and call this FM, which displays the F1 help.

[↓] Import

DOKCLASS

Document class. Name of the class to which you can refer the document for help. Mandatory. Type: DSYSH-DOKCLASS.

DOKLANGU

Language in which help should be displayed. This should always be SY_LANGU for self-programmed help. Type: DSYSH-DOKLANGU.

DOKNAME

Name of document. When used within the self-programmed help, the name of the data element should be in the form of *xxyyy*, where *xx* defines the segment, and *yyy* defines the message number. Mandatory. Type: DOKIL-OBJECT.

DOKTITLE

Window title. Type: DSYST-DOKTITLE.

CALLED_BY_PROGRAM

Program from where the help is called. Type: HELP_INFO-PROGRAM.

CALLED_BY_DYNP

Screen from where the help is called. Type: HELP_INFO-DYNPRO.

CALLED_FOR_TAB

F1 on a screen field corresponds with the reference structure to which this field is referenced. The information is required when displaying data element documentation for the TECHNICAL INFORMATION button. Type: HELP_INFO-TABNAME.

CALLED_FOR_FIELD

F1 corresponds to the name of the field on the screen and to the field name in the ABAP Data Dictionary reference referred structure. The information is required when displaying data element documentation for the TECHNICAL INFORMATION button. Type: HELP_INFO-FIELDNAME.

CALLED_FOR_TAB_FLD_BTCH_INPUT

Batch input relevance structure/field. This parameter defines the name of the ABAP Data Dictionary structure and referred reference in the field name, separated by "-". Type: HELP_INFO-DYNPROFLD.

MSG_VAR_1

Message: Variable 1. Value by which the first "&" in a message should be replaced. Type: SYMSGV.

MSG_VAR_2

Message: Variable 2. Value by which the second "&" in a message should be replaced. Type: SYMSGV.

MSG_VAR_3

Message: Variable 3. Value by which the third "&" in a message should be replaced. Type: SYMSGV.

MSG_VAR_4

Message: Variable 4. Value by which the fourth "&" in a message should be replaced. Type: SYMSGV.

CALLED_BY_CUAPROG

This parameter is used to fill the table HELP_INFOS, which helps transfer values to the help processor. Type: HELP_INFO-PROGRAM.

CALLED_BY_CUASTAT

This parameter is used to fill the table HELP_INFOS, which helps transfer values to the help processor. Type: HELP_INFO-PFKEY.

SHORT_TEXT

Short text that you want to display. Type: SY-BATCH.

CLASSIC_SAPSCRIPT

If this parameter is checked, the PERFORMANCE ASSISTANT screen isn't called, but help is displayed in a popup window. Type: CHAR1.

Tables

LINKS

This table isn't used. Type: TLINE.

[!] Exceptions

OBJECT_NOT_FOUND

Object isn't found.

SAPSCRIPT_ERROR

Error occurred while calling SAPscript.

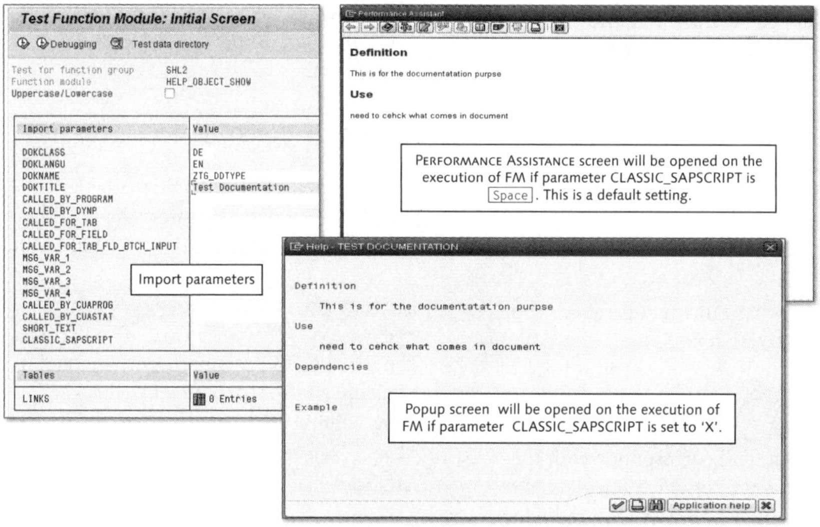

F1 Help Displayed from Data Element's Document

62 HELP_OBJECT_SHOW_FOR_FIELD

The FM shows any help document in a popup. Self-programmed help defined from the program is only acceptable if the standard F1 isn't used. This FM displays the documentation maintained in the data element for the database table or structure field. However, FM HELP_OBJECT_SHOW displays the documentation maintained in SAPscript by document class and document name.

Example: You want to display the F1 help of the data element that is maintained for the parameter field P_PARA1 on event `AT SELECTION SCREEN ON HELP REQUEST`. You can fill in the parameter details and call this FM, which displays the F1 help of the data element.

[↓] Import

DOKLANGU

Language in which help should be displayed. This should always be SY-LANGU for self-programmed help. Default: SY-LANGU. Type: DSYSH-DOKLANGU.

DOKTITLE

Window title. Type: DSYST-DOKTITLE.

CALLED_BY_TCODE

Transaction code from where the help is called. Type: HELP_INFO-TCODE.

CALLED_BY_PROGRAM

Program from where the help is called. Type: HELP_INFO-PROGRAM.

CALLED_BY_DYNP

Screen from where the help is called. Type: HELP_INFO-DYNPRO.

CALLED_FOR_TAB

F1 on a screen field corresponds with the reference structure to which this field is referenced. The information is required when displaying data element documentation for the TECHNICAL INFORMATION button. Type: HELP_INFO-TABNAME.

CALLED_FOR_FIELD

F1 corresponds to the name of the field on the screen and to the field name in the ABAP Data Dictionary reference referred structure. The information is required when displaying data element documentation for the TECHNICAL INFORMATION button. Type: HELP_INFO-FIELDNAME.

CALLED_FOR_TAB_FLD_BTCH_INPUT

Batch input relevance structure/field. This parameter defines the name of the ABAP Data Dictionary structure and referred reference in the field name, separated by "-". Type: HELP_INFO-DYNPROFLD.

CALLED_BY_CUAPROG

This parameter is used to fill the table HELP_INFOS, which helps transfer values to the help processor. Type: HELP_INFO-PROGRAM.

CALLED_BY_CUASTAT

This parameter is used to fill the table HELP_INFOS, which helps transfer values to the help processor. Type: HELP_INFO-PFKEY.

MERGE_DZ_IF_AVAILABLE

The DE and DZ descriptions are merged if available and this parameter is checked. Type: SY-BATCH.

MEMORYID

If the memory ID is attached to the field, you can pass it to the FM. Type: DFIES-MEMORYID.

EXPLICIT_MEMORYID

Memory ID is checked and returned if this parameter is set to X. Type: C.

Tables

LINKS

This table isn't in use. Type: TLINE.

EXCLUDEFUN

This table returns the list of function codes that are excluded from the help display.

[!] Exceptions

OBJECT_NOT_FOUND

Object isn't found.

SAPSCRIPT_ERROR

Error occurred while calling SAPscript.

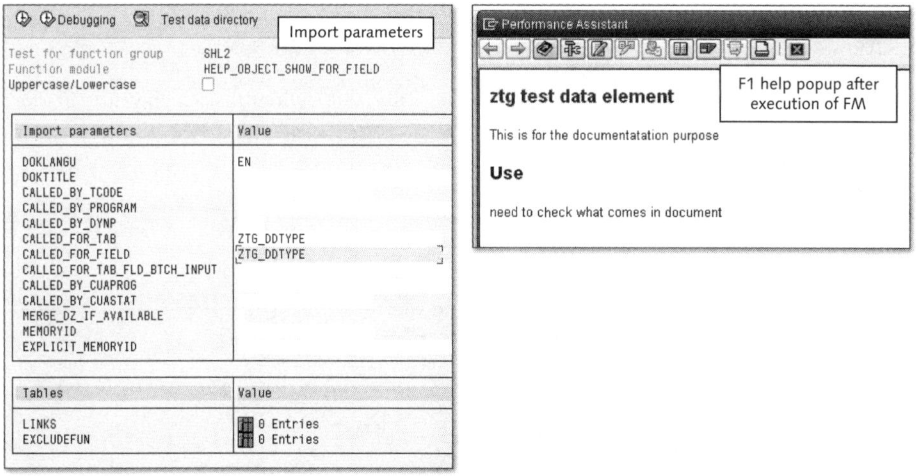

F1 Help Called from FM HELP_OBJECT_SHOW_FOR FIELD

63 HELP_START

The FM calls F1 help for Dynpro. This FM is called every time you press F1 help on the selection screen elements; that is, parameter and select options. You can also fill in the parameters of the table and call this FM within your program.

Example: You have a parameter P_MATNR in the selection screen that is linked to the data element MATNR. When you press the F1 help for that field, you are actually calling FM HELP_START in the background. After the FM is executed, you'll get the F1 help (PERFORMANCE ASSISTANT) screen.

[↓] Import

HELP_INFOS

Structure with transfer values for the help processor. This is a mandatory parameter. Type: HELP_INFO.

PROPERTY_BAG

This parameter defines the property set for the help field for the help processor.

[↑] Export

SELECTION

If the user has chosen data, this flag is set as X. Type: PHELP_PBAG.

SELECT_VALUE

Value that is selected comes in this field. Type: HELP_INFO-FLDVALUE.

RSMDY_RET

This is an interface between the screen and the selection report. Type: RSMDY.

Tables

DYNPSELECT

This table defines the screen selection fields in the process logic. Type: DSELC.

DYNPVALUETAB

This table defines the screen values in the process logic. Type: DVAL.

64 RESET_ENQUEUE

The FM deletes all of the collected locks from the local lock container. Locks that are already set aren't affected.

Example: You've collected four locks in the lock container, but you don't want to send the lock to the lock server; instead, you can reset or delete the collected lock from FM RESET_ENQUEUE. Executing this FM deletes all of the collected locks. This FM doesn't have any parameters.

65 VIEW_AUTHORITY_CHECK

This FM checks the authorization to a table or view. The authorization check depends on the access type (that is, whether the view is read or changed) table attributes, and the system configuration, that is, whether the data is allowed to be maintained and involves the authorization objects S_TABU_DIS, S_TABU_NAM, S_TABU_CLI, and S_TABU_LIN. The following other FMs are used in this process:

- VIEW_ENQUEUE (FM 66)
- VIEW_GET_DDIC_INFO (FM 67)
- VIEW_MAINTENANCE (FM 68)

Example: You have a maintenance view Z_VIEW1, which is used to maintain data in three tables: ZTAB1, ZTAB2, and ZTAB3. You've created a maintenance program for your view. To maintain this view, first you should check if you are authorized to make changes to this view with this FM. See FM VIEW_ENQUEUE (FM 66).

[↓] Import

VIEW_ACTION

The table access type is passed in this parameter. S is for read access, U is for change access, T is for transport table entries, which is handled like a change access. Default: 0.

VIEW_NAME

Name of the table/view to be checked. This is a mandatory parameter. Type: DD25V-VIEWNAME.

NO_WARNING_FOR_CLIENTINDEP

If this parameter is set to X, the warning popup to maintain client-specific tables is suppressed.

CHECK_ACTION_ALTERNATIVE

If the parameter CHECK_ACTION_ALTERNATIVE = X is set, when the authority check for the selected access type fails, the other access type is checked (display/change). If this check also fails, the exception NO_AUTHORITY is raised. Type: FLAG.

[↑] Export

GRANTED_ACTVT

If parameter CHECK_ACTION_ALTERNATIVE is active, the allowed activity is returned in this parameter. Possible return values are 00: No Authorization, 02: Change Authorization, and 03: Display Authorization. Type: ACTIV_AUTH.

[Δ] Changing

ORG_CRIT_INST

This defines the instantiated table organizational criteria. Type: VIMTY_OC_TYPE.

[!] Exceptions

INVALID_ACTION

The specified action isn't allowed.

NO_AUTHORITY

The specified table/view action isn't allowed.

NO_CLIENTINDEPENDENT_AUTHORITY

No authorization for maintaining cross-client tables/views.

TABLE_NOT_FOUND

The specified view/table doesn't exist.

NO_LINEDEPENDENT_AUTHORITY

No row authorization.

66 VIEW_ENQUEUE

This FM locks/unlocks data in the view structure; that is, this FM locks and unlocks all of the database tables that are located behind the view.

It's possible both to lock the entire table and to lock only a portion of the data records, which is defined by the given table with selection conditions. In the default case, a shared lock is set where other users can access the read data. When you are authorized to maintain the view, it's recommended to lock the view or table for which you want to maintain, change, or update the records so that no other user can change the same records you are changing.

Example: When you're modifying the record in the view, these tables should be locked so that other users can't make any modification on the table/record you are modifying. You should fill the appropriate parameters and call this FM to lock the underlying tables.

[↓] Import

ACTION

This parameter defines whether to lock or unlock the view tables. Value E is for lock, and D is for unlock. The default value is E. Type: C(1).

ENQUEUE_MODE

This parameter defines the lock mode: shared or exclusive. S is for shared, and E is for exclusive. Default: S. Type: C(1).

VIEW_NAME

Name of the view to be processed. This is a mandatory parameter. Type: OCUS-TABLE.

ENQUEUE_RANGE

This parameter defines whether to lock the entire table or only a portion of the table, which is defined by the selection criteria of the table. Only a portion of the table is locked if value X is passed. The default value is [Space]. Type: C(1).

Tables

SELLIST

This table defines the selection conditions that define the range of values to be locked. If this table is passed, and parameter ENQUEUE_RANGE is set to X, only values in this table are locked. Type: VIMSELLIST.

[!] Exceptions

CLIENT_REFERENCE

View is connected to another client.

FOREIGN_LOCK

View or subview is locked by another user.

INVALID_ACTION

Invalid value to parameter action is passed.

INVALID_ENQUEUE_MODE

Invalid enqueue mode is passed.

SYSTEM_FAILURE

System lock error.

TABLE_NOT_FOUND

Table not found in DDIC.

67 VIEW_GET_DDIC_INFO

The FM fills the view maintenance control block table; that is, it obtains the details of the view defined in the parameter VIEWNAME.

Example: To change the data in the view, you can get the view or table information by using this FM. Based on the table information you receive, you can maintain the view via the next FM, VIEW_MAINTENANCE (FM 68).

[↓] Import

VIEWNAME

Name of the view/table to be edited. This is a mandatory parameter. Type: OCUS-TABLE.

VARIANT_FOR_SELECTION

This parameter passes the selection condition variant. Type: TVIMV-VARIANT.

Tables

SELLIST

Table SELLIST contains the definition of the selection criteria and their values. Enter the name of the field in field VIEWFIELD of this table. Type: VIMSELLIST.

X_HEADER

The table contains the table or view header information from the ABAP Data Dictionary, such as the subset, selection conditions, maintenance status, and delivery class. The table also contains information about the table or view generation and event times for user form routines. Type: VIMDESC.

X_NAMTAB

This table contains the table or view field information from the ABAP Data Dictionary, such as structure field positions, key information, and maintenance characteristics of the field. Type: VIMNAMTAB.

[!] Exceptions

NO_TVDIR_ENTRY

View/table isn't entered in TVDIR.

TABLE_NOT_FOUND

View/table isn't found in the DDIC.

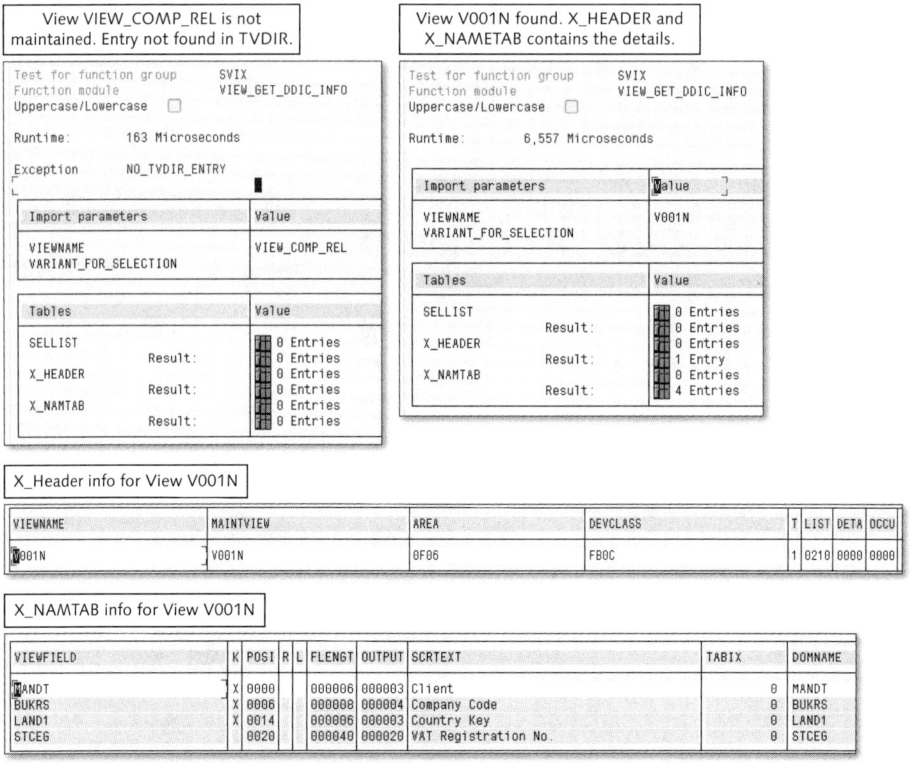

View Information

68 VIEW_MAINTENANCE

The FM calls extended table maintenance. It performs the selection, processing, and saving of the data and the interface layout.

The extended table maintenance dialog provides the possibility of processing table data in a consistent maintenance dialog, independent of whether access is to be made directly via the table or via a view defined in the ABAP Data Dictionary. Extended table maintenance can be called at three levels:

- **Highest level**
  Complete maintenance dialog for a table or a view is called. You can restrict the table entries that are read from the database using selections. FM VIEW_MAINTENANCE_CALL (FM 69) is used in this level.
- **Middle level**
  This level contains the actual maintenance dialog with the steps READ, EDIT, SAVE in this order. You must perform program

preparatory actions, such as locking the data, reading the ABAP Data Dictionary, and so on, yourself. FM VIEW_MAINTENANCE is used in this level.

- **Lowest level**
  In this level, you control all maintenance procedures individually from your program. You pass the actions to be performed to the FM as parameters. In this way, you can easily integrate individual maintenance steps (READ, EDIT, SAVE) in your program. FM VIEW_MAINTENANCE_LOW_LEVEL (FM 70) is used in this level.

Example: You can make modifications to the view or table entries by calling the table maintenance of the view or table. When you change the records, you should save the data back to the database. You should use this FM to call the table maintenance and saving the data to the view. See FM VIEW_ENQUEUE (FM 66).

[↓] Import

CORR_NUMBER

Correction number for changes made. This parameter is required if a parameter action has the value TRANSPORT. If you've passed a correction number, the user is prompted for a correction number in a dialog box when he calls the transport function. Type: E070-TRKORR.

VIEW_ACTION

This parameter defines the processing action to be performed. Value S is for display, U is for change, and T is for transport. This is a mandatory parameter.

VIEW_NAME

Name of the view to be edited. This is a mandatory parameter. Type: DD02V-TABNAME.

RFC_DESTINATION_FOR_UPGRADE

RFC destination of system for comparison/adjustment. Type: RFCDES-RFCDEST.

CLIENT_FOR_UPGRADE

This parameter checks the client where some comparison or adjustments are required. Type: SY-MANDT.

COMPLEX_SELCONDS_USED

This parameter defines whether DBA_SELLIST contains complex selection conditions. Type: TVDIR-FLAG.

NO_WARNING_FOR_CLIENTINDEP

If this parameter is set to X, the warning popup to maintain the client-specific table is suppressed.

OC_INST

This defines the organizational criteria object of class CL_VIEWFIELDS_ORG_CRIT. Type: CL_VIEWFIELDS_ORG_CRIT.

Tables

DBA_SELLIST

Table SELLIST contains the definition of the selection criteria and their values. Enter the name of the field in field VIEWFIELD of this table. Type: VIMSELLIST.

EXCL_CUA_FUNCT

CUA functions to be deactivated dynamically in the GUI at runtime. This table contains the functions that shouldn't be executed during the data processing. The FUNCTION field contains the function codes that are to be dynamically deactivated, and the MESSAGE field defines the online help auxiliary field. This isn't yet supported. Type: VIMEXCLFUN.

X_HEADER

This table contains the table or view header information from the ABAP Data Dictionary, such as the subset, selection conditions, maintenance status, and delivery class. The table also contains information about the table or view generation and event times for user form routines. Type: VIMDESC.

X_NAMTAB

This table contains the table or view field information from the ABAP Data Dictionary, such as structure field positions, key information, and maintenance characteristics of the field. Type: VIMNAMTAB.

DPL_SELLIST

This table display selection conditions. Type: VIMSELLIST.

[!] Exceptions

MISSING_CORR_NUMBER

Correction number missing.

NO_DATABASE_FUNCTION

Data retrieval function module missing.

NO_EDITOR_FUNCTION

Editor function module missing.

NO_VALUE_FOR_SUBSET_IDENT

No default value for SUBSET field. A structure field flagged as SUBSET is a field that can't be changed in maintenance and isn't considered part of the entries to be maintained. It restricts the data to be processed.

69 VIEW_MAINTENANCE_CALL

The FM calls the extended table maintenance at the highest level (for level descriptions, see FM VIEW_MAINTENANCE (FM 68)). This FM performs the following activities: authorization check, locking, fetching and formatting the necessary information from the ABAP Data Dictionary, selection, processing and saving data, restricting the data selection in the subset field dialog, and modifying dynamic interfaces such as menus and functions.

Example: You have maintenance view Z_VIEW1, which is used to maintain data in three tables: ZTAB1, ZTAB2, and ZTAB3. You want the complete maintenance dialog of the view to be called. You want to call the maintenance dialog programmatically via Transaction SM30. You can fill the parameter and call this FM in your program. The maintenance dialog is displayed after the execution of the FM.

[↓] Import

ACTION

This parameter defines the processing action to be performed. Value S is for display, U is for change, and T is for transport. This is a mandatory parameter.

CORR_NUMBER

Correction number for changes made. This parameter is required if the parameter action has the value TRANSPORT. If you have passed a correction number, the user is prompted for a correction number in a dialog box when he calls the transport function. Type: E070-TRKORR.

GENERATE_MAINT_TOOL_IF_MISSING

This isn't used anymore.

This parameter was used to create a maintenance dialog with the table maintenance generator for quick, temporary access to a table.

SHOW_SELECTION_POPUP

If this flag is set, it displays the popup that is passed in table DBA_SELLIST. The selection condition can be revised again by the user.

If the parameter is [Space], the user can't modify the data passed in the DBA_SELLIST selection condition.

VIEW_NAME

Name of the view to be edited. This is a mandatory parameter. Type: DD02V-TABNAME.

NO_WARNING_FOR_CLIENTINDEP

If this parameter is set to X, the warning popup to maintain the client-specific table is suppressed.

RFC_DESTINATION_FOR_UPGRADE

RFC destination of system for comparison/adjustment. Type: RFCDES-RFCDEST.

CLIENT_FOR_UPGRADE

This parameter checks the client where some comparison or adjustments are required. Type: SY-MANDT.

VARIANT_FOR_SELECTION

Selection conditions variant. Type: TVIMV-VARIANT.

COMPLEX_SELCONDS_USED

This parameter defines whether DBA_SELLIST contains complex selection conditions. Type: TVDIR-FLAG.

CHECK_DDIC_MAINFLAG

This parameter checks whether maintenance is allowed for the view.

SUPPRESS_WA_POPUP

This parameter suppresses the SPECIFY WORK AREA dialog box if a value is passed as X in this parameter. Type: XFELD.

Tables

DBA_SELLIST

Table SELLIST contains the definition of the selection criteria and their values. Enter the name of the field in field VIEWFIELD of this table. Type: VIMSELLIST.

EXCL_CUA_FUNCT

CUA functions to be deactivated dynamically in the GUI at runtime. This table contains the functions that shouldn't be executed during the data processing. The FUNCTION field contains the function codes that are to be dynamically deactivated, and the MESSAGE field defines the online help auxiliary field. This isn't yet supported. Type: VIMEXCLFUN.

[!] Exceptions

CLIENT_REFERENCE

View is chained to another client.

FOREIGN_LOCK

View/table is locked by another user.

INVALID_ACTION

ACTION contains invalid values.

NO_CLIENTINDEPENDENT_AUTH

No authorization for maintaining cross-client tables/views.

NO_DATABASE_FUNCTION

No data capture/disposal function module.

NO_EDITOR_FUNCTION

Editor function module missing.

NO_SHOW_AUTH

No display authorization.

NO_TVDIR_ENTRY

View/table isn't in TVDIR.

NO_UPD_AUTH

No maintenance or display authorization.

ONLY_SHOW_ALLOWED

Display but not maintenance authorization.

SYSTEM_FAILURE

System lock error.

UNKNOWN_FIELD_IN_DBA_SELLIST

Selection table contains unknown fields.

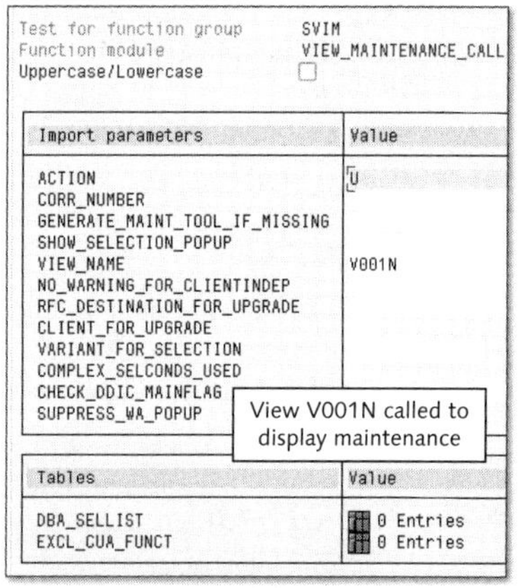

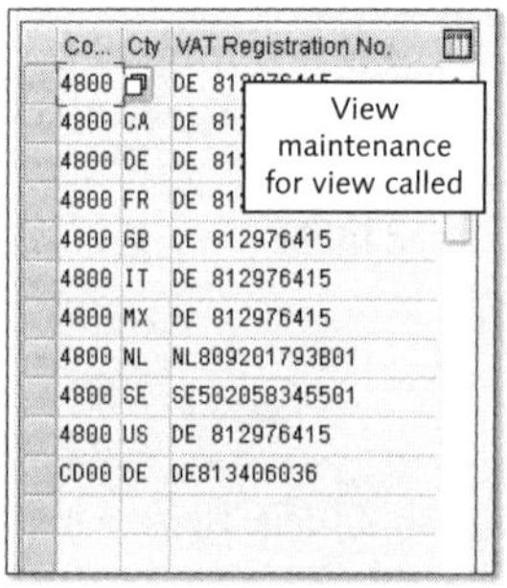

Called Maintenance Dialog for the View

VIEW_NOT_FOUND

View/table not found in DDIC.

MAINTENANCE_PROHIBITED

View/table can't be maintained according to DDIC.

70 VIEW_MAINTENANCE_LOW_LEVEL

The FM calls the extended table maintenance dialog at the lowest level (for level descriptions, see FM VIEW_MAINTENANCE [FM 68]). You must also perform all of the activities that the maintenance dialog otherwise performs; that is, authorization checking; locking, fetching and formatting the required information from the ABAP Data Dictionary; selecting, editing, and saving the data; restricting the data selection in the dialog for subset fields; and modifying dynamic interfaces (menus and functions).

Example: You have a maintenance view Z_VIEW1 that is used to maintain data in three tables: ZTAB1, ZTAB2, and ZTAB3. You want to call the maintenance dialog programmatically via Transaction SM30. You want everything to be called manually and the maintenance dialog to be called by the lowest level. Pass the actions to be performed in the FM as parameters and call this FM. Maintenance dialog is displayed.

[↓] Import

CORR_NUMBER

Correction number for changes made. This parameter is required if the parameter action has a value of TRANSPORT. If you have passed a correction number, the user is prompted for a correction number in a dialog box when he calls the transport function. Type: E070-TRKORR.

FCODE

Required function of the module. Valid codes are READ (read the data from the database), EDIT (edit the data), RDED (read and edit the data), SAVE (write the data in the database), ORGL (reset all marked entries in EXTRACT), and ORGD (reset the header entry in EXTRACT). The default value is RDED.

VIEW_ACTION

This parameter defines the processing action to be performed. Value S is for display, U is for change, and T is for transport. The default value is S.

VIEW_NAME

Name of the view to be edited. This is a mandatory parameter. Type: DD02V-TABNAME.

RFC_DESTINATION_FOR_UPGRADE

RFC destination of system for comparison/adjustment. Type: RFCDES-RFCDEST.

CLIENT_FOR_UPGRADE

This parameter checks the client where some comparison or adjustments are required. Type: SY-MANDT.

COMPLEX_SELCONDS_USED

This parameter defines whether DBA_SELLIST contains complex selection conditions. Type: TVDIR-FLAG.

NO_WARNING_FOR_CLIENTINDEP

If this parameter is set to X, the warning popup to maintain client-specific tables is suppressed.

OC_INST

This defines the organizational criteria object of class CL_VIEWFIELDS_ORG_CRIT. Type: CL_VIEWFIELDS_ORG_CRIT.

[↑] Export

LAST_ACT_ENTRY

Index of cursor position in the display table. On return, this field contains the last position of the cursor in the table EXTRACT (important for user processing). Type: SY-TABIX.

UCOMM

Last user command in view maintenance. On return from the FM, the parameter UCOMM contains the last user maintenance command. The following commands must be handled by the calling program:

- SAVE: Write the data in the database.
- ORGL: Reset all marked entries in EXTRACT.
- ORGD: Reset the header entry in EXTRACT.
- ANZG: Change the processing mode by choosing CHANGE • DISPLAY.
- AEND: Change the processing mode display/transport to change.
- TRSP: Change the processing mode to change to transport.
- ENDE: End processing.
- ATAB: Get another view for processing.
- BACK: Return to calling point, XXXX: User-specific command.

UPDATE_REQUIRED

This parameter on return indicates whether data have been changed and must therefore be saved. In this case, the parameter should have the value X.

CORR_NUMBER

Correction number for changes made. This correction number is only required in connection with the action transport (parameter VIEW_ACTION).

If you haven't passed a correction number, the user is prompted in a popup to enter a correction number when the transport function is called. Type: E070-TRKORR.

ACTION_OUT

This returns a value when the valid action is found. Values include S, U, or T. Type: SYCHAR01.

Tables

CORR_KEYTAB

Table with keys of entries to be transported. This table passes the table entry keys for the current correction. This table is significant for the transport mode (parameter VIEW_ACTION = T). Type: E071K.

DBA_SELLIST

Table SELLIST contains the definition of the selection criteria and their values. Enter the name of the field in field VIEWFIELD of this table. Type: VIMSELLIST.

DPL_SELLIST

Display selection conditions. This table contains the selection conditions for the creation of a subwork area for further processing. Type: VIMSELLIST.

EXCL_CUA_FUNCT

CUA functions to be deactivated dynamically in the GUI at runtime. This table contains the functions that shouldn't be executed during the data processing. The FUNCTION field contains the function codes that are to be dynamically deactivated, and the MESSAGE field defines the online help auxiliary field. This isn't yet supported. Type: VIMEXCLFUN.

EXTRACT

Display table. This table contains the display data of the last selection.

TOTAL

Table with all data read from the database. This table contains all of the data of the table.

X_HEADER

The table contains the table or view header information from the ABAP Data Dictionary, such as a subset, selection conditions, maintenance status, and delivery class. The table also contains information about the table or view generation and event times for user form routines. Type: VIMDESC.

X_NAMTAB

This table contains the table or view field information from the ABAP Data Dictionary, such as structure field positions, key information, and maintenance characteristics of the field. Type: VIMNAMTAB.

[!] Exceptions

FUNCTION_NOT_FOUND

Function module not found for view/table.

MISSING_CORR_NUMBER

Correction number missing.

NO_VALUE_FOR_SUBSET_IDENT

No default value for a subset field.

SAVING_CORRECTION_FAILED

Correction entries save error.

71 VIEW_MAINTENANCE_SINGLE_ENTRY

This FM maintains an individual entry in the extended table maintenance.

Example: There is a button on the ALV output screen such as CHANGE AN ENTRY. You select an entry from the ALV output and click the button. You want to open the extended table maintenance dialog where you can change the selected entry and save the entry to the database. You can achieve this task by filling the parameters so that this FM is called when the button is pressed. This FM displays the extended table maintenance where you can edit the selected entry.

[↓] Import

ACTION

This parameter defines the processing action to be performed. Value S is for display, U is for change, and T is for transport. The default value is S. This is a mandatory parameter. Type: VIMTY_TCODE.

CORR_NUMBER

Correction number for changes made. This parameter is required if the parameter action has a value of TRANSPORT. If you've passed a correction number, the user is prompted for a correction number in a dialog box when he calls the transport function. Type: E070-TRKORR.

VIEW_NAME

Name of the view to be edited. This is a mandatory parameter. Type: DD02V-TABNAME.

NO_WARNING_FOR_CLIENTINDEP

If this parameter is set to X, the warning popup to maintain the client-specific table is suppressed. Type: C.

RFC_DESTINATION_FOR_UPGRADE

RFC destination of system for comparison/adjustment. Type: RFCDES-RFCDEST.

CLIENT_FOR_UPGRADE

This parameter checks the client where some comparison or adjustments are required. Type: SY-MANDT.

VARIANT_FOR_SELECTION

This parameter checks the selection conditions variant. Type: TVIMV-VARIANT.

NO_TRANSPORT

If this parameter is set to X, changes made aren't recorded in transport. Type: TVDIR-FLAG.

SUPPRESSDIALOG

If this parameter is set to X, it suppresses the popup dialogs. Type: TVDIR-FLAG.

INSERT_KEY_NOT_FIXED

If this parameter is X, the key for the insert can be changed. Type: TVDIR-FLAG.

COMPLEX_SELCONDS_USED

This parameter defines whether DBA_SELLIST contains complex selection conditions. Type: TVDIR-FLAG.

[↑] Export

CORR_NUMBER

Correction number for changes made. This correction number is only required in connection with the action transport (parameter VIEW_ACTION).

If you haven't passed a correction number, the user is prompted in a popup to enter a correction number when the transport function is called. Type: E070-TRKORR.

[Δ] Changing

ENTRY

Entry to be maintained.

Tables

DBA_SELLIST

Table SELLIST contains the definition of the selection criteria and their values. Enter the name of the field in field VIEWFIELD of this table. Type: VIMSELLIST.

EXCL_CUA_FUNCT

CUA functions to be deactivated dynamically in the GUI at runtime. This table contains the functions that shouldn't be executed during the data processing. The FUNCTION field contains the function codes that are to be dynamically deactivated, and the MESSAGE field defines the online help auxiliary field. This isn't yet supported. Type: VIMEXCLFUN.

[!] Exceptions

ENTRY_ALREADY_EXISTS

Entry exists, it cannot be created.

ENTRY_NOT_FOUND

Entry not found, it cannot be edited.

CLIENT_REFERENCE

View is chained to another client.

FOREIGN_LOCK

View/table is locked by another user.

INVALID_ACTION

ACTION contains invalid values.

NO_CLIENTINDEPENDENT_AUTH

No authorization for maintaining cross-client tables/views.

NO_DATABASE_FUNCTION

Fct. mod. for data capture/disposal is missing.

NO_EDITOR_FUNCTION

Editor function module missing.

NO_SHOW_AUTH

No display authorization.

NO_TVDIR_ENTRY

View/table is not in TVDIR.

NO_UPD_AUTH

No maintenance or display authorization.

SYSTEM_FAILURE

System lock error.

UNKNOWN_FIELD_IN_DBA_SELLIST

Selection table contains unknown fields.

VIEW_NOT_FOUND

View/table not found in DDIC.

72 VIEWCLUSTER_MAINTENANCE_CALL

The FM calls the view cluster maintenance. The view cluster maintenance maintains data across several tables. It combines the maintenance of several tables or views (objects) into one maintenance dialog. The objects are maintained by the extended table maintenance.

Example: You have a view cluster ZVC1, which maintains two views and two tables. You want to modify the view cluster and need to call its maintenance dialog to modify it from your ABAP program. You can do so by filling in the parameters and calling this FM.

[↓] Import

VIEWCLUSTER_NAME

View cluster name you want to maintain. This is a mandatory parameter. Type: VCLDIR-VCLNAME.

START_OBJECT

One object is flagged as the initial object in a view cluster definition. This initial object is presented to the user first when the view cluster is called. If you want to use another object as the initial object, specify it in the START_OBJECT parameter. If nothing is specified, the initial object specified in the view cluster definition is the default object. Type: VCLSTRUC-OBJECT.

MAINTENANCE_ACTION

This parameter defines the processing action to be performed. Value S is for display, U is for change, and T is for transport. The default value is S. This is a mandatory parameter.

READ_KIND

The parameter READ_KIND influences when the view cluster object data is read in.

- [Space]: Complete. All view cluster object data is read in immediately at the start of view cluster maintenance.
- T: Subtree. At first, only the data of objects that depend on the initial object are read in. The data of other root objects (objects with no predecessor) and their dependent objects are read in as required, by subtree.

If a view cluster has several root objects, READ_KIND = T improves performance at the start of view cluster maintenance.

For view clusters with only one root object, this parameter makes no difference. The default value is [Space].

SHOW_SELECTION_POPUP

If this parameter is set to X, it displays the selection condition dialog box.

CORR_NUMBER

Correction number for changes made. This parameter is required if the parameter action has a value of TRANSPORT. If you've passed a correction number, the user is prompted for a correction number in a dialog box when he calls the transport function. Type: E070-TRKORR.

NO_WARNING_FOR_CLIENTINDEP

When data in client-independent tables or views are to be maintained, there is usually a warning about the possible effects on other clients. If this parameter is set to X, the warning popup to maintain the client-specific table is suppressed.

RFC_DESTINATION

RFC destination of system for comparison/ adjustment. Type: RFCDES-RFCDEST.

SUPPRESS_WA_POPUP

Some maintenance views contain key fields with the maintenance attribute *Subset", which requires the user to specify a value for the field at the start of the view maintenance dialog. If the field has a parameter ID that contains previous values or user parameters, a proposal is offered.

If the flag SUPPRESS_WA_POPUP is set, the input dialog box is suppressed if the field already has a parameter ID value. The user is then not prompted to specify a value for the field more than once per SAP ECC logon.

- X: Suppress dialog box if all input fields already have values from the parameter ID.
- [Space]: Always display input dialog box.

Type: XFELD.

Tables

DBA_SELLIST

Table SELLIST contains the definition of the selection criteria and their values. Enter the name of the field in field VIEWFIELD of this table. Table: VIMSELLIST.

DBA_SELLIST_CLUSTER

The table DBA_SELLIST_CLUSTER contains database access selection conditions for any view cluster objects. They are passed on to dependent objects according to the relationships specified in the view cluster definition field dependencies.

Selection conditions for fields of a dependent object that are related to the higher-level object according to the view cluster definition field dependencies are ignored. Table: VCLTY_SELLIST_TABLE.

EXCL_CUA_FUNCT_ALL_OBJECTS

The table EXCL_CUA_FUNCT_ALL_OBJECTS must contain all of the functions that shouldn't be callable while processing view cluster objects. These functions are deactivated dynamically in the GUI of each view cluster object at runtime. This parameter is ignored if the parameter EXCL_CUA_FUNCT_CLUSTER is specified. Table: VIMEXCLFUN.

EXCL_CUA_FUNCT_CLUSTER

The table EXCL_CUA_FUNCT_CLUSTER contains all of the functions for any view cluster objects that aren't callable while processing this object. These functions are deactivated dynamically in the GUI of this object at runtime. Table: VCLTY_EXCLFUNCT_TABLE.

DPL_SELLIST_FOR_START_OBJECT

The table DPL_SELLIST_FOR_START_OBJECT contains selection conditions for entries to be displayed for the initial view cluster object, if it isn't a root object (object without predecessor). If you go directly into a dependent object when you enter the view cluster, the entries to be displayed in this object can't be selected via the selected entries in the higher-level object, as in normal navigation. The view cluster maintenance displays a popup by default to enter the missing conditions.

These conditions can be specified in the parameter DPL_SELLIST_FOR_START_OBJECT. The standard popup is then not called, if a condition is specified for each field in the initial object that depends on a predecessor field. Table: VIMSELLIST.

[!] Exceptions

CLIENT_REFERENCE

Object chained in another client.

FOREIGN_LOCK

Object is locked by another user.

VIEWCLUSTER_NOT_FOUND

View cluster does not exist.

VIEWCLUSTER_IS_INCONSISTENT

View cluster definition is inconsistent.

MISSING_GENERATED_FUNCTION

Generated view maintenance modules missing.

NO_UPD_AUTH

No maintenance authorization.

NO_SHOW_AUTH

No display authorization

OBJECT_NOT_FOUND

Object not found in DDIC.

NO_TVDIR_ENTRY

Object not in TVDIR.

NO_CLIENTINDEP_AUTH

No maintenance authorization for client-independent objects.

INVALID_ACTION

MAINTENANCE_ACTION contains invalid value.

SAVING_CORRECTION_FAILED

Correction entries save error.

SYSTEM_FAILURE

System locking error.

UNKNOWN_FIELD_IN_DBA_SELLIST

Selection table contains unknown fields.

MISSING_CORR_NUMBER

Correction number missing.

Part III
Conversions

This part includes function modules (FMs) that are related to conversions of values from one format to another format, such as from character to float, from one SI unit to another, from spool to PDF, and so on. FMs related to date, time, unit of measure, round off, and so on are also covered in this part.

This part also includes FMs of the CONVERSION_EXIT type such as CONVERSION_EXIT_ALPHA_INPUT and CONVERSION_EXIT_ALPHA_OUTPUT, which are used to convert the SAP internal format to user output format, and vice versa. Conversion exits are used to convert the SAP internal format to the user's display format and vice versa. The CONVERSION_EXIT FMs generally come in a pair:

- CONVERSION_EXIT_*_INPUT converts external formats to internal formats.
- CONVERSION_EXIT_*_OUTPUT converts internal formats to external formats.

The * represents a five-character conversion routine that is entered in the domain of the table field you want to convert.

The FMs that we've included in this part are very useful in your day-to-day programming where you're required to convert one form to another to achieve your business requirements.

73 ADD_TIME_TO_DATE

This FM is used to add a specified amount of time (i.e., day, month, week, or year) to the date you provide in the input parameters. You can also subtract the time by providing a negative value in parameter I_TIME.

Example: You're working on the production of a car. You have a deadline to produce the car in 10 days. Based on the start date, you want to calculate the end date so you can give it to the customer. Fill in the parameters to add time, execute the FM, and you'll get the end date.

[↓] Import

I_IDATE

Provide the input date on which you want to add the time. Mandatory. Type: SY-DATUM.

I_TIME

Time that you want to add. Mandatory. Type: MARA-MHDHB.

I_IPRKZ

This field defines the unit for the time such as day, week, month, or year. You need to call conversion exit CONVERSION_EXIT_PERKZ_INPUT to convert the format from output to input. Mandatory.

- Day has output format as D and input format as [Space].
- Week has output format as W and input format as 1.
- Month has output format as M and input format as 2.
- Year has output format as Y and input format as 3.

Type: MARA-IPRKZ.

I_RDMHD

Rounding rule for calculation.

- [Space]: No rounding off.
- -: Start of chosen period (week, month, or year).
- +: End of chosen period (week, month, or year).
- F: Start of the following period (week, month, year).

Type: MARA-RDMHD.

[↑] Export

O_IDATE

Return date after processing. Type: SY-DATUM.

[!] Exceptions

INVALID_PERIOD

Period is not valid.

INVALID_ROUND_UP_RULE

Round off rule is invalid.

INTERNAL_ERROR

Some internal error while executing FM.

74 C14B_ADD_TIME

This FM adds up times with date correction on the date and time given in the import parameter.

Example: You have a parameter P_TIME in your production order selection screen, and you have a requirement that whatever time the end user enters in this parameter, you want to get the production order up to 1 hour from that time from the database table. After you fill in the parameters and execute, you'll get the updated time and date in the export parameters. Based on the time, you can fetch the data from your production order database table and display it in the output.

[↓] Import

I_STARTTIME

Time to which import parameter I_ADDTIME is added. Type: T.

I_STARTDATE

Date that may be corrected. Type: D.

I_ADDTIME

Time to which I_STARTTIME is added. Type: T.

[↑] Export

E_ENDTIME

Output time. Type: T.

E_ENDDATE

Output date. Type: D.

Time Addition on the Given Time

75 CALCULATE_EXCHANGE_RATE

This FM finds currency rates based on amounts, currency keys, and dates. A translation rate is determined from the amounts entered in the import parameters. Because the rate is also dependent on the units of each individual currency, (e.g., a DEM/ITL rate of 1.67 with the units 1 DEM and 1,000 ITL means that 1 DEM is equal to 1,670 ITL), Table TCURR must also be read before the rate can be established. For this, you need both currency keys, an exchange rate type, and a validity date. These are the key fields of database table TCURR. The factors determined in this way are transferred to the calling program along with the determined rate. If exchange rate fixing is defined for exchange rate type TYPE_OF_RATE, this information is transferred to the calling program. If one amount is calculated from the other, even where a fixed exchange rate is used, this amount is returned instead of the rate calculated. All import parameters are mandatory.

Example: You're working on a program to display the exchange rate on a daily basis; that is, you want to know today's exchange rate from INR to USD ($). You can use this FM to convert the exchange rate.

[↓] Import

CLIENT

Client. Default: SY-MANDT. Type: SY-MANDT.

DATE

Conversion date. This date is read from the table TCURR. Type: SYST-DATUM.

FOREIGN_AMOUNT

Amount in foreign currency.

FOREIGN_CURRENCY

Currency key for foreign currency.

LOCAL_AMOUNT

Amount in local currency.

LOCAL_CURRENCY

Currency key for local currency.

TYPE_OF_RATE

Exchange rate type. Default: M.

- M: Average rate.
- G: Bank buying rate.
- B: Bank selling rate.

[↑] Export

EXCHANGE_RATE

Exchange rate calculated from the amount entered.

FOREIGN_FACTOR

Factor for the foreign currency units. Type: TCURR-FFACT.

LOCAL_FACTOR

Factor for the local currency units. Type: TCURR-TFACT.

DERIVED_RATE_TYPE

Exchange rate type derived more than once. Type: TCURR-KURST.

FIXED_RATE

Returns the fixed exchange rate if present.

[!] Exceptions

NO_RATE_COMPUTABLE

Translation not possible.

NO_RATE_FOUND

Entry not found in table TCURR.

RATE_TOO_BIG

The exchange rate determined exceeds acceptable limits.

NO_FACTORS_FOUND

No entry found in table TCURF.

NO_SPREAD_FOUND

Spread not entered in table TCURS.

DERIVED_2_TIMES

Exchange rate type derived more than once.

OVERFLOW

Arithmetical overflow.

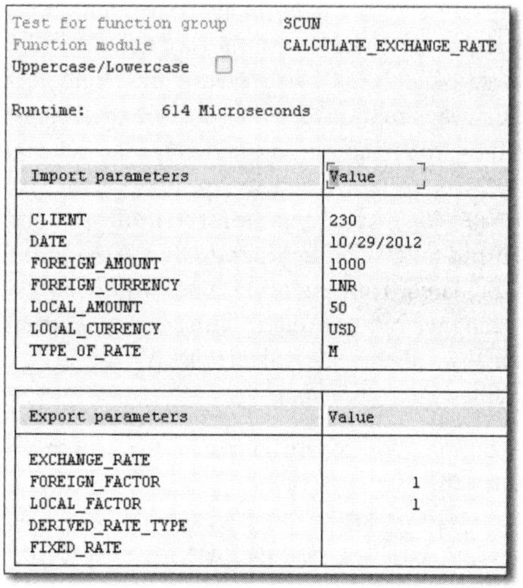

Test for function group SCUN
Function module CALCULATE_EXCHANGE_RATE
Uppercase/Lowercase

Runtime: 614 Microseconds

Import parameters	Value
CLIENT	230
DATE	10/29/2012
FOREIGN_AMOUNT	1000
FOREIGN_CURRENCY	INR
LOCAL_AMOUNT	50
LOCAL_CURRENCY	USD
TYPE_OF_RATE	M

Export parameters	Value
EXCHANGE_RATE	
FOREIGN_FACTOR	1
LOCAL_FACTOR	1
DERIVED_RATE_TYPE	
FIXED_RATE	

Conversion from INR to USD

---

76 CF_UT_UNIT_CONVERSION

This FM converts the quantity from one unit of measurement to another. You can also use it when a material has a base unit of measurement and an alternate unit of measure, or is capable of being converted to an alternative unit of measure per table T006.

Example: You have a material where you want to store the unit as centimeters for a material, but you want to display it in meters. In this case, you can fill in the parameters and can use this FM to convert the unit.

[↓] Import

MATNR_IMP

Material number. Type: MARA-MATNR.

MEINS_IMP

Material base unit of measure. Type: MARA-MEINS.

UNIT_NEW_IMP

New unit. This is mandatory parameter. Type: T006-MSEHI.

UNIT_OLD_IMP

Old unit. This is mandatory parameter. Type: T006-MSEHI.

VALUE_OLD_IMP

Old value. This is mandatory parameter. Type: PLFH-MGVGW.

[↑] Export

VALUE_NEW_EXP

New value of new unit. Type: PLFH-MGVGW.

[!] Exception

OVERFLOW

Field overflow.

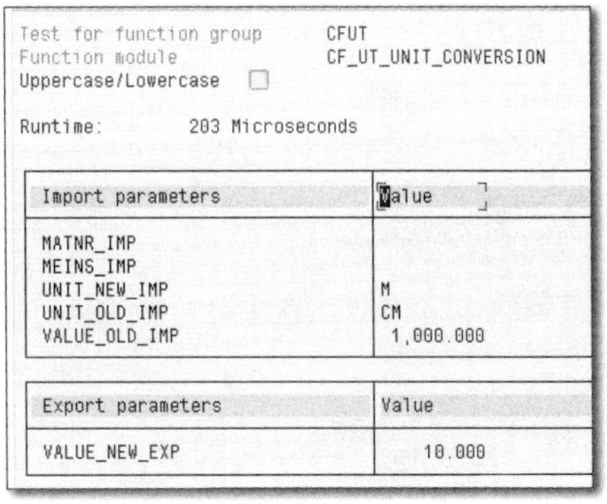

Converting Unit from Centimeters to Meters

77 CHAR_FLTP_CONVERSION

This FM converts a character types format field to a float type format. This FM checks whether the STRING parameter (format CHAR) can be interpreted as a floating point number. If not, an error message (exception STRING_NOT_FLTP) is output, so you can avoid an ABAP runtime error. If the check is successful, the floating point number is put in the parameter FLSTR (format FLTP).

Example: You have a field FLD1 of type character, but the field is used to save the numeric values. You need to add a value to that field say, `FLD1contains value = 100`, and you want to add 10.012 in the field. In this case, you have to first convert the character type into a float type variable by calling this FM, and then you have to add value 10.012 into the variable.

[↓] Import

DYFLD

Screen field name for cursor positioning. If there is a STRING parameter error, the cursor is positioned on the field whose name is in the DYFLD parameter before the error message. In step loops, SY-STEPL is used. The default value is [Space].

MASKN

This FM can create an input mask. The edit symbol used for a numerical value (e.g., "_") must be passed in the MASKN parameter so that it can be interpreted.

MAXDEC

Maximum number of input decimal places. The default value is 16. Type: T006-DECAN.

MAXEXP

Maximum gross power of 10. The default value is 59+. Type: T006-EXPON.

MINEXP

Minimum gross power of 10. The default value is 60-. Type: T006-EXPON.

STRING

CHAR input field for conversion to FLTP. The FM tries to interpret the contents of the STRING parameter as a floating point number. The maximum meaningful length of STRING is CHAR 22. This is a mandatory parameter.

MSGTYP_DECIM

Message type for TOO_MANY_DECIM. It can have the following values:

- Space: No message type conversion --> Normal message output.
- N: Message isn't output, not EXCEPTION.
- I: Message is output with type I, not EXCEPTION.
- W: Message is output with type W.
- E: Message is output with type E.
- A: Message is output with type A.
- S: Message is output with type S, not EXCEPTION.

The default value is W. Type: SREF-MSGTYP.

[↑] Export

DECIM

Number of decimal places determined in the mantissa in the interpretation of STRING. Type: DECIM.

EXPON

Gross power of 10 determined, that is, when the mantissa is presented so that there is one significant place before the decimal point, for example, 1.77E+20. The MAXEXP and MINEXP parameters use this format. Type: EXPON.

FLSTR

FLTP output field, from the CHAR conversion. This FM tries to interpret the STRING parameter as a floating point number, which is put in the FLSTR parameter. FLSTR must be assigned to a floating point field (FLTP). Type: FLSTR.

IVALU

Value in STRING flag. If STRING is 0 or a value that can be interpreted as a floating point value, the IVALU parameter is set to not initial (X). If STRING is empty or only the input mask, IVALU remains initial (Space). Type: IVALU.

[!] Exceptions

EXPONENT_TOO_BIG

Gross power of 10 is too big. If STRING has a greater gross exponent than specified in the MAXEXP parameter, an error message is issued.

EXPONENT_TOO_SMALL

Gross power of 10 is too small.

STRING_NOT_FLTP

STRING cannot be interpreted as a floating point. The floating point number in the STRING parameter is written incorrectly or in a syntax that cannot be interpreted.

TOO_MANY_DECIM

Too many decimal places. If the mantissa of the STRING parameter (without any power of 10) contains more decimal places than specified in the MAXDEC parameter, a warning is issued in a normal dialog (exception TOO_MANY_DECIM). This check isn't made in batch or batch input mode because it's only a warning.

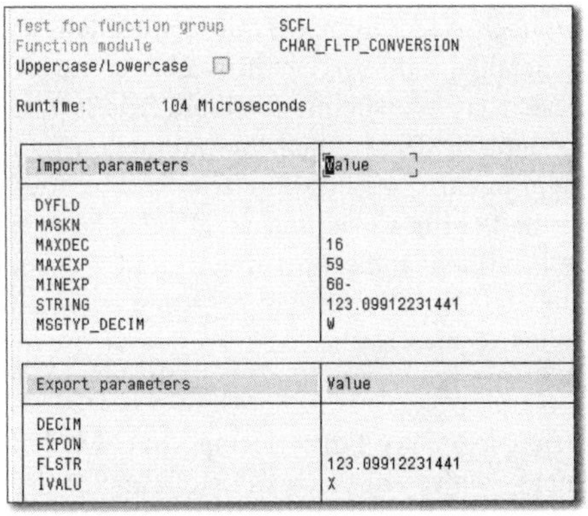

Character Format Converted into Float Format

---

78 COMPUTE_YEARS_BETWEEN_DATES

This FM is used to calculate the date difference.

Example: You have a program where after the user has given his date of birth in a selection screen, the program calculates the age based on today's date and displays it. You can pass the relevant parameter in this FM and call it to get the age in years.

[↓] Import

FIRST_DATE

Specifies the start date. Type: PREL-BEGDA.

MODIFY_INTERVAL

This parameter adds or subtracts the start date by 1 unit. Can have values + or –.

SECOND_DATE

Specifies the end date. Type: PREL-ENDDA.

[↑] Export

YEARS_BETWEEN_DATES

Returns the year in between the dates.

[!] Exception

SEQUENCE_OF_DATES_NOT_VALID

The first date is greater than the second date.

---

79 CONVERSION_EXIT_ALPHA_INPUT

This FM converts the user display format number to SAP format. Alpha conversions are generally used for account numbers. The SAP system checks whether the input field is purely numeric; if yes, then the number string is right justified, and leading zeros are inserted. If the field is alphanumeric, it's displayed as left justified, and spaces are inserted. Conversion exit alpha inputs are generally automatically triggered by the SAP server before saving the entry.

Example: Many times during your development, you've entered a value of a material in the screen, but you want to save the value in the database in SAP format with leading zeros. You can call this FM to convert the user format to SAP format.

[↓] Import

INPUT

Field in user format to be converted. Mandatory. Type: CLIKE.

[↑] Export

OUTPUT

Internal SAP converted field. Type: CLIKE.

80 CONVERSION_EXIT_ALPHA_OUTPUT

This FM converts the text from internal SAP format to user display format number.

Example: You've received the sales document number from table VBAK that has leading zeros and is in SAP internal format. You want to display it to the user and eliminate the leading zeros. You can pass the document number in the import parameter and call this FM to get the user-specific format for the sales document number.

[↓] Import

INPUT

Field in SAP format to be converted. Mandatory. Type: CLIKE.

[↑] Export

OUTPUT

Returned user format of the field. Type: CLIKE.

81 CONVERSION_EXIT_CUNIT_INPUT

This conversion exit converts the unit of measure from the user external format to the SAP internal format. This FM converts from the commercial unit to the business unit.

Example: A user has entered the unit of measure in the screen field, and you want to save it to the database table. When you save it to the table, this FM is called in the background to convert the user external format to the SAP format before saving the unit of measure to the table.

[↓] Import

INPUT

External display of unit of measurement. Mandatory. Default: SY-LANGU.

LANGUAGE

Language key. Default is login language. Type: SY-LANGU.

[↑] Export

OUTPUT

Internal display of unit of measurement.

[!] Exception

UNIT_NOT_FOUND

Unit of measurement not found.

---

82 CONVERSION_EXIT_CUNIT_OUTPUT

This conversion exit converts the internal format of a unit of measure to the external format. FM CUNIT converts a unit of measure from a business to a commercial unit.

Example: You've received the unit of measure from Table T006, which is in SAP internal format. You want to display it to the user. You can pass the parameter to the FM and call it to display the user format of the unit of measure.

[↓] Import

INPUT

Internal display of the unit of measurement. Mandatory. Default: SY-LANGU.

LANGUAGE

Language key. Default is login language. Type: SY-LANGU.

[↑] Export

LONG_TEXT

Long name of unit of measurement.

OUTPUT

External display of unit of measurement.

SHORT_TEXT

Short text of unit of measurement.

[!] Exception

UNIT_NOT_FOUND

Unit of measurement not found.

---

83 CONVERSION_EXIT_LUNIT_INPUT

This conversion exit converts the unit of measure from the user external format to the SAP internal format. This FM converts the unit from a technical unit to a business unit.

Example: A user has entered the unit of measure in the screen field, and you want to save it to the database table. When you save it to the table, this FM is called in the background to convert the user external format to the SAP format before saving the unit of measure to the table.

[↓] Import

INPUT

External display of unit of measurement. Mandatory. Default: SY-LANGU.

LANGUAGE

Language key. Default is login language. Type: SY-LANGU.

[↑] Export

OUTPUT

Internal display of unit of measurement.

[!] Exception

UNIT_NOT_FOUND

Unit of measurement not found.

---

84 CONVERSION_EXIT_LUNIT_OUTPUT

This conversion exit converts the internal format of the unit of measure to the user external format. This FM converts the unit of measure from a business to a technical unit. Details about units can be found in table T006A.

Example: You've received the unit of measure from Table T006, which is in SAP internal format. You want to display it to the user. You can pass the parameter to the FM and call it to display the user format of the unit of measure.

[↓] Import

INPUT

Internal display of the unit of measurement. Mandatory. Default: SY-LANGU.

LANGUAGE

Language key. Default is login language. Type: SY-LANGU.

[↑] Export

LONG_TEXT

Long name of unit of measurement.

OUTPUT

External display of unit of measurement.

SHORT_TEXT

Short text of unit of measurement.

[!] Exception

UNIT_NOT_FOUND

Unit of measurement not found.

85 CONVERSION_EXIT_GJAHR_INPUT

This conversion exit converts the fiscal year format from the user external format to the SAP internal format.

Example: A user has entered the fiscal year in the screen field, and you want to save it to the database table. When you save it to the table, this FM is called in the background to convert the user external format to the SAP format before saving the fiscal year to the table.

[↓] Import

INPUT

Fiscal year in the user format. Mandatory.

[↑] Export

OUTPUT

Fiscal year in the SAP format.

[!] Exception

WRONG_INPUT

Input is wrong.

---

86 CONVERSION_EXIT_GJAHR_OUTPUT

This conversion exit converts the internal format of the fiscal year to the user external format.

Example: You've received the fiscal year from the ABAP Data Dictionary table, which is in the SAP internal format. You want to display it to the user. You can pass the parameter to this FM to get the user display format of the fiscal year.

[↓] Import

INPUT

Fiscal year in SAP format. Mandatory.

[↑] Export

OUTPUT

Fiscal year in user format.

[!] Exception

WRONG_INPUT

Input is wrong.

---

87 CONVERSION_EXIT_ISOLA_INPUT

This conversion exit converts the two-digit ISO language field into a one-digit SAP language field.

Example: A user has entered the language in the screen field, and you want to save it to the database table. When you save it to the table, this FM is called in the background to convert the user external format to the SAP format before saving the language to the table.

[↓] Import

INPUT

Input language in user format. Mandatory.

[↑] Export

OUTPUT

Output language in SAP format.

[!] Exception

UNKNOWN_LANGUAGE

Language entered is unknown.

---

88 CONVERSION_EXIT_IDATE_INPUT

This FM converts the external date format to the internal date format.

Example: A user has entered the date in the screen field, and you want to save it to the database table. When you save it to the table, this FM is called in the background to convert the user external format to the SAP format before saving the date to the table.

[↓] Import

INPUT

Input date in user format. Mandatory.

[↑] Export

OUTPUT

Output date in SAP format.

---

89 CONVERSION_EXIT_IDATE_OUTPUT

This FM converts the internal date format to the external date format.

Example: You've received the date from the ABAP Data Dictionary table, which is in the SAP internal format. You want to display it to the user. You can pass the parameter to the FM to get the user display format of the date.

There are four similar FMs for conversion of dates with minute differences in the conversion:

- **CONVERSION_EXIT_IDATE_OUTPUT:** Date returned isn't based on user settings. For example, if your input date is 20121010, this FM returns the output date as Oct102012.
- **CONVERSION_EXIT_SDATE_OUTPUT:** Date returned is based on user settings. For example, if your input date is 20121010, this FM returns the output date as Oct/10/2012.
- **CONVERSION_EXIT_LDATE_OUTPUT:** Date returned isn't based on user settings. For example, if your input date is 20121010, this FM returns the output date as October 10, 2012.

- **CONVERT_DATE_TO_EXTERNAL:** Date returned isn't based on user settings. For example if your input date is 20121010, this FM returns the output date as 10/10/2012.

Refer to program RSSCA1T0 for the functionality of these IDATE and SDATE FMs.

[↓] Import

INPUT

Input date in SAP format. Mandatory.

[↑] Export

OUTPUT

Returned date in user format.

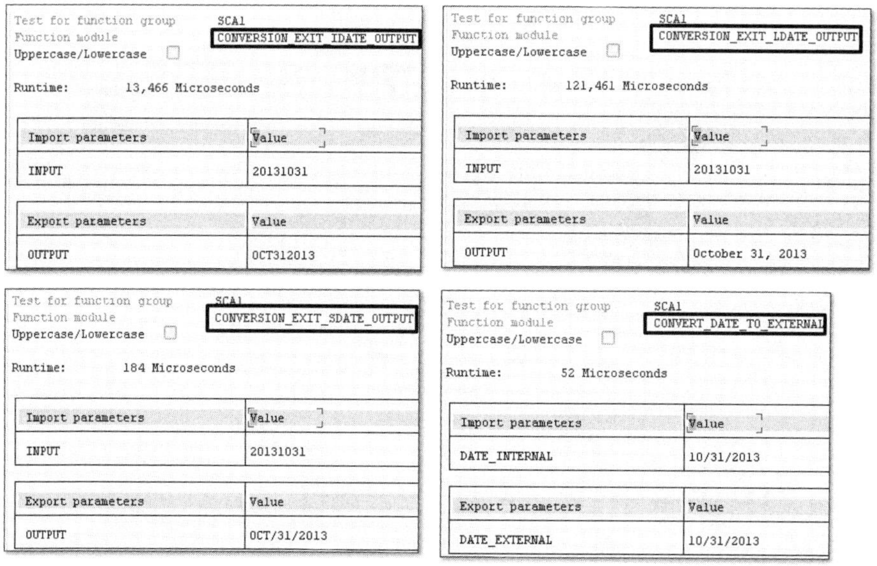

Difference in Date Output of Various Formats

90 CONVERSION_EXIT_ISOLA_OUTPUT

This conversion exit converts the one-digit language of the SAP format to a two-digit ISO format.

Example: You've received the language from the ABAP Data Dictionary table, which is in one-digit SAP internal format. You want to display it to the user. You can pass the parameter to the FM to get the user display format of the language.

[↓] Import

INPUT

Input language in SAP format. Mandatory.

[↑] Export

OUTPUT

Returned language in user format.

---

91 CONVERSION_EXIT_LDATE_OUTPUT

This FM converts the internal date format to the external date format.

Example: You've received the date from the ABAP Data Dictionary table, which is in the SAP internal format, and you want to display it to the user. You can pass the parameter to this FM to get the user display format of the date.

For differences between other date conversion FMs, refer to FM CONVERSION_EXIT_IDATE_OUTPUT (FM 89).

[↓] Import

INPUT

Input date in SAP format. Mandatory.

[↑] Export

OUTPUT

Returned date in user format.

---

92 CONVERSION_EXIT_PERI5_INPUT

This conversion exit converts the calendar quarter of user settings to the internal format calendar quarter (YYYYQ).

Example: A user has entered the calendar quarter in the screen field, and you want to save it to the database table. When you save it to the table, this FM is called in the background to convert the user external format to the SAP format before saving the calendar quarter to the table.

[↓] Import

INPUT

Input calendar quarter in the user format. Mandatory.

NO_MESSAGE

If this parameter has the value X, the error message is returned in the RETURN table; otherwise, exception INPUT_NOT_VALID is raised. Type: SY-DATAR.

[↑] Export

OUTPUT

Output language in SAP format.

RETURN

Return table for success or error message. Type: BAPIRETURN.

[!] Exception

INPUT_NOT_VALID

Input entered is not valid.

93 CONVERSION_EXIT_PERI5_OUTPUT

This conversion exit converts the calendar quarter of SAP format (YYYYQ) to the calendar format based on user settings.

Example: You've received the calendar format from the ABAP Data Dictionary table, which is in YYYYQ SAP internal format, and you want to display it to the user. You can pass the parameter to the FM to get the user display format of the calendar format.

[↓] Import

INPUT

Input calendar quarter in SAP format. Mandatory.

[↑] Export

OUTPUT

Returned calendar quarter in user format.

94 CONVERSION_EXIT_SDATE_INPUT

This FM converts the external date format to the internal date format

Example: A user has entered the date in the screen field, and you want to save it to the database table. When you save it to the table, this FM is called in the background to convert the user external format to the SAP format before saving the date to the table.

[↓] Import

INPUT
Input date in user format. Mandatory.

[↑] Export

OUTPUT
Output date in SAP format.

95 CONVERSION_EXIT_SDATE_OUTPUT

This FM converts the internal date format to the external date format.

Example: You've received the date from the ABAP Data Dictionary table, which is in the SAP internal format, and you want to display it to the user. You can pass the parameter to this FM to get the user display format of the date.

For the difference between other date conversion FMs, refer to FM CONVERSION_EXIT_IDATE_OUTPUT (FM 89).

[↓] Import

INPUT
Input date in SAP format. Mandatory.

[↑] Export

OUTPUT
Returned date in user format.

96 CONVERSION_FACTOR_GET

This FM determines the factors for the conversion of a measurement unit into another using the measurement unit table. Only measurement units in the same dimension can be converted. Units without a dimension can't be converted using the measurement unit table. This FM also provides the number of decimal places to which the values in the unit UNIT_OUT are to be rounded. This also applies to units without a dimension, if UNIT_IN = [Space]. The module also provides the dimension of the unit UNIT_OUT.

The following formula applies to the conversion:

> *(UNIT_OUT unit value) = (UNIT_IN unit value) * numerator/denominator + additive constant*

Example: You have a field that indicates length. Data is saved here in centimeters, but you want to display it as feet in output. You can fill in the parameters and call this FM to get the conversion factor, which you can use to convert the value from centimeters to feet.

[↓] Import

NO_TYPE_CHECK

The output parameters ADDITIVE CONSTANTS, COUNTER, and DENOMINATOR type check can be deactivated with this flag.

This is useful when the FM is only used to determine the number of decimals for value rounding in UNIT_OUT.

UNIT_IN

Unit from which to convert. Type: T006-MSEHI.

UNIT_OUT

Unit into which to convert. Type: T006-MSEHI.

[↑] Export

ADD_CONST

Additive constant in the conversion formula.

DECIMALS

The parameter specifies the number of decimal places to which the measurement unit UNIT_OUT should be rounded, according to the measurement unit table.

DENOMINATOR

Denominator in the conversion formula.

DIMENSION

Dimension key. Type: T006-DIMID.

NUMERATOR

Numerator in the conversion formula.

[!] Exceptions

CONVERSION_NOT_FOUND

The two units UNIT_IN and UNIT_OUT either have different dimensions or both have no dimensions. Conversion with the help of the measurement units table is therefore not possible.

OVERFLOW

One of the fields NUMERATOR, DENOMINATOR, or ADD_CONST is of type P or I and too small for the determined value. A field overflow occurs.

TYPE_INVALID

One of the output parameters ADD_CONST, DENOMINATOR, or NUMERATOR is not of the type F, P, or I; that is, is not a number or was not specified at the call. The type check that leads to this conclusion can be switched off with the input parameter NO_TYP_CHECK.

UNITS_MISSING

A unit was entered for neither UNIT_IN nor UNIT_OUT.

UNIT_IN_NOT_FOUND

UNIT_IN is not maintained in the measurement unit table.

UNIT_OUT_NOT_FOUND

UNIT_OUT is not maintained in the measurement unit table.

Test for function group SCV0
Function module CONVERSION_FACTOR_GET
Uppercase/Lowercase ☐

Runtime: 140 Microseconds

Import parameters	Value
NO_TYPE_CHECK	X
UNIT_IN	CM
UNIT_OUT	FT

Export parameters	Value
ADD_CONST	
DECIMALS	
DENOMINATOR	0000000000000000000000000000000000000000000000000000000000000000000000000000000000000000000000000000038100
DIMENSION	LENGTH
NUMERATOR	0000000000000000000000000000000000000000000000000000000000000000000000000000000000000000000000000000001250

Conversion from Centimeters to Feet

97 CONVERT_ABAPSPOOLJOB_2_PDF

This FM converts the ABAP list spool request into a PDF.

Example: You have to report REP1, and you want to display the output of the report in a PDF for mailing. You can generate the spool request of that output and can display it in PDF by filling in the appropriate parameters and calling this FM.

[↓] Import

SRC_SPOOLID

Pass the spool request number. This is a mandatory parameter. Type: TSP01-RQIDENT.

NO_DIALOG

If this parameter is set to [Space], a dialog box appears stating that the spool is very large and asking if you want to process in the background. Type: C.

DST_DEVICE

This parameter defines the spool output device. Type: TSP03-PADEST.

PDF_DESTINATION

This parameter defines the PDF destination whether it should be a PDF table, a spool request, or should be converted as an XString for the Unicode system. Type: C.

NO_BACKGROUND

If this parameter is set to X, conversion won't take place in the background. Type: C.

GET_SIZE_FROM_FORMAT

This parameter reads the size of treatment. If this parameter is [Space], rows and column from the spool request are received. Type: C.

USE_CASCADING

If x, cascading will be used. Default: Space. Type: C.

[↑] Export

PDF_BYTECOUNT

Returns the number of bytes in a PDF file. Type: I.

PDF_SPOOLID

Returns the spool request number. Type: TSP01-RQIDENT.

LIST_PAGECOUNT

Returns the number of pages in the ABAP spool job. Type: I.

BTC_JOBNAME

Returns the name of a background job. Type: TBTCJOB-JOBNAME.

BTC_JOBCOUNT

Returns the number of a batch job. Type: TBTCJOB-JOBCOUNT.

BIN_FILE

Returns the bin file that is of XString format. This is used for the Unicode system. Type: XSTRING.

Table

PDF

PDF file table. Type: TLINE.

[!] Exceptions

ERR_NO_ABAP_SPOOLJOB

SRC_SPOOLID parameter is not an ABAP list.

ERR_NO_SPOOLJOB

SRC_SPOOLID parameter does not exist.

ERR_NO_PERMISSION

No permission for reading SRC_SPOOLID.

ERR_CONV_NOT_POSSIBLE

Conversion to spool is not possible.

ERR_BAD_DESTDEVICE

DST_DEVICE parameter is invalid.

USER_CANCELLED

Operation is canceled by user.

ERR_SPOOLERROR

Error in reading spool request.

ERR_TEMSEERROR

Error in TEMSE server.

ERR_BTCJOB_OPEN_FAILED

Error in JOB_OPEN.

ERR_BTCJOB_SUBMIT_FAILED

Error in SUBMIT.

ERR_BTCJOB_CLOSE_FAILED

Error JOB_CLOSE.

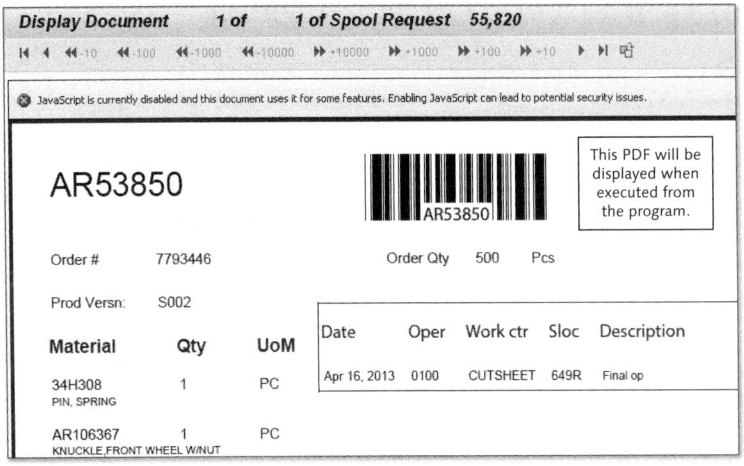

PDF Displayed When Executed from the Program

98 CONVERT_DATE_TO_EXTERNAL

The function block converts an internal date (e.g., 03262013) into the external preparation date. The conversion corresponds to the date of the conversion Dynpro interpreters. The date is processed according to the user master record (e.g., 03/26/2013).

Example: You have a screen parameter field P_ERSDA (created on date of material in the MARA table). In your program, when you retrieve the date, you get it in an internal format such as 20130101, but in output, you want to display it in the format as defined in user settings. You can do so by filling the date in the import parameter and executing the FM.

[↓] Import

DATE_INTERNAL

Internal date that needs to be converted. Type: SY-DATUM.

[↑] Export

DATE_INTERNAL

Date retuned in the external format according to the user settings. Type: SY-DATUM.

[!] Exception

DATE_INTERNAL_IS_INVALID

The internal date is invalid.

99 CONVERT_DATE_TO_INTERNAL

The function block converts an external date (e.g., 03/26/2013) into the internal date. The conversion corresponds to the date of the conversion Dynpro interpreters. The date is processed according to the user master record (e.g., 03/26/2013).

Example: You have a screen parameter field P_ERSDA (created on date of material in the MARA table). You enter the date in the user display format in this field, but you want to save the date. Saving occurs in the SAP internal format. You can do so by filling the date in the import parameter and executing the FM. You'll receive the date in the internal format, which you can use to save in the database.

[↓] Import

DATE_EXTERNAL

External date formatting. This is a mandatory parameter.

ACCEPT_INITIAL_DATE

Single-character indicator. Type: TADIR-EDTFLAG.

[↑] Export

DATE_INTERNAL

Internal date formatting. Type: SY-DATUM.

[!] Exception

DATE_EXTERNAL_IS_INVALID

The external date is invalid; that is, the date does not exist.

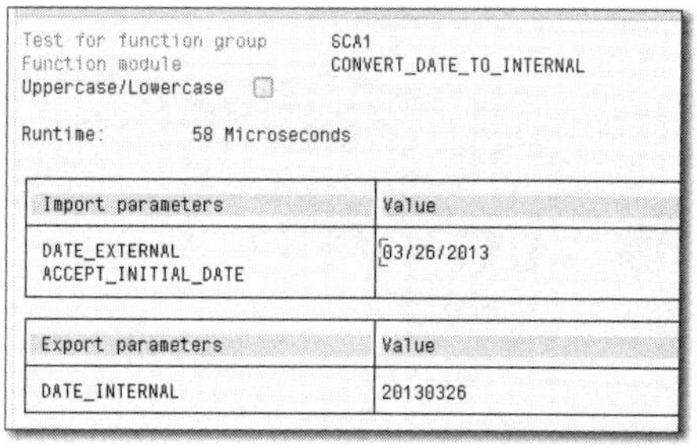

Converting Date from External Format to Internal Format

100 CONVERT_ITF_TO_HTML

This FM converts SAPscript text to HTML to allow it to be displayed using any web browser on the Internet. Both the text header (ABAP Data Dictionary structure THEAD) and the text lines (ABAP Data Dictionary structure TDLINES) must be entered in the import and table parameters I_HEADER and T_ITF_TEXT.

After a successful conversion, the FM provides you with a table of corresponding HTML text (DDIC structure T_HTML_TEXT). SAPscript commands, including texts, as well as standard, system, and text symbols are dissolved during conversion as long as the dissolution requires no form processing.

Example: You want to send the standard paragraphs texts by email in HTML format. In that case, you can read those standard paragraphs by FM READ_TEXT (see FM 283), which returns the paragraph in ITF format and then passes THEADER and TLINE in this FM. Execute this FM, and the output converts the ITF to HTML format.

[↓] Import

I_CODEPAGE

Target character set. This parameter isn't supported now. Type: TCP02-CPCODEPAGE.

I_HEADER

Text header of input text. The text header contains descriptive information on a text module such as a short text, creator, last person to make changes, and so on. The structure of the text header is specified in the DDIC structure THEAD. This is a mandatory parameter. Type: THEAD.

I_PAGE

Page specification for page window format. Use the parameter to determine a particular page of the form on which the input text is based. This page contains form-specific attributes such as sizes or the standard paragraph format of the text. Usually, you need not specify this parameter. The default value is [Space]. Type: ITCTH-TDPAGE.

I_WINDOW

Window specification for page window format. Use the parameter to determine a particular page window of the form on which the input text is based. This page window contains the form-specific attributes, such as sizes or the standard paragraph of the text. Usually, there is no need to specify this parameter. The default value is [Space]. Type: ITCTH-TDWINDOW.

I_SYNTAX_CHECK

Activating the ITF syntax check. Switches the SAPscript syntax check on for checking the syntax of the input text in ITF format. Because conversion always runs without corresponding print programs, the syntax check can't test program symbols. The default value is [Space]. Type: C.

I_REPLACE

Expanding symbols and includes. This parameter determines whether or not simple control commands, include texts, and standard, system, and text symbols should be dissolved in the input text. The converter uses the FMs TEXT_COMMAND_REPLACE, TEXT_SYMBOL_REPLACE, and TEXT_INCLUDE_REPLACE during this process. Program symbols that are provided with values from the print program can't, however, be dissolved because their values aren't available until after text formatting (form processor). The same is true for processing logic control using control commands. The default setting is symbols and control commands are dissolved. The default value is X. Type: C.

I_PRINT_COMMANDS

Outputting commands and text elements. This parameter determines whether or not the control commands that can't be edited or the names of text elements should appear in HTML output. The default setting is control commands are suppressed. The default value is [Space]. Type: C.

I_HTML_HEADER

Output of the HTML header tag. This parameter determines whether or not an HTML header is normally added to a converted text. The default setting is display with header. The default value is X. Type: C.

I_FUNCNAME

Exit module for link interpretation. The parameter allows you to specify the name of a FM that you want to use to interpret SAPscript links (e.g., for documentation texts). The text returned by this FM is then included in the HTML data stream. Converter checks whether the specified text exists and whether the call interface is correct at runtime. The call interface of the FM must look like this:

```
call function function_name exporting
itf_link = l_text (type c)
 importing
html_link = l_text. (type c)
```

From this, an HTML link key of the form `<A HREF="' l_text '">` is generated. The default value is [Space]. Type: TFDIR-FUNCNAME.

I_TITLE

Title in HTML header. Use the parameter to set in the HTML header the title of the document. The default setting is output of the text name as title. The default value is [Space]. Type: C.

I_BACKGROUND

File name for the HTML image as background. Use the parameter to set a background picture in the BODY tag (attribute BACKGROUND) of the converted text. You must specify the background picture as URL. The default setting is no background picture set. The default value is [Space]. Type: C.

I_BGCOLOR

Background color of text. Use the parameter to set a background color in the BODY tag (attribute BGCOLOR) of the converted text. You must specify the background color as RGB value, for example, #FF0000 for the color red. The default setting is no background color set. The default value is [Space]. Type: C.

I_UNESCAPE_FORMATS

Exception formats for masking special characters. Use this parameter to specify a table of character and paragraph formats within which you don't want to mask HTML special characters (<, >, &). Usually, such special characters are described using escape sequences, for example, < for the < character.

During text output into the data stream, the converter first checks whether within the current character format, the masking is deactivated. Then it checks the current paragraph format. If it finds the current font within the table, it doesn't mask the special characters.

Example: Suppressing masking characters in the description of HTML links (`<A HREF="url">text<(></>`) in workflow tasks. Type: TEDFORMAT.

I_ESCAPE_SPACES

Masking of blank characters. The parameter can control whether multiple blank characters in sequence are to be masked in HTML. The HTML converter always outputs blank characters correctly, but browsers only ever display one character. In this case, the blank characters are replaced by the character entities ` ` in HTML (nonbreakable spaces).

Also that alignment with blank characters is only useful with nonproportional fonts. Exact alignment isn't possible with proportional fonts because the characters are different widths. Default setting is no masking of blank characters. The default value is [Space]. Type: C.

I_ENCODING

Define character set coding in header. The parameter specifies the character set code of the HTML text (e.g., UTF-8, ISO-8859-15, etc.). The code is defined in the HTML header as a <META>-tag. In earlier HTML versions, the character set code ISO-8859-1 was automatically used, if none was explicitly specified in the header. In HTML 4.01, the browser can accept any code if none is specified. A complete list of the valid character set codes, including their official names, can be found in the HTML standard (under Character Encoding). Type: C.

[↑] Export

E_HTML_TEXT

This parameter returns the text content in HTML as XString. Type: XSTRING.

Tables

T_ITF_TEXT

Text lines in ITF (input). This table contains text lines belonging to a specific text module. The line's structure is determined by the ABAP Data Dictionary table TLINE. Line contents are stored in ITF format. This is a mandatory parameter. Type: TLINE.

T_HTML_TEXT

Text lines in HTML (output). The table contains the HTML text returned after conversion. The line structure is specified in the ABAP Data Dictionary table HTMLLINE (maximum 132 lines). Type: HTMLLINE.

T_CONV_CHARFORMATS

Table for character formats. This is a mapping table for mapping ITF character formats to corresponding HTML character formats (HTML tags). An example of just such a mapping table for character formats can be found in report RSTXHTML. Type: TLINE.

T_CONV_PARFORMATS

Table for paragraph formats. This is a mapping table for mapping ITF paragraph formats to corresponding HTML paragraph formats (HTML tags). An example of just such a mapping table for paragraph formats can be found in report RSTXHTML. Type: TLINE.

[!] Exceptions

SYNTAX_CHECK

Incorrect input output text for ITF syntax.

REPLACE

Errors expanding includes and symbols.

ILLEGAL_HEADER

Text header incorrect.

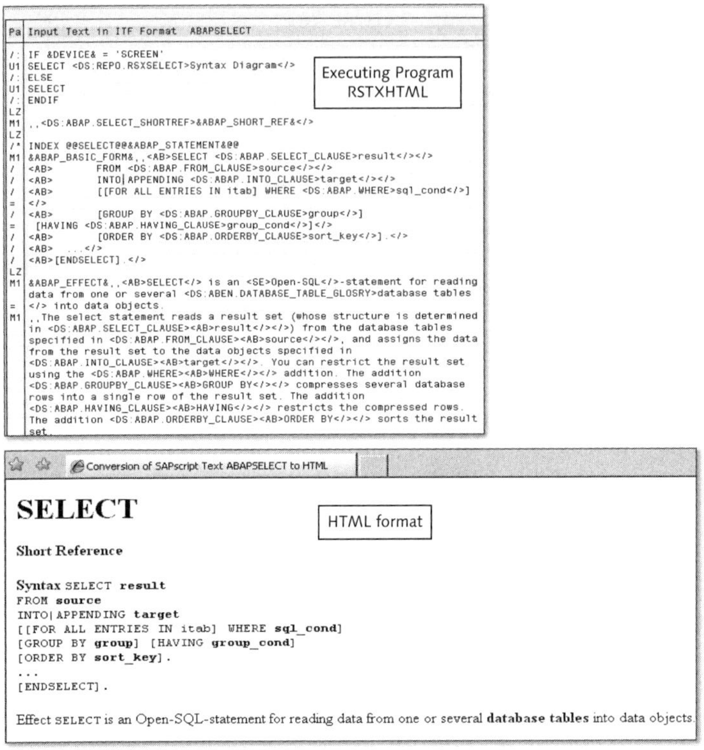

FM Call from Program RSTXHTML

---

101 CONVERT_OTF

This FM converts the OTF format to various formats, currently ASCII or PDF. The PDF is a binary format, although PDF files (or the included PDF instructions) are viewable in a text editor. For this reason, in PDF format, return parameter BIN_FILESIZE contains the number of bytes (in table LINES XSTRING or field) after the successful conversion. The PDF files are returned to the table LINES, although the structure TLINE has CHAR 134 but is treated as RAW134 (or RAW268 in Unicode). If a return parameter BIN_FILE (type XSTRING) is specified, the PDF data won't be returned in the table but in the field LINES BIN_FILE.

Example: You want to convert the content of your Smart Form into a PDF. You have a Smart Form ZSFORM_BOM that displays the BOM of an order material, and now you want it to be displayed in PDF so you can print/save it for future reference. You can do so by filling in the parameters and calling this FM.

[↓] Import

FORMAT

This parameter defines the target format for table LINES. Value can be ASCII or PDF. The default value is ASCII.

MAX_LINEWIDTH

In ASCII format: Number of characters in a LINES TDLINE. The default value is 132.

ARCHIVE_INDEX

This parameter defines the archive index at OTF to PDF. The default value is [Space]. Type: TOA_DARA.

COPYNUMBER

Number of copies to create. The default value is 0. Type: ITCPO-TDCOPIES.

ASCII_BIDI_VIS2LOG

BIDI conversion logical to visual order in ASCII format. The default value is [Space]. Type: C.

PDF_DELETE_OTFTAB

Delete OTF table in PDF conversion. The default value is [Space]. Type: C.

PDF_USERNAME

SY-UNAME to display in PDF document information. The default value is [Space]. Type: C.

[↑] Export

BIN_FILESIZE

This parameter returns the number of bytes in LINES in binary format.

BIN_FILE

Returns the PDF data. Type: XSTRING.

Tables

OTF

Input table with OTF format. Type: ITCOO.

LINES

Output table with target format. Type: TLINE.

[!] Exceptions

ERR_MAX_LINEWIDTH

This exception is raised if line width is not between 2 and 132 characters.

ERR_FORMAT

This exception is raised if the format is not supported.

ERR_CONV_NOT_POSSIBLE

This exception is raised when conversion is not possible.

ERR_BAD_OTF

This exception is raised for OTF faulty data.

102 CONVERT_OTFSPOOLJOB_2_PDF

This FM converts the OTF spool to PDF.

Example: You want to convert the content of your spool into a PDF. You have a spool number 12345 that displays the BOM of an order material, and now you want it to be displayed in a PDF. You can do so by filling in the parameters and calling this FM.

[↓] Import

SRC_SPOOLID

Spool request number you want to convert. Mandatory. Type: TSP01-RQIDENT.

NO_DIALOG

If this parameter is set to [Space], a dialog box appears stating that the spool is very large and asking if you want to process in the background. Type: C.

DST_DEVICE

This parameter defines the spool output device. Type: TSP03-PADEST.

PDF_DESTINATION

This parameter defines the PDF destination and whether it should be a PDF table, a spool request, or should be converted as XString for the Unicode system. Type: C.

NO_BACKGROUND

If this parameter is set to X, conversion won't take place in the background. Type: C.

[↑] Export

PDF_BYTECOUNT

Returns the number of bytes in a PDF file. Type: I.

PDF_SPOOLID

Returns the spool request number. Type: TSP01-RQIDENT.

LIST_PAGECOUNT

Returns the number of pages in the ABAP spool job. Type: I.

BTC_JOBNAME

Returns the name of a background job. Type: TBTCJOB-JOBNAME.

BTC_JOBCOUNT

Returns the number of a batch job. Type: TBTCJOB-JOBCOUNT.

BIN_FILE

Returns the bin file that is of XString format. This is used for the Unicode system. Type: XSTRING.

Table

PDF

PDF File table. Type: TLINE.

[!] Exceptions

ERR_NO_OTF_SPOOLJOB

SRC_SPOOLID parameter is not an ABAP list.

ERR_NO_SPOOLJOB

SRC_SPOOLID parameter does not exist.

ERR_NO_PERMISSION

No permission for reading SRC_SPOOLID.

ERR_CONV_NOT_POSSIBLE

Conversion to spool is not possible.

ERR_BAD_DESTDEVICE

DST_DEVICE parameter is invalid.

USER_CANCELLED

Operation is canceled by user.

ERR_SPOOLERROR

Error in reading spool request.

ERR_TEMSEERROR

Error in TEMSE server.

ERR_BTCJOB_OPEN_FAILED

Error in JOB_OPEN.

ERR_BTCJOB_SUBMIT_FAILED

Error in SUBMIT.

ERR_BTCJOB_CLOSE_FAILED

Error JOB_CLOSE.

---

103 CONVERT_TIME_INPUT

This FM converts a time from the externally formatted user input into the SAP system internal format.

Example: You have an input parameter field P_TIME where you enter the time. You want to save the time to the table. Fill in the parameter and call this FM to get the converted time.

[↓] Import

INPUT

Formatted time entry which can be six characters long.

PLAUSIBILITY_CHECK

Make plausibility check, that is, check whether the time entered in the input parameter is valid or not. The default value is X.

[↑] Export

OUTPUT

Output time as a database field in the format HHMMSS. Type: SY-UZEIT.

[!] Exceptions

PLAUSIBILITY_CHECK_FAILED

The plausibility check found that an invalid time was entered.

WRONG_FORMAT_IN_INPUT

Time in INPUT has incorrect format interpreted as a time.

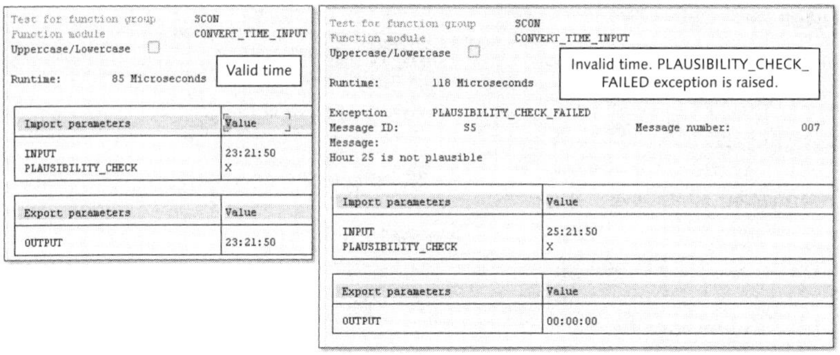

Time Conversion

---

104 CONVERT_TO_FOREIGN_CURRENCY

This FM translates a local currency amount into foreign currency. An amount in foreign currency is calculated from a specified local currency amount. For this, you may either specify the translation rate manually (parameter RATE) or have the system determine it from table TCURR on the basis of the rate type, date, and currency key. Because the ratios for the units of the currencies involved in the translation are significant for this translation, table TCURR is always read by the program, and there must be a valid entry there for the data specified. If exchange rate fixing is defined for exchange rate type TYPE_OF_RATE, or an alternative exchange rate type is defined for the currency pair, this information is transferred to the calling program. When table TCURR is read, the foreign currency key is always taken as the first part of the key and the local currency as the second part. If this entry isn't in the table, the key is reread in reverse sequence.

Example: You want to convert the currency from USD ($) to INR in your program. You can fill in the parameters and call this FM to display the rupees.

[↓] Import

CLIENT

Client. Type: SY-MANDT.

DATE

Currency translation date. This parameter is mandatory. Type: SYST-DATUM.

FOREIGN_CURRENCY

Currency key for foreign currency. This parameter is mandatory.

LOCAL_AMOUNT

Amount in local currency. This parameter is mandatory.

LOCAL_CURRENCY

Currency key for local currency. This parameter is mandatory.

RATE

Predefined exchange rate. Default: 0.

TYPE_OF_RATE

Exchange rate type:

- M: Average rate.
- G: Bank buying rate.
- B: Bank selling rate.

The default value is M.

READ_TCURR

Read exchange rate from table TCURR. If this flag has been deleted and the IMPORT parameter RATE has been given a value greater than zero, the following applies:

- If the exchange rate type transferred or diverted to TYPE_OF_RATE has been fixed (TCURV-XFIXD = X), then the corresponding TCURR entry is read and the fixed exchange rate in the FIXED_RATE parameter is returned. If there is no TCURR entry, the error message SG105 is issued.
- For exchange rates that aren't fixed, the TCURR isn't read.

Otherwise, the TCURR is read, and the error message SG105 is issued if the required TCURR entry has not been found.

The currency factor (TCURF) must have been maintained. Default: X.

[↑] Export

EXCHANGE_RATE

Exchange rate calculated from the amount entered.

FOREIGN_AMOUNT

Amount returned in foreign currency.

FOREIGN_FACTOR

Factor for the foreign currency units. Type: TCURR-FFACT.

LOCAL_FACTOR

Factor for the local currency units. Type: TCURR-TFACT.

EXCHANGE_RATEX

Exchange rate 4.11.

DERIVED_RATE_TYPE

Alternative exchange rate. Type: TCURR-KURST.

FIXED_RATE

Returns the fixed exchange rate if present.

[!] Exception

NO_RATE_FOUND

Entry not found in table TCURR.

OVERFLOW

Field FOREIGN_AMOUNT is too small.

NO_FACTORS_FOUND

No entry found in table TCURF.

NO_SPREAD_FOUND

Spread not entered in table TCURS.

DERIVED_2_TIMES

Exchange rate type derived more than once.

---

105 CONVERT_TO_FRACTION

This FM returns the numerator and denominator conversion factor of the fraction when you're building logic.

Example: You have a floating number 1.021344, which you get when you divide numerator = 31,917 to denominator = 31,250. When you provide the value as 1.021344 in the INPUT parameter of this FM, you get numerator = 31,917 and denominator = 31,250.

[↓] Import

INPUT

Input value of type float. Mandatory.
Type: F.

[↑] Export

NOMINATOR

Numerator value of the conversion factor.
Type: T006-ZAEHL.

DENOMINATOR

Denominator value of the conversion factor.
Type: T006-NENNR.

[!] Exception

CONVERSION_OVERFLOW

Number passed cannot be represented.

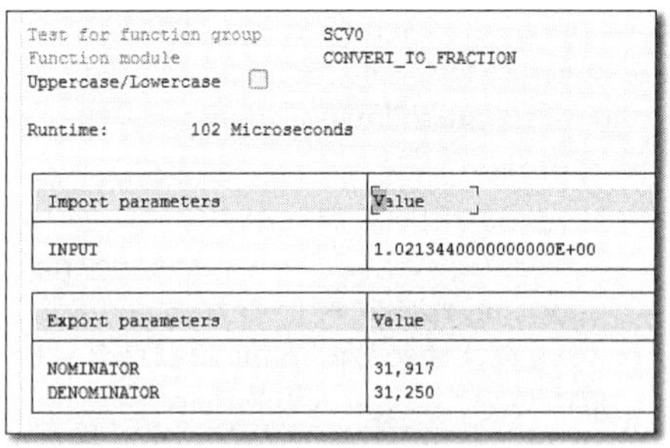

Conversion of Fraction

---

106 CONVERT_TO_LOCAL_CURRENCY

This FM translates foreign currency amount into local currency. An amount in local currency is calculated from a specified foreign currency amount. FM CONVERT_TO_FOREIGN_CURRENCY (FM 104) performs the opposite functionality; refer to it for details.

Example: You want to convert the currency from INR to USD ($) in your program. You can fill in the parameters and can call this FM to display the USD.

[↓] Import

CLIENT

Client. Type: SY-MANDT.

DATE

Currency translation date. This parameter is mandatory. Type: SYST-DATUM.

FOREIGN_CURRENCY

Currency key for foreign currency. This parameter is mandatory.

FOREIGN_AMOUNT

Amount in foreign currency. This parameter is mandatory.

LOCAL_CURRENCY

Currency key for local currency. This parameter is mandatory.

RATE

Predefined exchange rate. The default value is 0.

TYPE_OF_RATE

Exchange rate type:

- M: Average rate.
- G: Bank buying rate.
- B: Bank selling rate.

The default value is M.

READ_TCURR

Read exchange rate from table TCURR. If this flag has been deleted, and the IMPORT parameter RATE has been given a value greater than zero, the following applies:

- If the exchange rate type transferred or diverted to TYPE_OF_RATE has been fixed (TCURV-XFIXD = X), then the corresponding TCURR entry is read, and the fixed exchange rate in the FIXED_RATE parameter is returned. If there is no TCURR entry, the error message SG105 is issued.
- For exchange rates that aren't fixed, the TCURR isn't read.

Otherwise, the TCURR is read, and the error message SG105 is issued if the required TCURR entry hasn't been found.

The currency factor (TCURF) must have been maintained. Default: X.

[↑] Export

EXCHANGE_RATE

Exchange rate calculated from the amount entered.

FOREIGN_FACTOR

Factor for the foreign currency units. Type: TCURR-FFACT.

LOCAL_AMOUNT

Amount returned in local currency.

LOCAL_FACTOR

Factor for the local currency units. Type: TCURR-TFACT.

EXCHANGE_RATEX

Exchange rate 4.11.

FIXED_RATE

Returns the fixed exchange rate if present.

DERIVED_RATE_TYPE

Alternative exchange rate. Type: TCURR-KURST.

[!] Exception

NO_RATE_FOUND

Entry not found in table TCURR.

OVERFLOW

Field FOREIGN_AMOUNT is too small.

NO_FACTORS_FOUND

No entry found in table TCURF.

NO_SPREAD_FOUND

Spread not entered in table TCURS.

DERIVED_2_TIMES

Exchange rate type derived more than once.

---

107 COUNTRY_CODE_ISO_TO_SAP

This FM translates an ISO country code (International Organization for Standardization) into the SAP internal country key. If several internal country codes are assigned the same ISO code, the indicator UNIQUE isn't set (value = [Space]). Otherwise, it is set (value= X).

Example: To get the SAP country code for ISO code AG, you can use this FM. The export parameter returns the SAP internal country code as AG as maintained in table T005.

[↓] Import

ISO_CODE

ISO code for the currency. Mandatory. Type: T005-INTCA.

[↑] Export

SAP_CODE

SAP internal code for the currency. Type: T005-LAND1.

UNIQUE

If several internal country codes are assigned the same ISO code, the indicator UNIQUE isn't set (value = Space). Otherwise, it is set (value= X). Type: BDWFAP_PAR-CALLTRANS.

[!] Exception

NOT_FOUND

ISO code not found in the system.

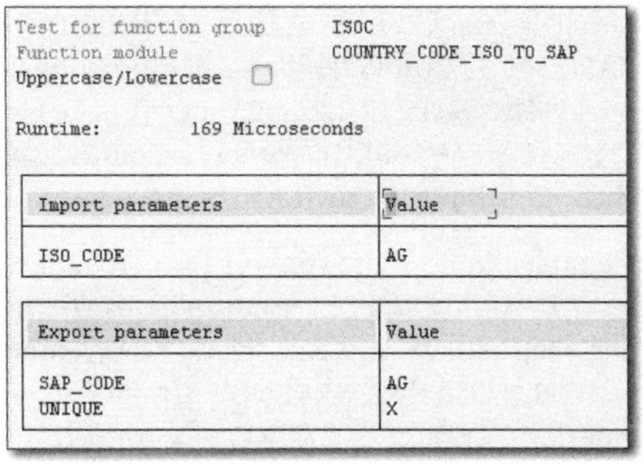

Test for function group ISOC
Function module COUNTRY_CODE_ISO_TO_SAP
Uppercase/Lowercase

Runtime: 169 Microseconds

Import parameters	Value
ISO_CODE	AG

Export parameters	Value
SAP_CODE	AG
UNIQUE	X

Conversion of Country Code from ISO to SAP Internal Country Code

---

108 COUNTRY_CODE_SAP_TO_ISO

This FM translates an SAP country code into the ISO internal country key.

Example: To get the ISO country code for SAP code AG, you can use this FM. The export parameter returns the ISO country code as AG as maintained in table T005.

[↓] Import

SAP_CODE

SAP internal code for the country. Mandatory. Type: T005-LAND1.

[↑] Export

ISO_CODE

ISO code for the country. Type: T005-INTCA.

[!] Exception

NOT_FOUND

ISO code not found in the system.

---

109 CURRENCY_AMOUNT_SAP_TO_BAPI

This FM converts an amount of currency from the SAP system internal format to the user external format, which can also be used for external languages such as Java and .NET via BAPI. This FM provides the similar functionality as FM BAPI_CURRENCY_CONV_TO_EXTERNAL (FM 560). Decimal amounts are checked from table TCURX.

Example: You want to convert $100.10 of SAP to BAPI so that you can use it in external system such as Java, and you can do so by entering "USDN" in the CURRENCY parameter and entering amount "100.10" in the SAP_AMOUNT parameter. In table TCURX, USDN is maintained as decimal 5, so ".1001" is returned in the BAPI_AMOUNT when you execute this FM.

[↓] Import

CURRENCY

Currency you want to convert. Mandatory. Type: TCURC-WAERS.

SAP_AMOUNT

Amount in SAP you want to convert. Mandatory. Type: BAPICURR-BAPICURR.

[↑] Export

BAPI_AMOUNT

Converted amount in BAPI. Type: BAPICURR-BAPICURR.

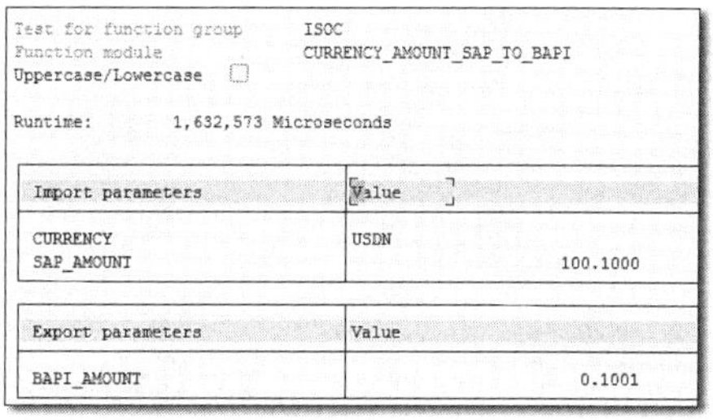

Conversion of Currency Amount from SAP to BAPI

---

110 CURRENCY_AMOUNT_SAP_TO_IDOC

This FM converts an amount of currency from an SAP system to be used for an external system by an IDoc. This FM determines the decimal amounts factor received from table TCURX.

Example: You want to convert $100.10 from the SAP system to an IDoc so that you can use it in an external system such as Java. You can do so by entering "USDN" in the CURRENCY parameter and entering amount "100.10" in the SAP_AMOUNT parameter. In table TCURX, USDN is maintained as decimal 5, so ".1001" is returned in the BAPI_AMOUNT when you execute this FM.

[↓] Import

CURRENCY

Currency you want to convert. This is a mandatory parameter. Type: TCURC-WAERS.

SAP_AMOUNT

Amount in SAP you want to convert. This is a mandatory parameter.

[↑] Export

IDOC_AMOUNT

Converted amount in IDoc.

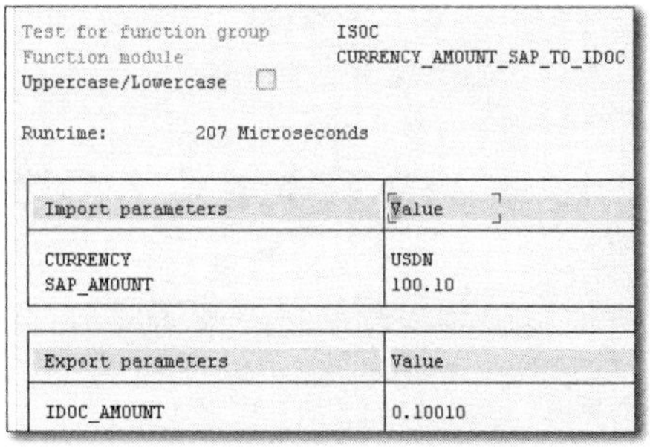
Test for function group ISOC
Function module CURRENCY_AMOUNT_SAP_TO_IDOC
Uppercase/Lowercase

Runtime: 207 Microseconds

Import parameters	Value
CURRENCY	USDN
SAP_AMOUNT	100.10

Export parameters	Value
IDOC_AMOUNT	0.10010

Currency Conversion from SAP to IDoc

111 CURRENCY_CODE_ISO_TO_SAP

This FM translates the international standard ISO currency code into the SAP internal currency key.

In release 4.6C of SAP ERP, a flag (field XPRIMARY) has been added to table TCURC which specifies that this currency doesn't conform to ISO and therefore can't be used for external communication (e.g., ALE). This FM uses only the TCURC entries for which the flag hasn't been set. If several SAP currency codes that refer to an ISO currency code don't have the flag set, the FM terminates with the exception MORE_THAN_ONE (and a T100 message).

Example: You know the ISO currency code of US dollar is USD, and you want to use the currency in your SAP program, but you don't know the SAP currency code. You can fill in the import parameter and can execute this FM to return the SAP currency code of USD. Note that there are USD and USDN currency codes in table TCURC, but the primary code is USD, which has XPRIMARY field checked in the table entry of TCURC.

[↓] Import

ISO_CODE

ISO code for the currency you want to check. Type: TCURC-ISOCD.

[↑] Export

SAP_CODE

SAP internal code for the currency. Type: TCURC-WAERS.

UNIQUE

When this parameter is set to X, it identifies that the ISO to SAP currency code is unique. Type: BDWFAP_PAR-CALLTRANS.

[!] Exception

NOT_FOUND

ISO code not found in the system.

---

112 CURRENCY_CODE_SAP_TO_ISO

This FM translates the SAP internal currency code into international standard ISO currency code.

In release 4.6C of SAP ERP, a flag (field XPRIMARY) has been added to table TCURC which specifies that this currency doesn't conform to ISO and therefore can't be used for external communication (e.g., ALE). This FM uses only the TCURC entries for which the flag hasn't been set. If the SAP currency code just transferred has a set flag, the FM terminates with the exception NOT_FOUND.

Example: You're internally using the currency code USDN, but you want the ISO code because you want to send that data to an external Java system that understands ISO code. You can fill in the parameter and call this FM to get the ISO value as USD. Because the currency USDN is for internal SAP use, parameter NOT_ALLOWED is also checked when the SAP currency is USDN. This parameter isn't checked if the currency is USD.

[↓] Import

SAP_CODE

SAP internal code for the country. Mandatory. Type: TCURC-WAERS.

[↑] Export

ISO_CODE

Returned ISO code for the country. Type: TCURC-ISOCD.

NOT_ALLOWED

ISO code isn't allowed due to one of the following reasons:

- The primary flag of the found ISO code is assigned more than once.
- The ISO code isn't unique, and the primary flag isn't assigned for associated SAP currency codes.
- The ISO code isn't unique, and the current SAP currency code hasn't assigned the primary flag.

[!] Exception

NOT_FOUND

SAP code not found in the system.

---

113 DATE_CHECK_PLAUSIBILITY

This FM is used to check the date plausibility; that is, the FM checks that the date has been entered in the correct format.

Example: You've entered a date in the parameter PARA1, and you want to see if the date is valid or not. You can pass the entered date in the import parameter of the FM and execute the FM to return whether the date is valid or not. If the date isn't valid, it returns with the exception.

[↓] Import

DATE

Transfer of date to be checked. Mandatory. Type: SY-DATUM.

[!] Exception

PLAUSIBILITY_CHECK_FAILED

Date is not plausible.

---

114 DATE_CHECK_WORKINGDAY

This FM defines whether the date that the user has entered is a working day. This FM calls the FM DATE_CONVERT_TO_FACTORYDATE (FM 117).

Example: You're working in a factory and you want to identify the factory date for normal working date 04/21/2013. You can do so by filling in the parameters and executing this FM. The factory date that corresponds to the entered date is returned.

[↓] Import

DATE

Date to be tested. Mandatory. Type: SY-DATUM.

FACTORY_CALENDAR_ID

Factory calendar Mandatory. ID. Type: SCAL-FCALID.

MESSAGE_TYPE

Type: of message if no date business working day is found. Mandatory. Type: SY-MSGTY.

[!] Exception

DATE_AFTER_RANGE

Date is after the end of the factory calendar.

DATE_BEFORE_RANGE

Date is prior to the start of the factory calendar.

DATE_INVALID

Date is invalid.

DATE_NO_WORKINGDAY

Date is not a business working day.

FACTORY_CALENDAR_NOT_FOUND

Factory calendar is unknown.

MESSAGE_TYPE_INVALID

Invalid message type.

115 DATE_COMPUTE_DAY

This FM determines what day of the week corresponds to the date you enter in the import parameters. The date must be passed with the valid value.

Example: A selection screen parameter has a date field where you want to enter the date and process the logic further if the day is on a weekend. You can pass the value of the date in the import parameter and execute the FM to get the day corresponding to the date in the export parameter. If the returned value is Saturday or Sunday, you can execute your logic.

[↓] Import

DATE

Date of weekday to be calculated. Date must lie between 01.01.1901 and 31.12.2098. Mandatory. Type: SCAL-DATE.

[↑] Export

DAY

Returned weekday value. Possible values are:

- 1: Monday.
- 2: Tuesday.
- 3: Wednesday.
- 4: Thursday.
- 5: Friday.
- 6: Saturday.
- 7: Sunday.

Type: SCAL-INDICATOR.

116 DATE_CONV_EXT_TO_INT

This FM converts the date from the external format of the user master record to the internal date format.

Example: You have a screen that contains a date parameter field. You've entered the date in MMDDYYYY format per your user settings, and you want to save the date field in the database in YYYYMMDD format. The system calls this FM to change the date format from the user format to the internal format automatically.

[↓] Import

I_DATE_EXT

This parameter inputs the date in the external user format. This is a mandatory parameter.

[↑] Export

E_DATE_INT

This parameter returns the date in the internal SAP format. Type: D.

[!] Exception

ERROR

Error in the conversion.

---

117 DATE_CONVERT_TO_FACTORYDATE

This FM returns the factory calendar date for the entered date in FM DATE_CHECK_WORKINGDAY (FM 114). This FM has the option of CORRECT_OPTION, which returns the previous date or following date if the entered date isn't a working day.

Example: You're working in a factory and want to identify the factory date for the normal date 04/21/2013. You can do so by filling in the parameters and executing this FM. The factory date is returned that corresponds to the entered date.

[↓] Import

CORRECT_OPTION

This parameter defines how the workday should be calculated. It can have the following values:

- +: If the specified date isn't a working day, the first working day after the date is returned. (This is the default.)
- –: If the specified date isn't a working day, the first working day before the date is returned.

Type: SCAL-INDICATOR.

DATE

Date to be converted into the factory calendar date. Mandatory. Type: SCAL-DATE.

FACTORY_CALENDAR_ID

Factory calendar ID. Mandatory. Type: SCAL-FCALID.

[↑] Export

DATE

Date returned after conversion. Mandatory. Type: SCAL-DATE.

FACTORYDATE

Number of the workday in the specified calendar. Type: SCAL-FACDATE.

WORKINGDAY_INDICATOR

Indicator that defines whether a date is a workday. The following values are possible:

- [Space]: The specified date is a working day.
- +: If the specified date isn't a working day, the following working day is returned.
- –: If the specified date isn't a working day, the previous working day is returned. Type: SCAL-INDICATOR.

[!] Exception

CALENDAR_BUFFER_NOT_LOADABLE

Factory calendar could not be buffered.

CORRECT_OPTION_INVALID

CORRECT_OPTION is not + or –.

DATE_AFTER_RANGE

Date is later than the factory calendar definition.

DATE_BEFORE_RANGE

Date is earlier than the factory calendar definition.

DATE_INVALID

Invalid date format.

FACTORY_CALENDAR_NOT_FOUND

Factory calendar is not in the buffer.

---

118 DATE_GET_WEEK

This FM calculates the week in which the date passed in the input parameter lies.

Example: You want to display the total sales of a specific week for your store. You can give the date, say, as 31 October 2013 and can get the week for the date by passing the parameter in this FM and executing it. The FM returns 201344. The first four digits of the return parameter correspond to the year, and the last two digits correspond to the week in which the date falls.

[↓] Import

DATE

Date for which the week should be calculated. Date should be between 01.01.1901 and 31.12.2098. Mandatory. Type: SCAL-DATE.

[↑] Export

WEEK

Returns week for date in YYYYWW format. Type: SCAL-WEEK.

[!] Exception

DATE_INVALID

Date does not lie between 01/01/1991 and 31/12/2098.

---

119 DATE_IN_FUTURE

This FM is used to give the future or past date (i.e., with the given date, a certain number of days is added or subtracted). The result is a valid calendar in the future or the past.

Example: You want to get the report of total sales from the past 45 days after the date that is specified. The user provides the start date in a parameter on the screen, and you want to calculate the date before 45 days. Based on the date range, you want to process your report. You can pass 45 days and the start date in the importing parameter of the FM and execute the FM to get the date after 45 days.

[↓] Import

ANZAHL_TAGE

Number of days that are added/subtracted. Mandatory.

IMPORT_DATUM

Date to which NUMBER_DAYS is added in the internal format of YYYYMMDD. Mandatory. Type: RM06B-EEIND.

[↑] Export

EXPORT_DATUM_EXT_FORMAT

Determined date in external format (according to users).

EXPORT_DATUM_INT_FORMAT

Determined date in the internal format (YYYYMMDD). Type: SY-DATUM.

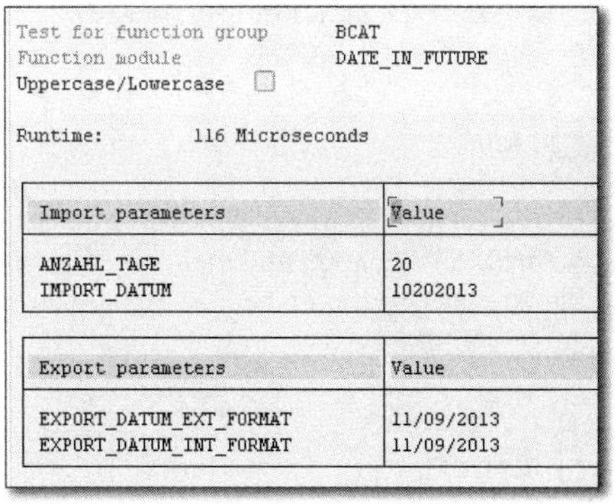
Test for function group BCAT
Function module DATE_IN_FUTURE
Uppercase/Lowercase
Runtime: 116 Microseconds

Import parameters	Value
ANZAHL_TAGE	20
IMPORT_DATUM	10202013

Export parameters	Value
EXPORT_DATUM_EXT_FORMAT	11/09/2013
EXPORT_DATUM_INT_FORMAT	11/09/2013

Date in Future

---

120 DATE_TO_PERIOD_CONVERT

This FM determines the period from the posting date on fiscal year variant tables T009 and T009B.

Example: You're required to convert the calendar month and the year to a fiscal period for your company finance documents. For example, 07 month of 2013 year for period version 24 is fiscal period 013 and fiscal year 2013. In this scenario, pass the details in to the parameter and execute the FM to get the posting period and fiscal year.

[↓] Import

I_DATE

Posting date. This is a mandatory parameter. Type: SY-DATUM.

I_MONMIT

Mid-month at half monthly period. Default: 00. Type: GJVAR-MONMIT.

I_PERIV

Period version. The period variant must be defined in table T009 and maintained in table T009B. This is a mandatory parameter. Type: T009B-PERIV.

[↑] Export

E_BUPER

Determined posting period. Type: T009B-POPER.

E_GJAHR

Determined fiscal year. Type: T009B-BDATJ.

[!] Exception

INPUT_FALSE

Input entered is not correct.

T009_NOTFOUND

Version period determination not in table T009.

T009B_NOTFOUND

Version period determination not in table T009B.

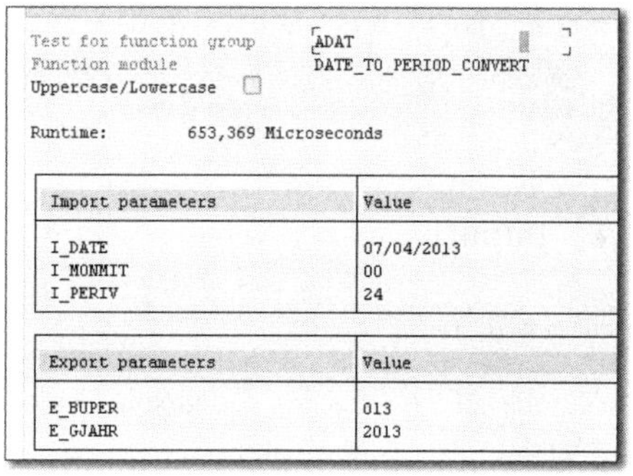

Test for function group ADAT
Function module DATE_TO_PERIOD_CONVERT
Uppercase/Lowercase

Runtime: 653,369 Microseconds

Import parameters	Value
I_DATE	07/04/2013
I_MONMIT	00
I_PERIV	24

Export parameters	Value
E_BUPER	013
E_GJAHR	2013

Fiscal Period Conversion

121 DAY_ATTRIBUTES_GET

This FM returns the specified period of time and the calendar (factory or holiday calendar) days with all their attributes. These attributes include, for example, the labeling; or whether it's a working day or a nonworking holiday. If it is a public holiday, the information of the holiday still is added to read.

Example: You're working on a leave and working day application for your company that includes a select option in the screen where you can give the start date and end date for which you want to get the holidays details. Fill in the parameter and call this FM to get the details of the date entered.

[↓] Import

FACTORY_CALENDAR

Factory calendar whose details are required. Type: SCAL-FCALID.

HOLIDAY_CALENDAR

Public holiday calendar whose details are required. Type: SCAL-HCALID.

DATE_FROM

Date from which details are required. Default: SY-DATUM. Type: SCAL-DATE.

DATE_TO

Date to which details are required. Default: SY-DATUM. Type: SCAL-DATE.

LANGUAGE

Language in which details are required. Type: SY-LANGU.

NON_ISO

Defines the ISO or non-ISO. Type: OS_BOOLEAN.

[↑] Export

YEAR_OF_VALID_FROM

This parameter returns the date from when the calendar is valid. Type: SCAL-YEAR.

YEAR_OF_VALID_TO

This parameter returns the date until the calendar is valid. Type: SCAL-YEAR.

RETURNCODE

This parameter is set to 0 if the specified time is within the validity period of the calendar. If the specified period, however, is partially or entirely outside of the validity period of the calendar, then the RETURN_CODE is returned with the value 4. Type: SY-SUBRC.

Table

DAY_ATTRIBUTES

This table contains the attributes of the date passed in the import parameter. Type: CASDAYATTR.

[!] Exception

FACTORY_CALENDAR_NOT_FOUND

Factory calendar is not in the buffer.

HOLIDAY_CALENDAR_NOT_FOUND

Holiday calendar is not in the buffer.

DATE_HAS_INVALID_FORMAT

Beginning or the end date has an invalid format.

DATE_INCONSISTENCY

Date is inconsistent.

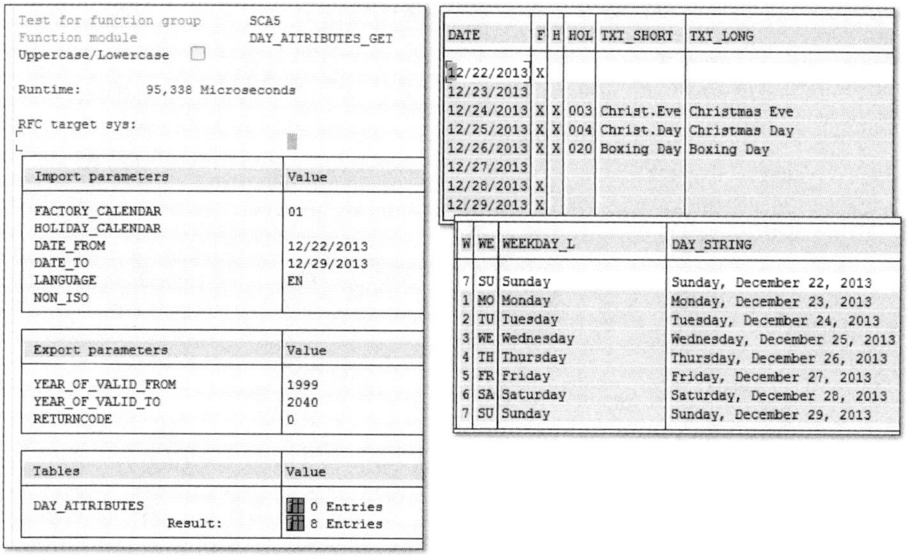

Calender Days Returned from Date Interval

122 DAY_IN_WEEK

This FM returns the day number of the week when a date is passed as an import parameter to this FM.

Example: Your selection screen parameter has a date field where you want to enter the date and process the further logic if the day is on a weekend. You can pass the value of the date in the import parameter and execute the FM to get the day corresponding to the date in the export parameter. If the returned value is Saturday or Sunday, you can execute your logic.

[↓] Import

DATUM

Date of weekday to be calculated. Date must lie between 01.01.1901 and 31.12.2098. Mandatory. Type: D.

[↑] Export

WOTNR

Returned weekday value. Possible values are:

- 1: Monday.
- 2: Tuesday.
- 3: Wednesday.
- 4: Thursday.
- 5: Friday.
- 6: Saturday.
- 7: Sunday.

Type: P.

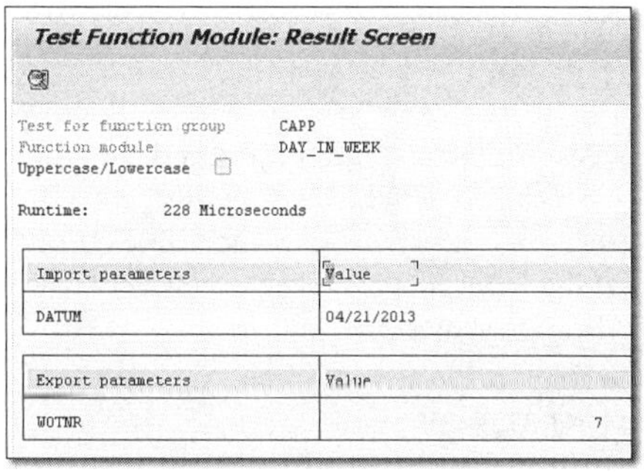

Day Returned from the Date Entered

---

123 DAYS_BETWEEN_TWO_DATES

This FM determines the number of days from a given start and end date based on the interest rate set by the selected calculation method. The interest calculation method is based on either the Gregorian calendar or the German interest calculation method.

Example: You've entered two dates in the input parameters of a selection screen, and you want to find out the number of days in between the start and end date in the Gregorian calendar. Fill in the parameter I_STGMETH and I_SZBMETH and call this FM to get the days between the dates passed.

[↓] Import

I_DATUM_BIS

This parameter defines the end date. It's assumed that the end date is a useful calendar date in the Gregorian calendar. This is a mandatory parameter. Type: VTBBEWE-DBERBIS.

I_DATUM_VON

This parameter defines the start date. It's assumed that the start date is a useful calendar date in the Gregorian calendar. This is a mandatory parameter. Type: VTBBEWE-DBERVON.

I_KZ_EXCL_VON

This parameter defines if the start date is to be included or to be taken exclusively. If the value is 0, the start date is included in the search criteria. If the value is 1, the start date is taken exclusively. The default value is 0.

I_KZ_INCL_BIS

This parameter defines if the end date is to be included or to be taken exclusively. If the value is 1, it's included in the search criteria. If the value is 0, it's taken exclusively. The default value is 0.

I_KZ_ULT_BIS

This parameter defines the ULTIMO indicator for the end date. If the value is [Space], the end date isn't regarded as the end date, which is the default setting. If this parameter has value X, the end date is viewed as the month end date. This and I_KZ_ULT_VON indicators are important for the interest calculation method and the Gregorian calendar without the lead days.

I_KZ_ULT_VON

This parameter defines the ULTIMO indicator for the start date. If the value is [Space], the start date isn't regarded as the end date, which is the default setting. If this parameter has the value X, the start date is viewed as the month end date.

I_STGMETH

This parameter defines the daily interest method. The default value is 0.

I_SZBMETH

This parameter defines the interest calculation method. This parameter supports the calculation for the Gregorian calendar, the German interest calculation method, and the Gregorian calendar without leap days. The default value is 1. It can have the following values.

- 1: 360/360 German-count method (default setting).
- 2: Act/360 French interest calculation method.
- 3: Actual/365.

[↑] Export

E_TAGE

This parameter returns the number of days in the time interval. Taking into account the included labels and marking the start and end of the end date, the number lying between the start and end date is determined in dependence of the interest calculation method.

[!] Exception

DAYS_METHOD_NOT_DEFINED

This exception is raised if the calculation method defined in I_SZBMETH is not defined.

124 DIMENSION_CHECK

This FM checks whether a unit is of a specified dimension. For a business unit, this FM checks if the internal measurement unit MSEHI has a dimension DIMID.

Example: You have a parameter for a unit and dimension in your selection screen and want to validate whether the unit matches with the dimension. You can fill in the parameter and execute this FM. If the execution is successful, it means the unit and dimension match. However, if an exception is raised, it means the unit and dimension don't match.

[↓] Import

BUSINESS_UOM

Commercial unit of measurement flag. If this parameter has the value X, it checks whether the specified unit is a business unit. If not, the exception DIMENSION_CHECK_FAILED is raised. If this parameter has the value [Space], only the validity of the dimension is checked. The default value is [Space]. Type: T006-KZKEH.

DIMID

Dimension key. You can get the dimension key from the FM DIMENSION_GET (FM 125) or FM DIMENSION_GET_FOR_UNIT (FM 126). The default value is [Space]. Type: T006-DIMID.

LANGUAGE

Language key. The default value is logon language. Type: SY-LANGU.

MSEHI

This parameter defines the internal unit of measurement. This is a mandatory parameter. Type: T006-MSEHI.

UNIT_EXTERN_LONG

This parameter determines whether a three-character (UNIT_EXTERN_LONG = [Space]) or six-character (UNIT_EXTERN_LONG NE [Space]) external measurement unit is used. The default value is UNIT_EXTERN_LONG = [Space].

[!] Exception

DIMENSION_CHECK_FAILED

Internal unit of measurement. MSEHI does not correspond to the dimension DIMID; that is, there is no corresponding entry in the table T006. If the exception is handled by the calling program, the corresponding external measurement unit comes in the field SY-MSGV1. SY_MSGV1 contains a three-character or six-character measurement unit, depending on the UNIT_EXTERN_LONG parameter.

UNIT_NOT_VALID

Internal unit of measurement is not maintained in table T006.

125 DIMENSION_GET

This FM determines the dimension key and text depending on the base unit exponents. According to the International System of Units, there are seven base units or dimensions. Those are meter for length, kilogram for mass, second for time, ampere for current, kelvin for temperature, mol for molarity, and candela for light intensity. Based on these units, all other units can be derived.

Example: Unit for speed (m/s) can be derived from the base unit, which is meter per second (m/s). If you enter value "1" in LENGTH and value "-1" in parameter TIME, you'll get the returned value as SPEED in the export parameter DIMID. You can use this FM to get the DIMID parameter, which is the import parameter of FM DIMENSION_CHECK (FM 124).

[↓] Import

AMOUNT_OF_SUBSTANCE

Mole quantity exponent.

ELECTRIC_CURRENT

Exponent for the electric current.

LANGUAGE

Language key. The default value is logon language. Type: SY-LANGU.

LENGTH

Length exponent.

LUMINOUS_INTENSITY

Luminosity exponent.

MASS

Mass exponent.

TEMPERATURE

Temperature exponent.

TIME

Current exponent.

USE_BUFFER

This parameter is used to determine whether the FM should work with a buffer (USE_BUFFER = X) or without a buffer (USE_BUFFER = [Space]). The buffered work method is recommended if the module is to be used several times. The default value is X.

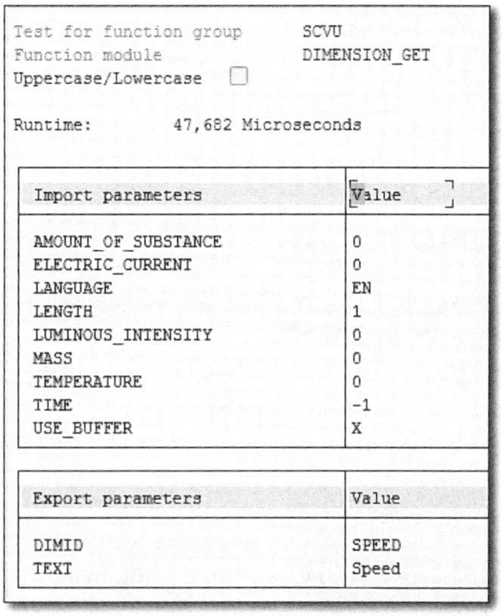

Dimension of Speed Returned from Base Units Meter and Second

[↑] Export

DIMID

This parameter returns the dimension key. Type: T006D-DIMID.

TEXT

This parameter returns the dimension text. If no text is maintained, [Space] is returned. Type: T006T-TXDIM.

[!] Exception

DIMENSION_NOT_FOUND

No dimension is maintained for specified exponent combination.

126 DIMENSION_GET_FOR_UNIT

This FM returns the dimension key and its text based on the unit entered in the importing parameter. You can use this FM to get the DIMID parameter, which is used as an import parameter for FM DIMENSION_CHECK (FM 124).

Example: You enter "M/S" in the UNIT parameter and execute this FM to get the dimension key, which is SPEED for M/S in the DIMID parameter and its text.

[↓] Import

LANGUAGE

Language key. The default value is logon language. Type: SY-LANGU.

UNIT

Unit of measurement. Mandatory. Type: T006-MSEHI.

USE_BUFFER_FOR_TEXT

This parameter determines whether the FM should work with a buffer (USE_BUFFER = X) or without a buffer (USE_BUFFER = [Space]). The buffered work method is recommended if the module is to be used several times. The default value is X.

[↑] Export

DIMID

This parameter returns the dimension key. Type: T006D-DIMID.

TEXT

This parameter returns the dimension text. If no text is maintained, [Space] is returned. Type: T006T-TXDIM.

[!] Exception

DIMENSION_NOT_FOUND

No dimension is maintained for the specified exponent combination.

127 DURATION_DETERMINE

This FM determines the length of time between a start time and an end time. The times you can enter are a date and time of day.

Example: You have a requirement to determine the number of operational hours of your organization between two dates; for example, between April 30th, 2013, and May 2nd, 2013. Considering May 1st as a holiday, you don't reduce the time for that day. Fill in the parameters and call this FM, and the DURATION parameter is returned, which gives you the duration.

[↓] Import

UNIT

This parameter defines the output time unit. Type: T006-MSEHI.

FACTORY_CALENDAR

A factory calendar from which the working days should be calculated if you don't want to use the Gregorian calendar. Type: SCAL-FCALID.

[↑] Export

DURATION

With import parameters, the FM calculates the duration and transfers it to this time unit field. Type: F.

[Δ] Changing

START_DATE

This parameter passes the start date of the time interval. Type: SY-DATUM.

START_TIME

This parameter passes the start time of the time interval. Type: SY-UZEIT.

END_DATE

This parameter passes the end date of the time interval. Type: SY-DATUM.

END_TIME

This parameter passes the end time of the time interval. Type: SY-UZEIT.

[!] Exceptions

FACTORY_CALENDAR_NOT_FOUND

Factory calendar not found.

DATE_OUT_OF_CALENDAR_RANGE

Date outside factory calendar validity.

DATE_NOT_VALID

Date is invalid.

UNIT_CONVERSION_ERROR

Time unit conversion error.

SI_UNIT_MISSING

Time unit missing.

PARAMETERS_NOT_VALID

Input parameters invalid.

128 END_TIME_DETERMINE

This FM determines the end date and time based on the specified start date, time, duration, and unit of time.

Example: You're working on the project management program where you have a list of company projects and durations in months. You want to determine the end date of these projects so that you can forecast the work load. Provide the start date and start time and duration, and then execute the FM to get the end date of the project. You can use this start date and end date to get the data for the report.

[↓] Import

DURATION

This parameter defines the duration that determines the end date. Default: 0.

UNIT

This parameter defines for which unit the calculation of the end point is to be made. Type: T006-MSEHI.

FACTORY_CALENDAR

If this parameter is defined, the end time is based on the factory calendar working days. Type: SCAL-FCALID.

[↑] Export

END_DATE

This parameter returns the end date. Type: SY-DATUM.

END_TIME

This parameter returns the end time. Type: SY-UZEIT.

[Δ] Changing

START_DATE

Start date. The default value is the current date. Type: SY-DATUM.

START_TIME

Start time. The default value is the current time. Type: SY-UZEIT.

[!] Exceptions

FACTORY_CALENDAR_NOT_FOUND

Factory calendar not found.

DATE_OUT_OF_CALENDAR_RANGE

Date outside factory calendar validity.

DATE_NOT_VALID

Date is invalid.

UNIT_CONVERSION_ERROR

Time unit conversion error.

SI_UNIT_MISSING

Time unit missing.

PARAMETERS_NOT_VALID

Input parameters invalid.

129 EVAL_FORMULA

This FM evaluates literals and variables, as well as analyzes the syntactical correction of the formula. This FM accepts arithmetic formulas, logical formulas (with results TRUE and FALSE), conditions of the IF...THEN...ELSE, or expressions with field of structure SY. This FM also evaluates user-defined functions that are defined in a table.

Example: You want to create a program where you enter any formula, such as a*b*c–d and then get back a value. If a user enters values such as a = 2, b =3, c=4, and d = 4, he should get the value 20. You can pass the relevant parameter in this FM and call this FM to get the value.

[↓] Import

DEGREES

This parameter defines the angle function operators in GRAD.

FORMULA

Formula to be calculated. This is a mandatory parameter.

PROGRAM

The parameter PROGRAM specifies a user-specific program that assigns values to the variables in the formula.

ROUTINE

The parameter ROUTINE specifies a user-specific routine that assigns values to variables in the formula.

UNIT_OF_MEASURE

The parameter UNIT_OF_MEASURE determines the measurement unit of the value that the formula calculates.

NO_EXISTENCE_CHECK

If this parameter is X, the existence of a program/routine isn't checked.

[↑] Export

VALUE

Calculated value of the formula returned.

[!] Exceptions

DIVISION_BY_ZERO

Division by zero in function evaluation.

EXP_ERROR

Invalid argument for exponential function.

FORMULA_TABLE_NOT_VALID

The specified formula set is empty or invalid. The Table TFKT contains the standard formula set.

INVALID_EXPRESSION

The formula contains the invalid expression and cannot be evaluated.

INVALID_VALUE

Invalid value occurred during function evaluation.

LOG_ERROR

Invalid argument for logarithmic function.

PARAMETER_ERROR

Parameter required for the function evaluation is invalid.

SQRT_ERROR

Invalid argument for square root function argument.

UNITS_NOT_VALID

Invalid unit of measure.

MISSING_PARAMETER

Program or routines missing for variables.

130 FACTORYDATE_CONVERT_TO_DATE

This FM returns the date for a factory calendar date entered. In the factory calendar, the working days are numbered sequentially from the first working day. The numbers of the working days are called factory dates. You can find the factory calendar definition from table TFACD.

Example: You're working on the factory plant and using a specific factory calendar. You're required to calculate the actual date based on the factory calendar date. Fill in the parameters and call this FM to get the actual date.

[↓] Import

FACTORYDATE

This parameter defines the serial number of the workday in the factory calendar. Mandatory. Type: SCAL-FACDATE.

FACTORY_CALENDAR_ID

Factory calendar ID. Mandatory. Type: SCAL-FCALID.

[↑] Export

DATE

Returned date of factory calendar date. Type: SCAL-DATE.

[!] Exceptions

CALENDAR_BUFFER_NOT_LOADABLE

Factory calendar could not be buffered into the calendar buffer.

FACTORYDATE_AFTER_RANGE

Factory date is larger than factory calendar defined.

FACTORYDATE_BEFORE_RANGE

Factory date is smaller than factory calendar defined.

FACTORYDATE_INVALID

Factory date has invalid format.

FACTORY_CALENDAR_ID_MISSING

Factory calendar ID was transferred with [Space].

FACTORY_CALENDAR_NOT_FOUND

Factory calendar is not in buffer.

131 FIMA_DAYS_AND_MONTHS_AND_YEARS

This FM determines the days, months, or years between two calendar dates.

Example: You have two dates in the selection screen and want to know the date difference between the two dates. You can fill in the parameter and execute this FM to get the date difference in the export parameter.

[↓] Import

I_DATE_FROM

This parameter defines the From date you want of the calendar. This is a mandatory parameter. Type: VTBBEWE-DBERVON.

I_KEY_DAY_FROM

This parameter defines the deadline for updating the calculation from date. Type: TFM_SVKEYDAY.

I_DATE_TO

This parameter defines the To date you want of the calendar. This is a mandatory parameter. Type: VTBBEWE-DBERBIS.

I_KEY_DAY_TO

This parameter defines the deadline for updating the calculation from date. Type: TFM_SVKEYDAY.

I_FLG_SEPARATE

If this flag is set, the value returned is absolute based on days, months, and year. Type: TRFF_FLG.

[↑] Export

E_DAYS

This parameter returns the days in between I_DATE_FROM and I_DATE_TO. Type: VTBBEWE-ATAGE.

E_MONTHS

This parameter returns the month in between I_DATE_FROM and I_DATE_TO. Type: VTBBEWE-ATAGE.

E_YEARS

This parameter returns the year in between I_DATE_FROM and I_DATE_TO. Type: VTBBEWE-ATAGE.

Both screens have the same import parameters except parameter I_FLG_SEPERATE. Please see the difference in the value returned in the export parameter.

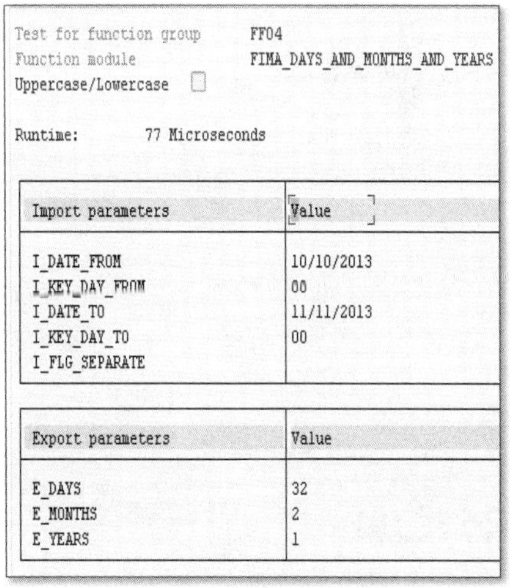

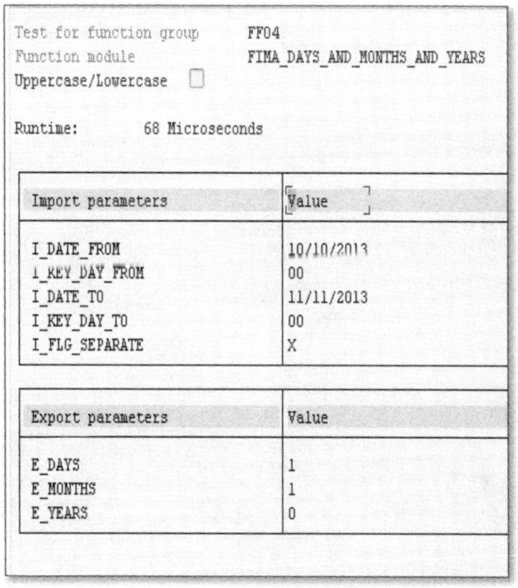

FM Parameter I_FLG_SEPARATE Explained

132 FIRST_DAY_IN_PERIOD_GET

This FM takes the given period and fiscal year and determines the first calendar date of the period.

Example: You may need to get the start date of a particular period based on the date given by the user in the selection screen. You can fill in parameters and call this FM to get the first date of the period.

[↓] Import

I_GJAHR

Value for fiscal year is passed in this variant. Mandatory. Type: T009B-BDATJ.

I_MONMIT

Mid-month for half-month periods. Type: GJVAR-MONMIT.

I_PERIV

Period variant. This period variant must be specified in table T009 and maintained in table T009B. Mandatory. Type: T009B-PERIV.

I_POPER

Fiscal period as defined in table T009B POPER. Mandatory. Type: T009B-POPER.

[↑] Export

E_DATE

First calendar day in fiscal period. Type: SY-DATUM.

[!] Exceptions

INPUT_FALSE

Incorrect input values.

T009_NOTFOUND

Period variant not found in table T009.

T009B_NOTFOUND

Period variant for fiscal year not found in table T009B.

133 FLTP_CHAR_CONVERSION

This FM converts the floating format to the character format.

Example: You have a floating type number that you get from a table, but you want to move it to another field in character format. You can fill in the parameter and convert your floating number to character format by executing this FM.

[↓] Import

DECIM

Number of decimal places in the floating point format in the export parameter FLSTR. Default: 15.

EXPON

Exponential, that is, power of 10, in floating point format in the export parameter FLSTR. Default: 0.

INPUT

A floating point field (FLTP) must be assigned to the INPUT parameter. The FM converts the contents of INPUT into a character string. Default: 0.

IVALU

This parameter distinguishes between initial and 0.

If the FLTP parameter INPUT is initial ([Space]), but the IVALU flag isn't initial (X),

0 is put in the CHAR parameter FLSTR instead of [Space].

MASKN

If the MASKN parameter isn't initial, an input mask can be created in the FLSTR parameter with the assigned character (e.g., "_").

[↑] Export

FLSTR

A character field maximum of length 22 characters must be assigned to the FLSTR parameter. The FM puts the formatted contents of the INPUT parameter in this field.

---

134 GET_CURRENT_YEAR

This FM returns the current and previous fiscal year and month based on the given date or company code.

Example: You've given the date in the selection screen, and you want to display the yearly expense report for the current fiscal year. You have to get the fiscal year to do this, so you fill in the parameters and execute this FM.

[↓] Import

BUKRS

Company code. Type: BKPF-BUKRS.

DATE

Date in the form YYYYMMDD. Type: SY-DATUM.

[↑] Export

CURRM

Fiscal month of the specified date. Type: BKPF-MONAT.

CURRY

Fiscal year of the specified date. Type: BKPF-GJAHR.

PREVM

Previous month of the specified date. Type: BKPF-MONAT.

PREVY

Previous year of the specified date. Type: BKPF-GJAHR.

---

135 HOLIDAY_CHECK_AND_GET_INFO

This FM checks if the given date is a holiday. If yes, its details are returned in the table.

Example: You're working on the leave management system of an employee, and you want to get the detail if a particular day is a holiday. You can give the date and check if the particular date is a holiday and its details by executing this FM.

[↓] Import

DATE

Date to be checked. Mandatory. Type: SCAL-DATE.

HOLIDAY_CALENDAR_ID

Public holiday calendar ID. Mandatory. Type: SCAL-HCALID.

WITH_HOLIDAY_ATTRIBUTES

This parameter determines where to return the holiday attributes if the date entered is a holiday. [Space] determines attributes not to be returned. (This is the default setting). X determines attributes to be returned. Type: SCAL-INDICATOR.

[↑] Export

HOLIDAY_FOUND

This parameter returns whether the specified date is a public holiday. [Space] specifies that it's not a holiday, and X specifies that it is a holiday. Type: SCAL-INDICATOR.

Tables

HOLIDAY_ATTRIBUTES

If the date entered is a holiday, this table returns the attribute of the holiday. Because several holidays can fall in the same date, this is a table parameter. Type: THOL.

[!] Exceptions

CALENDAR_BUFFER_NOT_LOADABLE

Public holiday calendar could not be buffered into the buffer area.

DATE_AFTER_RANGE

Date later than public holiday calendar defined.

DATE_BEFORE_RANGE

Date earlier than public holiday calendar defined.

DATE_INVALID

Invalid date format.

HOLIDAY_CALENDAR_ID_MISSING

Public holiday calendar was transferred with [Space].

HOLIDAY_CALENDAR_NOT_FOUND

Holiday calendar is not in buffer.

136 HOLIDAY_GET

This FM determines all of the holidays and all nonworking days that lie within a certain period. The FM can refer to the given calendar (holiday and factory calendar).

If both the holiday and factory calendar are passed in the import parameters, then holidays and nonworking days are determined based on both calendars. However, if only the holiday calendar is specified, only the associated holidays are determined (all other days are therefore working days).

Example: You've created a report that displays a list of holidays for the dates as specified by the user. You can pass the parameter in this FM and call this FM.

[↓] Import

HOLIDAY_CALENDAR

This parameter contains the public holiday calendar ID. Type: SCAL-HCALID.

FACTORY_CALENDAR

This parameter contains the factory calendar ID. Type: SCAL-FCALID.

DATE_FROM

This parameter determines the date from which holidays and nonworking days are to be determined. Type: SCAL-DATE.

DATE_TO

This parameter determines the date to which holidays and nonworking days are to be determined. Type: SCAL-DATE.

[↑] Export

YEAR_OF_VALID_FROM

This parameter returns the Valid From of the calendar, which is when the calendar is valid from. If both the holiday and the factory calendar are given in import the parameter, the factory calendar Valid From date is returned. Type: SCAL-YEAR.

YEAR_OF_VALID_TO

This parameter returns the Valid To of the calendar, which is until when the calendar is valid. If both the holiday and factory calendar are given in the import parameter, the factory calendar Valid Todate is returned. Type: SCAL-YEAR.

RETURNCODE

This parameter returns the selection period. Type: SY-SUBRC.

Table

HOLIDAYS

This table contains all of the days that are holidays or nonworking. Type: ISCAL_DAY.

[!] Exceptions

FACTORY_CALENDAR_NOT_FOUND

Factory calendar is not in the buffer.

HOLIDAY_CALENDAR_NOT_FOUND

Holiday calendar is not in the buffer.

DATE_HAS_INVALID_FORMAT

Start date or end has an invalid format.

DATE_INCONSISTENCY

Date is inconsistent.

137 HR_HK_DIFF_BT_2_DATES

This FM calculates the difference between two dates. The result can be in the form of year, month, or date depending on the output format specified.

Example: You want to find your age. You can put today's date in parameter Date1 and birth date in Date2. You'll get your age in years, months, and days.

[↓] Import

DATE1

Begin date. This is a mandatory parameter. Type: P0001-BEGDA.

DATE2

End date. This is a mandatory parameter. Type: P0001-BEGDA.

OUTPUT_FORMAT

Format to display the output. The default value is 01. Type: CHAR2.

[↑] Export

YEARS

Years between the begin date and the end date. Type: PEA_SCRYY.

MONTHS

Months between the begin date and the end date. Type: PEA_SCRMM.

DAYS

Days between the begin date and the end date. Type: PEA_SCRDD.

[!] Exceptions

INVALID_DATES_SPECIFIED

Dates specified are not valid.

OVERFLOW_LONG_YEARS_BETWEEN

Arithmatic overflow occurred while calculating year difference.

138 HR_SEN_CALE_DAYS_DATE

This FM is used to add or subtract calendar days from the given date. Depending on the operator specified in the parameter ID_DATE, the FM either adds the value to the date or subtracts the value from the import parameter ID_DATE.

Example: You want to get the total sales of your product for the number of days entered by the end user in the screen parameter. You have date and days parameter fields in the selection screen. The user enters the date and day for which he wants to get the total sales. You can pass these parameters as an import parameter in this FM and call this FM to get the end date. Based on the dates, you can fetch the details of sales from your sales database table.

[↓] Import

ID_DATE

Start or end date (depending on operator). Mandatory. Type: BEGDA.

ID_OPERATOR

Operator (+, –). Default: +. Type: ADSUB.

IS_DURATION

Duration in years, months, and days. Mandatory. Type: PSEN_DURATION_DEC.

[↑] Export

ED_DATE

Start or end date (depending on operator). Type: ENDDA.

[!] Exceptions

CONVERSION_NOT_SPECIFIED

Conversion not planned.

CONVERSION_NOT_POSSIBLE

Translation not possible.

---

139 HR_SEN_CALE_DAYS_DURATION

This FM calculates the duration between the calendar days for an interval that is defined by a start date and an end date.

Example: You have two dates—for example, October 31st, 2013, and November 16th, 2013—and you want to get the duration between those two dates. You can fill in the parameters and call this FM to get the duration between the dates, that is, the number of days, which is 17 in this case (inclusive of October 31st and November 16th).

[↓] Import

ID_BEGDA

Start date of duration. Mandatory. Type: BEGDA.

ID_ENDDA

End date of range. Mandatory. Type: ENDDA.

[↑] Export

ES_DURATION

Duration of interval in years, months, and days. Type: PSEN_DURATION_DEC.

[!] Exceptions

CONVERSION_NOT_SPECIFIED

Conversion not planned.

CONVERSION_NOT_POSSIBLE

Translation not possible.

---

140 HR_SEN_CRULE_0100_DATE

This FM is used for the calculation of a date; that is, you enter the start date and number of days, and you get the end date. Years, months, and days are used for the calculation. The required conversions take place based on the assumption that 1 year = 12 months and 1 month = 30 days. This FM is generally used for the calculation of employment periods.

Example: You joined on January 10, 2012, and served the company for 1 year and 2 months, and you want to know the date after 1 year and 2 months. Fill in the parameter and execute this FM to get the end date.

[↓] Import

ID_DATE

Start or end date you want to calculate. Mandatory. Type: BEGDA.

ID_OPERATOR

Operator (+, –). The default value is +. Type: ADSUB.

IS_DURATION

Duration in years, months, and days. Mandatory. Type: PSEN_DURATION_DEC.

[↑] Export

ED_DATE

Returned date. Type: ENDDA.

[!] Exceptions

CONVERSION_NOT_SPECIFIED

Conversion not planned.

CONVERSION_NOT_POSSIBLE

Translation not possible.

---

141 HRCM_AMOUNT_TO_STRING_CONVERT

This FM converts a number or currency amount to string format.

Example: You want to concatenate the string and currency amount and send it as a message such as "Total Currency amount is $100" to the end user. This $100 is a dynamic currency coming from database table. For the `CONCATENATE` statement, you need to convert the number in string format. To do so, pass the parameter and call this FM.

[↓] Import

BETRG

Parameter to pass amount or currency. Type: BETRG.

WAERS

Currency key. Type: WAERS.

NEW_DECIMAL_SEPARATOR

Decimal separator field. Type: C.

NEW_THOUSANDS_SEPARATOR

Thousand separator field.

[↑] Export

STRING

String where the converted amount will be returned.

---

142 HRCM_STRING_TO_AMOUNT_CONVERT

This FM converts the string amount to the actual amount, which can be stored in tables.

Example: You have a parameter field of character format. The user enters currency on this field, for example, $100. You want to convert this character format to currency format and save it to the database table. You can pass the relevant parameter in this FM and then call this FM.

[↓] Import

STRING

String that contains the amount to be converted. Mandatory.

DECIMAL_SEPARATOR

Decimal separator field. Type: C.

THOUSANDS_SEPARATOR

Thousand separator field.

WAERS

Currency key of the amount. Type: WAERS.

[↑] Export

BETRG

Converted amount or currency. Type: BETRG.

---

143 LANGUAGE_CODE_ISO_TO_SAP

This FM converts the ISO (International Standard of Organization) code into the SAP language code.

Example: To save a language externally, there is an ISO standard that is different from the way language is saved in SAP. You have an IDoc that passes the data from an external system to the SAP system. You have a language field in the IDoc that is saved in ISO format. Now, you want to save it in the SAP format. You can pass the ISO code language in the FM and execute this FM to get the language in SAP format.

[↓] Import

ISO_CODE

ISO language code to be converted. Mandatory. Type: T002-LAISO.

[↑] Export

SAP_CODE

SAP language code returned. Type: T002-SPRAS.

[!] Exception

NOT_FOUND

ISO code entered does not exist.

---

144 LANGUAGE_CODE_SAP_TO_ISO

This FM converts the SAP language code into ISO code.

Example: To save a language externally, there is an ISO standard that is different from the way language is saved in the SAP system. You have an IDoc that sends the data from the SAP system to an external system. You have a language field in the IDoc that is saved in SAP format. Now, you want to convert it in ISO format to send it. You can pass the SAP language in the FM and execute this FM to get the language in ISO format.

[↓] Import

SAP_CODE

SAP language code returned. Mandatory. Type: T002-SPRAS.

[↑] Export

ISO_CODE

ISO language code to be converted. Type: T002-LAISO.

[!] Exception

NOT_FOUND

ISO code entered does not exist.

---

145 LAST_DAY_IN_PERIOD_GET

You use this FM to return the last calendar date of the period from a given period and fiscal year.

Example: You need to get the end date of a particular period based on the date given by a user in the selection screen. You can fill in parameters and call this FM to get the first date of the period.

[↓] Import

I_GJAHR

Value for fiscal year is passed in this variant. Type: T009B-BDATJ.

I_MONMIT

Mid-month for half-month periods. Type: GJVAR-MONMIT.

I_PERIV

Period variant. This period variant must be specified in table T009 and maintained in table T009B. Type: T009B-PERIV.

I_POPER

Fiscal period as defined in table T009B POPER. Type: T009B-POPER.

[↑] Export

E_DATE

First calendar day in fiscal period. Type: SY-DATUM.

[!] Exceptions

INPUT_FALSE

Incorrect input values.

T009_NOTFOUND

Period variant not found in table T009.

T009B_NOTFOUND

Period variant for fiscal year not found in table T009B.

---

146 MATERIAL_UNIT_CONVERSION

The FM converts a quantity from one unit of measurement to another. Units of measure must relate to a material. One of them must be the base unit of measure of the material, while the other must be defined as an alternative unit of measure for the material or be capable of being converted to an alternative unit of measure as per table T006. This FM is similar to FM CF_UT_UNIT_CONVERSION (FM 76).

Example: You have a requirement where you want to convert units in the material documents to kilograms in the output, irrespective of the base unit of the material in ABAP Data Dictionary table. You can pass the parameters and call this FM.

[↓] Import

INPUT

Quantity to be converted.

KZMEINH

This parameter defines the relation of the input quantity to the base or alternative quantity. Value X denotes that the quantity to be converted is related to the alternative unit of measure. Value [Space] denotes that the quantity to be converted is related to the base unit of measure. Type: SY-MARKY.

MATNR

Material. Mandatory. Type: MARA-MATNR.

MEINH

Alternative unit of measure for the material. Mandatory. Type: MARA-MEINS.

MEINS

Base unit of measure of the material. Type: MARA-MEINS.

EXTENDED_CONV

If this flag is set, the alternative unit takes account not only of the material but also the volume and/or weight unit to the base unit. Possible values are:

- [Space]: No consideration of volume or weight.
- D: Considers existing gross weight.
- E: Considers net weight.
- F: Considers existing volume.

Type: SY-MARKY.

TYPE_UMR

This parameter specifies which type has the conversion factor. This information is used to improve the performance. This parameter can have following values:

- 1: Specifies a five-digit field of type P (e.g., MARM-UMREZ).
- 2: Specifies a 10-digit field of type I (e.g., T006-COUNT).
- 3: Field of type F.
- [Space]: The system determines the type.

Type: SY-MARKY.

NO_OUTPUT

Indicator: No OUTPUT is to be calculated. Type: SY-MARKY.

CHARGE

Charge. Type: MCHA-CHARG.

CHARG_CUOBJ

Internal object. Type: INOB-CUOBJ.

WERKS

Plant. Type: MARC-WERKS.

[↑] Export

ADDKO

Additive constant. Type: T006-ADDKO.

MEINS

Base unit of measure of the material. Type: MARA-MEINS.

OUTPUT

Output value.

UMREN

Conversion factor denominator.

UMREZ

Conversion factor numerator.

BATCH_CONVERSION_ERROR

Error flag parameter when converting in batch. Type: C.

KZWSO

Unit/ME variable is active. Type: MARM-KZWSO.

[!] Exceptions

CONVERSION_NOT_FOUND

Conversion factor could not be determined.

INPUT_INVALID

The quantity to be converted is not a number.

MATERIAL_NOT_FOUND

Material does not exist.

MEINH_NOT_FOUND

Alternative unit of measure not allowed.

MEINS_MISSING

No base unit of measure exists for the material.

NO_MEINH

No alternative unit of measure specified.

OUTPUT_INVALID

Quantity to be calculated is not a number.

OVERFLOW

Field overflow.

---

147 MD_CONVERT_MATERIAL_UNIT

This FM converts the amount of source unit to target unit that's linked to a material. This FM is generally used in Materials Management. All import parameters are mandatory.

Example: You have a material that is linked with a unit of measure, say kilograms. Now, in your program, you want to display the unit in grams and convert the quantity as well. You can pass the parameter in the FM and call the FM to get the converted output unit.

[↓] Import

I_MATNR

Material number. Type: MARA-MATNR.

I_IN_ME

Output unit. Type: MARA-MEINS.

I_OUT_ME

Target unit. Type: MARA-MEINS.

I_MENGE

Output quantity. Type: EKPO-MENGE.

[↑] Export

E_MENGE

Target amount quantity. Type: EKPO-MENGE.

[!] Exceptions

ERROR_IN_APPLICATION

Error occurred while executing the FM.

ERROR

A generic error occurred.

148 MONTH_NAMES_GET

This FM gives the name of the month based on the language entered.

Example: To get the list of all of the months in Portuguese to display in the screen, you can call this FM.

[↓] Import

LANGUAGE

Language in which month name is required. Type: SY-LANGU.

[↑] Export

RETURN_CODE

Return code. If the value returned is 0, it means successful execution of the FM. Type: SY-SUBRC.

Table

MONTH_NAMES

This table contains the names of all of the months for the specified language. Type: T247.

[!] Exception

MONTH_NAMES_NOT_FOUND

Month name is not found exception.

149 MONTH_PLUS_DETERMINE

This FM is used to determine the new date by adding months to the date the user specifies in the import parameters.

Example: Based on the selection screen date, you want to add three months to determine the quarter and execute the report to get the quarterly data. Pass the parameters in the FM and execute the FM to get the new date.

[↓] Import

MONTHS

Number of months to be added. Mandatory.

OLDDATE

From date. Mandatory. Type: SY-DATUM.

[↑] Export

NEWDATE

New date returned. Type: SY-DATUM.

---

150 PERIOD_DAY_DETERMINE

This FM determines the first and last day of the accounting period. This FM is a part of the Financial Accounting (FI) module. All import parameters are mandatory.

Example: You want to know the details of the fiscal year, such as its first and last day, and display these details in the program output. You can fill in the parameter and call this FM to determine the first and last period day.

[↓] Import

I_GJAHR

Fiscal year. Type: BKPF-GJAHR.

I_MONAT

Posting period. Type: BKPF-MONAT.

I_PERIV

Period version. Type: T001-PERIV.

[↑] Export

E_FDAY

First period day returned. Type: BKPF-BUDAT.

E_LDAY

Last period day returned. Type: BKPF-BUDAT.

E_SPERIOD

Special period indicator. Type: BKPF-BSTAT.

[!] Exceptions

ERROR_PERIOD

Period is not valid.

ERROR_PERIOD_VERSION

Period version not defined in T009.

FIRSTDAY_NOT_DEFINED

Previous period does not exist.

PERIOD_NOT_DEFINED

Period not defined in T009B.

YEAR_INVALID

Year is not valid.

---

151 READ_EXCHANGE_RATE

This FM determines the exchange rate from table TCURR. The exchange rate valid on the specified date is read from table TCURR according to the currency key and the exchange rate type.

Example: To write a program in which you have a currency in USD and can determine the rupees based on particular day exchange rate, fill in the parameters and call this FM.

[↓] Import

CLIENT

Client in which you're logged in. Type: SY-MANDT.

DATE

Currency translation date. Mandatory. Type: SYST-DATUM.

FOREIGN_CURRENCY

Currency key for foreign currency. Mandatory. Type: TCURR-FCURR.

LOCAL_CURRENCY

Currency key for local currency. Mandatory. Type: TCURR-TCURR.

TYPE_OF_RATE

Type of rate. It can have following values:

- M: Average rate.
- G: Bank buying rate.
- B: Bank selling rate.

Type: TCURR-KURST.

EXACT_DATE

If exchange rate for exact day is required, you need to set this parameter to X. Type: XFELD.

[↑] Export

EXCHANGE_RATE

Exchange rate returned.

FOREIGN_FACTOR

Factor for the foreign currency units.

LOCAL_FACTOR

Factor for the local currency units.

VALID_FROM_DATE

Date of read entry in TCURR. Type: SYST-DATUM.

DERIVED_RATE_TYPE

Alternative exchange rate type, where used. Type: TCURR-KURST.

FIXED_RATE

Fixed exchange rate if present.

OLDEST_RATE_FROM

Date of oldest exchange rate calculation. Type: SYST-DATUM.

[!] Exceptions

NO_RATE_FOUND

No exchange rate entered in table TCURR.

NO_FACTORS_FOUND

No conversion factors in TCURF.

NO_SPREAD_FOUND

No spread entered in table TCURS.

DERIVED_2_TIMES

Two times derived.

OVERFLOW

Value returned is overflowed.

ZERO_RATE

Indicator: No exchange rate exists.

---

152 RH_GET_DATE_DAYNAME

This FM determines the day name and number based on the date that's provided in the input parameters.

Example: You want to display the day of the date October 31st, 2013, in German language in the output ALV. You can pass in the parameters and call this FM to get the corresponding day and day number.

[↓] Import

LANGU

Language for weekday name. Mandatory. Type: SY-LANGU.

DATE

Date. Mandatory. Type: SY-DATUM.

CALID

Factory calendar ID. Type: P1027-CALID.

[↑] Export

DAYNR

Day number returned. 1 is for Monday, 2 is Tuesday, and so on, until 7 for Sunday. Type: HRVSCHED-DAYNR.

DAYTXT

Weekday name in the language given in the import parameter. Type: HRVSCHED-DAYTXT.

DAYFREE

Day off work according to factory calendar. Type: HRVSCHED-NODAY.

[!] Exceptions

NO_LANGU

Language is not entered or not valid.

NO_DATE

Date is not entered.

NO_DAYTXT_FOR_LANGU

Weekday name is not available for the language.

INVALID_DATE

Date is not valid.

---

153 RKE_TIMESTAMP_CONVERT_INPUT

This FM converts local time or date to GMT time stamp. This program supports the date from year 1990 to year 2057. All import parameters are mandatory.

Example: You have a requirement where you want to save current date and time to GMT time stamp format in your table. You can pass in the date and time in the import parameter and execute this FM. This FM saves your inputs to GMT time stamp.

[↓] Import

I_DATE

Date. Type: SY-DATUM.

I_DAYST

Active daylight savings time indicator. X represents daylight saving is on. Type: SY-DAYST.

I_TIME

Time. Type: SY-UZEIT.

I_TZONE

Time difference to GMT in seconds. Type: SY-TZONE.

[↑] Export

E_TIMESTMP

This parameter returns the time stamp in GMT. Type: CEST1-TIMESTMP.

[!] Exception

DATE_INVALID

If the date does not lie between 1990 and 2057, this exception is raised.

154 RKE_TIMESTAMP_CONVERT_OUTPUT

This FM converts Greenwich Mean Tme (GMT) to local date and time. All import parameters are mandatory.

Example: You have a requirement where you get the GMT time stamp from the ABAP Data Dictionary table. You want to display it in your ALV. You can pass the GMT time stamp along with other import parameter in this FM. You'll get back the corresponding date and time in your format after execution of this FM.

[↓] Import

I_DAYST

Active daylight savings time indicator. X represents daylight saving is on. Type: SY-DAYST.

I_TIMESTMP

Point in time in GMT. Type: CEST1-TIMESTMP.

I_TZONE

Time difference to GMT in seconds. Type: SY-TZONE.

[↑] Export

E_DATE

Returned date. Type: SY-DATUM.

E_TIME

Returned time. Type: SY-UZEIT.

155 ROUND

This FM rounds off the number to the decimal places specified in the import parameter DECIMALS.

Example: You've received a total amount to be paid by your customer from the sales database table. You want to round off the amount, print, and send it to the customer. You can pass the amount in this FM and execute this FM to round it off.

[↓] Import

DECIMALS

Number decimal places, to which rounding is to be carried out. Negative values are also allowed. For positive values, count to the right from the decimal places; for negative values, count to the left. For example 123.123 rounded to –2 decimal places returns 100, and rounding to 2 decimal places returns 123.12.

INPUT

Value to be rounded. Mandatory.

SIGN

Flag for rounding up, rounding down, and commercial round off. This parameter can have the following values:

- [Space]: No rounding.
- +: Round up.
- –: Round down.
- X: Commercial rounding.

[↑] Export

OUTPUT

Rounded value.

[!] Exceptions

INPUT_INVALID

Value to be rounded is not a number.

OVERFLOW

Field overflow.

TYPE_INVALID

Output value is not a number.

---

156 RP_CALC_DATE_IN_INTERVAL

This FM adds or subtracts a day, month, or year on the given input date.

Example: Based on the selection screen date, you want to add three months to determine the quarter and execute the report to get the quarterly data. Pass the date in the DATE parameter, pass "3" in the MONTHS parameter, and execute the FM to get the new date.

[↓] Import

DATE

Input date. Type: P0001-BEGDA.

DAYS

Days to add in the DATE. Type: T5A4A-DLYDY.

MONTHS

Months to add in the DATE. Type: T5A4A-DLYMO.

SIGNUM

Sign + or – for addition or subtraction. The default value is +. Type: T5A4A-SPLIT.

YEARS

Year to add in the DATE. Type: T5A4A-DLYYR.

[↑] Export

CALC_DATE

Output DATE. Type: P0001-BEGDA.

---

157 RP_LAST_DAY_OF_MONTHS

This FM determines the last day of the month based on a given date.

Example: You're working on an expense report for the company. You've written a program where all of the charges of the month are cumulated together and then charged to the department on the last day of the month. You can pass the import parameter as a date to get the last day. Based on this last day, you can cumulate the amount and then send this report in output in ALV to the user.

[↓] Import

DAY_IN

Input date. Type: SY-DATUM.

[↑] Export

LAST_DAY_OF_MONTH

Date of last day of the month from input date. Type: SY-DATUM.

[!] Exception

DAY_IN_NO_DATE

Date entered is not a date.

---

158 RS_CONV_EX_2_IN

This FM converts the field contents from an external format (INPUT_EXTERNAL) into an internal format (OUTPUT_EXTERNAL). The parameter OUTPUT_EXTERNAL must have the technical attributes of the ABAP Data Dictionary field specified in TABLE_FIELD. The field INPUT_EXTERNAL must be a character field. It may have any length but may not be filled to a greater length than that of OUTPUT_INTERNAL (this leads to the exception INPUT_TOO_LONG).

Example: You've given the value on a field in the screen output, but you want that field to be saved as an SAP internal format. You can do so by passing the relevant parameters and calling this FM.

[↓] Import

INPUT_EXTERNAL

Input (external format). Mandatory. Type: C.

TABLE_FIELD

Dictionary field name. Mandatory. Type: TABFIELD.

CURRENCY

If you've specified a currency field, you need to pass the currency field in this parameter. The currency key field is calculated based on the currency field from table TCURX.

[↑] Export

OUTPUT_INTERNAL

Field contents in internal format.

[!] Exceptions

INPUT_NOT_NUMERICAL

Nonnumeric input.

TOO_MANY_DECIMALS

Too many places after the decimal point.

MORE_THAN_ONE_SIGN

+/– sign input multiple times.

ILL_THOUSAND_SEPARATOR_DIST

Separation between thousands not = 0 mod 3.

TOO_MANY_DIGITS

Too many characters in whole number field.

SIGN_FOR_UNSIGNED

No +/– signs allowed.

TOO_LARGE

Value too large.

TOO_SMALL

Value too small.

INVALID_DATE_FORMAT

Invalid date format.

INVALID_DATE

Invalid date.

INVALID_TIME_FORMAT

Invalid time format.

INVALID_TIME

Invalid time.

INVALID_HEX_DIGIT

Incorrect hex character.

UNEXPECTED_ERROR

Unintercepted error.

INVALID_FIELDNAME

Field not in dictionary.

FIELD_AND_DESCR_INCOMPATIBLE

Output field does not match dictionary description.

INPUT_TOO_LONG

Input longer than output length of output field.

NO_DECIMALS

Decimal separator without subsequent decimal places.

INVALID_FLOAT

Invalid floating point entry.

CONVERSION_EXIT_ERROR

Error in conversion exit.

---

159 SCMS_BINARY_TO_XSTRING

This FM converts the binary format to XString text format. You'll also use FM SCMS_XSTRING_TO_BINARY (FM 160) to continue this process.

Example: You want to send a file in binary format as an attachment to an email. Call this FM, which converts the binary format text to XString, and then use the FM SCMS_XSTRING_TO_BINARY.

[↓] Import

INPUT_LENGTH

Length of the file to be converted. Mandatory. Type: I.

FIRST_LINE

First line number. Type: I.

LAST_LINE

Last line number. Type: I.

[↑] Export

BUFFER

Returned content in XString. Type: XSTRING.

[!] Exception

FAILED

Conversion failed because of technical reasons.

Table

BINARY_TAB

Input table that contains the binary format data to be converted.

160 SCMS_XSTRING_TO_BINARY

This FM converts the XString format to binary text format.

Example: After using FM SCMS_BINARY_TO_XSTRING, convert the XString format into a binary format table with this FM. Then call method ADD_ATTACHMENT of class CL_DOCUMENT_BCS to attach this table and send the attachment.

[↓] Import

BUFFER

XString content to be converted. Mandatory. Type: XSTRING.

APPEND_TO_TABLE

This parameter appends the binary table existing records to the new records if the value of this parameter isn't [Space]. Type: C.

[↑] Export

OUTPUT_LENGTH

Returned output length. Type: I.

Table

BINARY_TAB

Output table that contains the binary format data.

161 SCP_CODEPAGE_FOR_LANGUAGE

This FM determines the code page or character set that is to be used for a given language by SAP. As you know, all of the data in the database are stored as a sequence of bytes. For characters, a code page defines a mapping between a byte sequence and a character. If the system can't determine a code page, it sends an error message.

Example: You're developing a program to execute in Chinese language. There are some special characters that may not display properly in Chinese. Use this FM to find and check the code page that the Chinese language is using in detail.

[↓] Import

LANGUAGE

Language for which character set is required. Type: SPRAS.

[↑] Export

CODEPAGE

Character set to be used. Type: CPCODEPAGE.

[!] Exception

NO_CODEPAGE

No character set known for this language.

162 SCP_REPLACE_STRANGE_CHARS

This FM processes a text so that it contains only simple characters. Special characters and national characters are replaced in such a way that the text remains reasonably legible.

The character set 1146 is used by default. Following examples can be seen as replacement after this FM: á ==> AE (AE).

After the replacement, new text can be longer than the old.

Example: You have a text that contains some special and regional characters in the output of the program. You can edit it to be readable for all users by passing the parameters and executing this FM. You'll receive the replaced text.

[↓] Import

INTEXT

This parameter contains the text to be processed. Blank characters at the end of INTEXT aren't processed. Mandatory. Type: C.

INTEXT_LG

Number of bytes in the input text can be specified in this parameter. Type: I.

INTER_CP

Character set that you want to use instead of default character set 1146. Type: TCP00-CPCODEPAGE.

INTER_BASE_CP

The FM makes use of the fact that 1146 is a subcode page of 1100. If you set a code page with INTER_CP that is neither a subcode page of 1100 nor belongs to the same code page family as IN_CP, you can use INTER_BASE_CP to specify the code page that incorporates INTER_CP. Type: TCP00-CPCODEPAGE.

IN_CP

This parameter specifies that the data in INTEXT isn't in the system character set but is coded in the character set IN_CP. Type: TCP00-CPCODEPAGE.

REPLACEMENT

If there are characters that don't occur in one or the other of the two character sets, they are replaced by a period (.). This can be changed with the parameter REPLACEMENT. An SAP character number must be specified. A 0 means that no replacements should be made. In such a case, the exception CANNOT_CONVERT is generated. Default: 46.

[↑] Export

OUTTEXT

Processed output text. Type: C.

OUTUSED

Length of the output text is written in this parameter.

OUTOVERFLOW

This parameter has the value as X if the output text did not fit in OUTTEXT. Otherwise, the value is [Space].

[!] Exceptions

INVALID_CODEPAGE

If you specify the character set that does not exist, you'll get this exception.

CODEPAGE_MISMATCH

If the two specified character sets are not compatible, this exception is raised.

INTERNAL_ERROR

Internal error.

CANNOT_CONVERT

There was no replacement for a character in INTEXT.

FIELDS_NOT_TYPE_C

Fields passed must be type C for parameters INTEXT and OUTTEXT.

163 SD_DATETIME_DIFFERENCE

This FM determines the time difference between two dates and times that are specified.

Example: You've received an order September 20th, 2013, to produce a car, and you've actually delivered the product on September 27th, 2013. You can calculate the total time for the production by passing the parameter in this FM and calling this FM.

[↓] Import

DATE1

First input date. This is a mandatory parameter. Type: D.

TIME1

First input time. This is a mandatory parameter. Type: T.

DATE2

Second input date. This is a mandatory parameter. Type: D.

TIME2

Second input time. This is a mandatory parameter. Type: T.

[↑] Export

DATEDIFF

This parameter returns the date difference. Type: P.

TIMEDIFF

This parameter returns the time difference. Type: P.

EARLIEST

This parameter defines which date or time is earliest. 1 refers to the first date and time, 2 refers to the second date and time, and 0 means both the dates and times are the same. Type: C.

[!] Exception

INVALID_DATETIME

Date or time passed is invalid. None of the import parameter should be initial; otherwise, this exception is raised.

164 SI_UNIT_GET

This FM determines the SI unit (System of Unit (refer to FM DIMENSION_GET (FM 125) for more information on SI units) for the entered unit or dimension. If both the DIMENSION and UNIT parameters are passed, DIMENSION takes priority over UNIT, and the SI unit for the DIMENSION is determined.

Example: You want to find the unit for the dimension SPEED to display in the ALV report column. You can do so by passing the SPEED in the DIMENSION parameter and executing the FM. You'll receive its SI_UNIT, which is M/S.

[↓] Import

DIMENSION

Dimension key. Type: T006-DIMID.

UNIT

Unit of measurement. Type: T006-MSEHI.

[↑] Export

SI_UNIT

SI unit. Type: T006-MSEHI.

[!] Exceptions

DIMENSION_NOT_FOUND

Dimension has not been maintained.

UNIT_NOT_FOUND

Unit of measurement not maintained.

165 START_TIME_DETERMINE

This FM determines the start date and time based on the given end date and time, duration, unit, and factory calendar.

Example: You have an invoice showing a car sold to a customer. You have specified an order completion date and time, and you know how long it takes to make that car. To find the start date when the production of that order was started, you can fill in the parameters and call this FM.

[↓] Import

DURATION

Duration. The default value is 0.

UNIT

Time unit. Type: T006-MSEHI.

FACTORY_CALENDAR

If you want to refer to a factory calendar, you can specify it here. Type: SCAL-FCALID.

[↑] Export

START_DATE

Start date is returned. Type: SY-DATUM.

START_TIME

Start time is returned. Type: SY-UZEIT.

[Δ] Changing

END_DATE

You need to specify the end date. The default value is the current date. Type: SY-DATUM.

END_TIME

You need to specify the end time. The default value is the current time. Type: SY-UZEIT.

[!] Exceptions

FACTORY_CALENDAR_NOT_FOUND

Factory calendar not found.

DATE_OUT_OF_CALENDAR_RANGE

Date entered is outside factory calendar validity.

DATE_NOT_VALID

Date is invalid.

UNIT_CONVERSION_ERROR

Time unit conversion error.

SI_UNIT_MISSING

Time unit is missing.

PARAMETERS_NOT_VALID

Input parameters are invalid.

---

166 SUBTRACT_TIME_FROM_DATE

This FM is used to subtract the time; that is, day, month, week, or year, from the provided date. This FM is opposite to FM ADD_TIME_TO_DATE (FM 73) in functionality. You can also add the time by giving a negative value in the I_TIME field.

Example: You've produced a car in 14 days. You want to get the start date based on the end date by subtracting 14 from the end date. You can fill in the parameters and call this FM to get the start date.

[↓] Import

I_IDATE

Provide the input date on which you want to add the time. Mandatory. Type: SY-DATUM.

I_TIME

Time which you want to add. Mandatory. Type: MARA-MHDHB.

I_IPRKZ

This field defines the unit for the time such as day, week, month, or year. You need to call conversion exit CONVERSION_EXIT_PERKZ_INPUT to convert the format from output to input. Mandatory.

- Day has an output format of D and input format of [Space].
- Week has output format of W and input format of 1.
- Month has output format of M and input format of 2.

- Year has output format of Y and input format of 3.

Type: MARA-IPRKZ.

I_RDMHD

Rounding rule for calculation.

- [Space]: No rounding off.
- –: Start of chosen period (week, month, or year).
- +: End of chosen period (week, month, or year).
- F: Start of the following period (week, month, year).

Type: MARA-RDMHD.

[↑] Export

O_IDATE

Return date after processing. Type: SY-DATUM.

[!] Exceptions

INVALID_PERIOD

Period is not valid.

INVALID_ROUND_UP_RULE

Round off rule is invalid.

INTERNAL_ERROR

Some internal error occurred while executing the FM.

---

167 SWI_DURATION_DETERMINE

This FM determines the difference between the dates entered in the import parameters in seconds.

Example: Sometimes you need to calculate the date difference in seconds in your program development for any number of purposes. You can use this FM to calculate the difference.

[↓] Import

START_DATE

Start date. Type: DATS.

END_DATE

End date. Type: DATS.

START_TIME

Start time. Type: TIMS.

END_TIME

End time. Type: TIMS.

[↑] Export

DURATION

Duration returned in seconds. Type: SYTABIX.

---

168 SX_OBJECT_CONVERT_OTF_PDF

This FM converts the object in SAP output text format (i.e., OTF format) to Adobe PDF. Changing parameter CONTENT_BIN is returned with the PDF data.

Example: You're working on SAPscript, and you want to display the output in a PDF. You can fill in the parameters and call this FM.

[↓] Import

FORMAT_SRC

Current format, that is, OTF. This is a mandatory parameter. Type: SX_FORMAT.

FORMAT_DST

Required format, that is, PDF. This is a mandatory parameter. Type: SX_FORMAT.

ADDR_TYPE

This parameter isn't used in the FM. Type: SX_ADDR_TYPE.

DEVTYPE

SAPscript device type for conversion, such as a printer. Type: SX_DEVTYPE.

FUNCPARA

This parameter isn't used in the FM. Type: SX_FUNCPAR.

[Δ] Changing

TRANSFER_BIN

If this parameter is X, exception ERR_CONV_FAILED is raised. Type: SX_BOOLEAN.

CONTENT_TXT

Pass the OTF format text in this table. Type: SOLI_TAB.

CONTENT_BIN

This parameter returns the content of the PDF. Type: SOLIX_TAB.

OBJHEAD

This parameter defines the object header. This parameter is also not used. Type: SOLI_TAB.

LEN

Length of the object's contents (in characters/bytes). Type: SO_OBJ_LEN.

[!] Exception

ERR_CONV_FAILED

Conversion could not be executed. If parameter TRANSFER_BIN is X, this exception is raised.

169 SX_OBJECT_CONVERT_OTF_PRT

This FM converts the object in SAP output text format (i.e., OTF format) to printer format. Changing parameter CONTENT_BIN is returned with the PRT format data.

Example: You're working on SAPscript, and you want to send the output to the printer. You can fill in the parameters and call this FM to do so.

[↓] Import

FORMAT_SRC

Current format; that is, OTF. This is a mandatory parameter. Type: SX_FORMAT.

FORMAT_DST

Required format; that is, PRT. This is a mandatory parameter. Type: SX_FORMAT.

ADDR_TYPE

This parameter isn't used in the FM. Type: SX_ADDR_TYPE.

DEVTYPE

SAPscript device type for conversion, such as a printer. Type: SX_DEVTYPE.

FUNCPARA

This parameter isn't used in this FM. Type: SX_FUNCPAR.

[Δ] Changing

TRANSFER_BIN

If this parameter is X, exception ERR_CONV_FAILED is raised. Type: SX_BOOLEAN.

CONTENT_TXT

Pass the OTF format text in this table. Type: SOLI_TAB.

CONTENT_BIN

This parameter returns the content of PRT. Type: SOLIX_TAB.

OBJHEAD

This parameter defines the object header. This parameter is also not used. Type: SOLI_TAB.

LEN

Length of the object's contents (in characters/bytes). Type: SO_OBJ_LEN.

[!] Exception

ERR_CONV_FAILED

Conversion could not be executed. If parameter TRANSFER_BIN is X, this exception is raised.

170 SX_OBJECT_CONVERT_OTF_RAW

This FM converts the object in SAP output text format (OTF format) to the SAP internal text format (RAW format). Changing parameter CONTENT_TXT is returned with the RAW format data.

Example: You're working on SAPscript, and you want to convert the output in the SAP internal format; that is, RAW format. You can fill in the parameters and call this FM.

[↓] Import

FORMAT_SRC

Current format; that is, OTF. This is a mandatory parameter. Type: SX_FORMAT.

FORMAT_DST

Required format; that is, RAW. This is a mandatory parameter. Type: SX_FORMAT.

ADDR_TYPE

This parameter isn't used in the FM. Type: SX_ADDR_TYPE.

DEVTYPE

This parameter isn't used in this FM. Type: SX_DEVTYPE.

FUNCPARA

This parameter isn't used in this FM. Type: SX_FUNCPAR.

[Δ] Changing

TRANSFER_BIN

If this parameter is X, exception ERR_CONV_FAILED is raised. Type: SX_BOOLEAN.

CONTENT_TXT

You need to pass the OTF text in this parameter and You'll get the converted RAW text in this parameter itself. Type: SOLI_TAB.

CONTENT_BIN

This parameter isn't used in this FM. Type: SOLIX_TAB.

OBJHEAD

This parameter defines the object header. This parameter is also not used. Type: SOLI_TAB.

LEN

Length of the object's contents (in characters/bytes). Type: SO_OBJ_LEN.

[!] Exception

ERR_CONV_FAILED

Conversion could not be executed. If parameter TRANSFER_BIN is X, this exception is raised.

171 TIME_CHECK_PLAUSIBILITY

This FM is used to check the time plausibility; that is, this FM checks the validity of time.

Example: You've entered a time in the parameter PARA1 of the screen. You have to validate the time entered in that parameter by the user. You can pass the entered time in the import parameter of the FM and execute the FM to return whether the time entered is valid or not. If the time isn't valid, it returns with the exception.

[↓] Import

TIME

Time to be checked. Type: SY-UZEIT.

[!] Exception

PLAUSIBILITY_CHECK_FAILED

Time is not valid.

172 UNIT_CONVERSION_SIMPLE

This FM converts values from one measurement unit to another and rounds the result to the number of decimal places maintained in the measurement unit table.

Example: User has entered the value of distance in meters but you want to save the value in feet in your database. You can fill in the parameters and call this FM to get the converted value in feet.

[↓] Import

INPUT

Input value of measurement. Mandatory.

NO_TYPE_CHECK

If this parameter is [Space], the output parameters NUMERATOR, DENOMINATOR, and ADD_CONST are type checked and checked for overflow. If the value is X, no check occurs. The default value is X.

ROUND_SIGN

This parameter defines the rounding sign. The rounding can be Up (+), Down (–), commercial (X), or no rounding ([Space]).

UNIT_IN

Unit of input value comes from the measurement table. Type: T006-MSEHI.

UNIT_OUT

Unit of output value comes from the measurement table. Type: T006-MSEHI.

[↑] Export

ADD_CONST

Additive constant for conversion.

DECIMALS

Number of decimal places for rounding.

DENOMINATOR

Denominator for conversion.

NUMERATOR

Numerator for conversion.

OUTPUT

Output value.

[!] Exception

CONVERSION_NOT_FOUND

Conversion factor could not be determined.

DIVISION_BY_ZERO

Division by zero trapped.

INPUT_INVALID

Input value is not a number.

OUTPUT_INVALID

OUTPUT parameter is not a number.

OVERFLOW

Field overflow.

TYPE_INVALID

An output parameter is not a number.

UNITS_MISSING

No units specified.

UNIT_IN_NOT_FOUND

UNIT_IN is not maintained.

UNIT_OUT_NOT_FOUND

UNIT_OUT is not maintained.

---

173 UNIT_CONVERSION_WITH_FACTOR

This FM converts a unit from one measurement unit to another based on the following formula:

*Output value = (Numerator / Denominator) * input value + additive constant*

Example: You have a specific formula based on which you want to convert one unit to another. You can pass the parameters and call this FM to get the converted value.

[↓] Import

ADD_CONST

Additive constant.

DENOMINATOR

Denominator.

INPUT

Input value.

NUMERATOR

Numerator.

[↑] Export

OUTPUT

Output value.

[!] Exceptions

DIVISION_BY_ZERO

Zero was entered for the denominator. Division by zero is not defined.

OVERFLOW

The field OUTPUT is of type P or I and too small for the result.

TYPE_INVALID

The output parameter OUTPUT is not of type F, P, or I; that is, it is not a number.

---

174 UNIT_CORRESPONDENCE_CHECK

This FM checks whether two units have the same dimension.

Example: You've entered one unit of a length, for example, an inch on the screen, and you want to save the unit in the database table. Before saving the unit, you should validate whether the unit you've entered in the screen and the standard unit of the length (meter, in this example) have the same dimension and then convert it and save it. You can pass the screen unit in UNIT_IN parameter and standard unit in UNIT_OUT parameters and call this FM. You get the exception if the dimensions are different.

[↓] Import

UNIT_IN

Unit of measurement for input. Type: T006-MSEHI.

UNIT_OUT

Unit of measurement for output Type: T006-MSEHI.

[!] Exceptions

DIMENSIONS_ARE_DIFFERENT

The measurement units UNIT_IN and UNIT_OUT do not have the same dimension.

UNIT_IN_NOT_FOUND

UNIT_IN is not maintained.

UNIT_OUT_NOT_FOUND

UNIT_OUT is not maintained.

175 UNIT_GET

This FM searches for a unit based on the input parameters entered. The found unit is returned in export parameter UNIT.

Example: You have the dimensions of a unit from an ABAP table. Based on the dimensions, you want to display the unit on the ALV. You can pass the dimensions parameter and execute this FM to get the unit.

[↓] Import

ADD_CONST

Additive constant. Type: T006-ADDKO.

DECIMALS

Number of decimal places for rounding. If the decimals have a default value 88, the number of decimal places is ignored. Type: T006-ANDEC.

DENOMINATOR

Denominator. This is a mandatory parameter. Type: T006-NENNR.

DIMENSION

Dimension key. Valid dimension key can be obtained using the function block DIMENSION_GET or DIMENSION_GET_FOR_UNI. This is a mandatory parameter. Type: T006-DIMID.

EXPONENT

Exponent. Type: T006-EXP10.

NUMERATOR

Numerator. This is a mandatory parameter. Type: T006-ZAEHL.

[↑] Export

UNIT

Unit of measurement returned based on the export parameters. Type: T006-MSEHI.

[!] Exception

UNIT_NOT_FOUND

Unit of measurement with desired factor not maintained.

176 UNIT_OF_MEASURE_ISO_TO_SAP

This FM converts the internationally normed ISO code for measurement units into the SAP-internal measurement unit code.

Example: To get the SAP unit for the ISO unit meter, you can pass the ISO maintained unit MTR in parameter ISO_CODE and call this FM. The export parameter returns the SAP internal country code as M and the UNIQUE parameter as X.

[↓] Import

ISO_CODE

ISO code for the measurement unit. Type: T006-ISOCODE.

[↑] Export

SAP_CODE

SAP code for the measurement unit for SAP use. Type: T006-MSEHI.

UNIQUE

UNIQUE is a flag that states whether only one entry or several entries were found:

- X: Exactly one entry was found.
- [Space]: More than one entry was found. Type: ISOFIELDS-UNIQUE.

[!] Exception

NOT_FOUND

This exception is raised if no primary entry was found in the measurement units table with this ISO code.

---

177 UNIT_OF_MEASURE_SAP_TO_ISO

This FM converts the SAP-internal measurement unit code to the international measurement unit ISO code.

Example: To get the ISO unit for SAP unit meter, you can pass the SAP maintained unit M in parameter ISO_CODE and call this FM. The export parameter returns the SAP internal country code as M and the UNIQUE parameter as X.

[↓] Import

SAP_CODE

Internal SAP code for the measurement unit. Type: T006-MSEHI.

[↑] Export

ISO_CODE

ISO code for the measurement unit. Type: T006-ISOCODE.

[!] Exceptions

NOT_FOUND

ISO code not found in the system.

NO_ISO_CODE

SAP code has no ISO code assigned to it.

---

178 UNIT_OF_MEASUREMENT_HELP

This FM provides the entries help (i.e., F4 help) for units of a dimension. Either all of the units or all of the business measurement units for the specified dimension are displayed in a popup. If no dimension is specified, all units are displayed. If a dimension is specified, all business measurement units for that dimension are displayed.

Example: You have a parameter for a unit. You need to display the F4 help for the user. You can pass the parameter and call this FM to display the F4 help.

[↓] Import

BUSINESS_UOM

Commercial unit of measurement flag. Value X defines that only business measurement units are displayed for selection. Value [Space] defines that all of the measurement units are displayed for selection. The default value is [Space]. Type: T006-KZKEH.

CUCOL

Starting column popup. The parameters CUCOL and CUROW determine where the dialog window should appear. CUCOL = CUROW = 1 is the top left-hand corner of the screen. If both the parameters have the value 0, the dialog is positioned automatically. The default value is 0 (automatic positioning). Type: SY-CUCOL.

CUROW

Starting line popup. The default value is 0. Refer to parameter CUCOL for details. Type: SY-CUROW.

DIMID

Dimension key. You can get the valid dimension key from FM DIMENSION_GET or DIMENSION_GET_FOR_UNIT. The default value is [Space]. Type: T006-DIMID.

DISPLAY

Display flag. This flag specifies whether the popup is only used to display the possible entries (DISPLAY = X), or to select a value (DISPLAY = [Space]). The default value is [Space].

LANGUAGE

Language key (default: logon language). Type: SY-LANGU.

SELECT_UNIT_EXTERN

This flag determines whether the FM returns the internal measurement unit (SELECT_UNIT_FLAG = [Space]) or an external measurement unit (SELECT_UNIT_FLAG <> [Space]).

SELECT_UNIT_LONG

This flag determines whether three-character (SELECT_UNIT_LONG = [Space]) or six character (SELECT_UNIT_LONG <> [Space]) external measurement units are used.

Three-character external units are used by default.

[↑] Export

SELECT_UNIT

Selected unit of measurement is returned. If the CANCEL pushbutton is clicked, [Space] is returned in this parameter.

---

179 WDKAL_DATE_ADD_FKDAYS

This FM calculates the working date based on the input date and the number of working days for a given factory calendar.

Example: You want to produce an order in 10 working days and give the date to the customer. Pass the parameters and execute this FM to get the date when the order will be ready. You can then give this date to the customer.

[↓] Import

I_DATE

Input date. Type: SY-DATLO.

I_FKDAY

Number of working days. Type: MDCAL-FKDAY.

I_FABKL

Factory calendar. Type: TFACD-IDENT.

[↑] Export

E_DATE

Working date is returned. It is calculated as date + working days. Type: SY-DATLO.

E_FKDAY

This parameter returns the working date as a workday number for that factory calendar. Type: MDCAL-FKDAY.

[!] Exception

ERROR

Some error occurred.

---

180 WEEK_GET_FIRST_DAY

This FM returns the first day for the week.

Example: You have a date parameter in the screen and want to get the starting date of the week for the date the user has entered in the parameter field of the selection screen. For example, the user has entered November 13th, 2013, (a Wednesday), and he wants to get the start date of the week, that is, November 11th, 2013. First, you need to call the FM DATE_GET_WEEK (FM 118) and pass the selection screen date in the import parameter of this FM to get the week details. You need to pass this week detail as an import parameter to this FM to get the first day of the week and pass the parameter and call this FM to get the date.

[↓] Import

WEEK

Week for which the date is to be determined. Type: SCAL-WEEK.

[↑] Export

DATE

Date of the first day of the week. Type: SCAL-DATE.

[!] Exception

WEEK_INVALID

Week was incorrectly specified.

---

181 WEEKDAY_GET

This FM returns the day of the week for the given language.

Example: You may be developing the report for a Brazil factory and want to get the week start day. Based on the output, you want to put some logic in your program. You can call this FM for import parameter PT to get the weekday table that you can use for your logic.

[↓] Import

LANGUAGE

Language. Type: SY-LANGU.

[↑] Export

RETURN_CODE

Return code. It will have a value if the FM execution is successful. Type: SY-SUBRC.

Tables

WEEKDAY

Week number is returned in this table. For the English language, 1 for Monday, 2 for Tuesday, and so on are returned in this table. Type: T246.

[!] Exception

WEEKDAY_NOT_FOUND

Factory calendar is not found.

---

182 WLB3_GET_NUMBER_OF_WORKDAYS

This FM returns the number of working days in a week that are defined for the factory calendar.

Example: You're developing a report to provide the salaries for factory workers based on the number of days they are working per week. You may need to get the number of working days in a week for that factory calendar. You can pass the factory calendar as a parameter and call this FM to get the number of working days for that calendar.

[↓] Import

I_FABKL

Factory calendar. Type: T001W-FABKL.

[Δ] Changing

IE_WORKDAYS

Number of working days. Type: I.

[!] Exception

DEFAULT_SETTING

Factory calendar is not found.

Part IV
BDC, ALV, Report, and Screens

This part includes function modules (FMs) related to Batch Data Communication (BDC), ABAP List Viewer (ALV), report programming, and screen programming. BDC is a batch interfacing technique mainly used for uploading data from legacy systems to the SAP system. BDC works by simulating the user input from the transactional screen via an ABAP program.

Report and screen (dialog or module pool) programming are the main cruxes of the ABAP language. Reports are the executable programs that mainly deal with reporting, that is, fetching the data from the database, processing it, and displaying the result. Results are displayed by ALV.

Report programming isn't controlled by user-defined screens but by the runtime environment that calls the series of processing blocks in the fixed sequence, such as initialization, at selection screen, start of selection, end of selection, and so on.

Dialog or screen programing is another type of programming where screens and flow logic are designed by the user. Flow logic comes from creating different types of screens and logic built for navigating from one screen to another.

183 BDC_CLOSE_GROUP

After you're done inserting batch input data to a session, you can close the session with this FM. A batch input program may have only one session open at a time.

Example: You have transferred the data from Microsoft Excel to the BDC session. You need to pass the parameters in this FM and call this to close the BDC group.

[!] Exceptions

NOT_OPEN

Queue not opened (see SYSLOG).

QUEUE_ERROR

Error reading/writing the queue (see SYSLOG).

---

184 BDC_INSERT

This FM inserts the batch input data to the BDC session. For BDC using the session method, the BDC session needs to be created. The following FMs are used for data transfer using BDC:

1. BDC_OPEN_GROUP (FM 185)
2. BDC_INSERT
3. BDC_CLOSE_GROUP (FM 183)

After you've created a session, then you can insert batch input data into it with this FM.

Example: You're creating a BDC to send data from Microsoft Excel to the SAP system. You need to insert the data in the BDC session by calling this FM.

[↓] Import

TCODE

Transaction code of the transaction that is to be run. The default value is FILLER4. Type: TSTC-TCODE.

POST_LOCAL

Parameter to update data locally. If POST_LOCAL = X, data is updated locally. The default value is NOVBLOCAL. Type: BDCTH-MTYPE.

PRINTING

If this parameter is set to X, printing for the BDC data is switched on. The default value is NOPRINT. Type: BDCTH-STATE.

SIMUBATCH

If this parameter is set to X, BDC runs in the background, that is, SY-BATCH = X. The default value is [Space]. Type: SYBATCH.

CTUPARAMS

Call transaction using BDC method details. The default value is [Space]. Type: CTU_PARAMS.

Table

DYNPROTAB

This table contains the data that is to be processed by the transaction. Type: BDCDATA.

[!] Exceptions

INTERNAL_ERROR

Internal error of batch input.

NOT_OPEN

Queue was not opened.

QUEUE_ERROR

Error reading/writing the queue.

TCODE_INVALID

Transaction code is not known.

PRINTING_INVALID

Printing is not valid.

POSTING_INVALID

Posting is not valid.

---

185 BDC_OPEN_GROUP

BDC is used to transfer data from a legacy system such as a flat file or Microsoft Excel to a SAP system. This FM creates a new session for BDC. You can't reopen a session that already exists and has been closed. If you call this FM with the name of an existing session, then an additional session with the same name is created. Before opening a session, make sure that the opened session is already closed.

Example: You have materials in your file, and you want to create material data in SAP. You can do so by first recording and creating the BDC sessions to transfer the data. Execute this FM to create the session, then execute the following FMs:

1. BDC_INSERT (FM 184)
2. BDC_CLOSE_GROUP (183)

[↓] Import

CLIENT

Client in which the session is to be processed. If you don't provide a value for this parameter, the default is the client under which the batch input program runs when the session is created. Type: APQI-MANDANT.

DEST

This parameter isn't used now. Type: APQI-DESTSYS.

GROUP

Name of the session that is to be created. May be up to 12 characters long. There is no default. You must specify a session name. Type: APQI-GROUPID.

HOLDDATE

Lock date. The session is locked and may not be processed until after the date that you specify. Only a system administrator with the LOCK authorization for the authorization object BATCH INPUT AUTHORIZATIONS can unlock and run a session before this date. Type: APQI-STARTDATE.

KEEP

Retain session after successful processing. Set this option to the value X to have a session kept after it has been successfully processed. A session that is kept remains in the input/output queue until an administrator deletes it.

Sessions that contain errors in transactions are kept even if KEEP isn't set.

Default: If not set, then sessions that are successfully processed are deleted. Only the batch input log is kept. Type: APQI-QERASE.

USER

Authorizations user for background processing. This is the user name that is used for checking authorizations if a session is started in background processing. The user must be authorized for all of the transactions and functions that are to be executed in a session. Otherwise, transactions are terminated with "no authorization" errors. Type: APQI-USERID.

RECORD

If this parameter is set to X, batch input recording is activated. Type: APQI-PUTACTIVE.

PROG

Name of the creator program. Type: APQI-PROGID.

DCPFM

This parameter should be set if a decimal character is used. Type: USDEFAULTS-DCPFM.

DATFM

This parameter should be set if the date format is used. Type: USDEFAULTS-DATFM.

[↑] Export

QID

Unique database queue ID of the session is generated in this parameter. Type: APQI-QID.

[!] Exceptions

CLIENT_INVALID

Client is invalid.

DESTINATION_INVALID

Target system is invalid/no longer relevant.

GROUP_INVALID

Batch input session name is invalid.

GROUP_IS_LOCKED

Batch input session is protected.

HOLDDATE_INVALID

Lock date is invalid.

INTERNAL_ERROR

Internal error of batch input.

QUEUE_ERROR

Error reading/writing the queue (see SYSLOG).

RUNNING

Session is already being processed.

SYSTEM_LOCK_ERROR

System error when protecting batch input session.

USER_INVALID

Batch input user is not valid.

186 COMPLEX_SELECTIONS_DIALOG

You can use this FM to fill or display an internal table created with ranges or select options. This dialog contains the MULTIPLE SELECTIONS screen that you can call from the selection screen of the report.

Example: You have a requirement to create a SELECT option in your module pool screen. You want to have the same multiple selections behavior as in select options. For this, you need to create two fields—one field for low value and one for high value—and a button for multiple selections. You can call this FM to get the select option behavior.

[↓] Import

TITLE

Dialog box title. Type: SY-TITLE.

TEXT

Text for options dialog box. Type: RSSELINT-TEXT.

SIGNED

If this parameter is X, sign is allowed. Type: RSCONVERT-SIGN.

LOWER_CASE

If this parameter is X, lowercase isn't converted to uppercase. Type: RSCONVERT-LOWER.

NO_INTERVAL_CHECK

If this parameter is X, no interval check occurs. Type: RSVUVINT-NO_INT_CHK.

JUST_DISPLAY

If this parameter is X, then the field is display only. Type: C.

JUST_INCL

If this parameter is X, the field is select only. Type: C.

EXCLUDED_OPTIONS

Respective fields of this parameter should be marked as X for the options that are required to be excluded. For example, if you don't want BETWEEN BT, you can pass EXCLUDED-OPTIONS-BT as X. Type: RSOPTIONS.

DESCRIPTION

If the field attributes differ from those contained in RANGE-LOW and RANGE-HIGH, this should be explained in DESCRIPTION. If this is the case, the parameters SIGNED and LOWER_CASE are ignored. Entries with excessive length (DESCRIPTION-LENGTH) are cut off. Type: RSFLDESC.

HELP_FIELD

Dictionary field name for F1/F4 help. Type: RSSCR-DBFIELD.

SEARCH_HELP

Search help details for the field. Type: DDSHDESCR-SHLPNAME.

TAB_AND_FIELD

Table name, field name. Type: RSTABFIELD.

Table

RANGE

Contents table.

[!] Exceptions

NO_RANGE_TAB

Table not a RANGES table.

CANCELLED

Activity terminated.

INTERNAL_ERROR

Internal error.

INVALID_FIELDNAME

Incorrect field description.

---

187 DP_CONTROL_ASSIGN_TABLE

This FM sends the selected table information to a client. There is a special conversion object that allows you to access the table on the client using the RFC API or RFC table object. The data is stored on the DataProvider object on the client. You can either create the object by calling the FM DP_CREATE (FM 188), or it can be created implicitly by this FM.

Example: You want to send the internal table data to the text editor. You've created the DataProvider object, and you're required to send the internal table data to the DataProvider. You can do so by passing the relevant parameters in this FM and calling this FM.

[↓] Import

H_DP

If you pass this parameter, it contains a handle to a DataProvider object created using DP_CREATE. If you don't pass it, the necessary DataProvider object is created in the FM and automatically released by it when no longer required. Type: CNTL_HANDLE.

H_CNTL

Handle of the control to which the table DATA is to be assigned. You assign the

DATA table to the control using the PROPERTYNAME parameter. Mandatory. Type: CNTL_HANDLE.

TABNAME

Name of the structure of table DATA from the ABAP Data Dictionary. If you don't specify this parameter, you can fill the parameter TABFIELDS with the fields of table DATA (see DP_GET_FIELDS_FROM_TABLE [FM 189]). If you don't specify either FIELDS or TABFIELDS, the structure of table DATA is determined dynamically. In this case, the field names and the table name aren't available on the client. Type: X030L-TABNAME.

MEDIUM

This parameter represents the way in which data is stored and its location. The following are supported in Windows platform:

- CNDP_MEDIUM_R3TABLE: The data is stored in an internal table at the frontend. This medium is always available for data that you send from the ABAP program to the DataProvider. This medium is also available for data from other sources (HTTP, FILE, FTP), and you can read it into the ABAP program as well.
- CNDP_MEDIUM_OBJECT: The data is stored as an object at the frontend. The object type that is initialized with the data depends on the MIME type and the conversion components installed at the frontend. Not all mime types support this medium.
- CNDP_MEDIUM_STREAM: The data is presented as a byte stream at the frontend. This medium is recommended for all nontyped data such as bitmaps, OLE documents, text, and so on.
- CNDP_MEDIUM_DATA: The data is presented using the standard representation of its MIME type.

Type: C.

PROPERTYNAME

Name of the control property. Mandatory. Type: C.

DATE

Date from which data is valid. Type: SY-DATUM.

TIME

Time from which the data is valid. Type: SY-UZEIT.

DESCRIPTION

Additional description. Type: C.

Tables

DATA

Internal table containing the data you want to assign.

TABFIELDS

Structure of the table DATA. If you don't specify this parameter, you can pass the name of the structure of DATA in the TABNAME parameter. You can use the FM DP_GET_FIELDS_FROM_TABLE (FM 189) to fill the table TABFIELDS. If you don't specify either FIELDS or TABFIELDS, the system determines the structure of the DATA table dynamically. In this case, the field names and table name aren't available on the client. Type: RFC_FIELDS.

COLUMNS_TO_STRETCH

Columns to be widened in Unicode systems. This is for SAP internal use. Type: TABLE_OF_STRINGS.

[!] Exceptions

DP_ERROR_CREATE

DataProvider could not be created.

DP_ERROR_SEND_DATA

Error sending data.

DP_ERROR_ASSIGN

Could not make assignment.

DP_ERROR_INVALID_PARAM

One of the parameters is invalid.

DP_ERROR_TABNAME

Parameter TABNAME does not exist in the ABAP Data Dictionary.

---

188 DP_CREATE

This FM creates a DataProvider object. This object can be used to transfer internal tables in the SAP system to the client and from the client back to the ABAP/4 program. The DataProvider object must be created before it can be used by any other FM of a DataProvider (FM starting with DP_*). DataProvider is a standard interface for retrieving data used by controls in the SAP GUI. A DataProvider manages the data, maps the data in standard Windows data types and data objects, and controls the data conversion. The DataProvider is used in most ActiveX controls that are provided by SAP, such as text edit controls, HTML controls, tree controls, and so on.

Example: You want to use the text editor in your program. You'll make the reference object of text editor class CL_GUI_TEXTEDIT and use its methods. The constructor of this class calls this FM to create the DataProvider object.

Various methods of this class and other ActiveX classes call the other DataProvider FMs such as DP_CONTROL_ASSIGN_TABLE when the table information is required in the DataProvider. To fill in the table parameter TABFIELDS of FM DP_CONTROL_ASSIGN_TABLE (FM 187), you call the FM DP_GET_FIELDS_FROM_TABLE (FM 189).

[Δ] Changing

H_DP

If you pass this parameter, it contains a handle to a DataProvider object created using DP_CREATE. If you don't pass it, the necessary DataProvider object is created in the FM and automatically released by it when no longer required. Type: CNTL_HANDLE.

[!] Exceptions

DP_CREATE_ERROR

DataProvider could not be created.

DP_INSTALL_ERROR

DataProvider could not be installed.

DP_ERROR

DataProvider unknown error.

DP_ERROR_PLATFORM

DataProvider is not available on the platform.

189 DP_GET_FIELDS_FROM_TABLE

This FM dynamically analyzes the structure of an arbitrary table. The structure of the table must not be stored in the dictionary.

The return value DP_GET_FIELDS_FROM_TABLE fills the FIELDS table. This contains all significant information for access to individual elements within the table. The names of the columns and the table name itself can't be determined from this FM.

Example: You want to display the dynamic internal table data to Microsoft Excel. You've created the internal table and want to analyze the structure of the internal table. You can pass the relevant parameters in this FM and call it to analyze the structure of the internal table.

[↓] Import

TABNAME

Name of the structure of table DATA from the ABAP Data Dictionary. If you don't specify this parameter, you can fill the parameter TABFIELDS with the fields of table DATA. If you don't specify either FIELDS or TABFIELDS, the structure of table DATA is determined dynamically. In this case, the field names and the table name aren't available on the client. Type: X030L-TABNAME.

Tables

DATA

Any table whose structure is to be dynamically analyzed. FIELDS contains the structure of the DATA table after the function block. The field name and the table name aren't included here.

FIELDS

Table of type RFC_FIELDS. This table is filled with the structure of the table DATA. The field name and the table name can't be determined. Type: RFC_FIELDS.

[!] Exception

DP_INVALID_TABLE

Invalid table name entered.

190 DYNP_GET_STEPL

This FM defines the current line of step loop line. This FM is generally used at the time of the Process On Value-Request (POV) for step loop or table controls in a module pool. This FM should be called before calling the DYNP_VALUE_READ or DYNP_VALUE_UPDATE for step loop. The parameter value returned must be passed to the STEPL field of table parameter DYNPFIELDS of FM DYNP_VALUES_READ (FM 191) or FM DYNP_VALUES_UPDATE (FM 192).

Example: You have a module pool screen in which you've defined the table control with four rows. You have to assign the F4 help for the first row. To determine which line of the table control loop the F4 help should return to, call this FM.

[↑] Export

POVSTEPL

Current step-loop line. Type: SY-STEPL.

[!] Exception

STEPL_NOT_FOUND

Step loop line not found.

---

191 DYNP_VALUES_READ

This FM reads the screen field values before the PAI value transport. You want to use the value of one screen parameter for the F4 help of the other parameter, so you'll call this FM.

Example: You have two selection screen parameters: PLANT and WORK CENTER. Based on the value entered in PLANT by the end user, you want to get the value of only those work centers that are valid for that plant. For example, if the user entered "PLANT1", he should get only those work centers in F4 help that are valid for PLANT1. For this, you need to read the PLANT value entered by the end user. You can do so by passing the relevant parameter in this FM and calling it.

[↓] Import

DYNAME

Name of the program for which values are to be read from the screen. Variable SY-REPID must not be used in the call because it can be overwritten at runtime, for example, by the names of function groups. Either use the variable SY-CPROG, or enter the value as a literal instead. This is a mandatory parameter. Type: D020S-PROG.

DYNUMB

Number of the screen for which the values are to be read. If the variable SY-DYNNR is assigned to this parameter, it can in rare cases contain screen numbers of the current function groups SHLP and SHL2. Avoid this situation by assigning a four-character literal; for example, 0100, to the parameter. This is a mandatory parameter. Type: D020S-DNUM.

TRANSLATE_TO_UPPER

If this parameter is set to X, field values read are converted into uppercase. The default value is [Space].

REQUEST

This parameter defines whether the returned field is input enabled. The information can be found in the field FIELDINP of table parameter DYNPFIELDS. The default value is [Space]. Type: D020S-TYPE.

PERFORM_CONVERSION_EXITS

This parameter defines whether the output conversion exit stored for the screen field should be executed when the field values are read. The default value is [Space].

PERFORM_INPUT_CONVERSION

This parameter defines whether the input conversion exit stored for the screen field should be executed when the field values are read. The default value is [Space].

DETERMINE_LOOP_INDEX

Determine index of a line in the loop for all fields. The default value is [Space]. Type: CHAR1.

START_SEARCH_IN_CURRENT_SCREEN

The search for DYNAME/DYNUMB starts in the current (sub)screen. The default value is [Space]. Type: CHAR1.

START_SEARCH_IN_MAIN_SCREEN

The search for DYNAME/DYNUMB starts in the main Dynpro. The default value is [Space]. Type: CHAR1.

START_SEARCH_IN_STACKED_SCREEN

The search for DYNAME/DYNUMB starts in the specified main screen in the screen stack. The default value is [Space]. Type: DYNPRO_ID.

START_SEARCH_ON_SCR_STACKPOS

The search for DYNAME/DYNUMB starts at the specified position in the screen stack. The default value is [Space]. Type: SYTABIX.

SEARCH_OWN_SUBSCREENS_FIRST

The search through own screens before the stack is searched. The default value is [Space]. Type: CHAR1.

SEARCHPATH_OF_SUBSCREEN_AREAS

The path of the subscreen areas that is searched from the start screen. The default value is [Space]. Type: STRING.

Table

DYNPFIELDS

Table for reading current screen values. For step loops, step line values must be passed in this parameter. Type: DYNPREAD.

[!] Exceptions

INVALID_ABAPWORKAREA

Work area is not valid.

INVALID_DYNPROFIELD

Screen field is not valid.

INVALID_DYNPRONAME

Screen name is not valid.

INVALID_DYNPRONUMMER

Screen number is not valid.

INVALID_REQUEST

General request error.

NO_FIELDDESCRIPTION

No field description.

INVALID_PARAMETER

Internal system function parameters have incorrect values.

UNDEFIND_ERROR

Undefined error.

DOUBLE_CONVERSION

Double conversion required error.

STEPL_NOT_FOUND

Could not determine line in the loop.

---

192 DYNP_VALUES_UPDATE

This FM is used to change the screen field contents without calling the PBO event.

Example: You have a plant and work center and PARAMETER1 field in the selection screen. For one work center, there can be only one PARAMETER1 value. You want to get the F4 help on the work center based on the value entered in the PLANT field. Based on the value returned from the work center, you want to internally get the relevant PARAMETER1 entry and update the PARAMETER field at the same time. For this, you can fill the values in the table parameter DYNPFIELDS and call this FM to update the values in the screen without calling the PBO.

[↓] Import

DYNAME

Program name. Type: D020S-PROG.

DYNUMB

Screen number. Type: D020S-DNUM.

Table

DYNPFIELDS

Table for returning the field contents to the screen. For step loops, step line values must be passed in this parameter. Type: DYNPREAD.

[!] Exceptions

INVALID_ABAPWORKAREA

Work area is not valid.

INVALID_DYNPROFIELD

Screen field is not valid.

INVALID_DYNPRONAME

Screen name is not valid.

INVALID_DYNPRONUMMER

Screen number is not valid.

INVALID_REQUEST

General request error.

NO_FIELDDESCRIPTION

No field description.

UNDEFIND_ERROR

Undefined error.

---

193 FREE_SELECTIONS_DIALOG

This FM is used to enter dynamic selections and create a dynamic selection screen.

Example: After you use FM FREE_SELECTIONS_INIT (FM 194) and select which criteria you want to include in dynamic selection, you can call this FM to create the dialog box with the dynamic selection

[↓] Import

SELECTION_ID

The SELID returned by FREE_SELECTIONS_INIT. This parameter identifies the field list and the previous selections, so that several field lists can be processed in parallel. This is a mandatory parameter. Type: RSDYNSEL-SELID.

TITLE

Selection screen title. Type: SY-TITLE.

FRAME_TEXT

Text for the frame on the selection screen. If you don't fill the parameter, "Dynamic selections" is displayed. Type: SY-TITLE.

STATUS

Status number. Possible values:

- 0: Selections are copied with the F11 command.
- 1: Like a normal selection screen, status selections will be copied with F8.

Type: I.

AS_WINDOW

If the parameter has value X, it displays the selection screen as a dialog box. You can set the starting coordinates in the START_COL and START_ROW parameters. Type: C.

START_ROW

Start row for the dialog box. Type: SY-WINY1.

START_COL

Start column for the dialog box. Type: SY-WINX1.

NO_INTERVALS

If this parameter is set with X, select options won't come with intervals. Type: C.

JUST_DISPLAY

If this parameter is set to X, the selection screen is displayed in display mode. Type: C.

PFKEY

This parameter overwrites the GUI status set with the STATUS parameter. You should use this cautiously, taking care that the standard program flow still works. The default status belongs to program RSSYSTDB. Type: RSDSPFKEY.

ALV

This parameter is for internal use. Type: C.

TREE_VISIBLE

This parameter is used for the display of the field selection tree. Possible values:

- X: The field selection tree is displayed from the beginning and can be hidden by clicking a pushbutton.
- [Space]: The tree isn't displayed initially and, if required, must be activated by clicking the NEW FIELD SELECTION button. If no field is selected when the FM is called, this parameter is ignored, and the tree is always displayed. Type: C.

DIAG_TEXT_1

Replaces the default text in the warning: "Dynamic selections will be deleted." Type: C.

DIAG_TEXT_2

Replaces the default text in the warning: "Changing dynamic selections." Type: C.

WARNING_TITLE

Title for the warning dialog box displayed when you delete selections. Type: C.

AS_SUBSCREEN

Integrate dynamic selections as subscreen. This is for internal use.

NO_FRAME

This parameter isn't used in the FM.

[↑] Export

WHERE_CLAUSES

The selections entered by the user are returned for each table in the form of WHERE clauses (type RSDS_TWHERE ==> type pool RSDS). You can pass these directly to an SAP SELECT statement. Type: RSDS_TWHERE.

EXPRESSIONS

You can store the user selections here and pass them to the FM FREE_SELECTIONS_INIT again later. This allows you to access the selections entered earlier at a later stage. Type: RSDS_TEXPR.

FIELD_RANGES

Contains for each field a ranges table with the relevant selections (type RSDS_TRANGE ==> type pool RSDS). You can use these tables for the IN operator with SELECT, CHECK, and IF. Type: RSDS_TRANGE.

NUMBER_OF_ACTIVE_FIELDS

Contains the number of fields for which selections have actually been defined. Type: SY-TFILL.

Tables

FIELDS_TAB

Because the user can normally (i.e., apart from type F) change the field selection, the chosen fields are returned to this table. The table also contains the fields that have been chosen, but for which no selections have been entered. As with the export parameter EXPRESSIONS, you can store this field list and pass it to the FM FREE_SELECTIONS_INIT again later as a field preselection. Type: RSDSFIELDS.

FCODE_TAB

This table is for internal use. Type: RSDSFCODE.

FIELDS_NOT_SELECTED

The list of selections not currently displayed by the user is returned to the table. Type: RSDSFIELDS.

[!] Exceptions

INTERNAL_ERROR

Internal error.

NO_ACTION

Cancellation by user.

SELID_NOT_FOUND

Transfer nonexistent selection ID.

ILLEGAL_STATUS

Invalid status number.

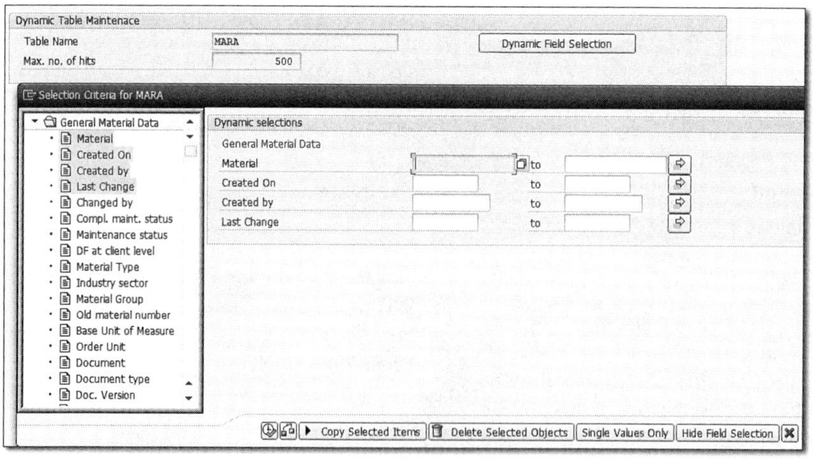

Dynamic Selection Called from FM FREE_SELECTION_DIALOG

194 FREE_SELECTIONS_INIT

This FM is used to initialize the capture of dynamic selections using the FM FREE_SELECTIONS_DIALOG (FM 193). The FM gets the information for the dynamic selection parameters that should be allowed and returns an ID in parameter SELECTION, which is used as input for the function block FREE_SELECTIONS_DIALOG. The function block can be called multiple times (with different field lists) during the course of a transaction, and each provides a new selection ID.

Example: You need to give a table name as an input parameter in the selection screen. Based on the table name given, you need to display the list of fields for dynamic selection. When you click EXECUTE on the dynamic selection, you should display the ALV with the fetched records. To enable dynamic selection, you need to first call this FM, which initializes the selection to be used.

[↓] Import

KIND

Type: of field list. Possible values:

- T: All fields of the tables specified in TABLES_TAB.
- G: A selection view.
- F: The fields passed in FIELDS_TAB.

The default value is T.

EXPRESSIONS

Allows you to store selections entered during an earlier call to FREE_SELECTIONS_DIALOG and pass them to FREE_SELECTIONS_INIT. The next time FREE_SELECTIONS_DIALOG is called, the old selections are displayed. These selections are returned apart from any that belong to tables no longer defined (parameter TABLES_TAB). Type: RSDS_TEXPR.

FIELD_RANGES_INT

Initial selections in the form of ranges tables. Type: RSDS_TRANGE.

FIELD_GROUPS_KEY

Selection view key. For KIND = G. You can create and maintain selection views in the Development Workbench. Type: RSDSQCAT.

RESTRICTION

You use this parameter to restrict the permitted sign and/or permitted selection options for some or all selections. For example, certain interval entries aren't supported, or for certain selections, the option of excluding values isn't available. For these types of cases, you need to restrict the selection for those selections. Type: SSCR_RESTRICT_DS.

ALV

This parameter is for internal use.

CURR_QUAN_PROG

The current currency (along with its corresponding number of decimal places) is retrieved from the appropriate work area in the program and supplied to all amount fields that have a currency key field (default program: the main program for the transaction). This also holds true for quantity fields and unit fields. Type: SY-REPID.

CURR_QUAN_RELATION

Assignment of currency key field or unit field to an amount field or a quantity field, where all of the fields are in the dynamic selection field selection. Type: SSCR_CURR_QUAN_T.

[↑] Export

SELECTION_ID

This parameter returns the selection IDs that should be passed as import parameters in FM FREE_SELECTIONS_DIALOG. Type: RSDYNSEL-SELID.

WHERE_CLAUSES

The selections entered by the user are returned for each table in the form of WHERE clauses (type RSDS_TWHERE ==> type pool RSDS). You can pass these directly to an SAP SELECT statement. Type: RSDS_TWHERE.

EXPRESSIONS

You can store the user selections here and pass them to the FM FREE_SELECTIONS_INIT again later. This allows you to access the selections entered earlier at a later stage. Type: RSDS_TEXPR.

FIELD_RANGES

Contains for each field a ranges table with the relevant selections (type RSDS_TRANGE ==> type pool RSDS). You can use these tables for the IN operator with SELECT, CHECK, and IF. Type: RSDS_TRANGE.

NUMBER_OF_ACTIVE_FIELDS

Contains the number of fields for which selections have actually been defined. Type: SY-TFILL.

Tables

TABLES_TAB

Table contains the names of those tables for which enhanced selections are supported. Type: RSDSTABS.

TABFIELDS_NOT_DISPLAY

Table field mentioned in this table won't be displayed. Type: RSDSFIELDS.

FIELDS_TAB

The meaning of this table parameter depends on the value of the import parameter KIND. If KIND = T or KIND = G, the listed fields are a preselection of fields. When FM FREE_SELECTIONS_DIALOG is first called, the dialog branches at once to the input screen for the preselected fields (here, fields that don't belong to any of the desired tables are deleted). If KIND = F, FIELDS_TAB simply contains the list of fields for which dynamic selections are required. In this case, you can use parameter FIELDS_NOT_SELECTED to exclude fields from the preselection. Type: RSDSFIELDS.

FIELD_DESC

To enable dynamic selection for fields without dictionary references, you have to specify field attributes for the fields. In ordinary cases, you can do this in table parameter FIELDS_TAB. If necessary, you can specify detailed descriptions in FIELD_DESC. Type: FLDCONVERT.

FIELD_TEXTS

You can overwrite the text that you've received from dictionary fields or mentioned in the FIELD_DESC table if you pass the text for the field in this table. Type: RSDSTEXTS.

EVENTS

Events for which control is required by the user. Possible events:

- O: AT SELECTION-SCREEN OUTPUT.
- A: AT SELECTION-SCREEN.
- S: AT SELECTION-SCREEN ON <selection>.
- E: AT SELECTION-SCREEN ON END OF <selection>.
- 1: AT SELECTION-SCREEN ON HELP-REQUEST FOR <selection>.
- 4: AT SELECTION-SCREEN ON HELP-REQUEST FOR <selection>-LOW/HIGH.

Type: RSDSEVENTS.

EVENT_FIELDS

List of the selections for which the control should be passed back to the called in selection-specific events (EVENTs S, E, 1, 4 in parameter EVENTS). Type: RSDSEVFLDS.

FIELDS_NOT_SELECTED

Kind F: Fields not initially selected. Type: RSDSFIELDS.

NO_INT_CHECK

This parameter is interpreted only where KIND = F. The selections listed aren't displayed when FREE_SELECTIONS_DIALOG is next called but can be activated by choosing New Field Selection. Type: RSDSTABS.

ALV_QINFO

Field for internal use. Type: LVC_S_QINF.

[!] Exceptions

FIELDS_INCOMPLETE

Only PRIM_FNAME filled.

FIELDS_NO_JOIN

Field assignment without join.

FIELD_NOT_FOUND

Dictionary field not found.

NO_TABLES

Table P_TABLES is empty.

TABLE_NOT_FOUND

A table from P_TABLES was not found.

EXPRESSION_NOT_SUPPORTED

Expression not (yet) supported.

INCORRECT_EXPRESSION

Incorrect logical expression.

ILLEGAL_KIND

KIND not equal to T, G, F, or G without P_RSDSQCAT.

AREA_NOT_FOUND

Invalid key FIELD_GROUPS_KEY.

INCONSISTENT_AREA

Inconsistent selection view.

KIND_F_NO_FIELDS_LEFT

KIND = F, but no field left after cleanup.

KIND_F_NO_FIELDS

KIND = F, but no field passed.

TOO_MANY_FIELDS

Too many entries in FIELDS_TAB.

DUP_FIELD

Field doubles in FIELDS_TAB.

FIELD_NO_TYPE

No field type in the field description.

FIELD_ILL_TYPE

Nonallowed field type.

DUP_EVENT_FIELD

Field doubled in EVENT_FIELDS.

NODE_NOT_IN_LDB

Node not part of logical database.

AREA_NO_FIELD

Selection view has no fields.

195 FREE_SELECTIONS_RANGE_2_EX

This FM converts the ranges format (RSDS_TRANGE) to the expression format (RSDS_TEXPR). The FMs FREE_SELECTIONS_INIT (FM 194) and FREE_SELECTIONS_DIALOG (FM 193) return the selections entered by the user in three different formats (types RSDS_TWHERE, RSDS_TEXPR, RSDS_TRANGE). Only the type RSDS_TEXPR can be used to call FREE_SELECTIONS_INIT later. If it's essential that the selections are stored in the format RSDS_TRANGE, you can use this FM to convert to type RSDS_TEXPR before you call FREE_SELECTIONS_INIT again.

Example: You've received and processed the selection in the format RSDS_TRANGE, and you want to call the FM FREE_DIALOG_INIT. To do this, you need to convert the format from ranges to expression by passing parameters and calling this FM.

[↓] Import

FIELD_RANGES

You can pass selections with the type RSDS_TRANGE, that is, as tables of RANGES. Type: RSDS_TRANGE.

[↑] Export

EXPRESSIONS

The selections are returned as an object with type RSDS_TEXPR—a table of logical expressions in Polish notation. You can pass the selections to the FM FREE_SELECTIONS_INIT in this form. Type: RSDS_TEXPR.

196 FREE_SELECTIONS_RANGE_2_WHERE

This FM converts dynamic selections from the ranges (RSDS_TRANGE) format to the where (RSDS_TWHERE) format.

Example: You've received ranges data from the FM FREE_SELECTIONS_DIALOG (FM 193) and processed it further, and you want to send the data to the `SELECT` statement in the `WHERE` clause. For this, you need to convert the data from the ranges format to the where format by passing the parameters and calling this FM.

[↓] Import

FIELD_RANGES

You can pass selections with the type RSDS_TRANGE, that is, as tables of RANGES. Type: RSDS_TRANGE.

[↑] Export

WHERE_CLAUSES

Selections in the form RSDS_TWHERE. Type: RSDS_TWHERE.

197 LIST_FREE_MEMORY

This FM deletes the last saved list from memory.

Example: After you're done, you need to free the list in memory to avoid causing system performance issues. You can do so by calling this FM.

Tables

LISTOBJECT

Deletes container for list object. Type: ABAPLIST.

198 LIST_FROM_MEMORY

This FM prepares a list import from memory. This FM retrieves a list from ABAP memory exported using `'LIST_TO_MEMORY'` or `SUBMIT... EXPORTING LIST TO MEMORY` commands and places it in the internal table LISTOBJECT.

Example: You want to submit a report program from another report program. In the submitted report, you'll prepare the SAP list and you want to return the list back to the parent report. You can do so by first submitting the report by using the `SUBMIT` command. Next, you can call this FM, which gives you the list data from the memory.

Tables

LISTOBJECT

Container to receive the list object. Type: ABAPLIST.

[!] Exceptions

NOT_FOUND

No list found in memory.

---

199 LIST_TO_ASCI

This FM converts the saved list object to the ASCII format. The purely text part of the list is copied line by line to the prepared internal table (LISTASCI) without any attributes (color, icon, symbol, etc.). There is no line break in this case; that is, if the internal table line isn't long enough, the list line is truncated.

Example: After using FM LIST_FROM_MEMORY (FM 198), call this FM to convert the list into ASCII format so that you can read the data into your internal table and display the ALV.

[↓] Import

LIST_INDEX

List level. The default value is –1. Type: SY-LSIND.

WITH_LINE_BREAK

If this parameter is set to X, line breaks are used for long lines. Type: C.

[↑] Export

LIST_STRING_ASCII

Table type returned with the values. Type: LIST_STRING_TABLE.

LIST_DYN_ASCII

Table type with generic table width. Type: DATA.

Table

LISTASCI

Table for receiving list.

LISTOBJECT

Container for list object. This table is taken into account only if the parameter LIST_INDEX = –1. Type: ABAPLIST.

[!] Exceptions

EMPTY_LIST

The list object transported is empty.

LIST_INDEX_INVALID

Invalid list index (LIST_INDEX).

---

200 LVC_FIELDCATALOG_MERGE

This FM creates the field catalog for the ALV FMs. The structure is based on ABAP Data Dictionary structure. If the field catalog contains more than 90 fields, the first 90 fields are displayed in the list; the remaining fields aren't displayed but are available in the field list.

Example: You want to display the ALV output. First you need to generate the field catalog that defines the fields of the ALV output. You can pass the parameters and can call this FM to generate the field catalog.

[↓] Import

I_BUFFER_ACTIVE

This parameter checks if the special buffer is active.

I_STRUCTURE_NAME

Structure name from which the field catalog is to prepare. Type: DD02L-TABNAME.

I_CLIENT_NEVER_DISPLAY

If this parameter is X, the CLIENT field won't be displayed. The default value is X. Type: SLIS_CHAR_1.

I_BYPASSING_BUFFER

If this parameter is set to X, the buffer is bypassed while reading. Type: CHAR01.

I_INTERNAL_TABNAME

Name of the internal output table in the CALLBACK_PROGRAM. Parameter CALLBACK_PROGRAM isn't present in this FM, so this program gives a problem if the internal table is passed in this parameter. Type: DD02L-TABNAME.

[Δ] Changing

CT_FIELDCAT

Field catalog with field descriptions. Type: LVC_T_FCAT.

[!] Exceptions

INCONSISTENT_INTERFACE

Call parameter combination error.

PROGRAM_ERROR

Program errors.

---

201 LVC_TRANSFER_FROM_SLIS

This FM is used to transfer the field catalog from SLIS to LVC; that is, the field catalog generated from REUSE_ALV_FIELDCATALOG_MERGE can be transferred to the LVC type.

Example: You want to display the ALV based on the LVC format (because you're using object-oriented programming) based on the internal table data. FM LVC_FIELDCATALOG_

MERGE (FM 200) can't be used with internal tables. You can create the field catalog from SLIS FM REUSE_ALV_FIELDCATALOG_MERGE (FM 209) and then convert it to LVC format by passing the parameter and calling this FM. The LVC method is used in object-oriented programming, whereas the SLIS method is used in normal programming.

[↓] Import

IT_FIELDCAT_ALV

Field catalog of SLIS. Type: SLIS_T_FIELDCAT_ALV.

IT_SORT_ALV

Sort of SLIS. Type: SLIS_T_SORTINFO_ALV.

IT_FILTER_ALV

Filter of SLIS. Type: SLIS_T_FILTER_ALV.

IS_LAYOUT_ALV

Layout structure of SLIS. Type: SLIS_LAYOUT_ALV.

[↑] Export

ET_FIELDCAT_LVC

Field catalog of LVC. Type: LVC_T_FCAT.

ET_SORT_LVC

Sort of LVC. Type: LVC_T_SORT.

ET_FILTER_LVC

Filter of LVC. Type: LVC_T_FILT.

ES_LAYOUT_LVC

Layout Structure of LVC. Type: LVC_S_LAYO.

Table

IT_DATA

Data table.

[!] Exception

IT_DATA_MISSING

Data missing.

---

202 LVC_VARIANT_DEFAULT_GET

This FM reads the default variant of an ALV.

Example: You want to give the default variant of ALV at the selection screen when you execute a transaction. You can pass the parameters and call this FM for the same at the INITIALIZATION event. You'll get the default variant that is set for the ALV so you can display it.

[↓] Import

I_SAVE

Variants can be saved.

[Δ] Changing

CS_VARIANT

Default variant information is returned. Type: DISVARIANT.

[!] Exceptions

WRONG_INPUT

Inconsistent input parameters.

NOT_FOUND

Variant not found.

PROGRAM_ERROR

Program errors.

203 LVC_VARIANT_EXISTENCE_CHECK

This FM checks the existence of the display variant in the database.

Example: You can validate the selection screen variant by passing the parameters in this FM. Variants have some user settings (if there are 10 output columns of ALV, one user doesn't want to see 2 of those columns) and save it as a variant; say Var1. When he executes the report again with this variant, he won't see those columns.

[↓] Import

I_SAVE

Variants can be saved.

[Δ] Changing

CS_VARIANT

Default variant information is returned. Type: DISVARIANT.

[!] Exceptions

WRONG_INPUT

Inconsistent input parameters.

NOT_FOUND

Variant not found.

PROGRAM_ERROR

Program errors.

204 LVC_VARIANT_F4

This FM displays the F4 values help if a variant is explicitly defined as an input field on a screen. The selection must be specified by at least partially filling in the parameter structure IS_VARIANT.

Example: To get the F4 help for the variant, pass the parameters and call this FM.

[↓] Import

IS_VARIANT

Variant information. This is a mandatory parameter. Type: DISVARIANT.

IT_DEFAULT_FIELDCAT

Field catalog with field descriptions. Type: LVC_T_FCAT.

I_SAVE

Variants can be saved.

[↑] Export

E_EXIT

Dialog canceled by the user.

ES_VARIANT

Variant information returned after F4. Type: DISVARIANT.

[Δ] Changing

CS_VARIANT

Default variant information is returned. Type: DISVARIANT.

[!] Exceptions

NOT_FOUND

Variant not found.

PROGRAM_ERROR

Program errors.

---

205 REUSE_ALV_BLOCK_LIST_APPEND

This FM is used to append the block list data to the blocked ALV.

Example: You want to display the ALV blocked list for the material number in one block and with the material description in another block. After you call FM REUSE_ALV_BLOCK_LIST_INIT (FM 207) to initiate the blocks, call this FM to append the data to the block for the material number, and then call this FM again to append the data of material text in another block.

[↓] Import

IS_LAYOUT

List layout specifications. This is a mandatory parameter. Type: SLIS_LAYOUT_ALV.

IT_FIELDCAT

Field catalog with field descriptions. This is a mandatory parameter. Type: SLIS_T_FIELDCAT_ALV.

I_TABNAME

Name of internal output table. This is a mandatory parameter. Type: SLIS_TABNAME.

IT_EVENTS

Table of events to perform. This is a mandatory parameter. Type: SLIS_T_EVENT.

IT_SORT

This parameter determines the sorting and/or the subtotals of the basic list. Type: SLIS_T_SORTINFO_ALV.

I_TEXT

Text of this list for the overview screen. The default value is [Space]. Type: SLIS_TEXT40.

Table

T_OUTTAB

Table with data to be displayed.

[!] Exceptions

PROGRAM_ERROR

Program errors.

MAXIMUM_OF_APPENDS_REACHED

Up to 19 lists can be appended in block mode. If the number exceeds this limit, this exception is raised.

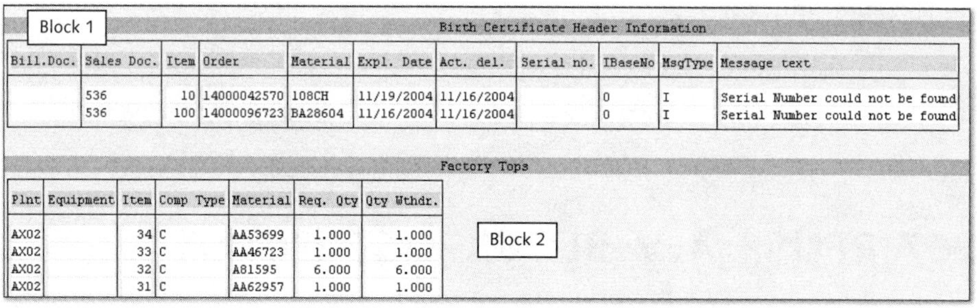

Block 1

Birth Certificate Header Information

Bill.Doc.	Sales Doc.	Item	Order	Material	Expl. Date	Act. del.	Serial no.	IBaseNo	MsgType	Message text
	536	10	14000042570	108CH	11/19/2004	11/16/2004		0	I	Serial Number could not be found
	536	100	14000096723	BA28604	11/16/2004	11/16/2004		0	I	Serial Number could not be found

Factory Tops

Plnt	Equipment	Item	Comp Type	Material	Req. Qty	Qty Wthdr.
AX02		34	C	AA53699	1.000	1.000
AX02		33	C	AA46723	1.000	1.000
AX02		32	C	A81595	6.000	6.000
AX02		31	C	AA62957	1.000	1.000

Block 2

ALV Block Display

206 REUSE_ALV_BLOCK_LIST_DISPLAY

This FM is used to display the block list to the output screen.

Example: You want to display the ALV blocked list for the material number in one block and with the material description in another block. Once you have appended the data from FM 205, use this FM to display the ALV block in the output.

[↓] Import

I_INTERFACE_CHECK

Check if the interface is available.

IS_PRINT

Print information is passed in this parameter. Type: SLIS_PRINT_ALV.

I_SCREEN_START_COLUMN

Specify the starting column of the screen.

I_SCREEN_START_LINE

Specify the starting line of the screen.

I_SCREEN_END_COLUMN

Specify the end column of the screen.

I_SCREEN_END_LINE

Specify the starting line of the screen.

[↑] Export

E_EXIT_CAUSED_BY_CALLER

Delete list in CALLBACK_USER_COMMAND.

ES_EXIT_CAUSED_BY_USER

EXIT or CANCEL button clicked by the user. Type: SLIS_EXIT_BY_USER.

[!] Exception

PROGRAM_ERROR

Program errors.

---

207 REUSE_ALV_BLOCK_LIST_INIT

This FM is used to initialize the ALV block. This FM, along with FMs REUSE_ALV_BLOCK_LIST_APPEND (FM 205) and REUSE_ALV_BLOCK_LIST_DISPLAY (FM 206), is used to display the block ALV.

Example: You have a report output must be displayed in block ALV format. You need to initialize the block ALV first by passing the relevant parameters in this FM.

[↓] Import

I_CALLBACK_PROGRAM

Name of the calling program. Type: SY-REPID.

I_CALLBACK_PF_STATUS_SET

Name of the form routine for setting the PF status. Type: SLIS_FORMNAME.

I_CALLBACK_USER_COMMAND

Name of form routine for handling USER COMMAND. Type: SLIS_FORMNAME.

IT_EXCLUDING

Table of inactive function codes. Type: SLIS_T_EXTAB.

---

208 REUSE_ALV_COMMENTARY_WRITE

This FM is used to assign the TOP-OF-PAGE event to the ALV display. List header information is output according to its type. The output information is put in an internal table. Output attributes are assigned to each line via the TYP field.

Example: To display the heading of the ALV data in the ALV TOP OF PAGE, pass the parameters and call this FM.

[↓] Import

IT_LIST_COMMENTARY

Comment block. This internal table has the TYP field with the following values:

- H (header: list header): In this option, only the contents of INFO are output, and KEY is ignored. Usually one line has one entry. Attributes are left-justified, bold, and followed by an empty line.
- S (selection): Important header information, such as currency of amounts in the list are displayed. KEY contents are bold, and INFO contents are normal. There can be several entries with one entry per line in this type. Attributes are left-justified and are followed by empty line.
- A (action): Action info. In this type, only the contents of INFO are output, and KEY is ignored. Usually there is one entry per line. Attributes are left-justified with normal intensity.

Besides TYP, there are two other possible fields in this table:

- KEY: String with key word character output in combination with TYP = S.
- INFO: Information output in header, formatted by type. Type: SLIS_T_LISTHEADER.

I_LOGO

This parameter holds the logo for display.

I_END_OF_LIST_GRID

This parameter is not used.

I_ALV_FORM

This parameter is not used. Type: SAP_BOOL.

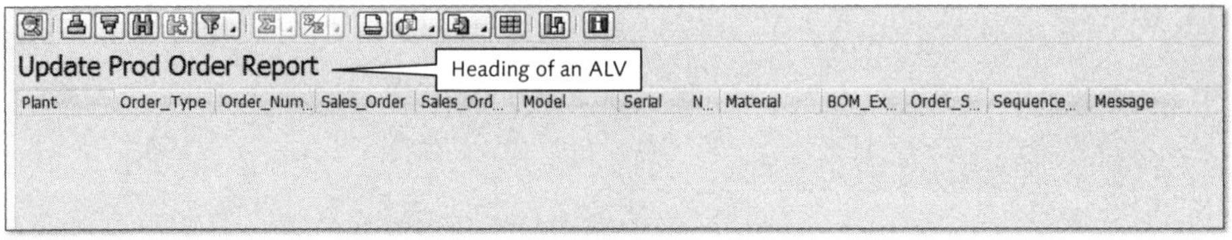

Heading of ALV

209 REUSE_ALV_FIELDCATALOG_MERGE

This FM creates a field catalog from the ABAP Data Dictionary structure or internal table. If the field catalog contains more than 90 fields, the first 90 fields are output in the list by default, whereas the remaining fields are only available in the field selection.

Example: You want to display the ALV output with normal classical programming. First you need to generate the field catalog that defines the fields of the ALV output by passing the parameters and calling this FM.

[↓] Import

I_PROGRAM_NAME

Program from which the FM is called and which contains the exit routines. The program should always be a report, function group, module pool, or form routine pool. Type: SY-REPID.

I_INTERNAL_TABNAME

Name of the internal output table in the CALLBACK_PROGRAM. An ABAP DDIC structure is preferable to a data object in a CALLBACK program include. Type: SLIS_TABNAME.

I_STRUCTURE_NAME

Name of the dictionary structure, table, or view. Type: DD02L-TABNAME.

I_CLIENT_NEVER_DISPLAY

If this parameter is X, the CLIENT field isn't displayed. The default value is X. Type: SLIS_CHAR_1.

I_INCLNAME

If the list output structure data declaration isn't in the assigned TOP include, the include name can be specified here if no structure is specified. Type: TRDIR-NAME.

I_BYPASSING_BUFFER

Ignore buffer while reading. Type: CHAR01.

I_BUFFER_ACTIVE

Special buffering active. Type: CHAR01.

[Δ] Changing

CT_FIELDCAT

Field catalog with field descriptions. Type: SLIS_T_FIELDCAT_ALV.

[!] Exceptions

INCONSISTENT_INTERFACE

Call parameter combination error.

PROGRAM_ERROR

Program errors.

210 REUSE_ALV_GRID_DISPLAY

This FM outputs the internal tables in the ALV grid format. You need to pass the internal table information that you want to output, field catalog information that will define the columns you want to display in the output, and layout information according to user preferences.

Example: Your business asks you to create a report of the business data and display the report in the ALV grid format. You need to create the field catalog from the FM REUSE_ALV_FIELDCATALOG_MERGE (FM 209) then call this FM to display the data in the output as ALV grid format.

[↓] Import

I_INTERFACE_CHECK

If this parameter is set to X, interface consistency is checked when the FM is called, and an error log is displayed. You should only set this parameter for testing purposes during the development process (e.g., for debugging). You can also perform the interface check on the result list by entering function code "&SOS". This parameter is currently not supported in full screen mode. As a workaround, go to the print preview from within the list, and enter function code "&SOS" there.

I_BYPASSING_BUFFER

If this parameter is set to X, buffers are ignored to check. Type: CHAR01.

I_BUFFER_ACTIVE

If this parameter is set to X, it checks whether the buffers are active.

I_CALLBACK_PROGRAM

Name of the calling program from which the FM is called. The program name should not be the include program name and should be a report, function group, module pool, or form routine. You should not pass SY-REPID directly in this parameter because this SY-REPID can be changed sometime by the current name of the internal intermediate program names. It's recommended to either pass SY-CPROG or store the current program name in the variable and pass it in this parameter. Type: SY-REPID.

I_CALLBACK_PF_STATUS_SET

Form name must be passed in this parameter. This parameter indicates that the caller wants to set a self-defined user status. As a result, the default status of the ALV isn't set. The interface of the form routine specified must be defined as follows:

```
FORM set_pf_status USING rt_extab
TYPE slis_t_extab
```

Table RT_EXTAB contains the function codes that are hidden on the standard user interface. If no EXIT routine is specified, the ALV sets a status that corresponds to status STANDARD_FULLSCREEN of function group SLVC_FULLSCREEN. Type: SLIS_FORMNAME.

I_CALLBACK_USER_COMMAND

Passing an EXIT routine indicates to the ALV that the application wants to respond to certain function codes. Generally, these are function codes that are unknown to the ALV (i.e., aren't standard ALV functions) and that were defined and set by a user status.

The interface of the form routine specified must be defined as follows:

```
FORM user_command  USING r_ucomm LIKE
sy-ucomm
     rs_selfield TYPE slis_selfield.
```

The EXIT routine is called whenever a function unknown to the ALV is triggered, or the routine call before/after the execution of a standard function code has been defined by interface parameter IT_EVENT_EXIT. Type: SLIS_FORMNAME.

I_CALLBACK_TOP_OF_PAGE

EXIT routine for handling TOP-OF-PAGE. The following form must be defined: FORM TOP_OF_PAGE. FM REUSE_ALV_COMMENTARY_WRITE can then be called within the EXIT routine.

If module REUSE_ALV_COMMENTARY_WRITE can't be used, you must use two parameters instead. Parameter I_CALLBACK_TOP_OF_PAGE passes the form routine that is responsible for normal formatting in batch mode or in the print preview mode, whereas parameter I_CALLBACK_HTML_TOP_OF_PAGE is required for online formatting. If one of these parameters isn't filled, TOP-OF-PAGE isn't output in the respective mode. Type: SLIS_FORMNAME.

I_CALLBACK_HTML_TOP_OF_PAGE

If FM REUSE_ALV_COMMENTARY_WRITE isn't used in the form for I_CALLBACK_TOP_OF_PAGE, the form routine must be passed in parameter I_CALLBACK_HTML_TOP_OF_PAGE for the online mode. The form should then have the following format:

```
form top_of_page using cl_dd type ref
to cl_dd_document.
```

Type: SLIS_FORMNAME.

I_CALLBACK_HTML_END_OF_LIST

In this parameter, you can pass a form for the online handling of end-of-list. The form must have the following format:

```
form end_of_list using cl_dd type ref
to cl_dd_document.
```

Type: SLIS_FORMNAME.

I_STRUCTURE_NAME

Internal output table structure name. If the internal output table is defined through an ABAP Data Dictionary structure (INCLUDE STRUCTURE structure or LIKE structure), you can automatically set up the field catalog by passing the structure name. Type: DD02L-TABNAME.

I_BACKGROUND_ID

You can use this parameter to pass a wallpaper for the output in TOP-OF-PAGE.

Caution: Because wallpaper-based formatting doesn't ensure a standards-based display of the page area in the new design, you should only use this parameter in exceptional cases. Type: SDYDO_KEY.

I_GRID_TITLE

Specifies the title of the control. This text is displayed above the grid. Type: LVC_TITLE.

I_GRID_SETTINGS

If TOP-OF-PAGE or END-OF-LIST are output online, these areas are displayed in a splitter above or below the list. Using I_GRID_SETTINGS, you can reduce the default size to 0%. To do this, you use two fields:

- COLL_TOP_P: Sets TOP-OF-PAGE to 0%.
- COLL_END_L: Sets END-OF-LIST to 0%.

Type: LVC_S_GLAY.

IS_LAYOUT

This parameter describes the layout specifications. These parameters are grouped based on the following sections: display options, exceptions, totals, interactions, detail screen, color, and other. Type: SLIS_LAYOUT_ALV.

IT_FIELDCAT

Field catalog containing the field descriptions of the fields to be considered for the list output (usually, this is a subset of the fields in the internal output table). Basically, you need a field catalog for each list output that uses the ALV. The field catalog associated with the output table is generated in the code of the caller. You can generate the field catalog automatically or semi-automatically by calling FM REUSE_ALV_FIELDCATALOG_MERGE. Type: SLIS_T_FIELDCAT_ALV.

IT_EXCLUDING

Table of inactive function codes. You must only fill this table if the caller uses the standard interface of the list tool but doesn't need certain interface functions and therefore wants to disable them. You must enter the function codes of these standard functions into the table. Type: SLIS_T_EXTAB.

IT_SPECIAL_GROUPS

If fields were grouped in the field catalog based on a common value for parameter SP_GROUP, an assignment is made in this internal table between the technical key of the group (FIELDCAT-SP_GROUP) and the associated field group text. Type: SLIS_T_SP_GROUP_ALV.

IT_SORT

Using internal table IT_SORT, the caller determines the sort order and/or the subtotaling of the basic list. Type: SLIS_T_SORTINFO_ALV.

IT_FILTER

Filter criteria for the first list output. Using this parameter, you can pass on filter criteria resulting from explicitly loading a display variant in advance, for example, to list output. This table should never be set up manually. Type: SLIS_T_FILTER_ALV.

IS_SEL_HIDE

This parameter isn't supported. Type: SLIS_SEL_HIDE_ALV.

I_DEFAULT

Value X defines that the initial variant is allowed, and value [Space] defines that the initial variant isn't allowed. The default value is X.

I_SAVE

This parameter controls saving the variant. Possible values:

- [Space]: Display variants can't be saved. Defined display variants (e.g., delivered display variants) can be selected for presentation independently of this flag. Changes can't be saved.
- X: Standard save. Display variants can be saved as standard display variants. User-specific saving isn't possible.
- U: Only user-specific saving. The user can only save display variants user-specifically.
- A: Standard and user-specific saving. The user can save a display variant user-specifically and as a standard display variant. The user chooses in the display variant save popup.

The default value is [Space].

IS_VARIANT

This structure is only relevant if display variants are to be saved and/or read. To allow display variants to be read within the ALV, you must specify the access path using fields REPORT (required field), HANDLE (optional field), and/or LOG_GROUP (optional field). If you also want to allow display variants to be saved, you must additionally fill in parameter I_SAVE accordingly. Type: DISVARIANT.

IT_EVENTS

This table is used to notify the ALV of all events that are processed by means of a CALLBACK through the caller. Type: SLIS_T_EVENT.

IT_EVENT_EXIT

Standard function code exit requests table. This table is used to pass the standard ALV function codes for which the user wants to get control before and/or after their execution by means of the USER_COMMAND callback event. Type: SLIS_T_EVENT_EXIT.

IS_PRINT

This parameter is relevant for printing. Type: SLIS_PRINT_ALV.

IS_REPREP_ID

Initialization key for report or report interface. Type: SLIS_REPREP_ID.

I_SCREEN_START_COLUMN

Start column coordinates for the list in the dialog box.

I_SCREEN_START_LINE

Start line coordinates for the list in the dialog box.

I_SCREEN_END_COLUMN

End column coordinates for the list in the dialog box.

I_SCREEN_END_LINE

End line coordinates for the list in the dialog box.

I_HTML_HEIGHT_TOP

Height for HTML_TOP_OF_PAGE. Type: I.

I_HTML_HEIGHT_END

Height for HTML_END_OF_PAGE. Type: I.

IT_ALV_GRAPHICS

Parameter for ALV graphic. Type: DTC_T_TC.

IT_HYPERLINK

Parameter for hyperlinks. Type: LVC_T_HYPE.

IT_ADD_FIELDCAT

Additional field catalog options. Type: SLIS_T_ADD_FIELDCAT.

IT_EXCEPT_QINFO

This parameter isn't used. Type: SLIS_T_QINFO_ALV.

IR_SALV_FULLSCREEN_ADAPTER

For internal use. Type: CL_SALV_FULLSCREEN_ADAPTER.

[↑] Export

E_EXIT_CAUSED_BY_CALLER

Delete list in CALLBACK_USER_COMMAND. This parameter specifies if the caller in CALLBACK USER_COMMAND has forced the system to exit the list.

ES_EXIT_CAUSED_BY_USER

If one of the default ALV function codes for Back, Cancel, or Exit is triggered, the system destroys the current list and quits the ALV FM. In this case, exporting structure ES_EXIT_CAUSED_BY_USER is filled accordingly:

- ES_EXIT_CAUSED_BY_USER-BACK = X. Function Back triggered.
- ES_EXIT_CAUSED_BY_USER-CANCEL = X. Function Cancel triggered.
- ES_EXIT_CAUSED_BY_USER-EXIT = X. Function Exit triggered.

Type: SLIS_EXIT_BY_USER.

Table

T_OUTTAB

This internal table contains the data to be displayed as a list.

[!] Exception

PROGRAM_ERROR

Program errors.

211 REUSE_ALV_GRID_LAYOUT_INFO_GET

This FM is used to get the current information of a list. This FM returns the information about the current layout, current field catalog, sort sequence, filter criteria, and information on the current scroll state of the list.

Example: You have a requirement to allow a user to download the results of a report to Microsoft Excel according to the ALV layout variant the user chooses on the selection screen; that is, output columns of the layout should be retrieved and downloaded into Excel in the background. You can call this FM to get the desired layout and the details of the ALV.

[↑] Export

ES_LAYOUT

Current list layout specifications are returned. Type: SLIS_LAYOUT_ALV.

ET_FIELDCAT

Current field catalog with field descriptions is returned. Type: SLIS_T_FIELDCAT_ALV.

ET_SORT

Current list output sort criteria are returned. Type: SLIS_T_SORTINFO_ALV.

ET_FILTER

Current list output filter criteria are returned. Type: SLIS_T_FILTER_ALV.

ES_GRID_SCROLL

Returns the ALV scroll information. Type: LVC_S_SCRL.

ES_VARIANT

Current layout. Type: DISVARIANT.

ET_MARKED_COLUMNS

Returns the selected columns. Type: SLIS_T_FIELDCAT_ALV.

ET_FILTERED_ENTRIES

Returns the filtered entries of the grid. Type: SLIS_T_FILTERED_ENTRIES.

ET_FIELDCAT_BACKEND

Returns the backend field catalog. Type: SLIS_T_FIELDCAT_ALV.

ES_PRINT

Returns the print settings. Type: SLIS_PRINT_ALV.

[!] Exceptions

NO_INFOS

No information available.

PROGRAM_ERROR

Program errors.

212 REUSE_ALV_GRID_LAYOUT_INFO_SET

This FM is used to change the current information of a list.

Example: When a user clicks a certain button on the output screen to do some processing on the selected record, a traffic light should be displayed to green if successful or red if an error occurs. You can make the changes in layout. Pass the parameter to change the layout and call this FM to change the ALV.

[↓] Import

IS_LAYOUT

List layout specifications. Type: SLIS_LAYOUT_ALV.

IT_FIELDCAT

Field catalog with field descriptions. Type: SLIS_T_FIELDCAT_ALV.

IT_SORT

List output sort criteria. Type: SLIS_T_SORTINFO_ALV.

IT_FILTER

List output filter criteria. Type: SLIS_T_FILTER_ALV.

IS_GRID_SCROLL

Current list scroll status. Type: LVC_S_SCRL.

IS_PRINT

ALV print details. Type: SLIS_PRINT_ALV.

213 REUSE_ALV_HIERSEQ_LIST_DISPLAY

This FM outputs two internal tables as a formatted hierarchical sequential list. In this ALV format, you need to pass an internal table containing the set of header information to be output, pass the internal table containing the set of item information to be output, pass a structure containing the general list layout details, and pass the field catalog information.

Example: You want to display the output of sales order headers and sales order items in the report ALV data based on the given selection condition. You can pass the header data, item data, layout, field catalog, and so on and call this FM to get the ALV in the form of a hierarchy list.

[↓] Import

I_INTERFACE_CHECK

Interface consistency check log output.

I_CALLBACK_PROGRAM

Name of the calling program. Type: SY-REPID.

I_CALLBACK_PF_STATUS_SET

Set EXIT routine to status. Type: SLIS_FORMNAME.

I_CALLBACK_USER_COMMAND

Name of form routine for handling USER COMMAND. Type: SLIS_FORMNAME.

IS_LAYOUT

List layout specifications. Type: SLIS_LAYOUT_ALV.

IT_FIELDCAT

Field catalog with field descriptions. Type: SLIS_T_FIELDCAT_ALV.

IT_EXCLUDING

Table of inactive function codes. Type: SLIS_T_EXTAB.

IT_SPECIAL_GROUPS

Grouping fields for column selection. Type: SLIS_T_SP_GROUP_ALV.

IT_SORT

Sort criteria for first list display. Type: SLIS_T_SORTINFO_ALV.

IT_FILTER

Filter criteria for first list output. Type: SLIS_T_FILTER_ALV.

IS_SEL_HIDE

Selection information modification. Type: SLIS_SEL_HIDE_ALV.

I_SCREEN_START_COLUMN

Coordinates for the list in the dialog box.

I_SCREEN_START_LINE

Coordinates for the list in the dialog box.

I_SCREEN_END_COLUMN

Coordinates for the list in the dialog box.

I_SCREEN_END_LINE

Coordinates for the list in the dialog box.

I_DEFAULT

Initial variant active/inactive logic.

I_SAVE

Variants can be saved.

IS_VARIANT

Variant information. Type: DISVARIANT.

IT_EVENTS

Table of events to perform. Type: SLIS_T_EVENT.

IT_EVENT_EXIT

Standard function code exit requests table. Type: SLIS_T_EVENT_EXIT.

I_TABNAME_HEADER

Header table name. This is a mandatory parameter. Type: SLIS_TABNAME.

I_TABNAME_ITEM

Item table name. This is a mandatory parameter. Type: SLIS_TABNAME.

I_STRUCTURE_NAME_HEADER

Internal output table (header) structure name. Type: DD02L-TABNAME.

I_STRUCTURE_NAME_ITEM

Internal output table (position) structure name. Type: DD02L-TABNAME.

IS_KEYINFO

Header/position table field links. This is a mandatory parameter. Type: SLIS_KEYINFO_ALV.

IS_PRINT

Print information. Type: SLIS_PRINT_ALV.

IS_REPREP_ID

Initialization key for a report or report interface. Type: SLIS_REPREP_ID.

I_BYPASSING_BUFFER

Skip all buffers? Type: CHAR01.

I_BUFFER_ACTIVE

User buffer A. Type: CHAR01.

IR_SALV_HIERSEQ_ADAPTER

Adapter for the hierseq. lists. Type: CL_SALV_HIERSEQ_ADAPTER.

IT_EXCEPT_QINFO

This table contains external tooltip values. Type: SLIS_T_QINFO_ALV.

I_SUPPRESS_EMPTY_DATA

If no data exists, no table is output. Type: SAP_BOOL.

[↑] Export

E_EXIT_CAUSED_BY_CALLER

Delete list in CALLBACK_USER_COMMAND.

ES_EXIT_CAUSED_BY_USER

How the user left the list. Type: SLIS_EXIT_BY_USER.

Tables

T_OUTTAB_HEADER

Header table with data to be output.

T_OUTTAB_ITEM

Position table with data to be output.

[!] Exception

PROGRAM_ERROR

Program errors.

Verkaufsbelege mit Storno oder Gutschrift zur Bestellung anzeigen

Beleg								
Position	Eint	Bestellung	Position	Gutschriftsmenge	Gutschriftswert	Lastschriftsmenge	Lastschriftswert	Storno
2144								
80	1	4500010489	10	1.000	102.54	1.000	102.54	
3449								
1165	1	4500009416	30	1.000	361.25	2.000	722.50	
4760	1	4500009416	20	1.000	32.02	2.000	64.04	
4149								
45	1	4500010008	10	1.000	372.47	1.000	372.47	
4576								
45	1	4500010640	10	1.000	370.20	1.000	370.20	
4667								
3270	1	4500011215	10	1.000	137.88	2.000	274.75	
3290	1	4500011215	20	3.000	120.93	4.000	153.70	
3340	1	4500011215	30	1.000	372.42	2.000	742.09	

Example of Hierarchy List Displayed by This FM

214 REUSE_ALV_LIST_DISPLAY

This FM outputs the internal tables in the ALV list format. You need to pass the internal table information that you want to output, field catalog information that will define the columns you want to display in output, and layout information according to the user preferences.

Example: Your business asks you to create a report of the business data and display the report in the ALV list format. You need to create the field catalog from the FM REUSE_ALV_FIELDCATALOG_MERGE (FM 209) and then call this FM to display the data in the output as ALV Grid format.

Note: For a list of parameters, please refer to FM REUSE_ALV_GRID_DISPLAY (FM 210).

215 REUSE_ALV_LIST_LAYOUT_INFO_GET

This FM is used to get the current information of the selected ALV list. This FM returns the information about the current layout, current field catalog, sort sequence, filter criteria and current scroll state of the list.

Example: You need to allow the user to download the results of a report to Microsoft Excel according to the ALV layout variant the user chooses on the selection screen; that is, output columns of layout should be retrieved and downloaded into Excel in the background.

You can call this FM to get the desired layout and the details of the ALV.

[↑] Export

ES_LAYOUT

Current list layout specifications are returned. Type: SLIS_LAYOUT_ALV.

ET_FIELDCAT

Current field catalog with field descriptions is returned. Type: SLIS_T_FIELDCAT_ALV.

ET_SORT

Current list output sort criteria are returned. Type: SLIS_T_SORTINFO_ALV.

ET_FILTER

Current list output filter criteria are returned. Type: SLIS_T_FILTER_ALV.

ES_LIST_SCROLL

Returns the ALV scroll information. Type: SLIS_LIST_SCROLL.

ES_VARIANT

Current layout. Type: DISVARIANT.

E_WIDTH

Current width of output list. Type: INT4.

ET_MARKED_COLUMNS

Returns the selected columns. Type: SLIS_T_FIELDCAT_ALV.

ET_FILTERED_ENTRIES

Returns the filtered entries. Type: SLIS_T_FILTERED_ENTRIES.

ET_FILTERED_ENTRIES_HEADER

This parameter isn't used. Type: SLIS_T_FILTERED_ENTRIES.

ET_FILTERED_ENTRIES_ITEM

This parameter returns items with filtered entries. This parameter isn't widely used. Type: SLIS_T_FILTERED_ENTRIES.

Tables

ET_OUTTAB

If this table parameter is requested before calling the FM, the tables with its values are populated. This parameter isn't widely used. Type: TABLE.

ET_OUTTAB_HEADER

If this table parameter is requested before calling the FM, the tables with its values are populated. This parameter isn't widely used. Type: TABLE.

ET_OUTTAB_ITEM

If this table parameter is requested before calling the FM, the tables with its values are populated. This parameter isn't widely used. Type: TABLE.

ET_COLLECT00

If this table parameter is requested before calling the FM, the tables with its values are populated. This parameter isn't widely used. Type: TABLE.

ET_COLLECT01

If this table parameter is requested before calling the FM, the tables with its values are populated. This parameter isn't widely used. Type: TABLE.

ET_COLLECT02

If this table parameter is requested before calling the FM, the tables with its values are populated. This parameter isn't widely used. Type: TABLE.

ET_COLLECT03

If this table parameter is requested before calling the FM, the tables with its values are populated. This parameter isn't widely used. Type: TABLE.

ET_COLLECT04

If this table parameter is requested before calling the FM, the tables with its values are populated. This parameter isn't widely used. Type: TABLE.

ET_COLLECT05

If this table parameter is requested before calling the FM, the tables with its values are populated. This parameter isn't widely used. Type: TABLE.

ET_COLLECT06

If this table parameter is requested before calling the FM, the tables with its values are populated. This parameter isn't widely used. Type: TABLE.

ET_COLLECT07

If this table parameter is requested before calling the FM, the tables with its values are populated. This parameter isn't widely used. Type: TABLE.

ET_COLLECT08

If this table parameter is requested before calling the FM, the tables with its values are populated. This parameter isn't widely used. Type: TABLE.

ET_COLLECT09

If this table parameter is requested before calling the FM, the tables with its values are populated. This parameter isn't widely used. Type: TABLE.

[!] Exceptions

NO_INFOS

No information available.

PROGRAM_ERROR

Program errors.

216 REUSE_ALV_LIST_LAYOUT_INFO_SET

This FM is used to change the current information of the ALV list.

Example: A user presses a button on the output screen and does some processing on the selected record. Based on the processing, a traffic light should be displayed to green if successful or red if an error occurs. You can do the changes in layout. You can pass the parameter to change the layout and call this FM to change the ALV.

[↓] Import

IS_LAYOUT

List layout specifications. Type: SLIS_LAYOUT_ALV.

IT_FIELDCAT

Field catalog with field descriptions. Type: SLIS_T_FIELDCAT_ALV.

IT_SORT

List output sort criteria. Type: SLIS_T_SORTINFO_ALV.

IT_FILTER

List output filter criteria. Type: SLIS_T_FILTER_ALV.

IS_LIST_SCROLL

Current list scroll status. Type: SLIS_LIST_SCROLL.

IS_PRINT

ALV print details. Type: SLIS_PRINT_ALV.

217 REUSE_ALV_POPUP_TO_SELECT

This FM is used to display or select table entries from internal tables (single or multiple selection possible) in a popup.

Example: You need to display the output on the popup screen, and you need to edit one column. You can choose and display the only column you want on a popup by calling this FM to edit the cell.

[↓] Import

I_TITLE

Dialog box title.

I_SELECTION

If this parameter is set as X, entries can be selected; however, if the value is [Space], values are in display mode.

I_ALLOW_NO_SELECTION

If this parameter is set to X, copy is allowed even if nothing is selected. Type: CHAR1.

I_ZEBRA

Line output with alternating color. The default value is X.

I_SCREEN_START_COLUMN

Start column coordinates for the list in the dialog box.

I_SCREEN_START_LINE

Start line coordinates for the list in the dialog box.

I_SCREEN_END_COLUMN

End column coordinates for the list in the dialog box.

I_SCREEN_END_LINE

End line coordinates for the list in the dialog box.

I_CHECKBOX_FIELDNAME

If the table output in the popup has checkboxes at the beginning of the rows (e.g., for multiple selection), the internal table must contain a field containing the value of the checkbox.

Assign the name of this field to the parameter I_CHECKBOX_FIELDNAME.

I_LINEMARK_FIELDNAME

Name of the internal output table field containing the color code. A complete list row is colored individually by color coding in a column of the internal output table. The name of this field is assigned to this parameter. The coding must have the following syntax: Cxy:

- C: Color (each code must begin with C).
- x: Color number (1 – 9).
- y: Highlight (0 = off, 1 = on).

I_SCROLL_TO_SEL_LINE

Scroll to the default selection if necessary. A row has been color highlighted, and its index in the internal output table is greater than the number of entries that can be displayed in the popup without scrolling.

When the popup is called, the cursor should be immediately positioned on, or scrolled to, this row.

I_TABNAME

Table name with chosen values. This is a mandatory parameter.

I_STRUCTURE_NAME

Internal output table structure name. Type: DD02L-TABNAME

IT_FIELDCAT

Field catalog with field descriptions. Type: SLIS_T_FIELDCAT_ALV.

IT_EXCLUDING

Table of inactive function codes. Type: SLIS_T_EXTAB.

I_CALLBACK_PROGRAM

Name of the calling program. Type: SY-REPID.

I_CALLBACK_USER_COMMAND

This parameter contains the form name for the USER_COMMAND handling event. Type: SLIS_FORMNAME.

IS_PRIVATE

Internal private use only. Type: SLIS_DATA_CALLER_EXIT.

[↑] Export

ES_SELFIELD

Selection info for simple selection. This structure contains the TABNAME, TABINDEX, FIELDNAME, and VALUE details of the selected returned field. Type: SLIS_SELFIELD.

E_EXIT

This parameter is set to X, if the user cancels the popup by selecting the CANCEL button.

Table

T_OUTTAB

Internal table of any structure that contains the data to be output in list form. Only the fields named in the field catalog and in the I_CHECKBOX_FIELDNAME parameter are used for the list output. Other internal table fields are ignored.

[!] Exception

PROGRAM_ERROR

Program errors.

---

218 REUSE_ALV_VARIANT_DEFAULT_GET

This FM reads the default variant of an ALV.

Example: You want to give the default variant of ALV at the selection screen when you execute the transaction. You can pass the parameters and call this FM for the same at the `INITIALIZATION` event. You'll get the default variant, which is set for the ALV, and then you can display it.

[↓] Import

I_SAVE

This parameter controls saving of the variant. Possible values:

- Space: Display variants can't be saved. Defined display variants (e.g., delivered display variants) can be selected for presentation independently of this flag. Changes can't be saved.
- X: Standard save. Display variants can be saved as standard display variants. User-specific saving isn't possible.

- U: Only user-specific saving. The user can only save display variants user-specifically.
- A: Standard and user-specific saving. The user can save a display variant user-specifically and as a standard display variant. The user chooses in the display variant save popup.

The default value is [Space].

[Δ] Changing

CS_VARIANT

Default variant information is returned. Type: DISVARIANT.

[!] Exceptions

WRONG_INPUT

Inconsistent input parameters.

NOT_FOUND

Variant not found.

PROGRAM_ERROR

Program errors.

---

219 REUSE_ALV_VARIANT_EXISTENCE

This FM checks the existence of the display variant in the database.

Example: You can validate the selection screen variant by passing the parameters in this FM.

[↓] Import

I_SAVE

Refer to parameter I_SAVE of FM REUSE_ALV_VARIANT_DEFAULT_GET (FM 218). The default value is [Space].

[Δ] Changing

CS_VARIANT

Default variant information is returned. Type: DISVARIANT.

[!] Exceptions

WRONG_INPUT

Inconsistent input parameters.

NOT_FOUND

Variant not found.

PROGRAM_ERROR

Program errors.

---

220 REUSE_ALV_VARIANT_F4

This FM displays a variant selection dialog box. If a variant is explicitly defined as an input field on a screen, this FM provides the F4 help for the ALV variant. The selection must be specified by at least partially filling in the parameter structure IS_VARIANT.

Example: If you want the F4 help for the variant so that you can select the variant, pass the parameters and call this FM.

[↓] Import

IS_VARIANT

Variant information. This is a mandatory parameter. Type: DISVARIANT.

I_TABNAME_HEADER

Name of the internal table in the program containing the output data of the highest hierarchy level. Type: SLIS_TABNAME.

I_TABNAME_ITEM

Name of the internal table in the program containing the output data of the lowest hierarchy level. Type: SLIS_TABNAME.

IT_DEFAULT_FIELDCAT

Field catalog with field descriptions. Type: LVC_T_FCAT.

I_SAVE

Refer to parameter I_SAVE of FM REUSE_ALV_VARIANT_DEFAULT_GET (FM 218).

I_DISPLAY_VIA_GRID

If this parameter is set to X, the variant is displayed via the grid method, otherwise, via the list method. Type: CHAR01.

[↑] Export

E_EXIT

Dialog cancelled by user.

ES_VARIANT

Variant information returned after F4. Type: DISVARIANT.

[!] Exceptions

NOT_FOUND

Variant not found.

PROGRAM_ERROR

Program errors.

221 REUSE_ALV_VARIANT_SELECT

This FM is used to read an ALV display variant. This function can also be performed with a dialog box popup screen; that is, variants are displayed in a popup screen.

Example: You've created variant VAR1, which hides two columns from an ALV report. To read the variant so that you can implement the settings of that variant, pass the parameter and call this FM.

[↓] Import

I_DIALOG

If this parameter is set to X, a dialog popup is displayed. The default value is X.

I_USER_SPECIFIC

If this parameter is set to X, user-specific variants are included. The default value is Space.

I_DEFAULT

If this parameter is X, the default variant is read. The default value is X.

I_TABNAME_HEADER

Name of the internal table in the program containing the output data of the highest hierarchy level. Type: SLIS_TABNAME.

I_TABNAME_ITEM

Name of the internal table in the program containing the output data of the lowest hierarchy level. Type: SLIS_TABNAME.

IT_DEFAULT_FIELDCAT

Field catalog with field descriptions. Type: SLIS_T_FIELDCAT_ALV.

I_LAYOUT

This parameter defines the output list description structure. This parameter is grouped under display options, exceptions, totals, interactions, detail screen, display variant, color, and other functionalities. Type: SLIS_LAYOUT_ALV.

I_BYPASSING_BUFFER

If this parameter is set to X, all of the buffers are skipped. Type: CHAR01.

I_BUFFER_ACTIVE

This parameter checks if the buffer is active.

[↑] Export

E_EXIT

This parameter is set (X) if the FM was called with the dialog, and the dialog was canceled by the user.

ET_FIELDCAT

Field catalog with field descriptions. Type: SLIS_T_FIELDCAT_ALV.

ET_SORT

List output sort criteria. Type: SLIS_T_SORTINFO_ALV.

ET_FILTER

List output filter criteria. Type: SLIS_T_FILTER_ALV.

ES_LAYOUT

List output layout details. Type: SLIS_LAYOUT_ALV.

[Δ] Changing

CS_VARIANT

Variant information. Type: DISVARIANT.

[!] Exceptions

WRONG_INPUT

Inconsistent input parameters.

FC_NOT_COMPLETE

Field catalog incomplete.

NOT_FOUND

Variant not found.

PROGRAM_ERROR

Program errors.

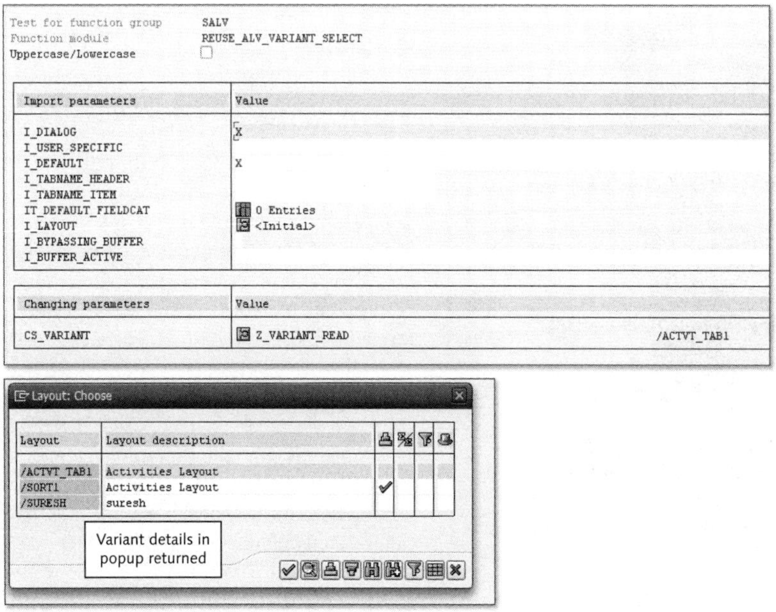

Variant Read Details

222 RPY_DYNPRO_READ

This FM gives you the details of the screen that exists in the repository browser. All of the information of the screen, such as flow logic, screen elements, and its properties, are returned back in the export and table parameters of this FM. The definitions and constants of the parameter are defined in the types group RPYTY. You can see the screen and its details in tables T020T, T021T, and D020S.

Example: You want the elements information of one screen in your dynamic programming. You can pass the program name and the screen number to this FM to get all of the parameters of the screen.

[↓] Import

PROGNAME

Program name of the screen. Type: D020S-PROG.

DYNNR

Screen number. Type: D020S-DNUM.

SUPPRESS_EXIST_CHECKS

If this parameter is set to [Space], an additional check is performed on table D020S to check if the screen exists. Exception NOT_FOUND is raised if the screen doesn't exist. Type: C.

SUPPRESS_CORR_CHECKS

If this parameter is set to [Space], an additional check is performed for authorization of the user. FM RS_ACCESS_PERMISSION is called for the authorization check. Exception CANCELED_IN_CORR or PERMISSION_FAILURE can be raised based on the exception returned by FM RS_ACCESS_PERMISSION. Type: C.

[↑] Export

HEADER

General information on the screen. Information that you see in the ATTRIBUTES section of the screen comes in this parameter. Type: RPY_DYHEAD.

Tables

CONTAINERS

Each entry in CONTAINERS describes a container. A container is a portion of a screen in which screen elements are grouped together logically and/or an inner container may be included such as a screen, subscreen, table control, custom control, and so on. Type: DYCATT_TAB.

FIELDS_TO_CONTAINERS

This table contains the individual elements of the screen and its details, such as buttons, text, radio buttons, checkboxes, frames, and other details. Type: DYFATC_TAB.

FLOW_LOGIC

Flow logic of the screen. Instructions that can be defined on the flow logic of the Screen Painter comes in this section. Type: RPY_DYFLOW.

PARAMS

Screen: parameter information for the screen. Type: RPY_DYPARA.

[!] Exceptions

CANCELLED

Canceled by the user in the correction dialog box.

NOT_FOUND

Screen is not found in table D020S.

PERMISSION_ERROR

No authorization to read.

---

223 RPY_TRANSACTION_READ

This FM gives you the details of the transaction code from ABAP Data Dictionary's transaction tables TSTC, TSTCT, and TSCC.

Example: You want to create a report for runtime error analysis where you want to display the transaction code, program name, and other information related to the runtime error to the user. You can call this FM to get the runtime error for a particular day. You'll get the program name from this FM as well. If you want to get the transaction code from the program name to be used for your report, you call pass the parameters and call this FM.

[↓] Import

TRANSACTION

Transaction code. Type: TSTC-TCODE.

PROGRAM

Program name. Type: TRDIR-NAME.

DYNPRO

Screen number. Type: D020S-DNUM.

TRANSACTION_TYPE

This parameter determines that the transaction code is attached to a type of object such as report, module pool, and so on. Based on this parameter, data is filtered from the table TSTC-CINFO field. Type: RGLIF-DOCUTYPE.

Tables

TCODES

Transaction code details returned. Type: TSTC.

GUI_ATTRIBUTES

Additional attributes for TSTC such as compatibility with HTML GUI, Windows GUI, Java GUI, and so on. Type: TSTCC.

[!] Exceptions

PERMISSION_ERROR

No authorization.

CANCELLED

Canceled by the user in the correction dialog box.

NOT_FOUND

If a transaction and program combination is entered or only the program name is entered and is not found in table TSTS, this exception is raised.

OBJECT_NOT_FOUND

If only the transaction parameter is entered and the transaction does not exist in TSTC, this exception is raised.

224 RS_COVERPAGE_SELECTIONS

This FM returns the parameters and select options of the selection screen. This FM is generally called when printing the cover sheet. This FM also calls FM RS_VARIANT_CONTENTS (FM 240) to get the values and process them further to return the values in export and tables parameters.

Example: You have CHANGE MODE or DISPLAY MODE radio button on the selection screen. The end user wants to know if he has executed the program in change mode or display mode. He also wants all of the other selection screen parameter values for which the report has been generated. Pass the parameters and call this FM to get the selection screen parameters and select options.

[↓] Import

REPORT

Report name. Type: RSVAR-REPORT.

VARIANT

Variant name ([Space] also valid). Type: RSVAR-VARIANT.

NO_IMPORT

If this parameter is set to X, import isn't allowed. The default value is [Space].

Table

INFOTAB

Table with parameter and selection options for printing.

[!] Exceptions

ERROR_MESSAGE

Variant entered is obsolete.

VARIANT_NOT_FOUND

Variant not found.

225 RS_CREATE_VARIANT

This FM creates the variant for the selection screen.

Example: You've created a selection screen as a subscreen and want to have the variant functionality; that is, when you give some values in parameters and select options, you want to save them and use them later. However, before you create the variant with this FM, you first need to call FM RS_REPORTSELECTIONS_INFO (FM 229).

[↓] Import

CURR_REPORT

Report name for which you want to create the variant. Type: RSVAR-REPORT.

CURR_VARIANT

Variant name. Type: RSVAR-VARIANT.

VARI_DESC

This parameter defines the variant details such as report name for variant, variant name, user who created the variant, if the variant is a system variant and is transported automatically, authorization group for the variant, variant create date and time, and so on. Type: VARID.

[↑] Export

VARI_CONTENTS_L

This parameter returns the general structure for parameters and select options. Type: RSPARAMSL.

Tables

VARI_CONTENTS

This parameter contains the selection screen field information, that is, select options and parameter information. Type: RSPARAMS.

VARI_TEXT

Variant text information such as the client in which the variant is to be created, language, program name in the variant key, name of the variant, and short text of the variant. Type: VARIT.

VSCREENS

This parameter defines the screen number and whether the screen is a normal screen or a dialog box. Type: RSDYNNR.

[!] Exceptions

ILLEGAL_REPORT_OR_VARIANT

Incorrect report or variant name entered in the VARI_DESC parameter.

ILLEGAL_VARIANTNAME

Variant name is [Space] or contains & or %.

NOT_AUTHORIZED

No authorization for variant creation.

NOT_EXECUTED

FM not executed properly.

REPORT_NOT_EXISTENT

Report name does not exist.

REPORT_NOT_SUPPLIED

Report is not type 1.

VARIANT_EXISTS

Variant already exists.

VARIANT_LOCKED

Variant is locked.

---

226 RS_EXTERNAL_SELSCREEN_STATUS

This FM sets the GUI status on the external selection screen defined in the external function group. This FM can be called at `INITIALIZATION` or at the `SELECTION-SCREEN OUTPUT` event.

Example: Instead of the selection screen GUI status, you want to use your own GUI status. You can do so either by calling this FM or by calling FM RS_SET_SELSCREEN_STATUS (FM 231). This FM displays the GUI status created in a function group, whereas FM RS_SET_SELSCREEN_STATUS displays the GUI status defined in the same ABAP program. Pass the relevant parameters in either of these FMs and execute. You'll get the desired GUI status; that is, PF status and title bar in the selection screen.

[↓] Import

P_FB

This parameter passes the name of the FM that sets its own GUI status when the system processes the selection screen. In the FM, you can set the SET PF-STATUS and SET TITLEBAR in the selection screen. Type: RS38L-NAME.

---

227 RS_FUNCTION_POOL_CONTENTS

This FM returns all of the names of the FMs in the function group that were entered in the import parameter of this FM.

Example: The end user doesn't have the authorization to view FMs via Transaction SE37, so he has requested to create a program to view FMs in a function group. You can do so by passing the function group name in the import parameter of this FM and calling this FM.

[↓] Import

FUNCTION_POOL

Function group name you want to find. Type: TLIBG-AREA.

Table

FUNCTAB

This table returns all FMs of the function group. Type: RS38L_INCL.

[!] Exception

FUNCTION_POOL_NOT_FOUND

Function group not found.

228 RS_REFRESH_FROM_SELECTOPTIONS

This FM returns the contents of the current selection of the report, that is, content of the parameter and select option of the selection screen in the table SELECTION_TABLE. The other FMs needed in this process are:

1. RS_REPORTSELECTIONS_INFO (FM 229)
2. RS_CREATE_VARIANT (FM 225)

Example: You need to call this FM to get the details values of the parameters and select options. You can get this information in the tables parameter CURR_REPORT. You need this information in the tables to build the table parameter VARI_CONTENTS of FM RS-CREATE_VARIANT. After you build this information and other import parameter details information, you need to call this FM RS_CREATE_VARIANT. On successful execution of this FM, your variant is created.

[↓] Import

CURR_REPORT

Program for which selections are to be displayed. Type: RSVAR-REPORT.

[↑] Export

SP

This is used for returning logical database values. Type: SYLDB_SP.

Tables

SELECTION_TABLE

Table with ranges structure that contains selections. Type: RSPARAMS.

SELECTION_TABLE_255

Table with ranges structure that contains selections of structure RSPARAMSL_255.

[!] Exceptions

NOT_FOUND

Program does not exist.

NO_REPORT

Program is not type 1.

---

229 RS_REPORTSELECTIONS_INFO

This FM returns information about parameters, select options, and default values for a report.

Example: To get the selection screen's field information; that is, its parameter and select options, call this FM and get this information in tables parameters FIELD_INFO and DEF_VALUES.

[↓] Import

REPORT

Report name. This is a mandatory parameter. Type: RSVAR-REPORT.

DEFAULT_VALUES

If this parameter is X, default values of the selection screen are also returned. The default value is X.

Tables

FIELD_INFO

Information about type, reference field, and so on. Type: RSEL_INFO.

DEF_VALUES

Defaults in RSPARAMS structure (if available). Type: RSPARAMS.

DEF_VALUES_255

Generic structure for parameters and select options (255 long). Type: RSPARAMSL_255.

[!] Exceptions

NO_SELECTIONS

Report has no selections.

REPORT_NOT_EXISTENT

Report does not exist.

SUBROUTINE_POOL

Program is subroutine pool.

230 RS_SELECTIONSCREEN_READ

The FM returns the current values of the field names specified in the table FIELDVALUES. On the standard selection screen, you can retrieve the values of any number of fields (select options and parameters). On the multiple selection screen, you can only retrieve the values of the LOW and HIGH fields in the current cursor line. This FM is similar to DYNP_VALUES_READ (FM 191).

Example: You have PLANT, WORK CENTER, and PARAMETER1 fields in the selection screen. For one work center, there can be only one PARAMETER1 value. You want to get the F4 help on the work center based on the value entered in the PLANT field. To get the PLANT field for the F4 help of the work center, you need to pass the relevant parameters and call this FM.

[↓] Import

PROGRAM

Current program name. Type: SY-REPID.

DYNNR

Screen number. Type: SY-DYNNR.

Table

FIELDVALUES

Table with field name, type, and position. In this table, you get the values of the field back. Type: RSSELREAD.

231 RS_SET_SELSCREEN_STATUS

This FM allows you to set a custom GUI status on the selection screen. With this FM, you can set a separate status in a selection screen.

The FM should be called only during the processing of the standard selection images; that is, either at `INITIALIZATION` or at the `AT SELECTION-SCREEN OUTPUT` event before the selection screen is created and displayed on the user SAP GUI.

Before you call this FM, FM RS_SUBMIT_INFO (FM 233) should be called to get the selection screen mode such as normal mode, variant maintenance mode, or job scheduling mode.

The value passed in parameter P_STATUS for GUI status must be defined in the report even if the parameter P_PROGRAM isn't supplied. Another way to set an externally defined status is from FM RS_EXTERNAL_SELSCREEN_STATUS (see FM 226).

Example: You want to remove the print function from the default GUI status. After you get this information, you can pass the parameter in P_STATUS and fill in the P_EXCLUDE table to remove the print function and execute this FM. You won't get the print function.

[↓] Import

P_STATUS

GUI status to be set on the selection screen. If you don't specify a value for P_PROGRAM, the system looks for the status in the program to which the selection screen belongs. Type: SY-PFKEY.

P_PROGRAM

Program from which GUI status P_STATUS is to be taken. The parameter is optional. If it's not filled, the system searches the status in the program to which the selection screen belongs. The default value is [Space]. Type: SY-REPID.

Table

P_EXCLUDE

Table of OK codes to be excluded. The table should have the structure RSEXFCODE. It contains all of the function codes that are deactivated when the selection screen is processed.

---

232 RS_SELECTIONSCREEN_UPDATE

The FM is the second step after using FM RS_SELECTIONSCREEN_READ (FM 230); it updates the values to selection screen fields on multiple selection screen. This FM allows you to supply values to any fields of the standard selection screen in the `AT SELECTION-SCREEN ON VALUE REQUEST...` event. You must pass field names and their values in the table FIELDVALUES. In the POSITION field, you specify whether you want to supply values to the LOW or the HIGH field of a select option. For parameters, the contents of the field should be [Space]. In the case of multiple selections, you can only change the LOW or HIGH field at the current cursor position. This FM is similar to DYNP_VALUES_UPDATE (FM 192).

Example: You have a main work center and next successor work center; based on the value entered on the main work center, you want to internally get the details of the next successor work center from the ABAP Data Dictionary table and update the parameter field of the next successor work center at the same time. Fill in the values in table parameter DYNPFIELDS, and call FM RS_SELECTIONSCREEN_UPDATE (FM 232) to update the values in the screen without calling the PBO.

[↓] Import

PROGRAM

Current program name. Type: SY-REPID.

DYNNR

Screen number. Type: SY-DYNNR.

Table

UPDATEVALUES

Table with field name, type, and position. In this table, you get the values of the field back. Type: RSSELREAD.

[!] Exceptions

NO_HIGH_FIELD

HIGH field of the select option is not specified.

233 RS_SUBMIT_INFO

This FM gives information about the mode of the current SUBMIT during the processing of the selection screen. The field of P_SUBMIT_INFO is X based on the mode of the selection screen. For example, if the selection screen is called from the normal mode, field MODE_NORML is set to X. Mode information is required when you want to set your own GUI status of the selection screen and call FM RS_SET_SELSCREEN_STATUS (FM 231) or FM RS_EXTERNAL_SELSCREEN_STATUS (FM 226).

Example: Before you can remove a function from the standard GUI status, you first need to call this FM to get the submit mode information of the selection screen; that is, normal mode, variant maintenance mode, or job scheduling mode.

[↑] Export

P_SUBMIT_INFO

Information according to RSSUBINFO about SUBMIT. This parameter have the following fields:

- MODE_NORML: Normal SUBMIT.
- MODE_VARI: Variant maintenance selection screen.
- MODE_JOB: SUBMIT VIA JOB. Background job scheduling selection screen is processed.
- VIA_SELSCR: Program was started with `SUBMIT ... VIA SELECTION-SCREEN`, that is, with the selection screen processed in the foreground. Type: RSSUBINFO.

234 RS_SUPPORT_SELECTIONS

This FM assigns the values stored in the variant to the parameter and select options of the report. The report must be loaded to load the variant.

Example: Whenever you execute the program on the selection screen you've created, you want to load the default values to the selection screen fields. First, call FM RS_VARIANT_EXISTS (FM 242), which gets the name of the default variant of that report. Next, pass the variant name received from RS_VARIANT_EXISTS to this FM to load the values to the selection screen fields.

[↓] Import

REPORT

Report name. Type: RSVAR-REPORT.

VARIANT

Variant name. Type: RSVAR-VARIANT.

[!] Exceptions

VARIANT_NOT_EXISTENT

Layout does not exist.

VARIANT_OBSOLETE

Obsolete variant.

235 RS_TOOL_ACCESS

This FM is used to find the where-used list of the particular object entered in the import parameter and calling the ABAP Workbench for the same.

Example: You want to create a program where you give the name of an object in the selection screen parameter and get the where-used list of that object when you execute the report. Pass the value as CROSSREF in the OPERATION parameter, pass relevant values to other parameters, and call this FM. You'll get the screen popup of the where-used list. On execution, you'll get the list of places where that object is used.

[↓] Import

OPERATION

Operation that can be performed. Possible operations:

- ADD: Creating.
- SHOW: Viewing.
- EDIT: Changing.
- COPY: Copying.
- RENAME: Renaming.
- DELETE: Deleting.
- MOVE: Moving.

- TRANSPOR: Adding to correction.
- DOCU: Documenting.
- CROSSREF: Cross reference for where-used list.
- PRINT: Printing.

This is a mandatory parameter.

OBJECT_NAME

Name of the object.

OBJECT_TYPE

Transport object type (R3TR/LIMU) or workbench object type (according to type group SWBM).

ENCLOSING_OBJECT

Program name must be passed if the object is a subobject. Function group name must be passed for FMs.

POSITION

Position of object. The default value is [Space].

DEVCLASS

Development class for the object. Type: TADIR-DEVCLASS.

INCLUDE

Include name. Type: TRDIR-NAME.

VERSION

Version of the object. Possible values:

- A: Active.
- I: Inactive.
- [Space]: Corresponds to work list.

The default value is [Space]. Type: CHAR1.

MONITOR_ACTIVATION

If this parameter is set to X, it activates the monitor. The default value is X.

WB_MANAGER

Interface reference to the workbench manager. This interface has methods of navigation back, navigation forward, callback for tools for PBO event and PAI event, and so on. Type: IF_WB_MANAGER.

IN_NEW_WINDOW

If this parameter has value X, the workbench is started in a new window. If [Space], the workbench is started in the same window. Type: SEU_BOOL.

WITH_OBJECTLIST

Additionally start object list. Type: SEU_BOOL

WITH_WORKLIST

Additionally show worklist. Type: SEU_BOOL.

[↑] Export

NEW_NAME

New program name (after rename).

WB_TODO_REQUEST

Returns the transport request class objects. Type: CL_WB_REQUEST.

[Δ] Changing

P_REQUEST

Request/task number for transport. Type: TRKORR.

Table

OBJLIST

Object list table. Type: RSEUAP.

[!] Exceptions

NOT_EXECUTED

Application function was not carried out.

INVALID_OBJECT_TYPE

Invalid object type.

236 RS_TREE_CONSTRUCT

This FM builds a new hierarchy or inserts a subtree. The resulting hierarchy can then be further processed and displayed by FM RS_TREE_LIST_DISPLAY (FM 237).

Example: You want to display the ALV tree hierarchy. First, you need to construct the hierarchy and insert its subtree by passing the parameters and calling this FM.

[↓] Import

INSERT_ID

Node at which you want to insert. The default value is 000000. Type: SNODE-ID.

RELATIONSHIP

Relationship. The default value is [Space].

LOG

Internal log entry is created if this parameter has the value X. Type: AS4FLAG.

Table

NODETAB

Structure information The two fields, NODETAB-NAME (i.e., node name) and NODETAB-TLEVEL (i.e., hierarchical level of the node), must be passed in this table. The root node must always be level 1. Type: SNODETEXT.

[!] Exceptions

TREE_FAILURE

Unable to generate hierarchy.

ID_NOT_FOUND

ID of node is not found.

WRONG_RELATIONSHIP

Relationship is wrong.

237 RS_TREE_LIST_DISPLAY

This FM displays the hierarchy. This FM also manipulates the hierarchies and graphs in which unused graphical links are displayed in a different color as link nodes. Before calling this FM, the hierarchy must exist. You can first construct the hierarchy by calling FM RS_TREE_CONSTRUCT (FM 236) and then displaying the tree with this FM.

Example: You need to display the tree. You can do so by passing the parameters in this FM and executing the FM.

[↓] Import

CALLBACK_PROGRAM

Program containing the EXIT routines. This program should either be the main program name or a function group name. Type: SY-REPID.

CALLBACK_USER_COMMAND

Name of the form routine for handling USER COMMAND. The form should have the following call:

```
FORM user_command
tables  node STRUCTURE seucomm
    using   command
            changing exit
            list_refresh.
```

CALLBACK_TEXT_DISPLAY

EXIT routine for node display. A subroutine of the form

```
FORM text_display
TABLES nodes STRUCTURE snodetext.
```

is called (if specified) each time the hierarchy is rebuilt.

CALLBACK_MOREINFO_DISPLAY

EXIT routine to node supplementary information. A subroutine with the following form:

```
FORM moreinfo_display
TABLES moreinfo STRUCTURE streeattr
USING node LIKE streenode.
```

CALLBACK_COLOR_DISPLAY

EXIT routine for color legend. The function code TRCL displays a color key with the various colors used in the function group.

Using the EXIT routine

```
form COLOR_DISPLAY tables COLORTAB
structure SEUCOLOR
```

the caller can add user-defined colors using the exit routine CALLBACK_TEXT_DISPLAY.

CALLBACK_TOP_OF_PAGE

EXIT routine for the TOP-OF-PAGE display.

CALLBACK_GUI_STATUS

EXIT routine for setting a GUI status.

CALLBACK_CONTEXT_MENU

EXIT routine for constructing the context menu. A subroutine with the form

```
FORM callback_context_menu
USING node LIKE snodetext
menu TYPE REF TO cl_ctmenu.
```

is called when the user requests a context menu (right-click).

STATUS

Status display parameter. It can have the following values:

- OWN: The caller has set its own status.
- STANDARD: The standard status of the function group is used.
- IMPLICIT: Implicit decision as to the status.

If the parameter CALLBACK_USER_COMMAND isn't set, the FM sets the STANDARD status. If the parameter is set, the status of the caller remains set (as with OWN).

The default value is IMPLICIT.

CHECK_DUPLICATE_NAME

Test for duplicate names. When you rename (TRRN) or create (TRAD) a node, the system checks that it's unique. Possible values:

- 0: No check.
- 1: Name must be unique within the entire hierarchy.
- 2: Name must be unique within the same level of the hierarchy (nodes with the same parent).
- 3: If a node with this name already exists, the system creates a node to it. Reference nodes are displayed in a different color (parameter COLOR_OF_LINK). You can't append further nodes to reference nodes.

The default value is 1.

COLOR_OF_NODE

Node color. Standard display for a node. You can override it in the exit routine CALLBACK_TEXT_DISPLAY. The default value is 4.

COLOR_OF_MARK

Color of a selected node. The default value is 3.

COLOR_OF_LINK

Color of a link node. This parameter is only effective if the parameter CHECK_DUPLICATE_NAME has the value 3. The default value is 1.

COLOR_OF_MATCH

Color of a found node. The default value is 5.

LOWER_CASE_SENSITIVE

X if case-sensitive. The default value is [Space].

MODIFICATION_LOG

X if the modification log is required. The default value is [Space].

NODE_LENGTH

Output length of a node. You can use this to set the general node length. The default value is 30.

TEXT_LENGTH

Length of text block 1. The default value is 75.

TEXT_LENGTH1

Length of text block 2. The default value is 0.

TEXT_LENGTH2

Length of text block 3. The default value is 0.

RETURN_MARKED_SUBTREE

Controls the nodes that should be passed with the exit routine CALLBACK_USER_COMMAND. Possible values:

- [Space]: Only the roots of selected subtrees are passed.

- X: All selected nodes of the hierarchy are passed

The default value is [Space].

SCREEN_START_COLUMN

Column where the list dialog box begins. The default value is 0.

SCREEN_START_LINE

Line where the list dialog box begins. The default value is 0.

SCREEN_END_COLUMN

Column where the list dialog box ends. The default value is 0.

SCREEN_END_LINE

Line where the list dialog box ends. The default value is 0.

SUPPRESS_NODE_OUTPUT

X suppresses the node display. Sometimes, only the long text of a node is relevant and not the node itself (e.g., if the node only contains key information). If this parameter is set to X, only the long text is displayed. The parameter NODE_LENGTH and the color and length details in the exit routine CALLBACK_TEXT_DISPLAY then refer to this text. The default value is [Space].

LAYOUT_MODE

Tree display normal/compressed. The default value is [Space].

USE_CONTROL

Display as a list or tree control. The default value is STREE_USE_LIST.

[↑] Export

F15

Function key [F15] is pressed.

238 RS_VARIANT_CATALOG

This FM returns the list of all report variants in a dialog box. The name of the variant selected from the list is returned in SEL_VARIANT.

Example: You've created a selection screen as a subscreen and want to display all of the variants of the selection screen on clicking the button. You can pass the parameters and call this FM. You'll get the list of variants created for that selection screen. After you select the variant, get the details of that variant by passing the parameter in FM RS_VARIANT_VALUES_TECH_DATA (FM 243) to get the selection screen fields values for that variant. Next, you need to copy the values you received from this FM to the selection screen parameters. The values of the variant are given on the selection screen.

[↓] Import

REPORT

Name of report for which the variant is to be changed. This is a mandatory parameter. Type: RSVAR-REPORT.

NEW_TITLE

Title, if desired. The default value is [Space].

DYNNR

Screen number for which the variant is required. Type: SY-DYNNR.

INTERNAL_CALL

If this parameter is set to X, the external call is processed, and the new environment is built. The default value is [Space]. Type: C.

MASKED

If this parameter has value X, the variant is masked. The default value is X. Type: C.

VARIANT

Name of the variant to be changed. The default value is [Space]. Type: RSVAR-VARIANT.

POP_UP

If this parameter is X, a popup dialog screen appears that asks for the variant name, environment, created by, changed by, and other details for filtering the variant values. The default value is [Space]. Type: C.

[↑] Export

SEL_VARIANT

Name of variant chosen in the list. Type: RSVAR-VARIANT.

SEL_VARIANT_TEXT

Text of the variant chosen from the list. Type: RSVAR-VTEXT.

Table

BELONGING_DYNNR

Belonging screen number and screen type. Type: RSDYNNR.

[!] Exceptions

NO_REPORT

Program type other than 1.

REPORT_NOT_EXISTENT

Report does not exist.

REPORT_NOT_SUPPLIED

No report name specified.

NO_VARIANTS

No variant exists.

NO_VARIANT_SELECTED

No variant selected.

VARIANT_NOT_EXISTENT

Layout does not exist.

239 RS_VARIANT_CHANGE

This FM is used to change the parameters, select options, and other properties of a variant.

Example: You've created the variant. Now you want to change the already-existing variant. You can do so by passing the parameters in this FM and calling this FM.

[↓] Import

REPORT

Report name. This is a mandatory parameter. Type: RSVAR-REPORT.

VARIANT

Variant name. This is a mandatory parameter. Type: RSVAR-VARIANT.

VALUE_OR_ATTR

This parameter differentiates between an external call (V) or one called on the initial screen (A). The default value is V. Type: RSVAR-FLAG1.

[↑] Export

VARIANT

Name of the variant to be changed. Type: RSVAR-VARIANT.

[!] Exceptions

NOT_AUTHORIZED

No authorization.

NOT_EXECUTED

Variant not changed.

NO_REPORT

Program type other than 1.

REPORT_NOT_EXISTENT

Report does not exist.

REPORT_NOT_SUPPLIED

No report name specified.

VARIANT_LOCKED

Variant locked.

VARIANT_NOT_EXISTENT

Layout does not exist.

VARIANT_PROTECTED

Variant protected.

240 RS_VARIANT_CONTENTS

This FM returns the value for the existing variant. This FM also checks whether any selection screen parameter or select options of the variant have changed. If so, then an error message is returned.

Example: You've created a program and want the variant values used in the selection screen to be displayed in the output. You can get the variant values by passing the parameters and calling this FM.

[↓] Import

REPORT

Report name. This is a mandatory parameter. Type: RSVAR-REPORT.

VARIANT

Variant name. This is a mandatory parameter. Type: RSVAR-VARIANT.

MOVE_OR_WRITE

This parameter determines in which format the values are output. This is mainly used for DATE fields. The values are W (external format), M (internal format). For example, MOVE_OR_WRITE = W for the DATE field depends on the user profile and may be dd.mm.yyyy. If MOVE_OR_WRITE = M, the DATE field is returned in the form yyyymmdd.

NO_IMPORT

If this parameter is set to X, import isn't allowed. The default value is [Space].

EXECUTE_DIRECT

This parameter should not be used.

[↑] Export

SP

Search help values returned. Type: SYLDB_SP.

Tables

L_PARAMS

Table that contains variant parameters. Type: VANZ.

L_PARAMS_NONV

Table that contains invisible parameters. Type: VANZ.

L_SELOP

Table that contains select options. Type: VANZ.

L_SELOP_NONV

Table that contains invisible select options. Type: VANZ.

VALUTAB

Table that contains the detail values of the parameter and select options such as low value, high value, and so on. Type: RSPARAMS.

VALUTABL

Generic structure for parameters and select options (132 characters long). Type: RSPARAMSL.

OBJECTS

Table that contains all of the selection screen objects such as parameter and select options, as well as details such as length, if the parameter is a radio button or checkbox, and so on. Type: VANZ.

FREE_SELECTIONS_DESC

Table with dynamic selection description. Type: RSDYNBREPI.

FREE_SELECTIONS_VALUE

Table with dynamic selection value. Type: RSSELDYN.

[!] Exceptions

VARIANT_NON_EXISTENT

Variant does not exist.

VARIANT_OBSOLETE

Report does not exist.

241 RS_VARIANT_DELETE

This FM is used to delete the variant. In a dialog box, you can choose whether to delete the variant in all clients or only in the current client.

Example: You've created the variant and now want to change it. You can do so by passing the parameters and calling this FM.

[↓] Import

REPORT

Report name. This is a mandatory parameter. Type: RSVAR-REPORT.

VARIANT

Variant name. This is a mandatory parameter. Type: RSVAR-VARIANT.

FLAG_CONFIRMSCREEN

If this parameter has the value X, you're not asked whether you really want to delete the variant for confirmation.

FLAG_DELALLCLIENT

If this parameter is set to X, the DELETE VARIANT popup for deleting in all clients or only in the current client isn't displayed, and variants are deleted only in the current client.

[↑] Export

VARIANT

Name of variant to be changed. Type: RSVAR-VARIANT.

[!] Exceptions

NOT_AUTHORIZED

No authorization.

NOT_EXECUTED

Variant not changed.

NO_REPORT

Program type other than 1.

REPORT_NOT_EXISTENT

Report does not exist.

REPORT_NOT_SUPPLIED

No report name specified.

VARIANT_LOCKED

Variant locked.

VARIANT_NOT_EXISTENT

Layout does not exist.

NO_CORR_INSERT

Object is not added to the correction.

VARIANT_PROTECTED

Variant protected.

---

242 RS_VARIANT_EXISTS

This FM checks whether the variant exists for a given report.

Example: You have a variant field in the selection screen. You've given the variant a name, and you need to validate it and check whether the variant exists. You can fill in the parameters and call this FM.

[↓] Import

REPORT

Report name. This is a mandatory parameter. Type: RSVAR-REPORT.

VARIANT

Variant name. This is a mandatory parameter. Type: RSVAR-VARIANT.

[↑] Export

R_C

Return code (= 0 if the variant exists). Type: SY-SUBRC.

[!] Exceptions

NOT_AUTHORIZED

No authorization.

NO_REPORT

Report name = [Space].

REPORT_NOT_EXISTENT

Report does not exist.

REPORT_NOT_SUPPLIED

Report is not type 1.

243 RS_VARIANT_VALUES_TECH_DATA

This FM provides the technical values to the selection screen fields, that is, select options and parameters. This FM returns the technical name of the select options and parameters, values, and description in the VARID format.

Example: You've created a selection screen as a subscreen and want to display the detail of the variant. After you select the variant, pass the parameter in this FM to get the selection screen fields values for that variant.

[↓] Import

REPORT

Report name. Type: RSVAR-REPORT.

VARIANT

Variant name. Type: RSVAR-VARIANT.

SEL_TEXT

If this parameter is set to X, additional tables with variant text are generated.

MOVE_OR_WRITE

This parameter determines in which format the values are output. This is mainly used for DATE fields. The values are W (external format), M (internal format). For example, MOVE_OR_WRITE = W for the DATE field depends on the user profile and may be dd.mm.yyyy. If MOVE_OR_WRITE = M, the DATE field is returned in the form yyyymmdd.

SORTED

Order as on the selection screen.

EXECUTE_DIRECT

This parameter should not be used.

[↑] Export

TECHN_DATA

Description of the variant in VARID format. Type: VARID.

Tables

VARIANT_VALUES

Selection screen field values. Type: RSPARAMS.

VARIANT_TEXT

Selection screen field texts. Type: RSVSELTEXT.

[!] Exceptions

VARIANT_NON_EXISTENT

Variant does not exist.

VARIANT_OBSOLETE

Variant is obsolete.

---

244 RZL_SUBMIT

This FM submits the program; that is, this FM performs similar operations as a `SUBMIT` statement.

Example: You have a button on the output screen, which when clicked, calls another report in the background. You can pass the report name in this FM and can call this FM to submit the report.

[↓] Import

REPID

Report name. Type: SY-REPID.

---

245 SAPGUI_PROGRESS_INDICATOR

This FM displays the SAP progress indictor in the current window.

Example: You've created a program that you know will take a lot of time to display so you want to display the progress indicator. You can pass the parameters and call this FM, which displays the SAP progress indicator.

[↓] Import

PERCENTAGE

Size of bar (0 <= PERCENTAGE <= 100).

TEXT

Text to be displayed.

---

246 SCROLLING_IN_TABLE

This FM supports scrolling in table structures according to the SAP ERP standards. Scrolling in internal tables is performed using an index and an internal table. This FM doesn't perform scrolling directly but determines the index of the first table line to be output. You can define the substructure within a table structure. A substructure is part of a table structure that allows scrolling.

The following scrolling modes are supported:

- Scrolling with overlap (OVERLAPPING = X)
- Scrolling without overlap (OVERLAPPING = [Space])
- Fill last page (LAST_PAGE_FULL = X)
- Fill last page according to page (LAST_PAGE_FULL = [Space])

Example: You have a table control in your screen. There are a few lines that are visible in the table control area to the user. This value is stored in SY-LOOPC. After all of the lines of the table control are filled up with the values and you add more records, you want the table control to automatically scroll and show the added entry as well. This functionality can be achieved by passing the relevant parameters and calling this FM. You'll see that whenever you add new records after the table control visible area is full, the table control is automatically scrolled.

[↓] Import

ENTRY_ACT

Starting index. The number of the current first output line is passed in this parameter. The current page can also be passed to the parameter PAGE_ACT. By default, both PAGE_ACT and ENTRY_ACT have the value 0, but at least one of the two parameters must be specified and not be zero. Values between ENTRY_FROM to ENTRY_TO are valid. Type: SY-TABIX.

ENTRY_FROM

Start of the table substructure. This parameter only has to be specified when substructures are used. It specifies the index at which the sub-structure begins. Values smaller than ENTRY_TO are valid. No check is made. The default value is 1. Type: SY-TABIX.

ENTRY_TO

End of table substructure. This parameter determines the index at which the tabular structure or sub-structure ends. Values greater than ENTRY_FROM are valid. No check is made. This is a mandatory parameter. Type: SY-TABIX.

LAST_PAGE_FULL

Indicator of whether the last page is full. X means the last page is full, and [Space] means the last page is in the paging sequence. The default value is X.

LOOPS

Number of entries per page, that is, the number of records that are to be output per page. This is a mandatory parameter. Type: SY-TABIX.

OK_CODE

Function code for the scroll operation. Possible values:

- P: First page.
- P—: Previous page.
- P+–: Next page.

- P++-: Last page.
- -: Paging by absolute page number. The absolute destination page number must be passed in the PAGE_GO parameter.

The default value is [Space].

OVERLAPPING

Indicator for overlapping scrolling. Overlapping means that the last entry on the current page is the first entry on the new page. Value X signifies overlapping, and value [Space] signifies without overlapping. The default value is [Space].

PAGE_ACT

Start page. This parameter passes the page number of the current page.

Alternatively, the current first output line can also be passed to the parameter ENTRY_ACT. At least one of these two parameters must be specified and not be zero. The default value is 0. Type: SY-TABIX.

PAGE_GO

Target page. This parameter contains the absolute destination page number. Alternatively, the paging operation can also be controlled via the parameter OK_CODE. The default value is 0. Type: SY-TABIX.

[↑] Export

ENTRIES_SUM

Total number of entries of the table (sub)structure. The parameter contains the number of entries of the table structure or substructure the user is scrolling through. Type: SY-TABIX.

ENTRY_NEW

Index of the first entry to be output. Type: SY-TABIX.

PAGES_SUM

Total number of pages. Type: SY-TABIX.

PAGE_NEW

The parameter contains the scroll destination page number. Type: SY-TABIX.

[!] Exceptions

NO_ENTRY_OR_PAGE_ACT

No start index or page specified.

NO_ENTRY_TO

End of table substructure is zero.

NO_OK_CODE_OR_PAGE_GO

No function code or target page specified.

247 SELECT_OPTIONS_RESTRICT

This FM simplifies the use of select options on the selection screen by restricting possible selection options and signs. You can call this FM only once in any program containing a selection screen. This means if the selection field appears on two different selection screens, you can't set different restrictions for it each time.

Example: You have a requirement where you want your selection screen select option SOPT1 to accept only equal-to values and not between, not equal to, contains, and so on. You can fill the parameters and restriction table and call this FM. Your select option now can accept only equal-to values.

[↓] Import

PROGRAM

Program name (default SY-CPROG or SY-LDBPG). Type: SY-REPID.

RESTRICTION

Description of restrictions. Type: SSCR_RESTRICT.

DB

X: Call using logical database. Type: RSSCR-DB.

[!] Exceptions

TOO_LATE

Call is too late. If you call this FM after PBO of the selection screen, this exception is raised.

REPEATED

Multiple calls of this FM on the program will raise this exception.

SELOPT_WITHOUT_OPTIONS

One of the select options contains no valid options.

SELOPT_WITHOUT_SIGNS

One of the select options does not have a valid sign.

INVALID_SIGN

Invalid sign.

EMPTY_OPTION_LIST

One of the options lists is empty.

INVALID_KIND

One line has a KIND value unequal to A, B, or S.

REPEATED_KIND_A

More than one line has KIND = A.

248 VIEW_SET_PF_STATUS

This FM sets the PF status for table maintenance, that is, the SAPLSVIM function group interface.

Example: You want to disable the DELETE button from the standard table maintenance generator screen of your table. You can pass the parameter and can call this FM to delete the function and set the PF status.

[↓] Import

STATUS

PF status to be set. This is a mandatory parameter. Type: SY-PFKEY.

TITLE

Space is set status, otherwise, set title. The default value is Space.

TITLE_TEXT

Text variable of the title. The default value is Space.

OBJIMP

If this parameter is set to X, the object can be adjusted; that is, the table maintenance generator for the object occurs. Type: OBJIMP.

Table

EXCL_CUA_FUNCT

Table of the UI functions not to be activated. Type: VIMEXCLFUN.

---

249 VRM_DELETE_VALUES

This FM clears or deletes the values set for the ID, which were added previously by the FM VRM_SET_VALUES (FM 250).

Example: You created a dropdown list on selection screen parameter. If you want to remove the already existing list form the dropdown list, you can do so by passing the parameter and calling this FM.

[↓] Import

ID

Name of the value set. This parameter contains the screen's field name for which you need to generate the dropdown. This is a mandatory parameter. Type: VRM_ID.

ID_CONTAINS_PROGNAME

If this parameter is X, the current program name isn't appended to the ID parameter.

[!] Exception

ID_NOT_FOUND

The value set (ID) does not exist.

250 VRM_SET_VALUES

This FM stores the values table VALUES under the name of ID. If you want to use the value table as the contents of a dropdown list box or dropdown combo box, you must call the name of the input field with value set. This FM should be called at PBO or at the `SELECTION SCREEN OUTPUT` event.

Example: You have a requirement where you want to show airline routes in a dropdown list. You can do so by passing the relevant ID as a parameter name for which you want to generate the dropdown and values to be displayed in the dropdown in the values table and executing this FM. You'll receive the list of values as a dropdown.

[↓] Import

ID

Name of the value set. This parameter contains the screen's field name for which you need to generate the dropdown. This is a mandatory parameter. Type: VRM_ID.

VALUES

Values returned in the dropdown. This is a mandatory parameter. Type: VRM_VALUES.

[!] Exception

ID_ILLEGAL_NAME

ID name is illegal.

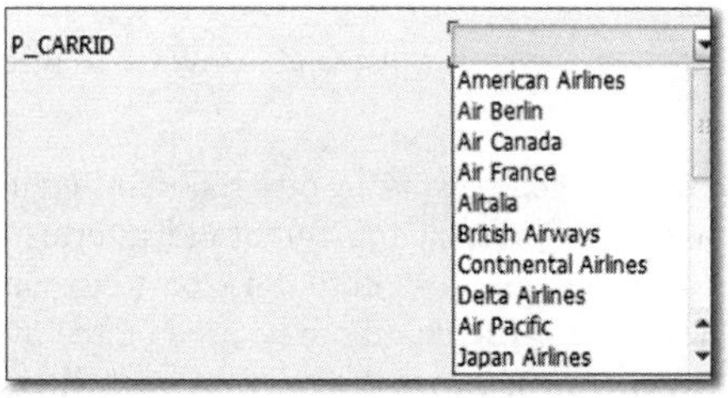

Dropdown List Generated from This FM

251 WRITE_LIST

This FM displays the saved list object; that is, the list saved in the container LISTOBJECT is displayed as a list. The list is displayed by the WRITE statement.

Example: You saved the output in the list format and want to display it in the output of your report program. You can call this FM for the same.

[↓] Import

WRITE_ONLY

If this parameter is set to X, the list is shown or displayed in another list. The default value is X. Type: C.

Table

LISTOBJECT

This table receives the list object. Type: ABAPLIST.

[!] Exception

EMPTY_LIST

The list object transported is empty.

252 WWW_LIST_TO_HTML

This FM saves the list and converts it into HTML format.

Example: You're working on a report program and want to display the output in list format. You want your list output to be saved in HTML format so that it can be used on the Internet. You can pass the relevant parameters in this FM for the same.

[↓] Import

LIST_INDEX

List index. Type: SY-LSIND.

Table

HTML

Contains the HTML page. Type: W3HTML.

Part V
Scripts and Forms

This part focuses on function modules (FMs) that are related to form printing. SAP's first form technology was the SAPscript technology. In release 4.6, SAP introduced Smart Forms, which eased the creation of form output by allowing modeling of logic and output using graphical tools. In release 6.40, SAP introduced Interactive Forms by Adobe, where the graphical editor was much easier to use, many more functionalities were available, and output was rendered exactly the same as designed in the editor (PDF).

This part contains the FMs related to working with SAPscript, Smart Forms, and Interactive Forms by Adobe. SAPscript also provides text modules that are used for writing and storing long texts. This part also contains the FMs related to SAPscript text modules.

253 CHECK_TEXT_AUTHORITY

This FM checks whether a user has the proper authorization to display or change the specified text module. This FM can only be used with texts of type TEXT; that is, texts that can be processed using Transaction SO10 (Standard Texts).

Example: An end user needs to use the standard text in your program. Before the user can do this, his authorization for this task needs to be checked. When this FM is called through the logic, the end user receives an exception if he doesn't have the authorization.

[↓] Import

ACTIVITY

This parameter specifies the functions in the application program with which the text module should be processed.

- SHOW: Text module should only be displayed.
- EDIT: Text module should be changed.

This is a mandatory parameter.

ID

Text ID of text to be checked. The text ID defines a further subdivision of the texts for a text object. Valid values for the text ID are defined in table TTXID. This is a mandatory parameter. Type: THEAD-TDID.

LANGUAGE

Text modules can exist in several languages, so this parameter checks for the language. This is a mandatory parameter. Type: THEAD-TDSPRAS.

NAME

Text name of a text to be checked. It can be 70 characters long. All characters except * and & are valid. This is a mandatory parameter. Type: THEAD-TDNAME.

OBJECT

Text objects of text to be checked. The text object is part of the text key and assigns the text module to the specified application object. Valid text is defined in table TTXOB. The default object is TEXT. Type: THEAD-TDOBJECT.

[!] Exceptions

NO_AUTHORITY

The user has no authorization to edit or display the text object.

---

254 CHECK_TEXT_ID

This FM checks whether the text ID is defined in the table TTXID for the text object.

Example: You want to read the SAP text in your program. You have a screen field to enter the text ID, but you need to first check if the text ID entered is valid. You can do that by passing the parameter in this FM. You'll get the exception if the text ID isn't valid.

[↓] Import

ID

Text ID to be checked. For details of this parameter, refer to FM CREATE_TEXT (FM 263). Type: TTXID-TDID.

OBJECT

Object for text ID. For details of this parameter refer to FM CREATE_TEXT (FM 263). Type: TTXID-TDOBJECT.

[↑] Export

ID_INFO

Information on text ID received from table TTXID. Type: TTXID.

[!] Exception

ID

Text ID invalid.

---

255 CHECK_TEXT_LANGUAGE

This FM checks if the language of the text is defined in table T002.

Example: You want to read an SAP text in your program only if the language is maintained in language key table T002. You need to first check for the language by passing the parameter in this FM. You'll get the exception if the text ID isn't valid.

[↓] Import

LANGUAGE

Language to be checked. Type: SYST-LANGU.

[!] Exception

LANGUAGE

Invalid language.

---

256 CHECK_TEXT_NAME

This FM checks if the name of the text module contains valid characters only.

Example: You can use this FM internally in other FMs such as READ_TEXT (FM 283) to check the validation of the text name entered by the user.

[↓] Import

NAME

Text name to be checked. Type: THEAD-TDNAME.

[!] Exception

NAME

Invalid text name.

---

257 CHECK_TEXT_OBJECT

This FM checks if the text object is valid and defined in table TTXOB.

Example: You want to read the SAP text in your program, but you need to first check if the text object is valid. You can do that by passing the parameter in this FM. You'll get the exception if the text object isn't valid.

[↓] Import

OBJECT

Text object to be checked. Type: TTXOB-TDOBJECT.

[↑] Export

OBJECT_INFO

Information on text object returned from table TTXOB. Type: TTXOB.

[!] Exception

OBJECT

Invalid text object.

258 CLOSE_FORM

The FM closes the form opened using FM OPEN_FORM (FM 278). The system executes the terminating processing steps for the last opened form. You must use this FM to close form printing; otherwise, no output appears at the printer or on the screen.

Example: To print an SAPscript, first you need to open the form from FM OPEN_FORM (FM 278), perform the write operations from FM WRITE_FORM (FM 297), and finally when you're done with the printing, close the form with this FM. To print the SAPscript form you've created, you need to use these three FMs in sequence.

[↑] Export

RESULT

This parameter returns the output, such as anything modified by the user in the print dialog option values. Type: ITCPP.

RDI_RESULT

This parameter structure contains information about the generated document if printed on the RAW data interface. For example, the parameter-RDI_RESULT-DOCNUM contains the document number in output mode. Type: RDIRESULT.

Tables

OTFDATA

OTF table. If the field TDGETOTF of import parameter OPTIONS in FM OPEN_FORM or PRINT_FORM is set to X, this means no spool or print output will be performed, but the formatted output will be returned in the format of OTF in this Table. Type: ITCOO.

[!] Exceptions

UNOPENED

Form printing has not been opened.

BAD_PAGEFORMAT_FOR_PRINT

Page format is not valid.

SEND_ERROR

Error when sending for print.

SPOOL_ERROR

Error at transfer to spool.

CODEPAGE

Code page not allowed.

---

259 COMMIT_TEXT

This FM transfers text creation text(s) from text memory and updates it into log files. All of the text modules with saving defined in the update task are first held in the text memory. This FM updates and commits the text modules present in the text memory. This FM generates a `CALL FUNCTION...IN UPDATE TASK` for every text module entered in the text memory corresponding to the action to be executed; that is, DELETE, CREATE, or CHANGE.

Example: When you make changes to the script text modules—that is, if you write anything in the text box in your screen—you need to save the text from FM SAVE_TEXT (FM 287) and then commit the saved text in the database by calling this FM.

[↓] Import

OBJECT

Text object. The default value is *. Type: THEAD-TDOBJECT.

NAME

Text name. The default value is *. Type: THEAD-TDNAME.

ID

Text ID. The default value is *. Type: THEAD-TDID.

LANGUAGE

Text language. The default value is *. Type: THEAD-TDSPRAS.

SAVEMODE_DIRECT

Save texts immediately indicator. The default value is [Space].

KEEP

Indicator to continue to keep texts in text memory. The default value is [Space].

LOCAL_CAT

Catalog stored locally and not in memory. The default value is [Space].

[↑] Export

COMMIT_COUNT

This parameter defines the number of text modules transferred to the update program. Type: SY-INDEX.

Tables

T_OBJECT

You can use table parameter T_OBJECT to pass a ranges table of text objects that are to be saved either through an update or directly. The text keys of the table can also be generic. In this case, all texts are selected whose text objects have the same characters up to the first occurrence of *. Parameter T_OBJECT generally overrides the text key field OBJECT that can be additionally specified as an import parameter in the function call of COMMIT_TEXT. Type: STXDROBJ.

T_NAME

You can use table parameter T_NAME to pass a ranges table of text names that are to be saved either through an update or directly. The text keys of the table can also be generic. In this case, all texts are selected whose text names have the same characters up to the first occurrence of *.

Parameter T_NAME is a table of reference type STXDRNAME. Parameter T_NAME generally overrides the text key field NAME that can be additionally specified as an import parameter in the function call of COMMIT_TEXT. Type: STXDRNAME.

T_ID

You can use table parameter T_ID to pass a ranges table of text IDs that are to be saved either through an update or directly. The text keys of the table can also be generic. In this case, all texts are selected whose text IDs are the same up to the first occurrence of *.

T_ID is a table of reference type STXDRID. Parameter T_ID generally overrides the ID text key field that can be additionally specified as an import parameter in the function call of COMMIT_TEXT. Type: STXDRID.

T_LANGUAGE

You can use table parameter T_LANGUAGE to pass a ranges table of text languages that are to be saved either through an update or directly. The text keys of the table can also be generic. In this case, all texts are selected whose text languages have the same characters up to the first occurrence of *.

Parameter T_LANGUAGE is a table of reference type STXDRLANG. Parameter T_LANGUAGE generally overrides the text key field LANGUAGE that can be additionally specified as an import parameter in the function call of COMMIT_TEXT. Type: STXDRLANG.

260 CONTROL_FORM

This FM is used to give the control commands to the layout set, such as page breaks or new window at runtime.

Example: When printing the SAPscript form, if you want the text to come in as a new page, you can pass the `PAGE BREAK` command in the parameter and call this FM.

[↓] Import

COMMAND

This parameter is an indication of the SAPscript commands executed in ITF format. The command is / without the paragraph attribute specified.

[!] Exceptions

UNOPENED

The current form function could not be performed because the form output has not been initialized with OPEN_FORM.

UNSTARTED

No form opened. The form processing was indeed initialized with OPEN_FORM without specifying a form, but there was no form still open with START_FORM.

261 CONVERT_OTF_MEMORY

This FM reads the memory buffer and converts the OTF into text format. This FM is used when the parameter DEVICE has the value OTF_MEM of FM OPEN_FORM (FM 278). In this case, SAPscript is outputted to the buffer instead of the spool. The value is returned in the table LINES.

Example: You don't want to send the SAPscript output for printing, but you want the output to be used within your program to process other logic. For this, you need to first set the DEVICE parameter with value OTF_MEM of FM OPEN_FORM (FM 278), and then you need to execute this FM to get the OTF text in the form of an internal table. You can use this internal table for your further logic.

[↓] Import

FORMAT

Target format for the LINES table. Value ASCII is supported, which is a default value.

MAX_LINEWIDTH

For ASCII format: Number of lines in LINES-TDLINE. The maximum line length in this parameter must not be smaller than 2 or greater than 132. The default value is 132.

[↑] Export

BIN_FILESIZE

For binary format: Number of bytes in table LINES.

Table

LINES

Output table with target format. The table contains all text lines belonging to a text module. The line structure is defined in table TLINE. The line contents are stored in ITF format. Type: TLINE.

[!] Exceptions

MEMORY_EMPTY

Memory for OTF data is empty.

ERR_MAX_LINEWIDTH

Line width must be between 2 and 132.

ERR_FORMAT

Format not supported.

ERR_CONV_NOT_POSSIBLE

Conversion not possible/supported.

262 COPY_TEXTS

This FM copies the texts that exist in the text catalog; that is, text memory can be copied. If the text is already stored on the database, the system doesn't overwrite these texts.

Example: You have one text and a COPY button that will copy the text. When you click that button, you need to copy the text that has been entered. You can do so by this passing the relevant parameters and calling this FM. Because the text is already stored in the database, it won't overwrite the same text again if the user clicks the COPY button again.

[↓] Import

SAVEMODE_DIRECT

This indicator checks whether the text needs to be copied immediately. If this parameter is X, text modules are saved immediately; otherwise, the save occurs according to the text object in the update task if this parameter has value [Space]. The default value is [Space].

INSERT

The parameter informs the word processor that it's a new text module. Normally, SAPscript reads the text file to check whether the text module is new. If this is already known in the calling program, the test read can be suppressed by the parameter and the performance therefore improved. If this parameter has the value X, it means the texts are new. The default value is [Space].

LOCAL_CAT

Catalog stored locally and not in memory. Default value is [Space].

[↑] Export

ERROR

This parameter indicates whether an error occurred during copying.

Possible values:

- [Space]: No error.
- X: Error copying.

This error flag only indicates that an error occurred copying a text. Further information can be obtained from the field SUBRC in table TEXTS.

Table

TEXTS

This table contains the texts to be copied. A copy function can be called per table line. The complete text key of the source text and target text must be specified. Generic entries aren't allowed.

The return code of the copy operation is returned for each text in the field SUBRC. Type: ITCTC.

263 CREATE_TEXT

This FM creates the SAPscript text modules for long text in your module pool screens. This FM internally calls the FMs INIT_TEXT (FM 277) and SAVE_TEXT (FM 287). After the texts are created and saved, an entry can be found in table STXH.

Example: You have a list box for a notes section on the screen. You need to create an SAP text module to allow an end user to write in and save the long text, which you can do with this FM.

[↓] Import

FID

Text ID of the text to be created. This is a mandatory parameter. The text ID defines a further subdivision of the texts for a text object. Valid values for the text ID are defined in table TTXID. Type: THEAD-TDID.

FLANGUAGE

Language of the text to be created. This is a mandatory parameter. Type: THEAD-TDSPRAS.

FNAME

Name of the text to be created. This is a mandatory parameter. It can be 70 characters long. All characters except * and & are valid. Type: THEAD-TDNAME.

FOBJECT

Object of the text to be created. This is a mandatory parameter. Text object is part of text key and assigns the text module to the specified application object. Valid text is defined in table TTXOB. The default object is TEXT. Type: THEAD-TDOBJECT.

SAVE_DIRECT

If this parameter is X, text modules are saved immediately; otherwise, save occurs according to the text object in the update task if this parameter has value [Space]. The default value is X.

FFORMAT

Format in which the text is to be created. The default value is *. Type: TLINE-TDFORMAT.

Tables

FLINES

Lines of the text to be created. The line structure is defined in table TLINE. The line contents are stored in ITF format. Type: TLINE.

[!] Exceptions

NO_INIT

Some exception occurred from FM INIT_TEXT (FM 277); that is, creating text did not work.

NO_SAVE

Some exception occurred from FM SAVE_TEXT (FM 287); that is, saving text did not work.

264 DELETE_TEXT

This FM is used to delete one or more text modules in the text file. Generic (wild card) entries are also possible in the fields ID and LANGUAGE.

Example: You want to delete the text created because you no longer require it. You can do so by passing the parameters in this FM and calling this FM.

[↓] Import

CLIENT

Client. Type: SY-MANDT.

ID

Text ID of text to be deleted. Type: THEAD-TDID.

LANGUAGE

Language of text to be deleted. Type: THEAD-TDSPRAS.

NAME

Name of text to be deleted. Type: THEAD-TDNAME.

OBJECT

Text object of text to be deleted. Type: THEAD-TDOBJECT.

SAVEMODE_DIRECT

Delete text immediately.

TEXTMEMORY_ONLY

You can specify that the delete function is only to apply to the text memory. If the parameter is assigned the value X, the text itself isn't deleted but only its entry in the text memory. All changes made in a text during a transaction can therefore be canceled. This parameter is only suitable for texts that are held in text memory.

LOCAL_CAT

Text catalog local.

[!] Exception

NOT_FOUND

Text not found.

---

265 DELETE_TEXT_FROM_CATALOG

This FM is used to delete text from a catalog; that is, from text memory.

Example: You don't want to actually delete the text but you want to delete the entry from the text memory so that you can store other text in memory. You can fill in the parameters and call this FM to delete text from text memory.

[↓] Import

OBJECT

Text object of text to be deleted. Type: THEAD-TDOBJECT.

NAME

Text name of text to be deleted. Type: THEAD-TDNAME.

ID

Text ID of text to be deleted. Type: THEAD-TDID.

LANGUAGE

Text language of text to be deleted. Type: THEAD-TDSPRAS.

LOCAL_CAT

Text catalog local. The default value is [Space].

Tables

T_OBJECT

You can use table parameter T_OBJECT to pass a ranges table of text objects that are to be saved either through an update or directly. The text keys of the table can also be generic. In this case, all texts are selected whose text objects have the same characters up to the first occurrence of *. Parameter T_OBJECT generally overrides the text key field OBJECT that can be additionally specified as an import parameter in the function call of COMMIT_TEXT. Type: STXDROBJ.

T_NAME

You can use table parameter T_NAME to pass a ranges table of text names that are to be saved either through an update or directly. The text keys of the table can also be generic. In this case, all texts are selected whose text names have the same characters up to the first occurrence of *.

Parameter T_NAME is a table of reference type STXDRNAME. Parameter T_NAME generally overrides the text key field NAME that can be additionally specified as an import parameter in the function call of COMMIT_TEXT. Type: STXDRNAME.

T_ID

You can use table parameter T_ID to pass a ranges table of text IDs that are to be saved either through an update or directly. The text keys of the table can also be generic. In this case, all texts are selected whose text IDs are the same up to the first occurrence of *.

T_ID is a table of reference type STXDRID. Parameter T_ID generally overrides the ID text key field that can be additionally specified as an import parameter in the function call of COMMIT_TEXT. Type: STXDRID.

T_LANGUAGE

You can use table parameter T_LANGUAGE to pass a ranges table of text languages that are to be saved either through an update or directly. The text keys of the table can also be generic. In this case, all texts are selected whose text languages have the same characters up to the first occurrence of *.

Parameter T_LANGUAGE is a table of reference type STXDRLANG. Parameter T_LANGUAGE generally overrides the text key field LANGUAGE that can be additionally specified as an import parameter in the function call of COMMIT_TEXT. Type: STXDRLANG.

[!] Exception

NOT_FOUND

Text not found.

---

266 EDIT_TEXT

This FM is used to call the full screen SAPscript editor. There, you're able to edit the lines of text with the functions available. The editor interface is set according to the interface assigned to the text object. When you exit the editor, the FM SAVE_TEXT (FM 287) is generally called implicitly with the function SAVE. This call can be deactivated via the parameter SAVE.

Example: To add the long text description for a field with the SAPscript editor, call this FM.

[↓] Import

DISPLAY

This parameter determines if the text editor is called in display mode or change mode. Value X determines display mode and value [Space] determines change mode.

EDITOR_TITLE

Text title specified on the title bar of the text editor window. Type: TTXIT-TDTEXT.

HEADER

Text header contains information that describes a text module such as short text, created by user, changed by user, and so on. The structure of the header is defined in table THEAD. Type: THEAD.

PAGE

Page. The default value is [Space]. Type: ITCTH-TDPAGE.

WINDOW

Window. The default value is [Space]. Type: ITCTH-TDWINDOW.

SAVE

The parameter determines whether the function SAVE_TEXT is called by the text editor when the text is saved. If saving by the editor isn't required, the application program must call the FM SAVE_TEXT (FM 287) itself. In both cases, the changed text lines are returned to the application program.

- X: Call SAVE_TEXT in this FM.
- [Space]: Do not call SAVE_TEXT.

LINE_EDITOR

This parameter determines which type of editor should be called: line editor or PC editor. Value X calls the line editor, and [Space] calls the PC editor. Type: C.

CONTROL

This parameter can be used to set certain attributes of the SAPscript editor. The attributes are described in the field of the structure ITCED. Type: ITCED.

PROGRAM

To replace program symbols, SAPscript has to know in which active program the work areas for those values that are to be adopted lie. SAPscript searches by default for the TABLES statements for program symbols in the program located at the root of the current call hierarchy. If this isn't what you want, you can also enter the name of the program with the TABLES work area whose program symbol values should replace the current ones. If no program name is entered, the system searches for field values in the first program called (SY-CPROG).

With this FM, program texts replacement takes place exclusively in print preview and not in the editor itself (TEXT • PRINT • PRINT PREVIEW). Type: SY-REPID.

LOCAL_CAT

Text catalog local.

[↑] Export

FUNCTION

The parameter returns the processing status of the text module for the current editor call.

- [Space]: No action.
- I: Text module was inserted.
- U: Text module was updated.
- D: Text module was deleted.

NEWHEADER

The parameter returns the text header with the fields changed according to the action carried out. Type: THEAD.

RESULT

The SAPscript editor returns status information to the calling program via this parameter. This information is described by the following fields of the structure ITCER:

RESULT-FUNCTION: The parameter returns the processing status of the text module for the current editor call:

- [Space]: No action
- I: Text module was inserted.
- U: Text module was updated.
- D: Text module was deleted.

RESULT-USEREXIT: The parameter indicates which function code was used by a user to leave the text editor:

- E: Exit: TEXT EXIT.
- B: Back: GOTO BACK.
- C: Cancel: EDIT CANCEL.
- P: Previous: PREVIOUS TEXT (only in interface TA).
- N: Next: NEXT TEXT (only in interface TA).

Type: ITCER.

Table

LINES

The table contains all text lines belonging to a text module. The line structure is defined in table TLINE. The line contents are stored in ITF format. Type: TLINE.

[!] Exceptions

ID

The text ID specified in the parameter ID or text header—field TDID—does not exist in table TTXID. It must be defined there with the text module object.

LANGUAGE

The language specified in the parameter LANGUAGE or in the text header—field TDSPRAS—is not defined in table T002.

LINESIZE

The line width specified in the text header—field TDLINESIZE—is invalid. The values for line widths must be greater than 0 and less than or equal to 132.

NAME

The text name specified in the parameter NAME or in the text header—field TDNAME—is blank or contains the invalid characters * or &.

OBJECT

The text object specified in the parameter OBJECT or in the text header—field TDOBJECT—is not contained in table TTXOB.

TEXTFORMAT

The text lines passed to the FM are available in a format that cannot be processed in the current environment. The format is stored in the text header in the field TDTEXTTYPE.

COMMUNICATION

The text format specified in the field TDTEXTTYPE of the text header requires that an external word processor is called. An error occurred when this word processor was called.

---

267 EDIT_TEXT_INLINE

This FM is useful if you want to edit text in the inline editor.

Example: You need to edit the first line of text on your module screen. Use this FM to open the editor, and then edit the text.

[↓] Import

DISPLAY

This parameter determines if the text editor is called in display mode or change mode. Value X determines display mode, and value [Space] determines change mode.

EDITOR_TITLE

Text title specified on the title bar of the text editor window. Type: TTXIT-TDTEXT.

HEADER

Text header contains information that describes a text module such as short text, created by user, changed by user, and so on. The structure of the header is defined in table THEAD. Type: THEAD.

INLINE_COUNT

The parameter specifies how many text lines are transferred from the beginning of the text line table LINES to table INLINES.

SAVE

The parameter determines whether the function SAVE_TEXT is called by the text editor when the text is saved. If saving by the editor isn't required, the application program must call the FM SAVE_TEXT (FM 287) itself. In both cases, the changed text lines are returned to the application program.

- X: Call SAVE_TEXT in this FM.
- [Space]: Do not call SAVE_TEXT.

The default value is X.

TEXTSCREEN

The parameter determines whether the text editor should be called after adjustment of the INLINE lines.

- [Space]: Text editor isn't called.
- X: Text editor is called.

LINE_EDITOR

This parameter determines which type of editor should be called: line editor or PC editor. Value X calls the line editor, and [Space] calls the PC editor. Type: C.

CONTROL

This parameter can be used to set certain attributes of the SAPscript editor. The attributes are described in the field of the structure ITCED. Type: ITCED.

SUPPRESS_END_SPACE

No [Space] at end in full line.

PROGRAM

To replace program symbols, SAPscript has to know in which active program the work areas for those values that are to be adopted lie. SAPscript searches by default for the TABLES statements for program symbols in the program located at the root of the current call hierarchy. If this isn't what you want, you can also enter the name of the program with the TABLES work area whose program symbol values should replace the current ones. If no program name is entered, the system searches for field values in the first program called (SY-CPROG).

With FM EDIT_TEXT (FM 266), program texts replacement takes place exclusively in print preview and not in the editor itself (Text • Print • Print Preview).

Type: SY-REPID.

LOCAL_CAT

Text catalog local.

[↑] Export

FUNCTION

The parameter returns the processing status of the text module for the current editor call.

- [Space]: No action.
- I: Text module was inserted.

- U: Text module was updated.
- D: Text module was deleted.

NEWHEADER

The parameter returns the text header with the fields changed according to the action carried out. Type: THEAD.

RESULT

The SAPscript editor returns status information to the calling program via this parameter. This information is described by the following fields of the structure ITCER:

RESULT-FUNCTION: The parameter returns the processing status of the text module for the current editor call.

- [Space]: No action.
- I: Text module was inserted.
- U: Text module was updated.
- D: Text module was deleted.

RESULT-USEREXIT: The parameter indicates which function code was used by a user to leave the text editor:

- E: Exit: TEXT EXIT.
- B: Back: GOTO BACK.
- C: Cancel: EDIT CANCEL.
- P: Previous: PREVIOUS TEXT (only in interface TA).
- N: Next: NEXT TEXT (only in interface TA).

Type: ITCER.

Tables

INLINES

This table contains the first lines of the read text module. The number is defined in the parameter INLINE_COUNT. The line structure of this table is described in TLINE. The line contents are stored in ITF format. Type: TLINE.

LINES

The table contains all text lines belonging to a text module. The line structure is defined in table TLINE. The line contents are stored in ITF format. Type: TLINE.

[!] Exceptions

ID

The text ID specified in the parameter ID or text header—field TDID—does not exist in table TTXID. It must be defined there with the text module object.

LANGUAGE

The language specified in the parameter LANGUAGE or in the text header—field TDSPRAS—is not defined in table T002.

LINESIZE

The line width specified in the text header—field TDLINESIZE—is invalid. The values for line widths must be greater than 0 and less than or equal to 132.

NAME

The text name specified in the parameter NAME or in the text header—field TDNAME—is blank or contains the invalid characters * or &.

OBJECT

The text object specified in the parameter OBJECT or in the text header—field TDOBJECT—is not contained in table TTXOB.

268 END_FORM

This FM closes the form that you've opened for printing.

Example: You've opened a form for printing by using FM START_FORM (FM 292). After you're done with the printing, you need to end the form by calling this FM.

[↑] Export

RESULT

This parameter returns the output, such as anything modified by the user in the print dialog option values. This parameter is of type ITCPP. Type: ITCPP.

[!] Exceptions

UNOPENED

Form printing has not been opened.

BAD_PAGEFORMAT_FOR_PRINT

Page format is not valid.

SPOOL_ERROR

Error at transfer to spool.

CODEPAGE

Code page not allowed.

269 EXCHANGE_ITF

This FM is interactive. You use it to convert the character and paragraph names of a style or form into those of another style or form. You specify the desired new style or form allocation in a dialog box. To specify which original names match which new names, use Transaction SE74 (Format Conversion) and column FORMAT CONVERSION IN SAPSCRIPT TEXTS.

Example: You've created the script with style STYLE1. To change the format of STYLE1 to STYLE2, call this FM and specify the style in the parameters.

[↓] Import

HEADER

Text header of original text. Type: THEAD.

FORMATWIDTH

Line width of resulting text. If an entry exists in table TTXOB for the text object, parameter FORMATWIDTH is overwritten with this value. Command lines are never broken up. The default value is 72. Type: ITCTK-TDLINESIZE.

[↑] Export

NEWHEADER

Text header of resulting text. Type: THEAD.

Table

ITF_LINES

Text that should be converted is imported into ITF_LINES, edited, and then returned to the same table. Type: TLINE.

[!] Exception

CANCELLED

Conversion was either cancelled by the user, or no conversion took place because the target format was the same as the source format.

---

270 EXPORT_TEXT

This FM converts an ITF text into the desired format FORMAT_TYPE and then saves it in a local file. FORMAT_TYPE can have values ITF, RTF, or ASCII. When exporting the formats ASCII and RTF, the system resolves text includes and control structures (/: IF, /: ELSE, /: ENDIF, /: CASE, /: WHEN, /: ENDCASE) and replaces text, standard, and system symbols.

Example: You need to update the long text of a transaction, and the information that you need to upload should be in a file with the extension *.rtf* (Rich Text Format). With this FM, you can convert the data from ITF to RTF format and then save the RTF to the file.

[↓] Import

CODEPAGE

This parameter specifies the character set numbers from the spool administration. The default value is [Space]. Type: TCP02-CPCODEPAGE.

FILE

Name of the file you want to create on the presentation server (if necessary, with leading directory name). If the directory doesn't exist or if the system can't open the file for writing, it triggers the respective exceptions. Type: RLGRAP-FILENAME.

FORMATWIDTH

Line width of the target file, which can be between 40 and 132 characters. The default value is 72. Type: ITCTK-TDLINESIZE.

FORMAT_TYPE

Format of the target file (ITF, RTF, or ASCII). The default value is RTF. Type: C.

HEADER

The text header contains a description of a text module, such as short text, creator, and so on. Type: THEAD.

SSHEET

Name of a template (**.dfv* or **.dot*). To trigger a format conversion, you must enter the symbolic name for which you maintained a conversion to the respective SAPscript style or form in Transaction SE74. In addition, you must set the parameter WITH_TAB to X. The default value is [Space]. Type: C.

WITH_TAB

If WITH_TAB is X and the parameter SSHEET is filled, the system converts character and paragraph formats. Source and target formats can be a SAPscript style, a SAPscript form, or a Microsoft Word template. Using Transaction SE74 (Format Conversion), you must specify which character or paragraph names or Word templates match. The default value is [Space]. Type: TDBOOL.

TAB_SUBSTITUTE

The parameter is used only for a conversion from ITF to ASCII. You define the substitution value in a character field of length 5. The first character defines the type of substitution. The following values are possible:

- C: Substitutes the tab with a character string of up to four characters. TAB_SUBSTITUTE = 'C<<>>' "before tab,,after tab " in SAPscript becomes "before tab <<>>after tab " in the ASCII file.
- X: Substitutes the tab with one or two binary characters. TAB_SUBSTITUTE = 'X09_ _' in SAPscript becomes the hexadecimal value 09.
- _: A blank substitutes the tab with 1 to 99 blanks. TAB_SUBSTITUTE = '_ 8_ _ _'. "before tab, after tab " in SAPscript becomes "before tab_ _ _ _ _ _ _ _after tab " in the ASCII file.

The default value is 'X09_ _'. Type: C.

REPLACE_SYMBOLS

If this parameter has value X, it evaluates symbols. Type: TDBOOL.

REPLACE_SAPCHARS

If this parameter has value X, it evaluates SAP characters. Type: TDBOOL.

Table

ITF_LINES

Text that should be converted is imported into ITF_LINES, edited, and then returned to the same table. Type: TLINE.

[!] Exceptions

DOWNLOAD_ERROR

Other error in downloading the file.

FILE_OPEN_ERROR

File cannot be opened.

FILE_WRITE_ERROR

Cannot write to file.

271 FORMAT_TEXTLINES

This FM formats the length of a line of SAPscript.

Example: To print your lines in output in 60 spaces, pass in the parameter and call this FM.

[↓] Import

CURSOR_COLUMN

Column where cursor is placed. The default value is 0.

CURSOR_LINE

Line where cursor is placed. The default value is 0.

ENDLINE

End line of formatting. The default value is 99999.

FORMATWIDTH

Width to be formatted. The default value is 72.

LINEWIDTH

Width of line to be formatted. The default value is 132.

STARTLINE

Start line for formatting. The default value is 1.

LANGUAGE

Language. The default value is SU-LANGU. Type: SY-LANGU.

[↑] Export

NEW_CURSOR_COLUMN

Returned new cursor column.

NEW_CURSOR_LINE

Returned new cursor line.

Table

Lines

The table contains all text lines belonging to a text module to be formatted. Type: TLINE.

[!] Exceptions

BOUND_ERROR

This exception is not raised in the FM.

272 FP_FUNCTION_MODULE_NAME

This FM returns the name of the FM that processes the logic written in the Interactive Forms by Adobe and executes the Adobe Interactive Form. This FM is required because the generated FM name is unique only in one system. If you don't generate the FM name by calling this FM, the generated FM might not exist in the other system and results into dump.

Example: You've written an Adobe Interactive Form. To execute it, you need the name of the FM. You can fill in the parameters and call this FM. On execution, the FM name is returned. When you execute this generated FM, you'll execute the Adobe Interactive Form in your program.

[↓] Import

I_NAME

Name of the Adobe Interactive Form. Type: FPNAME.

[↑] Export

E_FUNCNAME

Returned FM name that executes the Adobe Interactive Form. Type: FUNCNAME.

E_INTERFACE_TYPE

Interface Type. FPINTERFACETYPE.

EV_FUNCNAME_INBOUND

Name of the FM for converting XML data to DDIC data. Type: FUNCNAME.

[!] Exceptions

CX_FP_API_REPOSITORY

Exception API (repository).

CX_FP_API_USAGE

Exception API (use).

CX_FP_API_INTERNAL

Exception API (internal).

273 FP_GET_LAST_ADS_ERRSTR

This FM returns the text about the cause of the error.

Example: You want to display a PDF-based print form. However, the system displays the runtime errors FPRUNX001, FPRUNX002, or FPRUNX004 when the generated FM is called. To get the error text, you can execute this FM after the generated FM. Refer to FM FP_JOB_OPEN (FM 275).

[↑] Export

E_ADSERRSTR

Returned text that indicates the cause of the error. Type: STRING.

---

274 FP_JOB_CLOSE

This FM closes the PDF output job.

Example: You've printed the Adobe Interactive Form from FM FP_FUNCTION_MODULE_NAME (FM 272). You now need to close the Adobe Interactive Form processing by calling this FM. Refer to FM FP_JOB_OPEN (FM 275).

[↑] Export

E_RESULT

This parameter returns the form processing. Type: SFPJOBOUTPUT.

[!] Exceptions

USAGE_ERROR

Incorrect call.

SYSTEM_ERROR

System error.

INTERNAL_ERROR

Internal error.

---

275 FP_JOB_OPEN

This FM determines the output settings of the form and opens the spool job. You'll get the popup where you can set various print parameters such as PRINT IMMEDIATELY, PRINT OR PRINT PREVIEW, and so on.

Example: Whenever you want to execute the Adobe Interactive Form, first you need to execute this FM, which determines the output settings. You then use the following FMs to complete the process:

- FP_FUNCTION_MODULE_NAME (FM 272)
- FNAME (dynamically created)
- FP_GET_LAST_ADS_ERRSTR (FM 273)
- FP_CLOSE_JOB (FM 274)

[Δ] Changing

IE_OUTPUTPARAMS

Form processing output parameters. Type: SFPOUTPUTPARAMS.

[!] Exceptions

CANCEL

Abort by user.

USAGE_ERROR

Incorrect call.

SYSTEM_ERROR

System error.

INTERNAL_ERROR

Internal error.

---

276 IMPORT_TEXT

This FM uploads the file from a presentation server to an ITF-based format on the FORMAT_TYPE. FORMAT_TYPE can have values ITF, RTF, or ASCII.

Example: You have to update the long text of a transaction, and the data is in a file with extension *.rtf* (Rich Text Format) in your presentation server. You need to convert the file into the ITF format so that you can use it in SAPscript. You can fill in the parameters and call this FM. This FM converts the data from RTF format to ITF format and also uploads the file to SAP ERP.

[↓] Import

CODEPAGE

This parameter specifies the character set numbers from the spool administration. The default value is [Space]. Type: TCP02-CPCODEPAGE.

FILE

Name of the file you want to create on the presentation server (if necessary, with leading directory name). If the directory doesn't exist or if the system can't open the file for writing, it triggers the respective exceptions. Type: RLGRAP-FILENAME.

FORMAT_TYPE

Format of the target file (ITF, RTF, or ASCII). The default value is RTF. Type: C.

HEADER

The text header contains a description of a text module, such as short text, creator, and so on. Type: THEAD.

SSHEET

Name of a template (**.dfv* or **.dot*). To trigger a format conversion, you must enter the symbolic name for which you maintained a conversion to the respective SAPscript style or form in Transaction SE74. In addition, you must set the parameter WITH_TAB to X. The default value is [Space]. Type: C.

WITH_TAB

If WITH_TAB is X, and the parameter SSHEET is filled, the system converts character and paragraph formats. Source and target formats can be a SAPscript style, a SAPscript form, or a Microsoft Word template. Using Transaction SE74 (Format Conversion), you must specify which character or paragraph names or Word templates match. The default value is [Space]. Type: TDBOOL.

WORD_LANGU

Only for RTF conversion: language of the word processing program that created the RTF file.

The default value is SY-LANGU. Type: SPRAS.

MASK_BRACKETS

If this parameter is X, it masks special characters. Type: TDBOOL.

[↑] Export

NEWHEADER

Text header of resulting text. Type: THEAD.

Table

ITF_LINES

Text that should be converted is imported into ITF_LINES, edited, and then returned to the same table. Type: TLINE.

[!] Exceptions

FILE_OPEN_ERROR

File cannot be opened.

FILE_READ_ERROR

File cannot be read (completely).

UPLOAD_ERROR

Other error in uploading the file.

NO_CONTENTS

No contents to be uploaded.

277 INIT_TEXT

This FM initializes or clears the work area text header and line table, which is required for text module processing. These work areas must be initialized when a text is created, when the FM READ_TEXT (FM 283) or FM READ_TEXT_INLINE (FM 284) are terminated with an exception, or when the text is no longer required.

The text module must be specified fully with entries in the fields OBJECT, NAME, ID, and LANGUAGE. Generic entries aren't valid.

When INIT_TEXT is called up, the header is provided with initial values from table TTXOB. The line table is empty.

Example: You have read one text with READ_TEXT (FM 283) and now you want to read another text. You first need to initialize or clear the existing work area for the text header and line table. You can do so by calling this FM. A cleared work area and lines table are returned.

[↓] Import

ID

Text ID of text to be checked. The text ID defines a further subdivision of the texts for a text object. Valid values for the text ID are defined in table TTXID. This is a mandatory parameter. Type: THEAD-TDID.

LANGUAGE

Text module can exist in several languages so this parameter checks for the language. This is a mandatory parameter. Type: THEAD-TDSPRAS.

NAME

Text name of text to be checked. It can be 70 characters long. All characters except * and & are valid. This is a mandatory parameter. Type: THEAD-TDNAME.

OBJECT

Text objects of text to be checked. Text objects are part of text key and assign the text module to the specified application object. Valid text is defined in table TTXOB. The default object is TEXT. Type: THEAD-TDOBJECT.

[↑] Export

HEADER

Initialized work area for text header is returned. Type: THEAD.

Table

Lines

Initialized table for lines is returned. Type: TLINE.

[!] Exceptions

ID

The text ID specified in the parameter ID or text header—field TDID—does not exist in table TTXID. It must be defined there with the text module object.

LANGUAGE

The language specified in the parameter LANGUAGE or in the text header—field TDSPRAS—is not defined in table T002.

NAME

The text name specified in the parameter NAME or in the text header, field TDNAME, is blank or contains the invalid characters * or &.

OBJECT

The text object specified in the parameter OBJECT or in the text header—field TDOBJECT—is not contained in table TTXOB.

278 OPEN_FORM

This FM opens form printing. You must call this FM before you can use any other form function (WRITE_FORM (FM 297), START_FORM (FM 292), CONTROL_FORM (FM 260), etc.). Note that in this FM, you don't need to specify a form name. If you omit the name, you must use the FM START_FORM (FM 292) to open a form before starting the output.

You must end form printing by using the FM CLOSE_FORM (FM 258). Otherwise, the system doesn't print or display anything.

Within a program, you can use several OPEN_FORM/CLOSE_FORM pairs. This allows you to write output to several different spool requests from within one program.

Example: You want to print the SAPscript form FORM1. You first need to open the form by passing the parameters to this FM and calling this FM. After your form is opened, you can perform other processing in it.

[↓] Import

APPLICATION

Application code for interface. For the device type SCREEN, the system displays the text formatting on the screen. This requires an interface in which the different menu entries are defined. The same applies if the user chooses to display a print view on the screen for the other device types. The default value is TX. Type: TTXOB-TDAPPL.

ARCHIVE_INDEX

Archive index for the layout set to be archived. This information (DARA line) is moved to the layout set in the archive. You can then access the layout set in the archive directly. Type: TOA_DARA.

ARCHIVE_PARAMS

Archive parameters. The system interprets the settings passed in this parameter when archiving the output. The archive parameters have the ABAP Data Dictionary structure ARC_PARAMS. Type: ARC_PARAMS.

DEVICE

SAPscript can format a text for output on different device types. Enter the desired device type here.

Possible values:

- PRINTER: Print output
- TELEX: Telex output
- TELEFAX: Telefax output
- ABAP: Screen output as ABAP list (interface of the calling program).
- SCREEN: Screen output as ABAP list (interface controlled by SAPscript, can be set with parameter APPLICATION).

The user can display output formatted for PRINTER, TELEX, or TELEFAX as print view on the screen. From within the program, you can set the field TDPREVIEW (structure ITCPO) in the parameter OPTIONS, or the user can call the function on the print control screen.

The fields TDSENDTIME and TDSENDDATE designed for fax output (structure ITCPO) will be used for future enhancements; they aren't used at present.

The default value is PRINTER. Type: C.

DIALOG

Use parameter DIALOG to determine whether to display a dialog box before printing, in which the user can set several spool parameters for print formatting. Possible values:

- [Space]: Display no print parameter screen.
- X: Display print parameter screen.

The default value is X. Type: C.

FORM

The parameter contains the name of the form that is used for output. The characters * and & aren't allowed in the form name. Type: C.

LANGUAGE

This parameter specifies which language version of the form is to be used. If the form isn't available for the language defined in this parameter, the form in the original language is used. Type: THEAD-TDSPRAS.

OPTIONS

Use parameter OPTIONS to set several options for print formatting. The parameter has the structure ITCPO. The user can change some of the defined settings on the print control screen. Type: ITCPO.

MAIL_SENDER

Sender object ID of MAIL. Type: SWOTOBJID.

MAIL_RECIPIENT

Recipient object ID of MAIL. Type: SWOTOBJID.

MAIL_APPL_OBJECT

This parameter defines the application object number for the Device = MAIL. Type: SWOTOBJID.

RAW_DATA_INTERFACE

Indicator for Raw Data Interface (RDI). Type: C.

SPONUMIV

Number range number. Type: NRIV-NRRANGENR.

[↑] Export

LANGUAGE

The LANGUAGE parameter indicates which language version of the form is used for the output. Type: Form language.

NEW_ARCHIVE_PARAMS

The parameter contains results of the archiving process, including the archive parameters the user changed on the print control screen. The parameter has the ABAP Data Dictionary structure ARC_PARAMS. Type: ARC_PARAMS.

RESULT

The parameter contains results of the print formatting process. By comparing the corresponding fields of parameter OPTIONS with those of parameter RESULT, you can determine whether the user made changes to any settings on the print control screen. Type: ITCPP.

[!] Exceptions

CANCELED

When starting SAPscript print formatting, the system displayed a selection screen, on which the user can enter settings for the output, such as:

- Printer name
- Information on the cover page
- Page selection
- Number of copies

The user did not call any subsequent actions allowed on this screen but canceled the output formatting instead. The FM was ended without further action. No form is open for output anymore.

DEVICE

Invalid device type.

FORM

The parameter FORM contains the name of a form that the system could not find.

Possible reasons:

- The form doesn't exist.
- There is no active version of this form.

SAPscript first searches for the form in the current client and in the specified language. If the form doesn't exist there, it tries the original language of the form. If the form is still not found, it searches in client 0, first in the specified language and then in the original form language.

OPTIONS

The parameter OPTIONS contains invalid values for the formatting options. Possible errors:

- The output device specified in field TDDEST does not exist.
- The field TDPAGESLCT for selecting the pages to be printed contains invalid characters.

UNCLOSED

The system was told to open a new form even though an old form is still active. The old form must be closed first (CLOSE_FORM or END_FORM).

MAIL_OPTIONS

Error in mail options.

ARCHIVE_ERROR

Some error while archiving.

INVALID_FAX_NUMBER

Fax number entered is not valid.

MORE_PARAMS_NEEDED_IN_BATCH

More parameters are required.

SPOOL_ERROR

Error at transfer to spool.

CODEPAGE

Code page not allowed.

CANCELED

User cancelled print dialog.

279 PRINT_TEXT

Use this FM to prepare the text module specified in the parameters HEADER and LINES for an output device and to output it. The system takes the required style and form specifications from the fields TDSTYLE or TDFORM of the text header.

Example: To print the text TEXT1 that you've created or to see the text on the output screen of the program, call this FM.

[↓] Import

APPLICATION

Application code for the interface. This parameter can be used to select the interface be used for formatting for the screen (parameter DEVICE = SCREEN). Possible values:

- DS: Interface for hypertext environment.
- TO: Interface for Microsoft Office environment.
- TD: Interface for documentation environment.
- TX: Interface for all other texts.

The default value is TX. Type: TTXOB-TDAPPL.

ARCHIVE_INDEX

Archive index for the layout set to be archived. This information is moved to the layout set in the archive. You can then access the layout set in the archive directly. The default value is [Space]. Type: TOA_DARA.

ARCHIVE_PARAMS

Archive parameters that are evaluated when the output is archived (field TDARMOD in the options parameter). Type: ARC_PARAMS.

DEVICE

The device type for which the formatting process is to be performed. Possible values:

- PRINTER: Formatting for the specified printer.
- TELEFAX: Formatting for fax output.
- MAIL: Formatting for the Business Communication Interface (BCI).
- SCREEN: Formatting for the screen as an ABAP list. The UI is determined by the APPLICATION parameter.
- TELEX: Formatting for telex output.
- ABAP: Formatting for the screen as an APAB list. The UI is controlled by the calling program.

The print output for PRINTER, TELEX, and TELEFAX can also be displayed on the screen in print preview (field TDPREVIEW in option parameters). You can also choose the preview on the print dialog.

DIALOG

The parameter DIALOG can be used to control whether a dialog box appears before output to the printer. The user can set various spool parameters for the output in this window. The default value is X; that is, the dialog should come.

HEADER

Text header of text to be output. Type: THEAD.

OPTIONS

Different variants can be defined for the output via the parameter OPTIONS. The structure is described in table ITCPO. Some of the fields contained in this structure can be changed by the user via the print dialog. Type: ITCPO.

[↑] Export

NEW_ARCHIVE_PARAMS

The parameter contains results of the archiving, for example, the archive parameters changed by the user in the print dialog. Type: ARC_PARAMS.

RESULT

The parameter contains results of the output, for example, option values changed by the user in the print dialog. Type: ITCPP.

Tables

LINES

Lines of text to be output. Type: TLINE.

OTFDATA

OTF table (optional). If the field TDGETOTF of import parameter OPTIONS in FM OPEN_FORM or PRINT_FORM is set to X, this means no spool or print output, screen display, or fax/telex/teletex output will be performed, but the formatted output will be returned in the format of OTF in this table. Type: ITCOO.

[!] Exceptions

CANCELED

User terminated printing on print options dialog box.

DEVICE

Invalid device type.

FORM

The form specified in the parameter FORM does not exist, or there is no active version of this form.

OPTIONS

Invalid parameter values were passed in table ITCPO for the print options. Possible errors:

- An output device that does not exist was specified in field TDDEST.
- Invalid characters were used for the page selection in field TDPAGESLCT.

UNCLOSED

You tried to open a new form although a form was active. This form must first be closed with CLOSE_FORM or END_FORM.

UNKNOWN

Error not specified.

FORMAT

Only forms with the same page format can be used in a spool job defined by OPEN_FORM and CLOSE_FORM. The new form has a different page format to the previous forms used in this spool job.

TEXTFORMAT

The text lines passed to the FM are available in a format that cannot be processed in the current environment. The format is stored in the text header in the field TDTEXTTYPE.

COMMUNICATION

The text format specified in the field TDTEXTTYPE of the text header requires that an external word processor is called. An error occurred when this word processor was called.

BAD_PAGEFORMAT_FOR_PRINT

Page format is not valid.

280 PRINT_TEXT_ITF

This FM prints the SAPscript text passed in table LINES in ITF format, which means that it prints the text lines in the ways they appear in the lines table. The printout thus resembles the representation in the SAPscript editor.

Example: You want to see the output in the same format as in SAPscript, that is, ITF format. You can pass the parameter and call this FM to do so.

[↓] Import

HEADER

Text header of text to be output. Type: THEAD.

OPTIONS

Different variants can be defined for the output via the parameter OPTIONS. The structure is described in table ITCPO. Some of the fields contained in this structure can be changed by the user via the print dialog. Type: ITCPO.

[↑] Export

RESULT

The parameter contains results of the output, for example, option values changed by the user in the print dialog. Type: ITCPP.

Tables

LINES

Lines of text to be output. Type: TLINE.

281 READ_FORM_ELEMENTS

This FM determines all of the text elements of the table that occur in a layout set.

Example: You want to display a report that shows all of the text elements used in the layout set. You can do so by calling this FM. After you execute this FM, you'll receive the text elements defined in that form.

[↓] Import

FORM

This parameter specifies the form name that is used for output. The characters * and & aren't allowed in the form name.

LANGUAGE

Language for which the text elements are required to determine.

STATUS

Status of modification. Type: ITCTA-TDSTATUS.

Table

ELEMENTS

This table contains all of the windows defined in the form with its text elements. For each text element, the system specifies the number of text lines (editor lines) it consists of. Type: ITCWE.

[!] Exceptions

FORM

The parameter FORM contains the name of a form that could not be found. Possible reasons:

- The form does not exist.
- There is no active version of this form.

SAPscript first searches for the form in the current client and in the specified language. If the form does not exist there, it tries the original language of the form. If the form is still not found, it searches in client 0, first in the specified language and then in the original form language.

UNOPENED

The current form function could not be executed because the form output was not yet initialized using OPEN_FORM.

282 READ_FORM_LINES

This FM transfers the lines of a form element into an internal table. If you specify no form name, the system transfers the text lines of the currently open form. If you specify a form, the system uses the text lines of the active version of the form from the database.

Example: You've created a SAPscript form, and you want to read the line of its main window. You can do so by calling this FM. After you execute this FM, you'll receive the form elements in the internal table.

[↓] Import

ELEMENT

Enter the name of the form element whose text lines you want to read. If you leave the field blank, the system searches for the "nameless" element. The default value is Space.

FORM

Specify the name of the form whose element list you want to create. If you leave the field blank, the system uses the currently active form.

The default value is Space.

LANGUAGE

Form language. The default value is Space.

WINDOW

Enter the name of the window in which to find the form element. If you leave the field blank, the system searches for the form element in the main window.

The default value is MAIN.

Table

LINES

The table receives the lines of the specified element. Type: TLINE.

[!] Exceptions

ELEMENT

The parameter ELEMENT contains the name of a form element that the system could not find.

FORM

Form does not exist.

UNOPENED

Form could not be executed because the form output was no yet initialized using OPEN_FORM.

283 READ_TEXT

This FM reads the SAPscript text modules. You need to transfer all of the information of the text module to the internal work area of this FM to read the text. Text can be read from the text file or text memory.

Example: You have a list box in the notes section of the screen. Now you want to read the already created text when you enter into the program for the next time. You can pass the parameters and can call this FM to read the text.

[↓] Import

CLIENT

Client. Type: SY-MANDT.

ID

Text ID of the text to be created. This is a mandatory parameter. The text ID defines a further subdivision of the texts for a text object. Valid values for the text ID are defined in table TTXID. Type: THEAD-TDID.

LANGUAGE

Language of the text to be created. This is a mandatory parameter. Type: THEAD-TDSPRAS.

NAME

Name of the text to be created. This is a mandatory parameter. It can be 70 characters long. All characters except * and & are valid. Type: THEAD-TDNAME.

OBJECT

Object of the text to be created. This is a mandatory parameter. Text object is part of the text key and assigns the text module to the specified application object. Valid text is defined in table TTXOB. The default object is TEXT. Type: THEAD-TDOBJECT.

ARCHIVE_HANDLE

A handle is required to access an archive. If an archive is to be written to, the handle must be created via the FM ARCHIVE_OPEN_FOR_WRITE. If an archive is to be read from, the handle must be created with the FM ARCHIVE_OPEN_FOR_READ. Type: SY-TABIX.

LOCAL_CAT

Catalog stored locally and not in memory. The default value is [Space].

[↑] Export

HEADER

Text header of text read. Type: THEAD.

Table

LINES

Lines of the text to be created. The line structure is defined in table TLINE. The line contents are stored in ITF format. Type: TLINE.

[!] Exceptions

ID

The text ID specified in the parameter ID or text header—field TDID—does not exist in table TTXID. It must be defined there with the text module object.

LANGUAGE

The language specified in the parameter LANGUAGE or in the text header—field TDSPRAS—is not defined in table T002.

NAME

The text name specified in the parameter NAME or in the text header—field

TDNAME—is blank or contains the invalid characters * or &.

NOT_FOUND

The text module specified is not found.

OBJECT

The text object specified in the parameter OBJECT or in the text header—field TDOBJECT—is not contained in table TTXOB.

REFERENCE_CHECK

The text module to be read does not have any text lines but refers to the lines of another text module. This reference chain can have several levels. In the current text, the reference chain is interrupted; a text module referenced in this chain no longer exists.

WRONG_ACCESS_TO_ARCHIVE

This exception is raised if an archive is accessed with an incorrect or nonexistent archive handle or with an incorrect mode.

---

284 READ_TEXT_INLINE

This FM reads the text in the work area. To process text modules in an application program, information regarding the text modules is required in the internal work area. Text can be read from the text file or text memory. Text must be described fully by specifying OBJECT, NAME, ID, and LANGUAGE. Generic entries in these parameters aren't valid. When header information and text lines have been read successfully, they are transferred to the work areas HEADER and LINES.

Similar to what FM READ_TEXT (FM 283) does, in this FM, the number of lines defined in the parameter INLINE_COUNT is transferred from the table LINES to the table INLINES. This line table can be used to display the first text lines on a screen and change them. At PAI time, the FM EDIT_TEXT_INLINE (FM 267) should be called.

Example: You have a field called ACTIVITY that has long text. You've saved the long text description of the field by using FM SAVE_TEXT (FM 287). To see the long text for the field when you select the field, pass the parameters in this FM to read the long text and display it on screen.

[↓] Import

ID

Text ID of the text to be created. This is a mandatory parameter. The text ID defines a further subdivision of the texts for a text object. Valid values for the text ID are defined in table TTXID. Type: THEAD-TDID.

INLINE_COUNT

The parameter specifies how many text lines are transferred from the beginning of the text line table LINES to table INLINES.

LANGUAGE

Language of the text to be created. This is a mandatory parameter. Type: THEAD-TDSPRAS.

NAME

Name of the text to be created. This is a mandatory parameter. It can be 70 characters long. All characters except * and & are valid. Type: THEAD-TDNAME.

OBJECT

Object of the text to be created. This is a mandatory parameter. The text object is part of the text key and assigns the text module to the specified application object. Valid text is defined in table TTXOB. The default object is TEXT. Type: THEAD-TDOBJECT.

LOCAL_CAT

Catalog stored locally and not in memory. The default value is [Space].

[↑] Export

HEADER

Text header of text read. Type: THEAD.

Tables

INLINES

TLINE

This table contains the first lines of the read text module. The number is defined in the parameter INLINE_COUNT. The line structure of this table is described in TLINE. The line contents are stored in ITF format.

LINES

TLINE

Lines of the text to be created. The line structure is defined in table TLINE. The line contents are stored in ITF format.

[!] Exceptions

ID

The text ID specified in the parameter ID or text header—field TDID—does not exist in table TTXID. It must be defined there with the text module object.

LANGUAGE

The language specified in the parameter LANGUAGE or in the text header—field TDSPRAS— is not defined in table T002.

NAME

The text name specified in the parameter NAME or in the text header—field TDNAME—is blank or contains the invalid characters * or &.

NOT_FOUND

The text module specified is not found.

OBJECT

The text object specified in the parameter OBJECT or in the text header, field TDOBJECT, is not contained in table TTXOB.

REFERENCE_CHECK

The text module to be read does not have any text lines but refers to the lines of another text module. This reference chain can have several levels. In the current text, the reference chain is interrupted; a text module referenced in this chain no longer exists.

285 REFER_TEXT

This FM creates the SAPscript texts as a reference to the text lines in another text module. No text lines are saved; only a reference to the specified text module is created. Changes to the reference text module then affect the calling text as well. This reference chain can extend over several levels. The text module specified in the header is to reference another text module. This is specified via the parameters REF_OBJECT, REF_NAME, and REF_ID. The language of the referenced text is the same as that of the referencing text. The referenced text usually exists already in the database. The text lines of the referenced text can then be returned using the parameter READ_LINES in the LINES table. If the text exists in the text memory; that is, was not yet committed, the reference can be configured although the text lines of the referenced text can't be imported using READ_LINES. REFER_TEXT doesn't execute SAVE_TEXT. To save the reference text permanently, SAVE_TEXT must be called after REFER_TEXT.

Example: Two texts are used in two different screens of your program. Whenever one text is changed, the other text must be changed automatically. In this case, you can refer one text to the other by calling this FM. When you change one text, the other is also changed.

[↓] Import

CHECK_REFERENCE

This parameter defines whether the existence of the referenced text module should be checked.

- [Space]: No check is made.
- X: Checks whether referenced text exists.

The default value is X.

HEADER

The text header contains information that describes a text module, for example, short text, user who created the text, user who last changed it, and so on. Type: THEAD.

READ_LINES

The lines of the text to be referenced are held in the table LINES.

- [Space]: Don't read the text lines.
- X: Reads the text lines.

REF_ID

Text ID of the text module to be referenced. Valid values for the text ID are defined in table TTXID. Type: THEAD-TDID.

REF_NAME

Name of the text module to be referenced. The name can be up to 70 characters long. The structure depends on the text object used. All characters except for * and & are valid. Type: THEAD-TDNAME.

REF_OBJECT

Object name of the text to be referenced. Valid text objects are defined in the table TTXOB. Type: THEAD-TDOBJECT.

[↑] Export

NEWHEADER

The parameter returns the text header with the fields changed according to the action carried out. Type: THEAD.

Table

LINES

Lines of the text to be referred. The line structure is defined in table TLINE. The line contents are stored in ITF format. Type: TLINE.

[!] Exceptions

NOT_FOUND

Text module specified is not found.

REFERENCE_CHECK

The text module to be read does not have any text lines but refers to the lines of another text module. This reference chain can have several levels. In the current text, the reference chain is interrupted; a text module referenced in this chain no longer exists.

---

286 RENAME_TEXT

In application programs, the complete name of the master record or document is usually only fixed shortly before you use the `COMMIT WORK` command. In the meantime, a dummy name must be used to work with the related text modules. These dummy names must be replaced by the actual names before the FM COMMIT_TEXT (FM 259) is used with this FM. The texts concerned are given the name specified in the NEWNAME parameter. Only the name of a text module can be changed. Generic entries for ID and language are valid. Note that the text module doesn't perform a RENAME on the text file. If a text module in the text file is to be renamed, it must first be saved under the new text name, and the old text module must then be deleted.

Example: When you want to change the dummy name created by SAP to the name according to your standards, you can pass the parameters and call this FM.

[↓] Import

ID

Text ID of the text to be created. This is a mandatory parameter. The text ID defines a further subdivision of the texts for a text object. Valid values for the text ID are defined in table TTXID. Type: THEAD-TDID.

LANGUAGE

Language of the text to be created. This is a mandatory parameter. Type: THEAD-TDSPRAS.

NAME

Name of the text to be replaced. This is a mandatory parameter. It can be 70 characters long. All characters except * and & are valid. Type: THEAD-TDNAME.

NEWNAME

New name of text. This is a mandatory parameter. It can be 70 characters long. All characters except * and & are valid. Type: THEAD-TDNAME.

OBJECT

Object of the text to be created. This is a mandatory parameter. The text object is part of the text key and assigns the text module to the specified application object. Valid text is defined in table TTXOB. The default object is TEXT. Type: THEAD-TDOBJECT.

RENAME_REFERENCES

Also rename references in update task memory. Type: C.

LOCAL_CAT

Catalog stored locally and not in memory. The default value is [Space].

[!] Exception

NOT_FOUND

Text module specified is not found.

287 SAVE_TEXT

This FM writes a text module to the text file or text memory, depending on the specific text object. This FM is helpful to change existing texts and to create new texts. If you're sure it's a new text, then the parameter can be specified with parameter INSERT. This result is better for performance because a text read won't be performed.

Example: Whenever you create a text by using FM CREATE_TEXT (FM 263) or you change any text, you need to save the changed text. Pass the parameter and call this FM to save the text.

[↓] Import

CLIENT

Client. Type: SY-MANDT.

HEADER

The text header information to save. Type: THEAD.

INSERT

The parameter informs the word processor that it's a new text module. Normally, SAPscript reads the text file to check whether the text module is new. If this is already known in the calling program, the test read can be suppressed by the parameter and the performance therefore improved. Possible values:

- [Space]: Determines update mode.
- X: Determines text is new.

SAVEMODE_DIRECT

If this parameter is X, text modules are saved immediately; otherwise, save occurs according to the text object in the update task if this parameter has the value [Space]. The default value is X.

OWNER_SPECIFIED

When a new text is created, this parameter determines whether the information in the text header regarding the first creation is set automatically by SAPscript or whether the data passed in the header is to be used. The fields involved are TDFUSER, TDFDATE, TDFTIME, and TDFRELES in the text header. Possible values:

- [Space]: Information is set by SAPscript (default).
- X: Information is taken from the header.

If this option is activated and one of the fields listed above is initial, this field is still filled by SAPscript.

LOCAL_CAT

Catalog stored locally and not in memory. The default value is [Space].

[↑] Export

FUNCTION

The parameter returns the processing status of the text module for the current call. Possible values:

- [Space]: No action was taken.
- I: Text module was inserted.
- U: Text module was updated.
- D: Text module was deleted.

NEWHEADER

The parameter returns the text header with the fields changed according to the action carried out. Type: THEAD.

Table

LINES

Lines of the text to be saved. The line structure is defined in table TLINE. The line contents are stored in ITF format. Type: TLINE.

[!] Exceptions

ID

The text ID specified in the parameter ID or text header—field TDID—does not exist in table TTXID. It must be defined there with the text module object.

LANGUAGE

The language specified in the parameter LANGUAGE or in the text header—field TDSPRAS—is not defined in table T002.

NAME

The text name specified in the parameter NAME or in the text header—field TDNAME—is blank or contains the invalid characters * or &.

OBJECT

The text object specified in the parameter OBJECT or in the text header—field TDOBJECT—is not contained in table TTXOB.

288 SELECT_TEXT

This FM creates a table from the text headers of all text modules that match the selections specified in the fields OBJECT, NAME, ID, and LANGUAGE. The entries in these fields can also be generic. Usually, the FM searches for texts in the text file as well as in the text memory. To limit the search area, use the options TEXTMEMORY_ONLY or DATABASE_ONLY. If the option ARCHIV_HANDLE contains a value greater than 0, the system searches the archive for the specified texts.

Example: You want to display the header text module of the purchase order in the purchase order script. For that, you need to get the header text of purchase order in the table. You can pass the parameters and call this FM to get the table that contains the purchase order text. You can then display it in the script.

[↓] Import

CLIENT

Specify the client in which to search for the texts. If you omit this parameter, the system uses the current client by default. Type: SY-MANDT.

DATABASE_ONLY

Mark this field if you want to search for the text modules in the text file only. Possible values:

- [Space]: Search in text file and text memory.
- X: Search in text file only.

The default value is [Space].

TEXTMEMORY_ONLY

Mark this field if you want to search for the text modules in the text memory only. Possible values:

- [Space]: Search in text file and in text memory.
- X: Search in text memory only.

The default value is [Space].

OBJECT

Enter the name of the text object of the text you want to find. You can enter a generic value. Type: THEAD-TDOBJECT.

NAME

Enter the name of the text module you want to find. You can enter a generic value. Type: THEAD-TDNAME.

ID

Enter a text ID. You can enter a generic value. Type: THEAD-TDID.

LANGUAGE

Enter the language key of the text module you want to find. You can enter a generic value. The system then searches for the text in all languages that match the entry. Type: THEAD-TDSPRAS.

ARCHIVE_HANDLE

If you want the system to search for the text modules in the archive, you must enter a handle here. The system needs the handle to access the archive. You can use the FM ACHIVE_OPEN_FOR_READ to create the handle.

The value 0 indicates that you don't want to read the archive. The default value is 0. Type: SY-TABIX.

LOCAL_CAT

Catalog stored locally and not in memory. The default value is [Space].

[↑] Export

ENTRIES

This parameter contains the number of selected text modules. This value corresponds to the number of lines in table SELECTIONS. Type: SY-TFILL.

Tables

SELECTIONS

This table contains the text headers of all found texts that match the selection criteria specified in the call of FM SELECT_TEXT (FM 288). Type: THEAD.

T_OBJECT

You can use table parameter T_OBJECT to pass a ranges table of text objects that are to be saved either through an update or directly. The text keys of the table can also be generic. In this case, all texts are selected whose text objects have the same characters up to the first occurrence of *. Parameter T_OBJECT generally overrides the text key field OBJECT that can be additionally specified as an import parameter in the function call of COMMIT_TEXT. Type: STXDROBJ.

T_NAME

You can use table parameter T_NAME to pass a ranges table of text names that are to be saved either through an update or directly. The text keys of the table can also be generic. In this case, all texts are selected whose text names have the same characters up to the first occurrence of *.

Parameter T_NAME is a table of reference type STXDRNAME. Parameter T_NAME generally overrides the text key field NAME that can be additionally specified as an import parameter in the function call of COMMIT_TEXT. Type: STXDRNAME.

T_ID

You can use table parameter T_ID to pass a ranges table of text IDs that are to be saved either through an update or directly. The text keys of the table can also be generic. In this case, all texts are selected whose text IDs are the same up to the first occurrence of *.

T_ID is a table of reference type STXDRID. Parameter T_ID generally overrides the ID text key field that can be additionally specified as an import parameter in the function call of COMMIT_TEXT. Type: STXDRID.

T_LANGUAGE

You can use table parameter T_LANGUAGE to pass a ranges table of text languages that are to be saved either through an update or directly. The text keys of the table can also be generic. In this case, all texts are selected whose text languages have the same characters up to the first occurrence of *.

Parameter T_LANGUAGE is a table of reference type STXDRLANG. Parameter T_LANGUAGE generally overrides the text key field LANGUAGE that can be additionally specified as an import parameter in the function call of COMMIT_TEXT. Type: STXDRLANG.

I_NAME

Table is used to copy a large number of text names. Type: STEXTNAME.

[!] Exception

WRONG_ACCESS_TO_ARCHIVE

Archive handle invalid for access.

---

289 SSF_FUNCTION_MODULE_NAME

This FM returns the name of the function module, which, when executed, processes the logic written in the Smart Form and executes the Smart Form. This FM is required because the generated FM name is unique only in one system. The generated FM name has the prefix /1BCDWB/SF. If you don't generate the FM name by calling this FM, the generated FM might not exist in the other system and results into a dump.

Example: You've written a Smart Form. To execute the Smart Form, you need the name of the FM that will execute the Smart Form. You can fill in the parameters and call this FM. On execution, the FM name is returned. When you execute this generated FM, you'll execute a Smart Form in your program.

[↓] Import

FORMNAME

Name of the Smart Form. Type: TDSFNAME.

VARIANT

Name of the variant. This parameter isn't used for time variants and should remain blank. Type: TDVARIANT.

DIRECT_CALL

If this parameter is [Space], the variant won't be used; otherwise, if this parameter isn't [Space], the variant is used. Type: TDBOOL.

[↑] Export

FM_NAME

Name of the FM returned. This FM executes the Smart Form. Type: RS38L_FNAM.

[!] Exceptions

NO_FORM

Smart Form entered does not exist.

NO_FUNCTION_MODULE

FM does not return.

---

290 SSF_GET_DEVICE_TYPE

This FM returns the device type for the given language import parameter. The spool device type that is returned is passed in the Smart Form FM in the parameter CONTROL_PARAMETERS. The application and language are read from table TSFDEVTY to get the device type.

Example: If you want to convert Smart Form output in the PDF format, you need to generally use the following set of FMs:

- **CALL FUNCTION SSF_GET_DEVICE_TYPE**
  Returns the device type.
- **CALL FUNCTION SSF_FUNCTION_MODULE_NAME**
  Returns the Smart Form name FM_NAME.
- **CALL FUNCTION FM_NAME**
  Executes the Smart Form. The device type returned from the FM SSF_GET_DEVICE_TYPE is passed in the CONTROL_PARAMETERS parameter.
- **CALL FUNCTION CONVERT_OTF_2_PDF**
  Converts the OTF format to PDF format.
- **CALL METHOD CL_GUI_FRONEND_SERVICES->**FILE_SAVE_DIALOG
  Shows the file save dialog box.
- **CALL FUNCTION GUI_DOWNLOAD**
  Sends the data to the presentation server.

[↓] Import

I_LANGUAGE

Smart Form language. Type: SFLANGU.

I_APPLICATION

Smart Forms: Application identification to get the device type. The default value is SAPDEFAULT. Type: SFAPPL.

[↑] Export

E_DEVTYPE

Spool: Device type name returned. Type: RSPOPTYPE.

[!] Exceptions

NO_LANGUAGE

If language is not passed in the I_LANGUAGE parameter.

LANGUAGE_NOT_INSTALLED

Language is not installed.

NO_DEVTYPE_FOUND

No device type for language/application is available.

SYSTEM_ERROR

Error determining the installed languages.

291 SSF_READ_ERRORS

Each Smart Form can trigger a number of exceptions and errors during its execution. If an error (or a warning) occurs, an error table is filled internally. The application can read this table as soon as the FM for the form is processed completely. To read the table that contains all errors and warnings that occurred, call this FM.

Example: You've executed the FM to execute the Smart Form, and you get an error. To read the error table and correct the error, you need to call this FM.

[↑] Export

ERRORTAB

This table returns the errors and warning. Type: TSFERROR.

---

292 START_FORM

This FM starts the form and can change the layout set in the current print set. Therefore, it's possible to combine several different layout sets in one printer output with this FM. Before opening a new layout set with this FM, an already opened layout set must be closed with END_FORM (FM 268).

Example: You're printing the script form and want one layout to print one block of text and another layout set to print the other block of text. You can do that by calling START_FORM separately for both the layout sets.

[↓] Import

ARCHIVE_INDEX

Archive index for the layout set to be archived. This information (DARA line) is moved to the layout set in the archive. You can then access the layout set in the archive directly. Type: TOA_DARA.

FORM

The parameter contains the name of the form that is used for output. The characters * and & aren't allowed in the form name. Type: C.

LANGUAGE

This parameter specifies which language version of the form is to be used. If the form isn't available for the language defined in this parameter, the form in its original language is used. Type: THEAD-TDSPRAS.

STARTPAGE

First page of the form where the output is to be placed. If this parameter isn't specified, the output is displayed on the current page. Type: C.

PROGRAM

This parameter defines the program name for symbol replacement. This parameter is helpful for FMs TEXT_SYMBOL_REPLACE (FM 295) TEXT_INLCUDE_REPLACE (FM 294), or TEXT_CONTROL_REPLACE (FM 93), where text is replaced. Type: SY-REPID.

MAIL_APPL_OBJECT

This parameter defines the application object number for the Device = MAIL. Type: SWOTOBJID.

[↑] Export

LANGUAGE

This parameter indicates which language version of the form is used for the output. Type: FORM LANGUAGE.

[!] Exceptions

FORM

Form does not exist.

FORMAT

Page format is not allowed.

UNENDED

Previous form is not yet ended.

UNOPENED

Form output not opened.

UNUSED

No form opened yet.

SPOOL_ERROR

Spool is not generated properly.

CODEPAGE

Code page not allowed.

---

293 TEXT_CONTROL_REPLACE

This FM expands and replaces control statements such as IF, ELSE, or CASE structures in text.

Example: If you've used the `IF ELSE` statement in your script and want to replace the control statement, call this FM within the logic.

[↓] Import

HEADER

Text header information. This is a mandatory parameter. Type: THEAD.

PROGRAM

When interpreting control statements, SAPscript must determine the values of program symbols. To do this, it must know which active program contains the work areas for the values to be passed.

If you omit the parameter, the system searches for the field values in the program that was called first (SY-CPROG).

If you enter a program name, the system replaces the program symbols with the values from this program. This applies only for the current call of the FM.

The default value is [Space]. Type: SY-REPID.

REPLACE_COMMENT

This parameter can be used to define whether comment lines in the text lines should be removed. Value X removes comment lines, and value [Space] doesn't remove comment lines. The default value is X.

[↑] Export

CHANGED

Replace control structures indicator. If the value is X, statements are interpreted; if the value is [Space], statements aren't found.

NEWHEADER

The parameter returns the text header with the fields changed. Type: THEAD.

Table

LINES

The table contains the text lines in which to resolve the SAPscript statements. Type: TLINE.

---

294 TEXT_INCLUDE_REPLACE

This FM expands the `INCLUDE` commands in the text lines, replacing them with the lines of the text that are to be included. If a text module isn't available, or the user doesn't have the appropriate authorization, the `INCLUDE` command is retained.

Example: You've used the INCLUDE files in your script. To expand the INCLUDE command and process the text of the INCLUDE as well, call this FM.

[↓] Import

ALL_LEVEL

This parameter determines whether to expand the nested INCLUDE as well.

- [Space]: Expand existing INCLUDE only.
- X: Expand all includes.

The default value is X.

ENDLINE

End line for INCLUDE replacement. The default value is 99999. Type: SY-TABIX.

HEADER

Text header. This is a mandatory parameter. Type: THEAD.

STARTLINE

Start line for INCLUDE replacement. The default value is 1. Type: SY-TABIX.

PROGRAM

Program name for symbol replacement. The default value is [Space]. Type: SY-REPID.

[↑] Export

CHANGED

Replace control structures indicator. If the value is X, statements are interpreted; if the value is [Space], statements aren't found.

ERROR_TYPE

This parameter currently doesn't return any error messages. Type: SY-TABIX.

NEWHEADER

The parameter returns the text header with the fields changed. Type: THEAD.

Table

LINES

The table contains the text lines in which to resolve the SAPscript statements. Type: TLINE.

---

295 TEXT_SYMBOL_REPLACE

The FM replaces the symbols contained in the text lines with their respective values. The text isn't formatted for printing, but the symbols are replaced in the ITF format. Symbols that occur on comment or raw lines remain unchanged, as well as symbols enclosed by the character formats <(> and <)>.

The system interprets `DEFINE` statements in text lines and changes the values of text symbols accordingly. It also executes the control statements SET DATE MASK, SET TIME MASK, SET SIGN LEFT, and SET SIGN RIGHT.

Example: You've put some symbols in your text header of SAPscript. Your header is "Hi &1, please pay your bill promptly." You want to replace the text symbols &Name with the values you've mentioned, for example, "Tanmaya" in the program. You can call this FM to do so.

[↓] Import

ENDLINE

End line for symbol replacement. You can specify the index of the last text line on which you want to use the required function. If no value or an invalid value is specified, the last line of the text table is generally used. Type: SY-TABIX.

HEADER

Text header. This is a mandatory parameter. Type: THEAD.

INIT

This parameter can be used to specify whether the symbol administration in SAPscript should be initialized before replacement is carried out. Possible values:

- X: Initialize.
- [Space]: Don't initialize.

OPTION_DIALOG

This parameter specifies whether a dialog box is to be displayed before replacement. The user can make settings for the replacement in this box. Possible values:

- X: Display dialog box.
- [Space]: Don't display dialog box.

PROGRAM

Use this parameter to determine the program whose work areas the system uses to replace the values of the program symbols. If you omit this parameter, the system uses the first program called to search for the field values (SY-CPROG).

This assignment is valid for the current call of the FM only. The default value is [Space]. Type: SYST-REPID.

REPLACE_PROGRAM

The parameter specifies whether to resolve program symbols during replacing.

Possible values:

- X: Replace program symbols.
- [Space]: Don't replace program symbols.

The default value is X.

REPLACE_STANDARD

This parameter specifies whether to resolve standard symbols during replacing. Possible values:

- X: Replace standard symbols.
- [Space]: Don't replace standard symbols.

The default value is X.

REPLACE_SYSTEM

The parameter specifies whether to resolve system symbols during replacing. The system symbols &PAGE& and &NEXTPAGE& aren't replaced because the text isn't formatted for printing. Possible values:

- X: Replace system symbols.
- [Space]: Don't replace system symbols.

The default value is X.

REPLACE_TEXT

The parameter specifies whether to resolve text symbols during replacing. Possible values:

- X: Replace text symbols.
- [Space]: Don't replace text symbols.

STARTLINE

Enter the index of the table line from which to start the FM. If you omit the parameter or specify an invalid value, the system starts on the first text table line. The reference field is SY-TABIX. The default value is 1. Type: SY-TABIX.

[↑] Export

CHANGED

Replace control structures indicator. If the value is X, statements are interpreted; if the value is [Space], statements aren't found.

NEWHEADER

The parameter returns the text header with the fields changed. Type: THEAD.

Table

LINES

The table contains the text lines in which to resolve the SAPscript statements. Type: TLINE.

---

296 TEXT_SYMBOL_SETVALUE

This FM assigns a value to an SAPscript symbol. Note that you can only change the values of text symbols. System symbols, standard symbols, and program symbols can't receive new values. The system keeps the new value until you initialize the symbol administration tables in SAPscript. To do this, you can use; for example, OPEN_FORM (FM 278) or TEXT_SYMBOL_REPLACE (FM 295) with the parameter INIT = X.

Example: You've declared the placeholder in your text symbol. To assign the value to it so that the actual value will be displayed in the output, (for instance), you have a text symbol for the heading and you want to replace the text symbol with "Heading of Script" value. Pass the parameters to this FM and call it.

[↓] Import

NAME

The parameter contains the name of the symbol involved. Only the symbol name can be specified, without the text supplements or the possible formatting options. The symbol name can be passed with or without the delimiter &. This is a mandatory parameter.

VALUE

Define the value of the symbol. The system can use up to 80 characters as the symbol value. This is a mandatory parameter.

VALUE_LENGTH

If you omit this parameter or specify 0, the system automatically determines the length of the symbol value. The system then includes all characters up to the last blank into the value.

If you want to include trailing blanks into the assignment, specify the desired length of the symbol value here. The default value is 0. Type: SY-TABIX.

REPLACE_SYMBOLS

The value X in this parameter means that symbols contained in the field VALUE are replaced immediately with their current values; the assigned value no longer contains SAPscript symbols.

If the parameter is empty, the system includes the contents of the VALUE field without change. Any symbols occurring within are replaced with their current value only after the symbol specified in the parameter NAME is called in the text. The default value is [Space].

297 WRITE_FORM

This FM outputs the opened form elements into the current opened form.

Example: You want to print the SAPscript form, and you've created the window and different elements within that window. You can output the form elements into the current opened form (which you've opened from FM OPEN_FORM [FM 278]) by passing the parameters and calling this FM.

[↓] Import

ELEMENT

Text element name.

FUNCTION

The parameter defines how the text or text element is output in the respective window. The output method depends on the respective window type. Possible values:

- SET: In the case of the MAIN window, the text is appended to the previous output. For all other windows, the old window contents are deleted first.
- APPEND: For the MAIN window, the output is identical to that in the function SET. In the case of the other windows, the text element is appended to those preceding it.
- DELETE: The text element specified is deleted. The function can't be used in the MAIN window.

The default value is SET.

TYPE

Window area in MAIN window. The MAIN window is divided into three areas. This parameter can be used to control in which area the text is output. Possible values:

- TOP: Output occurs in the header area of the MAIN window.
- BODY: Output occurs in the main area of the MAIN window.
- BOTTOM: Output occurs in the footer area of the MAIN window.

The default value is BODY.

WINDOW

The parameter specifies the name of the window in which the current output is to occur. The window must be defined in the layout set. The default value is MAIN.

[↑] Export

PENDING_LINES

If corresponding text is output to the BOTTOM area of the main window (TYPE = BOTTOM), it's possible that no space will be left on the current output page. It's therefore marked internally only for the BOTTOM output of the next page. The actual output is therefore still due (hangs).

The parameter value X conveys these facts. For example, at the end of the text, this BOTTOM output which is still hanging could be implicitly shifted to the next page with an explicit page break (NEW PAGE).

[!] Exceptions

ELEMENT

Text element name does not exist.

FUNCTION

Invalid function specified in the parameter function.

TYPE

The window area type specified in the parameter TYPE is unknown or is not valid for the current window.

UNOPENED

The current form function could not be executed because the form output was not initialized with OPEN_FORM (FM 278).

UNSTARTED

A layout set was not opened. Possible reasons:

- The layout set processing was initialized with OPEN_FORM (FM 278) without specification of a layout set, but a layout set was not opened with START_FORM (FM 292).
- The layout set last used was concluded with END_FORM (FM 268), but a new layout set was not opened with START_FORM (FM 292).
- In the layout set used, no next page was defined. A layout print was therefore concluded automatically, without the explicit call from END_FORM (FM 268).
- In the layout set used, a window MAIN was not defined on any page. However, a call to output a text element in the MAIN window is nevertheless made.

WINDOW

The form window specified in the parameter WINDOW does not exist in the current form. Possible reasons:

- The window name was entered incorrectly.
- The form version with the specified window is not active.

BAD_PAGEFORMAT_FOR_PRINT

Page format given for print does not exist.

SPOOL_ERROR

Error at transfer to spool.

CODEPAGE

Code page is not allowed.

298 WRITE_FORM_LINES

This FM outputs the text lines specified in the table LINES in the specified layout set window. The text lines must have the SAPscript ITF format. From the data in the text header, the system uses only the field TDSTYLE to apply the formatting attributes defined in the specified style for this text. If the field is empty, the system uses the identically named formatting attributes (character and paragraph formats) of the form.

Example: You want to print the SAPscript form, and you've created the window and different elements within that window. To output the form elements into the current form so that you get the output, pass the parameters and call this FM.

[↓] Import

FUNCTION

The parameter defines how the text or text element is output in the respective window. The output method depends on the respective window type. Possible values:

- SET: In the case of the MAIN window, the text is appended to the previous output. For all other windows, the old window contents are deleted first.
- APPEND: For the MAIN window, output is identical to that in the function SET. In the case of the other windows, the text element is appended to those preceding it.
- DELETE: The text element specified is deleted. The function can't be used in the MAIN window.

The default value is SET.

HEADER

Text header of text to be output. Type: THEAD.

TYPE

Window area in the MAIN window. The MAIN window is divided into three areas. This parameter can be used to control into which area the text is output. Possible values:

- TOP: Output occurs in the header area of the MAIN window.
- BODY: Output occurs in the main area of the MAIN window.
- BOTTOM: Output occurs in the footer area of the MAIN window.

The default value is BODY.

WINDOW

The parameter specifies the name of the window in which the current output is to occur. The window must be defined in the layout set. The default value is MAIN.

[↑] Export

FROMPAGE

This parameter returns the page number where the first character of the current output was placed.

PENDING_LINES

If corresponding text is output to the BOTTOM area of the main window (TYPE = BOTTOM), it's possible that no space will be left on the current output page. It's therefore marked internally only for the BOTTOM output of the next page. The actual output is therefore still due (hangs).

The parameter value X conveys these facts. For example, at the end of the text, this BOTTOM output that is still hanging could be implicitly shifted to the next page with an explicit page break (NEW PAGE).

Table

LINES

Lines of text to be output. Type: TLINE.

[!] Exceptions

FUNCTION

Invalid function specified in the parameter function.

TYPE

The window area type specified in the parameter TYPE is unknown or is not valid for the current window.

UNOPENED

The current form function could not be executed because the form output was not initialized with OPEN_FORM (FM 278).

UNSTARTED

A layout set was not opened. Possible reasons:

- The layout set processing was initialized with OPEN_FORM (FM 278) without specification of a layout set, but a layout set was not opened with START_FORM (FM 292).
- The layout set last used was concluded with END_FORM (FM 268), but a new layout set was not opened with START_FORM (FM 292).
- In the layout set used, no next page was defined. A layout print was therefore concluded automatically, without the explicit call from END_FORM (FM 268).
- In the layout set used, a window MAIN was not defined on any page. However, a call to output a text element in the MAIN window is nevertheless made.

WINDOW

The form window specified in the parameter WINDOW does not exist in the current form. Possible reasons:

- The window name was entered incorrectly.
- The form version with the specified window is not active.

BAD_PAGEFORMAT_FOR_PRINT

Page format given for print does not exist.

SPOOL_ERROR

Error at transfer to spool.

CODEPAGE

Code page is not allowed.

Part VI
Archive, Number Ranges, Strings, and Text

This part of the book contains various functions such as archives, number ranges, texts, and strings that you'll use while developing your ABAP programs. You'll use archives when you want to store the old data and information (which isn't in use) in your secondary storage so that you can later refer to the information whenever required and free the current memory for usage.

Database records of the commercial objects, such as purchase documents, sales documents, addresses, and so on, are uniquely identified by a number or an ID. Number ranges assign the next available numbers to these commercial object database records and helps in their unique identification. This part also contains function modules (FMs) for text and string operations such as splitting the text based on some separator, concatenating string, and so on. Other functions such as opening, saving, closing an archive object, creating a new number from a number range, checking the number of an number range, word wrap functions, reversing a string, concatenating strings, splitting strings, and more can be found here.

299 ARCHIVE_CLOSE_FILE

This FM closes the open archive files and ends the data archiving action. The archive handle that is passed in parameter ARCHIVE_HANDLE is thereafter invalid and is no longer to be used. Further operations with this parameter ARCHIVE_HANDLE raises exception WRONG_ACCESS_TO_ARCHIVE.

Example: After you're done archiving files, call this FM to close the archive so that it's no longer available. Also refer to FM ARCHIVE_GET_CUSTOMIZING_DATA (FM 300).

[↓] Import

ARCHIVE_HANDLE

This parameter is used for temporary identification of group data objects and their archive files. In this case, this parameter is used to close a data archiving action. After this FM call, this parameter is invalid and can't be used further. Type: SY-TABIX.

[!] Exceptions

INTERNAL_ERROR

Internal error. It indicates the occurrence of an unforeseen event.

WRONG_ACCESS_TO_ARCHIVE

An archive handle was used that isn't valid (or is no longer valid). An archiving class may have called the FM and thereby triggered the exception.

ARCHIVING_STANDARD_VIOLATION

Violation of the archiving standard.

300 ARCHIVE_GET_CUSTOMIZING_DATA

This FM gives the customizing settings for an archiving object to the archiving program. Values that are important for the delete program are returned. Parameter COMMIT_COUNT_FOR_DELETE_PRG, which is returned from this FM, identifies the number of objects to be deleted. The deletion archiving program should delete this many objects and then `COMMIT WORK` to make the deletion permanent. The FMs you'll use in this process are as follows:

- ARCHIVE_OPEN_FOR_DELETE (FM 308)
- ARCHIVE_GET_CUSTOMIZING_DATA (FM 300)
- ARCHIVE_GET_INFORMATION (FM 301)
- ARCHIVE_GET_NEXT_OBJECT (FM302)
- ARCHIVE_GET_NEXT_RECORD (FM 303
- ARCHIVE_GET_TABLE (FM 305)
- ARCHIVE_GIVE_STATISTICS (FM 306)
- ARCHIVE_WRITE_STATISTICS (FM 316)
- ARCHIVE_GET_STATISTICS (FM 304)
- ACHIVE_CLOSE_FILE (FM 299)

Example: If you want to delete the data from the table, you can do so by using the listed FMs (refer to program SBOOKD in the SAP

ERP system). This FM gets the customizing data from the archiving object.

[↓] Import

OBJECT

This parameter specifies the archiving object that you want to open for archiving. Type: ARCH_OBJ-OBJECT.

[↑] Export

COMMIT_COUNT_FOR_DELETE_PRG

Number of objects for a `COMMIT WORK`. This parameter returns the value of the counter for the placement of a `COMMIT WORK`. The deletion programs must use this value to set a `COMMIT WORK` after this number of objects. The archiving classes settle any `COMMIT WORK`. Type: ARCH_USR-ARCH_COMIT.

MAINTAIN_INDEX

This parameter indicates whether a data object index is to be built. If the corresponding customizing option is checked, then the corresponding archive file identifier is written to an index table for each data object. Type: ARCH_USR-ARCH_INDEX.

[!] Exceptions

OBJECT_NOT_FOUND

If the archiving object is not known to the system, this exception is raised.

---

301 ARCHIVE_GET_INFORMATION

This FM returns information that shows you the current information for a handle.

Example: Refer to FM ARCHIVE_GET_CUSTOMIZING_DATA (FM 300). You are deleting the data from the table for an archiving object. You need to get all the records you want to delete. You need to use this FM to get this information.

[↓] Import

ARCHIVE_HANDLE

Open archive handle. This parameter is used for temporary identification of group data objects and their archive files. In this case, the handle is used to specify for which archive the information is needed. Type: SY-TABIX.

[↑] Export

ARCHIVE_CREATION_DATE

Date on which the archive was created. This value can change at runtime according to the handle because the archiving can manage several archive files per handle. The current value is always returned. Type: HEADA-DATUM.

ARCHIVE_CREATION_RELEASE

This parameter gives the release number under which the archive was created. Type: HEADA-SAPRL.

ARCHIVE_CREATION_SYSTEM

SAP system in which the archive was created. Type: HEADA-SYSID.

ARCHIVE_DOCUMENT

Archiving run according to archive management. This parameter returns the key of the archiving run, which is handled just under the archive handle. Type: ADMI_RUN-DOCUMENT.

ARCHIVE_NAME

Archive name that is currently being processed under the handle is returned in this parameter. This value can be changed at runtime depending on the handle because archiving each handle can manage multiple archive files. This parameter will always return current value of an archiving name. Type: HEADA-ARKEY.

OBJECT

Name of the object edited with the handle. Type: ARCH_OBJ-OBJECT.

ARCHIVE_CODE_PAGE

Name of the code page with which the archive file was written. The code page is required to achieve the ABAP command to `TRANSLATE` an implementation of the records on the current code page. Type: TCP00-CPCODEPAGE.

ARCHIVE_NUMBER_FORMAT

Description of the number format that allows the archive file to be written. The number format is needed to reach the ABAP command to `TRANSLATE` an implementation of the records on the current number format. Type: TCP00-CPCODEPAGE.

ADK_VERSION

Version of the archiving development kit. Type: HEADA-SAPRL.

ARCHIVING_CLASS

This parameter specifies the call context from the handle. If this call context is an archiving class, the name of the calling class archiving is issued. Otherwise, the parameter is left blank. Type: ARCH_CLAS_.

OBJECT_NUMBER_IN_FILE

This parameter returns the number of data objects that are in the archive file being edited under the handle. The significance of the value depends on the type of archiving program:

- Write program: Number of data objects written until the time of the call to the archive file data objects.
- Read program: Total number of data objects in the archiving file.
- Delete program: Total number of data object in the archiving file
- Reload program and conversion program: Call with write handle: Number of objects written until the time of the call to the new archive file objects.
- Call with read handle: Total number of data objects present in the archive file.

Type: ADMI_COUNT.

OBJECT_NUMBER_IN_RUN

This parameter returns the number of data objects that are in the archiving process that's currently being processed under the handle. The significance of the value depends on the type of archiving program:

- Write program: Number of data objects written in this session up to the point the parameter was called.
- Read program: Total number of data objects belonging to a archiving file in a archiving session.
- Delete program: Total number of data objects contained in ta archive file in an archiving session.
- Reload program and conversion program.
- Call with write handle: Number of data objects written until the time of the call in the new archiving session objects.
- Call with reading handle: Total number of data objects in the archiving files belonging to archive data objects.

Type: ADMI_COUNT.

Table

USED_CLASSES

This table returns the ADK class. Type: ADK_CLASSES.

[!] Exceptions

INTERNAL_ERROR

Internal error.

WRONG_ACCESS_TO_ARCHIVE

Incorrect archive access. An archive handle was used, which is not valid or is no longer valid.

---

302 ARCHIVE_GET_NEXT_OBJECT

This FM reads the next data object for an archive handle from an archive that has been opened for reading.

The data that belongs to an archiving class is forwarded automatically to the relevant archiving class. This means that the data can be retrieved from the archiving class immediately after this FM has been called. Typical archiving classes are long texts and change documents.

Example: you are deleting data from an archiving object. You need to call this FM to identify the next data object for an archive handle which you want to delete. Refer to FM ARCHIVEING_GET_CUSTOMIZING_DATA (FM 300).

[↓] Import

ARCHIVE_HANDLE

Open archive handle. In this case, the handle is used to specify which archive you want to read. Type: SY-TABIX.

[↑] Export

OBJECT_ID

You receive the ID of the data object read via this parameter. This ID was specified via the FM ARCHIVE_NEW_OBJECT (FM 307) when the archive was generated.

OBJECT_OFFSET

Offset of data object in the archive. The parameter OBJECT_OFFSET displays the address of the current data object in bytes. This address can be stored in the index table ARCH_IDX to enable direct access to the archived data object. Type: ARCH_IDX-OFFSET.

ARCHIVE_NAME

Archive key according to archive management. This value can change for each handle at runtime because the archiving can manage several archive files per handle. The current value is always returned. Type: HEADA-ARKEY.

COMPR_OBJECT_LENGTH

Length of a data object in bytes, which is in its compressed archive file format. Type: I.

SESSION

This parameter identifies the archiving session that archived the data object that has just been read. Type: ADMI_RUN_D.

[!] Exceptions

END_OF_FILE

End of the archive file reached.

FILE_IO_ERROR

Archive file access error.

INTERNAL_ERROR

Internal error.

OPEN_ERROR

Archive file open error.

WRONG_ACCESS_TO_ARCHIVE

Incorrect access to the archive.

---

303 ARCHIVE_GET_NEXT_RECORD

This FM reads the records of a data object sequentially. On the first call after the call of the function block ARCHIVE_GET_NEXT_OBJECT, the first record is automatically returned. From then on, more records are returned until no further record exists in the data object.

To read an archive file sequentially, it's enough to read in an outer DO loop with the function block ARCHIVE_GET_NEXT_OBJECT, which reads the next archive data object, and then read in an inner DO loop with the function block ARCHIVE_GET_NEXT_

RECORD, which reads the next records of the data object of the outer loop.

Example: In continuation of the deletion example, this is the next step you need to follow. You have identified the archiving object which you want to delete. Now you need to call this FM to get the records in this data object. Refer to FM ARCHIVE_GET_CUSTOMIZING_DATA (FM 300).

[↓] Import

ARCHIVE_HANDLE

Handle to the open archive files. In this case, the handle is used to specify which archive you want to read. This is a mandatory parameter. Type: SY-TABIX.

GET_REAL_STRUCTURE_NAME

Conversion of the structure name. This parameter checks whether a real structure name is used or its alias is passed in the parameter. Value X specifies that the actual structure name is passed. The default value is X.

AUTOMATIC_CONVERSION

Automatic conversion of the data record. If the value is X, an automatic adjustment of the data set to the current structure is made. The automatic conversion of the data set also includes an adjustment to the current code page and the current number format.

If this parameter has the value [Space], no automatic conversion is made. This can be useful if between writing and reading of the archive name, changes were made on columns of the table, but the length of the data set is the same. In this case, there is no automatic adjustment of the number format and code page. If archives with different hardware are read as they were written, then it can lead to problems if you don't make yourself an adaptation of the archives to the current code page and the current number format. The default value is X.

[↑] Export

RECORD

Data record. The parameter blocks the memory area in which a record of a data object is stored or to be stored.

The FMs of archiving don't check whether the size of the storage area is sufficient to accommodate the record.

RECORD_CURSOR

Pointer to data record. The parameters specified in the function blocks of the memory storage area determine where the record pointer of a data set is stored or to be stored. In this case, the record pointer to the current record should be placed in the specified memory area. Type: SY-TABIX.

RECORD_FLAGS

This parameters specified in the ADK storage area are stored in the control information to a data object or a data to be stored. Each set in a data object may include additional information in these eight characters (flags). In the context of sets of an archiving object, there are no rules for using these flags. However, in the context of an archiving class ARCH_FLAGS, the structure is used consisting of FLAG_1 to FLAG_8. FLAG_1 is for sentences belonging to a storage class, reserved as a deletion indicator.

You can, for example, also set an archiving object at about RECORD_FLAGS to control whether the data set deleting this after archiving should be considered or not. You only need to note here that all of the programs written to this object archiving understand the encryption in this field. Type: ARC_BUFFER-FLAGS.

RECORD_STRUCTURE

Name of returned structure. If you're not using the direct structure name but using an alias instead, the alias for the assignment of structure in Transaction AOBJ (Archiving Object Definition) must be made. Then pass the alias on RECORD_STRUCTURE. This alias is returned for subsequent read accesses. Type: ARC_BUFFER-RNAME.

RECORD_LENGTH

Length of the data record in the parameter RECORD. Type: ARCH_INT-INT4.

RECORD_REF

The RECORD_REF parameter has the same function as the parameter RECORD. However, when RECORD_REF is returned, it isn't the record itself but only a reference. Type: DATA.

[!] Exceptions

END_OF_OBJECT

No further data record exists.

INTERNAL_ERROR

Internal error.

WRONG_ACCESS_TO_ARCHIVE

Incorrect access to the archive.

---

304 ARCHIVE_GET_STATISTICS

This FM is used in write, delete, reload, and read programs for data archiving to transfer the statistics that have been collected by the Archive Development Kit (ADK); that is, this FM transfers the statistics data from the database to the ABAP archiving program.

When writing, deleting, reading, or reloading, the system automatically generates statistical data on each archiving run (such as the storage space that has been freed in the database by deletions, or the number of deleted data objects) and stores it persistently in the database. This data can be analyzed by data archiving administrators, and they can plan the future archiving projects. Statistics data is created in write, delete, read, and reload processes. In other archiving contexts—for example, execution of preprocessing, postprocessing, or conversion programs—no statistical data is collected. This is the reason these statistics FMs should not be used in conversions, preprocessing, and post processing programs.

Example: Refer to FM ARCHIVE_GET_CUSTOMIZING_DATA (FM 300). In all write, delete, reload, and read programs in data archiving; this FM should be called immediately before FM ARCHIVE_CLOSE_FILE (FM 299). In reload programs, it must be called each time with the read handle. In our delete program, you need to get the statistical data from the database to your program. You can pass the parameter in this FM to get the statistical data.

[↓] Import

ARCHIVE_HANDLE

Handle to the open archive files. This FM can only be called with the read handle. Type: SY-TABIX.

[↑] Export

ARCHIVE_DOCUMENT

Archiving session number or number of archive runs are stored in this parameter. Type: ADMI_RUN-DOCUMENT.

BINARY_STAT_DATA

ADK statistics. Type: XSTRING.

Table

STATISTICS_DATA

Table with statistical data. Type: ARCH_STATI.

[!] Exceptions

INTERNAL_ERROR

Internal error.

WRONG_ACCESS_TO_ARCHIVE

Incorrect access to the archive.

---

305 ARCHIVE_GET_TABLE

This FM reads several data records for the structure RECORD_STRUCTURE from the current data object and returns the value in the exporting parameter.

This FM can also be used together with logical databases. Logical databases are special ABAP programs that retrieve data and make it available to application programs. The most common use of logical databases is to read data from database tables by linking them to executable ABAP programs. To use this FM with logical databases, ensure that the parameter is ALL_RECORDS_OF_OBJECT = [Space].

Example: In continuation of our example for deleting records from table, if you have table structure associated with the data object. You need to pass the parameter in this FM to get the data of the table structure. Refer to FM ARCHIVE_GET_CUSTOMIZING_DATA (FM 300).

[↓] Import

ARCHIVE_HANDLE

Handle to the open archive files. In this case, the handle is used to specify which archive you want to read. This is a mandatory parameter. Type: SY-TABIX.

RECORD_STRUCTURE

Name of structure with data records. If you're not using the direct structure name but an alias instead, the alias for the assignment of structure in Transaction AOBJ (Archiving Object Definition) must be made. Then pass the alias on RECORD_STRUCTURE. This alias is returned for subsequent read accesses. Type: ARC_BUFFER-RNAME.

ALL_RECORDS_OF_OBJECT

Read all records of the structure in the data object.

If the value is [Space], the FM functions similarly to the FM ARCHIVE_GET_NEXT_STRUCT_SPECIF. All data records for the structure entered are read sequentially until the structure changes. All data records read are put into the internal table TABLE. The accompanying indicators are stored in the same file in the table RECORD_FLAGS_TABLE. The indicators from the last read record are returned in the parameter RECORD_FLAGS.

If this parameter has value X, the FM reads all data records of the specified structure in the entire data object. The data records are put into the internal table TABLE. It doesn't matter whether there are data records with another structure between the specified structure RECORD_STRUCTURE. All data records for the structure entered are collated in the current data object, regardless of where they are in the current data object.

The default value is [Space].

AUTOMATIC_CONVERSION

Automatic conversion of the data record. If the value is X, an automatic adjustment of the data set to the current structure is made. The automatic conversion of the data set also includes an adjustment to the current code page and the current number format.

If this parameter has the value [Space], no automatic conversion is made. This can be useful if between writing and reading of the archive name, changes were made on columns of the table, but the length of the data set is the same. In this case, there is no automatic adjustment of number format and code page. If archives with a different hardware are read as they were written, then it can lead to problems if you don't make yourself an adaptation of the archives to the current code page and the current number format. The default value is X.

ARCHIVING_CLASS

Archiving class name. The default value is [Space]. Type: ARCH_CLAS_.

[↑] Export

RECORD_CURSOR

Pointer to the data record. The parameters specified in the function blocks of the memory storage area determine where the record pointer of a data set is stored or to be stored. In this case, the record pointer to the current record should be placed in the specified memory area. Type: SY-TABIX.

RECORD_FLAGS

Specifies the area of the memory where control information on a data record for a data object is to be stored. Type: ARCH_FLAGS.

RECORD_LENGTH

Length of the data record in parameter RECORD. Type: ARCH_INT-INT4.

Tables

TABLE

Data records are returned in this table.

RECORD_FLAGS_TABLE

Table with indicator bars for all data records that are returned in the table TABLE.

Record TABLE[i] has the indicator bar RECORD_FLAGS_TABLE[i], which is returned in this table. Type: ARCH_FLAGS.

[!] Exceptions

END_OF_OBJECT

No further data record exists.

INTERNAL_ERROR

Internal error.

WRONG_ACCESS_TO_ARCHIVE

Incorrect access to the archive.

---

306 ARCHIVE_GIVE_STATISTICS

This FM transfers statistical data from your ABAP data archiving programs to the database. This FM is used in write, delete, read, and reload programs for data archiving. This FM must be called always at prescribed points. Not using or incorrectly using this FM can lead to incorrect statistical data.

Example: You are working on a delete program of data archiving. Just before saving and committing the deleted data to the database, you need to give the statistics of data you have deleted to ADK. Use this FM for the same. Refer to FM ARCHIVE_GET_CUSTOMIZING_DATA (FM 300).

[↓] Import

ARCHIVE_HANDLE

Handle to the open archive files. Type: SY-TABIX.

Table

TABLE

Table for transferring the statistics information. Type: ARCH_STAT.

[!] Exceptions

INTERNAL_ERROR

Internal error.

WRONG_ACCESS_TO_ARCHIVE

Incorrect access to the archive.

---

307 ARCHIVE_NEW_OBJECT

This FM must be called when a new data object is created that's used to write to an archive. Because there can only be one data object per handle, when this FM is called, all data for the current data object is discarded and overwritten, and all associated archiving classes are informed that all subsequent operations on this handle belong to a new data object. The archiving classes themselves discard all information for the current data object.

Example: You're creating an archiving program for your production orders; that is, you want all of the production orders older than 90 days to be archived. You need to create a data object for archiving. This can be done by passing relevant parameters in this FM and executing this FM. Refer to FM ARCHIVE_OPEN_FOR_WRITE (FM 311).

[↓] Import

ARCHIVE_HANDLE

Handle to the open archive. This is a mandatory parameter. Type: SY-TABIX.

OBJECT_ID

Identification of the new data object. You can assign an ID to the newly requested data object using this parameter. This ID is returned if you read this object from an available archive using the FM ARCHIVE_GET_NEXT_OBJECT (FM 302). If an index file is to be set up to read individual data objects from an archive file, this ID is used to set up the index.

[!] Exceptions

INTERNAL_ERROR

Internal error.

WRONG_ACCESS_TO_ARCHIVE

If you try to append a data record to a data object with the FM ARCHIVE_PUT_RECORD (FM 312) before a data object has been made available with this FM, the exception WRONG_ACCESS_TO_ARCHIVE is raised.

This exception occurs when you close a data object and write it to the archive file with the FM ARCHIVE_SAVE_OBJECT (FM 315) but then try to append another data record to the data object with the FM ARCHIVE_PUT_RECORD (FM 312).

308 ARCHIVE_OPEN_FOR_DELETE

This FM opens archives to delete their data in the database. This FM returns the parameter ARCHIVE_HANDLE, which is used in other FMs for archiving. Each call to this FM generates a new handle so it's possible to access several archives of different archive objects at the same time. The FMs you'll use in this process are as follows (in order):

- ARCHIVE_OPEN_FOR_DELETE (FM 308)
- ARCHIVE_GET_CUSTOMIZING_DATA (FM 300)
- ARCHIVE_GET_INFORMATION (FM 301)
- ARCHIVE_GET_NEXT_OBJECT (FM 302)
- ARCHIVE_GET_NEXT_RECORD (FM 303)
- ARCHIVE_GET_TABLE (FM 305)
- ARCHIVE_GIVE_STATISTICS (FM 306)
- ARCHIVE_WRITE_STATISTICS (FM 316)
- ARCHIVE_GET_STATISTICS (FM 304)
- ACHIVE_CLOSE_FILE (FM 299)

Example: Refer to FM ARCHIVE_GET_CUSTOMIZING_DATA (FM 300). You want to delete the data in the database for archives. Open the archiving object for deletion with this FM.

[↓] Import

ARCHIVE_NAME

Archive file name according to archive management. Type: ADMI_FILES-ARCHIV_KEY.

OBJECT

Archiving object name. You announce over the parameter to which you want to open archiving object archives. The archiving object must be known to the system; otherwise, exception OBJECT_NOT_FOUND is thrown. This is a mandatory parameter. Type: ARCH_OBJ-OBJECT.

TEST_MODE

This parameter controls whether the status information is to be adjusted in archive administration.

AINDFLAG

If this parameter is set to X, info structures are prepared from SAP AS. The default value is X. Type: C.

NO_STATISTICS

Parameter NO_STATISTICS controls whether, in a delete program, statistical data can be gathered using FM ARCHIVE_GIVE_STATISTICS (FM 306) and displayed using FM ARCHIVE_WRITE_STATISTICS (FM 316) or FM ARCHIVE_GET_STATISTICS (FM 304). If parameter NO_STATISTICS is [Space], statistics data can be collected and displayed by the previously mentioned FMs. The default value is [Space]. Type: C.

OUTPUT_SEL_SCREEN_WHEN_DIALOG

This parameter controls whether the selection screen of the archiving program is displayed or not, when the archiving program is running in the dialog mode. Value [Space] means the selection screen won't be displayed, and value X means it will be displayed. Type: C.

OUTPUT_SEL_SCREEN_WHEN_BATCH

This parameter controls whether the selection screen of the archiving program is displayed or not, when the archiving program is running in the background. Value [Space] means the selection screen won't be displayed, and value X means it will be displayed. Type: C.

[↑] Export

ARCHIVE_HANDLE

Handle to the open files. In this case, the handle is used to specify which archive you want to delete. The handle must be generated via FM ARCHIVE_OPEN_FOR_READ (FM 310), ARCHIVE_OPEN_FOR_MOVE (FM 309), or ARCHIVE_OPEN_FOR_DELETE (FM 308); otherwise, the exception WRONG_ACCESS_TO_ARCHIVE is raised. Type: SY-TABIX.

Table

SELECTED_FILES

This table contains the selected entries from the table ADMI_FILES. Type: ADMI_FILES.

[!] Exceptions

FILE_ALREADY_OPEN

The system tried repeatedly to open a file.

FILE_IO_ERROR

Archive file access error.

INTERNAL_ERROR

Internal error.

NO_FILES_AVAILABLE

No files exist for object.

OBJECT_NOT_FOUND

Object unknown.

OPEN_ERROR

Could not open file.

NOT_AUTHORIZED

No authorization for this FM.

ARCHIVING_STANDARD_VIOLATION

Violation of the archiving standard.

---

309 ARCHIVE_OPEN_FOR_MOVE

This FM writes and/or reloads data objects to a new archive file. This FM is used to read archive files and to write data objects from these archive files to new archive files. This facilitates the selective reloading of individual data objects from archives to the SAP system.

Two handles are returned as the result of this FM. You can use ARCHIVE_READ_HANDLE as if you had received this handle via the FM ARCHIVE_OPEN_FOR_READ (FM 310). You can therefore carry out all read operations. The handle that is returned via the parameter ARCHIVE_WRITE_HANDLE is used to write data objects that aren't to be written back to the database to the new archive. You can therefore only call the FM ARCHIVE_SAVE_OBJECT (FM 315) with this handle. Each time the FM ARCHIVE_SAVE_OBJECT (FM 315) is called with the handle ARCHIVE_WRITE_HANDLE, the data that was previously read with the FM ARCHIVE_GET_NEXT_OBJECT (FM 302), and the handle ARCHIVE_READ_HANDLE is written to the new archive file.

Example: You're creating an archiving program for sales orders. You want to read the existing archive file for archiving production orders, which you've already created, and you also want to reload some of the data to the new archiving program of sales orders. You can pass the parameter and call this FM to open the existing file for reading. You'll use these FMs in sequential order to complete the process:

- ARCHIVE_GET_NEXT_OBJECT (FM 302)
- ARCHIVE_OPEN_FOR_MOVE (FM 309)
- ARCHIVE_SAVE_OBJECT (FM 315)
- ARCHIVE_CLOSE_FILE (FM 299)

Also refer to program RSARCH07 in SAP ERP.

[↓] Import

OBJECT

Archiving object name. You announce the parameter to which you want to open archiving object archives. The archiving object must be known to the system; otherwise, exception OBJECT_NOT_FOUND is thrown. This is a mandatory parameter. Type: ARCH_OBJ-OBJECT.

TEST_MODE

This parameter controls whether the status information is to be adjusted in archive administration.

ARCHIVE_NAME

Archive file name according to archive management. This parameter is obsolete now. Don't use it. Type: ADMI_FILES-ARCHIV_KEY.

[↑] Export

ARCHIVE_READ_HANDLE

Handle to the open files for reading. This parameter provides the temporary identification to group data objects and their archive files that are to be processed in a specific sequence by a program of ADK FMs. In the reload program, it's possible to distinguish between the read handle (data volume read from the archive and reloaded into the database) and the write handle (data volume that that is written to a new archiving session.) Type: SY-TABIX.

ARCHIVE_WRITE_HANDLE

Handle to the files open for writing. This parameter provides the temporary identification to group data objects and their archive files that are to be processed in a specific sequence by a program of ADK FMs. In the reload program, it's possible to distinguish between the read handle (data volume read from the archive and reloaded into the database) and the write handle (data volume that is written to a new archiving session). Type: SY-TABIX.

[!] Exceptions

FILE_ALREADY_OPEN

The system tried repeatedly to open a file.

FILE_IO_ERROR

Archive file access error.

INTERNAL_ERROR

Internal error.

NO_FILES_AVAILABLE

No files exist for object.

OBJECT_NOT_FOUND

Object unknown.

OPEN_ERROR

Could not open file.

NOT_AUTHORIZED

No authorization for this FM.

ARCHIVING_STANDARD_VIOLATION

Violation of the archiving standard.

310 ARCHIVE_OPEN_FOR_READ

You can use this FM to open archives that you need to read. This FM returns the parameter ARCHIVE_HANDLE, which is used in other FMs for archiving. Each call to this FM generates a new handle, so it's possible to access several archives of different archive objects at the same time.

Example: To read an archive, you need to pass the parameters to this FM, which opens the archive for reading. The parameter ARCHIVE_HANDLE is returned, which you can use to call more archive FMs.

[↓] Import

ARCHIVE_DOCUMENT

Archive name according to archive management. This name is a six-digit number. The default value is 000000. Type: ADMI_RUN-DOCUMENT.

ARCHIVE_NAME

Archive file name according to archive management. The default value is [Space]. Type: ADMI_FILES-ARCHIV_KEY.

OBJECT

Archiving object name. You announce the parameter to which you want to open archiving object archives. The archiving object must be known to the system; otherwise, an exception is thrown OBJECT_NOT_FOUND. This is a mandatory parameter. Type: ARCH_OBJ-OBJECT.

MAINTAIN_INDEX

This parameter determines which archive files appear in a dialog box. If the flag is set to X, archive files that have an index for single access appear in the dialog. If the flag is empty, then all readable files appear in the dialog. The default value is [Space].

[↑] Export

ARCHIVE_HANDLE

Handle to the open files for reading. The handle must be generated via the FM ARCHIVE_OPEN_FOR_READ (FM 310), ARCHIVE_OPEN_FOR_MOVE (FM 309), or ARCHIVE_OPEN_FOR_DELETE (FM 308); otherwise, the exception WRONG_ACCESS_TO_ARCHIVE is raised. Type: SY-TABIX.

Tables

ARCHIVE_FILES

Table of the required archive files. Type: RNG_ARCHIV.

SELECTED_FILES

Table with the selected entries. Type: ADMI_FILES.

[!] Exceptions

FILE_ALREADY_OPEN

The system tried repeatedly to open a file.

FILE_IO_ERROR

Archive file access error.

INTERNAL_ERROR

Internal error.

NO_FILES_AVAILABLE

No files exist for the object.

OBJECT_NOT_FOUND

Object unknown.

OPEN_ERROR

Could not open file.

NOT_AUTHORIZED

No authorization for this FM.

ARCHIVING_STANDARD_VIOLATION

Violation of the archiving standard.

---

311 ARCHIVE_OPEN_FOR_WRITE

This FM is the first FM you need to call in a write program if you want to create a new archive file. The parameter ARCHIVE_HANDLE gives you the handle that you need to pass in write operations.

Example: You want to write table data to a new archive, so you call this FM after you open a new archiving session to archive data. You can pass the parameters and call this FM. You'll use the following FMs in sequential order:

- ARCHIVE_NEW_OBJECT (FM 307)
- ACHIVE_PUT_RECORD (FM 312)
- ARCHIVE_SAVE_OBJECT (FM 315)

- ARCHIVE_WRITE_STATISTICS (FM 316)
- ARCHIVE_CLOSE_FILE (FM 299)

Refer to program SBOOKA in the SAP ERP system for detailed implementation information.

[↓] Import

CALL_DELETE_JOB_IN_TEST_MODE

If this parameter is X, it runs the delete program in test mode only. The value [Space] runs the delete program in normal mode. The default value is X.

CREATE_ARCHIVE_FILE

If this parameter is set to X, archive files are created; otherwise, they aren't created. The default value is X.

OBJECT

Archiving object name. You announce the parameter to which you want to open archiving object archives. The archiving object must be known to the system; otherwise, an OBJECT_NOT_FOUND exception is thrown. This is a mandatory parameter. Type: ARCH_OBJ-OBJECT.

COMMENTS

Comment line for archiving session. The default value is [Space].

DO_NOT_DELETE_DATA

Do not delete archived data. This parameter controls whether regular archive files or archive files with the status Database Copy are created. Archive files with the status Database Copy can only be read. No deletion run is generated for these archive files. Value [Space] determines that archive files have the normal status, and value X determines archive files have the status of Database Copy. The default value is [Space].

OUTPUT_SEL_SCREEN_WHEN_DIALOG

This parameter controls whether the selection screen of the archiving program is displayed when the archiving program is running in the dialog mode. Value [Space] means the selection screen won't be displayed, and value X means it will be displayed. Type: C.

OUTPUT_SEL_SCREEN_WHEN_BATCH

This parameter controls whether the selection screen of the archiving program is displayed when the archiving program is running in the background. Value [Space] means the selection screen won't be displayed, and value X means it will be displayed. Type: C.

DESTROY

Indicator for archiving session for data destruction. The default value is ABAP_FALSE. Type: ADMI_DESTR.

[↑] Export

ARCHIVE_HANDLE

Handle to open files for writing. The handle must be generated via the FMs ARCHIVE_OPEN_FOR_READ (FM 310), ARCHIVE_OPEN_FOR_MOVE (FM 309), or ARCHIVE_OPEN_FOR_DELETE (FM 308); otherwise, the exception WRONG_ACCESS_TO_ARCHIVE is raised. Type: SY-TABIX.

[!] Exceptions

INTERNAL_ERROR

Internal error.

OBJECT_NOT_FOUND

Object unknown.

OPEN_ERROR

Could not open file.

NOT_AUTHORIZED

No authorization for this FM.

ARCHIVING_STANDARD_VIOLATION

Violation of the archiving standard.

---

312 ARCHIVE_PUT_RECORD

This FM is used to pass a data record to the data object that has already been requested with the FM ARCHIVE_NEW_OBJECT (FM 307). All data records you pass to the data object are written to the archive file together when you call the FM ARCHIVE_SAVE_OBJECT (FM 315).

Example: You're creating an archive program to archive production orders that are older than 90 days, so you need to pass the records to archive. You can pass the relevant parameters in this FM to write the records to archive the data objects. See FM ARCHIVE_OPEN_FOR_WRITE (FM 311).

[↓] Import

ARCHIVE_HANDLE

Handle to open archive files. This is a mandatory parameter. Type: SY-TABIX.

RECORD

Data record. The parameter blocks the memory area in which a record of a data object is stored or to be stored.

The FMs of archiving don't check whether the size of the storage area is sufficient to accommodate the record.

RECORD_FLAGS

The parameters specified in the ADK storage area are stored in the control information for a data object or a data object to be stored. Each set in a data object may include additional information in these eight characters (flags). In the context of sets of an archiving object, there are no rules for using these flags. However, in the context of an archiving class ARCH_FLAGS, the structure is used consisting of FLAG_1 to FLAG_8. FLAG_1 is for sentences belonging to a storage class, reserved as a deletion indicator.

You can, for example, also set an archiving object in which RECORD_FLAGS controls whether the data set deleting this after archiving should be considered or not. You only need to note here that all of the programs written to this object archiving understand the encryption in this field. Type: ARC_BUFFER-FLAGS.

RECORD_STRUCTURE

Name of returned structure. If you're not using the direct structure name, but with an alias, the alias for the assignment of structure in Transaction AOBJ (archiving object definition) must be made. Then pass the alias on RECORD_STRUCTURE. This alias is returned for subsequent read accesses. Type: ARC_BUFFER-RNAME.

[!] Exceptions

INTERNAL_ERROR

Internal error.

WRONG_ACCESS_TO_ARCHIVE

Incorrect access to the archive.

INVALID_RECORD_STRUCTURE

Structure of the record is unknown.

---

313 ARCHIVE_PUT_TABLE

This FM places a data record in a current data object. To do this, the FM transfers an internal table to the data object that was previously requested by FM ARCHIVE_NEW_OBJECT (FM 307). Each line of the internal table is placed in the data object as a separate data record. All data records that are transferred to the data object are written together to the archive file when FM ARCHIVE_SAVE_OBJECT (FM 315) is called.

Example: You're creating the archiving program for production orders that are older than 90 days, which are listed in an internal table. You need to pass this table data along with other parameter in this FM and call this FM to put the table data in the archiving data object. Also refer to program RSARCH10 in SAP ERP.

[↓] Import

ARCHIVE_HANDLE

Handle to the open archive files. This is a mandatory parameter. Type: SY-TABIX.

RECORD_FLAGS

Refer to parameter RECORD_FLAGS of FM ARCHIVE_PUT_RECORD (FM 312). Type: ARC_BUFFER-FLAGS.

RECORD_STRUCTURE

Refer to parameter RECORD_FLAGS of FM ARCHIVE_PUT_RECORD (FM 312). Type: ARC_BUFFER-RNAME.

Tables

TABLE

Internal table for the data object.

RECORD_FLAGS_TABLE

Flag bar for all data records in the table. Type: ARCH_FLAGS.

[!] Exceptions

INTERNAL_ERROR

Internal error.

WRONG_ACCESS_TO_ARCHIVE

Incorrect access to the archive.

INVALID_RECORD_STRUCTURE

Structure of the record is unknown.

---

314 ARCHIVE_READ_OBJECT

This FM reads a data object for an archiving object from an archive file and returns a handle that acts as an import parameter for other archiving FMs. After the data object has been read by ARCHIVE_READ_OBJECT, you can access the handle to the data via the FM ARCHIVE_GET_NEXT_RECORD (FM 303).

Example: You've created an archive file, and you want to read the file and display it. Pass the parameter in this FM and execute to read the archive file.

[↓] Import

OBJECT

Archiving object name. You announce the parameter to which you want to open archiving object archives. The archiving object must be known to the system; otherwise an OBJECT_NOT_FOUND exception is thrown. This is a mandatory parameter. Type: ARCH_IDX-OBJECT.

OBJECT_ID

Data object key. Using this parameter, you specify an object for which you want to import the corresponding data records. The default value is [Space]. Type: ARCH_IDX-OBJECT_ID.

USER_EXIT_PROGRAM

User exit, program name. Using this program, you can read the index from the table in this program. If no program name is given, the index is read from index table ARCH_IDX. The default value is [Space]. Type: ARCH_GENER-EXIT_PROG.

USER_EXIT_FORM

User exit, form routine name. The default value is [Space]. Type: ARCH_GENER-EXIT_FORM.

ARCHIVKEY

Logical name of the archive file. The default value is [Space]. Type: ARCH_IDX-ARCHIVEKEY.

OFFSET

Data object address. The default value is 0. Type: ARCH_IDX-OFFSET.

MOVEFLAG

Reloading control indicator. If this flag is X, the system calls the FM ARCHIVE_OPEN_FOR_MOVE (FM 309) to reload the data object. If the flag is [Space], then the system calls the FM ARCHIVE_OPEN_FOR_READ (FM 310) for reading an individual data object. The default value is [Space]. Type: ARCH_OBJ-INDXFLAG.

[↑] Export

ARCHIVE_HANDLE

Pointer to open file. Type: SY-TABIX.

COMPR_OBJECT_LENGTH

Length of a data object in bytes as it's stored in compressed form in the archive file. Type: I.

[!] Exceptions

NO_RECORD_FOUND

No data object found.

FILE_IO_ERROR

Archive file access error.

INTERNAL_ERROR

Internal error.

OPEN_ERROR

Archive file open error.

CANCELLED_BY_USER

User has cancelled the action.

ARCHIVELINK_ERROR

ArchiveLink error occurred.

OBJECT_NOT_FOUND

Unknown object.

FILENAME_CREATION_FAILURE

File name could not be created.

FILE_ALREADY_OPEN

The system tried repeatedly to open a file.

NOT_AUTHORIZED

No authorization.

FILE_NOT_FOUND

The archive file does not exist.

315 ARCHIVE_SAVE_OBJECT

This FM writes a data object in the archive file. If the data object no longer fits into the current archive file, the archive file is closed, and a new archive file is opened. The FM checks whether an interrupt request is received for the current writing program. If so, the current data object isn't written to the archive file and cause the TERMINATION_REQUESTED exception.

Example: You're creating an archiving program to archive production orders, and you've received all of the data to archive in the archive data object from FMs such as ARCHIVE-PUT_RECORDS (FM 312) and ARCHIVE_PUT_TABLE (FM 313). You need to save this information in the archiving file. You can pass the relevant parameters of this FM and execute it to archive the file. See FM ARCHIVE_OPEN_FOR_WRITE (FM 311).

[↓] Import

ARCHIVE_HANDLE

Handle to the open archive file. In this case, the handle is used to specify in which file you want to write. The handle must have been created using the FM ARCHIVE_OPEN_FOR_WRITE (FM 311); otherwise, a WRONG_ACCESS_TO_ARCHIVE exception is thrown. Type: SY-TABIX.

[↑] Export

OBJECT_OFFSET

Offset of data object in the archive. The parameter OBJECT_OFFSET displays the address of the current data object in bytes. This address can be stored in the index table ARCH_IDX to enable direct access to the archived data object. Type: ARCH_IDX-OFFSET.

ARCHIVE_NAME

Archive key according to archive management. This value can change for each handle at runtime because the archiving can manage several archive files per handle. The current value is always returned. Type: HEADA-ARKEY.

END_OF_RETENTION

End date of the minimum retention period. Type: LRM_RETENTION_PERIOD_END_MIN.

[!] Exceptions

FILE_IO_ERROR

Archive file access error.

INTERNAL_ERROR

Internal error.

OPEN_ERROR

Archive file open error.

TERMINATION_REQUESTED

Termination of archiving was requested.

WRONG_ACCESS_TO_ARCHIVE

Incorrect access to the archive.

DATA_OBJECT_NOT_SAVED

The data object was not written to the archive.

316 ARCHIVE_WRITE_STATISTICS

This FM writes the statistical data to the standard log that was created by FM ARCHIVE_READ_STATISTICS. This FM can be used in all write, delete, reload, and read programs in data archiving. It should be called immediately before the FM ARCHIVE_CLOSE_FILE (FM 299). In reload programs, it must be called using the read handle. In all other programs, such as conversion, preprocessing, and post processing programs, this FM can't be used.

Example: You've created an archive program that deletes production orders older than three years. You need to find how many data objects are deleted, maintain their statistics, and write that information to the system logs. You can pass the relevant parameters in this FM and call this FM for the same purpose. Refer to FM ARCHIVE_GET_CUSTOMIZING_DATA (FM 300).

[↓] Import

ARCHIVE_HANDLE

Handle to the open archive files. This is a mandatory parameter. Type: SY-TABIX.

STATISTICS_ONLY_PER_RUN

Do not output statistics for single files. If this parameter is X, statistics are generated only for the complete archiving session. If this parameter is [Space], detailed statistics are provided for each archive file, and statistics for the complete archiving session are also generated. However, if parameter STATISTICS_ON_PER_FILE is set to X, the system won't output statistics for the complete archiving session. The default value is [Space].

STATISTICS_ONLY_PER_FILE

This parameter controls how the standard logs are displayed. If X, no statistics are displayed for the whole archiving session. If [Space], the statistics are listed for all archive file and for the whole archiving session. The default value is [Space].

[!] Exceptions

INTERNAL_ERROR

Internal error.

WRONG_ACCESS_TO_ARCHIVE

Incorrect access from the archive.

317 AS_API_READ

This FM is used to read data from the archive information system. This FM returns the archive file and offsets the location that can be used for the index-based access of archived data. This FM won't execute the archive accesses but will find where in the archive a special data object can be found. The actual archives are read from FM ARCHIVE_READ_OBJECT (FM 314) and FM ARCHIVE_GET_TABLE (FM 305).

Example: You want to read the two-year-old production data of your business, which you've archived in an archive object. You can pass the relevant parameter and call this FM to get these details.

[↓] Import

I_FIELDCAT

Field catalog for which data is to be read. Type: AIND_FCAT.

I_SELECTIONS

Selections to be used for searching.

I_OBLIGATORY_FIELDS

Fields that have to appear in the InfoStructure. Type: TABLE.

I_MAXROWS

Maximum number of InfoStructure records to be selected. Type: I.

[↑] Export

E_RESULT

Results table with the archive information returned. Type: TABLE.

[!] Exceptions

PARAMETERS_INVALID

Parameters are incorrect.

NO_INFOSTRUC_FOUND

No suitable InfoStructure found.

FIELD_MISSING_IN_FIELDCAT

Field missing from field catalog.

318 NUMBER_CHECK

This FM checks whether the specified number lies in the number range interval.

Example: You have a field in your ZTABLE called ID, for which you've defined a number range. You want to check whether the ID specified is present in the number range or not by passing the parameter in this FM and executing it.

[↓] Import

NR_RANGE_NR

The number range number clearly identifies a number range interval for an object/subobject. For types 1 and 4, the number range number is allocated manually via the maintenance dialog. For types 2, 3, and 5–8, the allocation is automatic: The number range number is assigned to the elements via the group table. Type: INRI-NRRANGENR.

NUMBER

Number to be checked.

OBJECT

The name of the number range object must be passed as it was defined in number range object maintenance. Type: INRI-OBJECT.

SUBOBJECT

This parameter contains the value of subobjects. This is specified only for types 4–8. The default value is [Space].

TOYEAR

Value of "to" fiscal year. This parameter is specified only for objects with intervals that distinguish by years. The year is the "to" fiscal year; that is, the year when the interval ceases to be valid. The default value is 0000. INRI-TOYEAR.

NUMERIC_CHECK

If this parameter is set to X, the input string is checked for numeric values. This check is only performed if the upper- and lower limits of the external interval are numeric.

[↑] Export

RETURNCODE

If the number specified in the parameter NUMBER lies in the interval, this parameter has the value [Space], and if the number doesn't lie in the interval, this parameter has the value X. Type: INRI-RETURNCODE.

[!] Exceptions

INTERVAL_NOT_FOUND

Interval not found. The number range number could not be found for the object/subobject. The exception is also raised when no valid-to year is found for year-dependent objects.

NUMBER_RANGE_NOT_EXTERN

Number range is not external. The specified number range number interval is not external.

OBJECT_NOT_FOUND

Object not defined in number range table TNRO. A number range object was not defined under the object name.

319 NUMBER_GET_INFO

This FM provides the information on the interval of the number range based on the import parameters.

Example: If you want to get the information of the number range, you can pass the parameter in this FM and execute. The interval information of the number range and current number of the number range will be returned.

[↓] Import

NR_RANGE_NR

The number range number clearly identifies a number range interval for an object/subobject. For types 1 and 4, the number range number is allocated manually via the maintenance dialog. For types 2, 3, and 5–8, the allocation is automatic: The number range number is assigned to the elements via the group table. Type: INRI-NRRANGENR.

OBJECT

The name of the number range object must be passed as it was defined in number range object maintenance. Type: INRI-OBJECT.

SUBOBJECT

This parameter contains the value of the subobject. This is specified only for types 4–8. The default value is [Space].

TOYEAR

Value of the "to" fiscal year. This parameter is specified only for objects with intervals that distinguish by years. The year is the "to" fiscal year; that is, the year when the interval ceases to be valid. The default value is 0000. Type: INRI-TOYEAR.

[↑] Export

INTERVAL

This parameter contains the number range interval information. Type: NRIV.

NRIV

Returns number range interval. Type: NRIV_TT.

[!] Exceptions

INTERVAL_NOT_FOUND

Interval not found. The number range number could not be found for the object/subobject. The exception is also raised when no valid-to year is found for year-dependent objects.

OBJECT_NOT_FOUND

Object not defined in the number range table TNRO. A number range object was not defined under the object name.

320 NUMBER_GET_NEXT

This FM assigns the next free numbers for a number range interval to the specified object. By default, a number is assigned when no greater number is required in the parameter QUANTITY. If more than one number is required—that is, parameter QUANTITY is specified with the quantity of the number to be assigned and fewer numbers are available to the interval limit—only the numbers are assigned up to the interval boundary. If the last number of the interval range is assigned, the first interval number is delivered on the next call again. The length of the number depends on the definition of the number range object.

Example: You've created the number range for the EMPID. Whenever a new employee joins, you're required to fill in his details in your program and give him the Employee ID number (which is determined from the number range). To use the number range safely, it's recommended to first lock that number range. You can pass the parameters in the FM NUMBER_RANGE_ENQUEUE (FM 322) and execute this FM to lock the number range you want to use. You can generate the next number of the number range by passing the parameters in this FM and executing. You'll get the next number. After you're done generating the numbers, you should dequeue the lock on the number range by passing the object in the FM NUMBER_RANGE_DEQUEUE (FM 321).

[↓] Import

NR_RANGE_NR

The number range number clearly identifies a number range interval for an object/subobject. For types 1 and 4, the number range number is allocated manually via the maintenance dialog. For types 2, 3, and 5–8, the allocation is automatic: The number range number is assigned to the elements via the group table. This is a mandatory parameter. Type: INRI-NRRANGENR.

OBJECT

The name of the number range object must be passed as it was defined in number range object maintenance. This is a mandatory parameter. Type: INRI-OBJECT.

QUANTITY

Number of numbers. This parameter specified the number of numbers that should be assigned. If fewer than this number of numbers are available, this parameter contains the number of numbers assigned. The default value is 1. Type: INRI-QUANTITY.

SUBOBJECT

This parameter contains the value of the subobject. This is specified only for types 4–8. The default value is [Space].

TOYEAR

Value of the "to" fiscal year. This parameter is specified only for objects with intervals distinguish by years. The year is the "to" fiscal year, that is, the year when the interval ceases to be valid. The default value is 0000. Type: INRI-TOYEAR.

IGNORE_BUFFER

Ignore object buffering. The parameter IGNORE_BUFFER causes numbers to be taken directly from the database. Number range object buffering is ignored. The application programs must prevent numbers for a number range object from being procured both from the buffer and from the database during a LUW because this may cause a deadlock situation. The default value is [Space].

[↑] Export

NUMBER

Free numbers are returned. When more than one number is requested, the last number issued is passed. The free numbers then lie in the following interval: NUMBER - QUANTITY + 1 to NUMBER.

QUANTITY

Specifies the number of numbers that should be assigned. If fewer than this number of numbers are available, this parameter contains the number of numbers assigned. Type: INRI-QUANTITY.

RETURNCODE

If the number specified in the parameter NUMBER lies in the interval, this parameter has the value [Space], and if the number doesn't lies in the interval, this parameter has the value X. Type: INRI-RETURNCODE.

[!] Exceptions

INTERVAL_NOT_FOUND

Interval not found. The number range number could not be found for the object/subobject. The exception is also raised when no valid-to year is found for year-dependent objects.

NUMBER_RANGE_NOT_INTERN

Number range is not internal.

OBJECT_NOT_FOUND

Object is not defined in TNRO.

QUANTITY_IS_0

Number of numbers requested must be > 0.

QUANTITY_IS_NOT_1

Number of numbers requested must be 1.

INTERVAL_OVERFLOW

Interval used up. Change not possible.

321 NUMBER_RANGE_DEQUEUE

This FM is used to unlock the number range. You can lock the number range from FM NUMBER_RANGE_ENQUEUE (FM 322).

Example: You're creating a specific production order and using the number range to get the next number. Before using the number range, you've already locked the number range by calling the FM NUMBER_RANGE_ENQUEUE (FM 322) and generating the number from FM NUMBER_GET_NEXT (FM 320). Because you're done with the number range object creation, you need to unlock the number range by calling this FM.

[↓] Import

OBJECT

Name of number range object. Type: TNRO-OBJECT.

[!] Exception

OBJECT_NOT_FOUND

Object not defined in number range table TNRO. A number range object was not defined under the object name.

---

322 NUMBER_RANGE_ENQUEUE

This FM is used to lock the number range object.

Example: See FM NUMBER_GET_NEXT (FM 320). You've created the number range for the EMPID. Whenever a new employee joins, you're required to fill in his details in your program and give him the Employee ID number (which is determined from the number range). To use the number range safely, it's recommended to first lock that number range. You can pass the parameters in this FM and execute to lock the number range you want to use.

[↓] Import

OBJECT

Name of number range object. Type: TNRO-OBJECT.

[!] Exceptions

FOREIGN_LOCK

Number range object is locked for maintenance.

OBJECT_NOT_FOUND

Object not defined in number range table TNRO. A number range object was not defined under the object name.

SYSTEM_FAILURE

System error occurred during locking.

---

323 NUMBER_RANGE_INTERVAL_UPDATE

This FM is used to create, change, or delete number range intervals. The changes are transferred from the ABAP program through an internal table to the local memory where they are first written. Before calling this FM, you must call FM NUMBER_RANGE_ENQUEUE (FM 322) to lock the number range object. This is to avoid competing updates. In the case of an error, no change is copied, so that the local memory always has a consistent, error-free status. The error is transferred to the calling program via an error structure. As soon as an error is found, the FM cancels processing.

Example: You've created the number range for the employee ID. You want to modify the number range interval. You can pass the parameters and call this FM to change the interval of the number range. After you've made the changes, they go into the local memory. Next, you need to write the changes to the database, so you use FM NUMBER_RANGE_UPDATE_CLOSE (FM 324).

[↓] Import

OBJECT

Name of number range object. Type: TNRO-OBJECT.

CHECK_AT_ALL_EVENTS

Interval will be checked if value is X. Default: [Space]. Type C.

[↑] Export

ERROR

In case of error, the structure contains the error found or a warning. The error or warning number is in the field MSGNR of this structure. Type: INRER.

ERROR_OCCURED

If this parameter is set to X, it means an error occurred during testing.

WARNING_OCCURED

If this parameter is set to X, it means a warning occurred during testing.

Tables

ERROR_IV

The intervals for which overlaps were found are passed in the table. If no intervals overlap, this table is empty. Type: INRIV.

INTERVAL

Table with the intervals to be changed. The processing flag PROCIND of this table can have the following values:

- U: Change interval.
- I: Insert interval.
- Y: Insert year for an existing interval.
- D: Delete interval.
- [Space]: No changes.

Type: INRIV.

[!] Exception

OBJECT_NOT_FOUND

Object not defined in number range table TNRO. A number range object was not defined under the object name.

324 NUMBER_RANGE_UPDATE_CLOSE

This FM writes the changes that were placed in the local memory to the database in the previous FM. This FM doesn't execute the `COMMIT` nor does it unlock the number range object. Before the changes are copied into local memory, they are checked.

Example: Execute this FM to write your changes to the database to complete the process begun in FM NUMBER_RANGE_INTERVAL_UPDATE (FM 323).

[↓] Import

OBJECT

Name of the number range object. Type: TNRO-OBJECT.

INTERVAL

Number range interval. Type: INRIV_TT.

[!] Exceptions

NO_CHANGES_MADE

Nothing was changed.

OBJECT_NOT_INITIALIZED

The object was not initialized.

325 RKD_WORD_WRAP

This FM word wraps the text entered in a parameter.

Example: You have a long character string to show in the information box in output. You can word wrap the string so that after the length of the box is reached, data is displayed in the next line. You can pass the string in the parameter TEXTLINE and call this FM. This FM returns the output lines that you can display in your box.

[↓] Import

TEXTLINE

Source text line. This is a mandatory parameter. Type: C.

DELIMITER

Indicator, which is used as a separator. The default value is [Space]. Type: C.

OUTPUTLEN

Maximum output line width. The default value is 35. Type: I.

[↑] Export

OUT_LINE1

Output line 1. Type: C.

OUT_LINE2

Output line 2. Type: C.

OUT_LINE3

Output line 3. Type: C.

Table

OUT_LINES

All output lines comes in this table.

[!] Exception

OUTPUTLEN_TOO_LARGE

Output length is too large to convert.

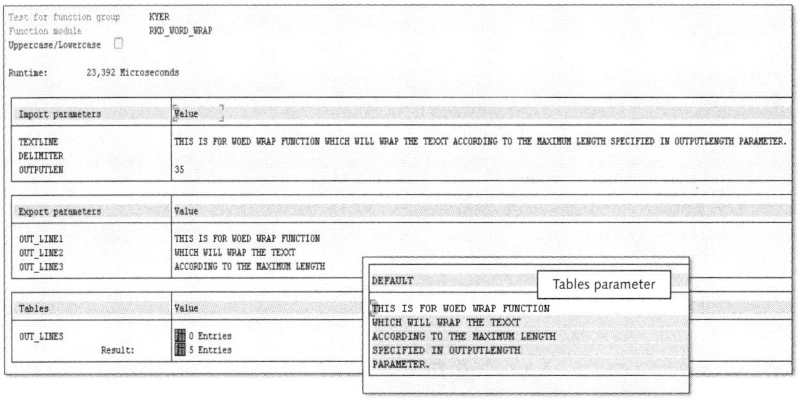

Word Wrap Function

---

326 SPELL_AMOUNT

This FM converts an amount or number into words. This FM can be used for the following reasons:

- **To convert a number into words**
  You need to pass the parameters LANGUAGE and AMOUNT.
- **To convert an amount into words**
  You need to pass the parameters LANGUAGE, CURRENCY, and AMOUNT.
- **To set the number of places before and after the decimal point for an amount**
  You need to enter parameters CURRENCY and AMOUNT. The LANGUAGE field must be transferred with the value [Space].

This FM uses table T015Z (Figures and Numbers in Words) for the conversion.

Example: You have an amount in a Figures column and an amount in a Word column. You want to display both the columns but the amounts are in number format only. You can pass the numbers in the parameter in this FM and execute this FM. Exporting parameter IN_WORDS returns the amount in the word format.

[↓] Import

AMOUNT

Amount/number to be spelled out.

CURRENCY

Currency for amounts, blank for numbers. The default value is [Space]. Type: SY-WAERS.

FILLER

Filler for padding the output field. The default value is [Space].

LANGUAGE

Language indicator. The default value is SY-LANGU. Type: SY-LANGU.

[↑] Export

IN_WORDS

This structure transfers the converted amounts. If you're calling the FM for entering the number of places before and after the decimal point, you need to pass the values to the NUMBER and DECIMAL fields of this structure. Type: SPELL.

[!] Exceptions

NOT_FOUND

Argument not found in T015Z.

TOO_LARGE

Amount too large to be spelled out.

327 STRING_CONCATENATE

The FM puts two character strings together in one character string. Blanks at the end of the first string are truncated, unless they are of the type STRING. Blanks at the beginning of the second character string are included. The behavior is the same as with the ABAP command `CONCATENATE .. IN CHARACTER MODE`. There are, however, some differences to the `CONCATENATE` command:

- The parameter STRING must not be of the type STRING, but rather of the type C.
- If the result is longer than the parameter STRING permits, the exception TOO_SMALL is thrown, and the parameter STRING remains unchanged.
- If the system code page is a multibyte code page (e.g., Chinese, Japanese, Korean), the `CONCATENATE` command checks whether the parameters STRING1 and STRING2 end with the first byte of a double-byte character. If this is the case, this byte is replaced by a blank. In

contrast to this, the FM always links the parameters STRING1 and STRING2 without performing a check or making changes.

Example: You want to concatenate two strings together and display the single string in the output screen as follows:

- String1: "Changes aren't allowed in Display Mode"
- String2: "Please reenter the program with change mode"

You want to concatenate both the strings. You can use this FM to pass the parameters as String1 and String2. Export parameter String will be the concatenation of the two import strings.

[↓] Import

STRING1

First string.

STRING2

Second string.

[↑] Export

STRING

Concatenated string.

[!] Exception

TOO_SMALL

Output field too short.

328 STRING_REVERSE

The FM returns the input string in reverse order. Leading spaces are taken into account, but not closing spaces.

Example: You want to reverse the string "ABC" and use it in your program. You can do so by passing the parameters in this FM and calling it. You'll receive the reversed string as "CBA". If the exporting parameter RSTRING is longer than the importing parameter STRING, the remaining digits are padded with spaces.

[↓] Import

STRING

Character string to be reversed.

LANG

Language in which you want to display the string

[↑] Export

RSTRING

Reverse character string.

[!] Exception

TOO_SMALL

Output field too short.

---

329 STRING_UPPER_LOWER_CASE

This FM eliminates the separator and makes the first letter just after the separator in uppercase and other letters of the word in lowercase.

Example: If you need to display the string without separators and you have the string "Space is a separator", execute this FM to get "SpaceIsASeparator".

[↓] Import

DELIMITER

Delimiter for separator.

STRING1

String to process.

[↑] Export

STRING

Processed string.

[!] Exceptions

NOT_VALID

Invalid delimiter string.

TOO_LONG

Delimiter is too long.

TOO_SMALL

Output field too short.

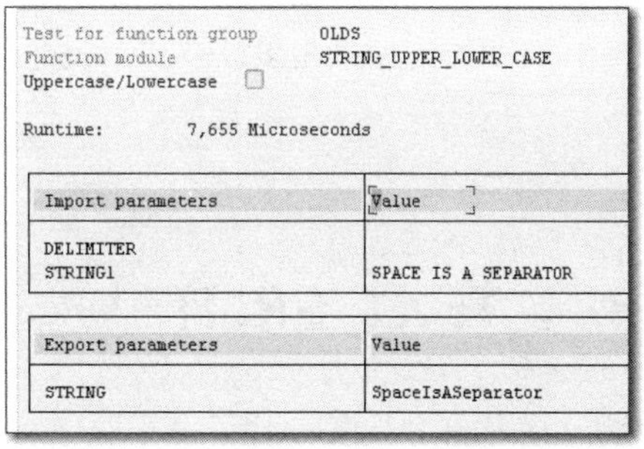

Execution of Function Module Removes the Separator

330 SWA_STRING_SPLIT

This FM is used to split the strings into substrings based on the import parameter TERMINATING_SEPARATORS.

Example: You've received a large text of a key such as 12345=121=23456, where each number string is separated by "=" and has a significance. To check if the first string number is of length 5 characters, you have to split the string into smaller strings based on the delimiter "-". Based on the split, you need to check if the length is 5 characters or not. You can pass the parameters in this FM and call this FM to receive the split string.

[↓] Import

INPUT_STRING

Input string that you want to split. Type: STRING.

MAX_COMPONENT_LENGTH

Maximum length of the segments. Type: I.

TERMINATING_SEPARATORS

Terminating delimiters such as "> =) }]" are given in this parameter. Type: STRING.

OPENING_SEPARATORS

Opening delimiters such as "< ({ [" are given in this parameter. Type: STRING.

Table

STRING_COMPONENTS

Table of the resulting substrings. Type: SWASTRTAB.

[!] Exception

MAX_COMPONENT_LENGTH_INVALID

Exceeded maximum supported length for subcomponents.

331 TEXT_SPLIT

This FM splits the texts. Separators for splitting can be [Space] or hyphen. The hyphen remains in the result, but the space doesn't. The result is filled with spaces from the end of the last word until the specified length. If the separator used ([Space] or hyphen) is followed by more separators, they are returned as leading separators in parameter REST.

Example: To split the string "I want to split this String" based on the separator [Space] when you reach the position of 30 characters, you can pass the parameters in this FM and call this FM. You can also do the same function from ABAP command `SPLIT`.

[↓] Import

LENGTH

The position where you want to split the string. The value is treated as a byte, not a character, even in the Unicode environment. If the value needs to be handled on a character basis in the Unicode environment, the other importing parameter AS_CHARACTER has to be set. This is a mandatory parameter.

TEXT

Input text. This is a mandatory parameter.

AS_CHARACTER

This flag is available only in the Unicode environment. In the non-Unicode environment, it doesn't function. If this flag is set (in the Unicode environment), the value of the other importing parameter LENGTH is treated as a character, not a byte.

[↑] Export

LINE

Results row.

REST

Remaining text that doesn't fits into line.

Part VII
Frontend Services and Microsoft Office Integration

This part contains the function modules (FMs) related to frontend services and Microsoft Office integration. Frontend services contain the FMs related to SAP GUI frontends such as GUI download, which downloads the FMs to your computer; GUI upload, which uploads the FMs; and FMs to check whether the GUI is available, active, and so on.

Office integration deals with the FMs related to Microsoft Office applications such as Word, Excel, and so on. You can use FMs in this topic to perform tasks such as opening the file in Word and Excel, downloading the data from SAP to Excel, and sending an email.

332 ALSM_EXCEL_TO_INTERNAL_TABLE

This FM is used to import data from Excel into an SAP internal table. This FM uses Object Linking and Embedding (OLE) technology, which is developed by Microsoft for embedding and linking the documents and other objects.

Example: You have data in your Excel sheet. You want to analyze the data and create the ALV report from the data. For that, you need to get the data from the Excel sheet. You can pass the parameters in this FM and then call it. You'll get the data imported from Excel to your internal table, which you can use to generate the ALV report.

[↓] Import

FILENAME

Excel file name you want to import the data from. Type: RLGRAP-FILENAME.

I_BEGIN_COL

Column from which you want to import the data. Type: I.

I_BEGIN_ROW

Row from which you want to import the data. Type: I.

I_END_COL

Column until which you want to import the data. Type: I.

I_END_ROW

Row until which you want to import the data. Type: I.

Tables

INTERN

Internal table returned with the Excel data. Type: ALSMEX_TABLINE.

[!] Exceptions

INCONSISTENT_PARAMETERS

Exception is raised if the parameters are inconsistent such as begin column/row is greater than end column/row.

UPLOAD_OLE

Exception while uploading from OLE such as when the file name entered is not valid.

---

333 EXCEL_OLE_STANDARD_DAT

This FM transfers the data from your system internal tables to your computer and formats it in Excel format, standard formatting via OLE 2.0. Note that for this to work, all columns in the worksheet must have a column heading. This FM internally calls FM MS_EXCEL_OLE_STANDARD_DAT (FM 344).

Example: You have an SAP report that analyzes the weekly data of your work, and you want the SAP data to be downloaded into your PC in Excel format so that you can use Excel functions for analysis. To do this, fill in the parameters and call this FM.

[↓] Import

FILE_NAME

Name of download file. Type: RLGRAP-FILENAME.

CREATE_PIVOT

Creating a pivot table. Possible values:

- 0: Don't create a pivot table.
- 1: Create a pivot table in addition to the data sheet.
- 2: Create a pivot table and suppress data sheet.

DATA_SHEET_NAME

Name of the data sheet. Type: RLGRAP-FILENAME.

PIVOT_SHEET_NAME

Name of the pivot sheet. Type: RLGRAP-FILENAME.

PASSWORD

Password for XLS file.

PASSWORD_OPTION

Regulates the use of the password. Possible values:

- 0: XLS file is saved without a password.
- 1: XLS file with the specified password (in the parameter PASSWORD) secured.
- 2: It creates its own password. This password is used to open the sheet.

Type: I.

Tables

PIVOT_FIELD_TAB

If the CREATE_PIVOT parameter has a value greater than 0, then data from this table goes to the pivot sheet. Type: EXCELPIVOT.

DATA_TAB

Data from this table goes to the data sheet.

FIELDNAMES

Table that contains column header details.

[!] Exceptions

FILE_NOT_EXIST

Specified file does not exist.

FILENAME_EXPECTED

No file name transferred.

COMMUNICATION_ERROR

SAPGUI receive error.

OLE_OBJECT_METHOD_ERROR

Error while calling OLE method.

OLE_OBJECT_PROPERTY_ERROR

Error while calling OLE attribute.

INVALID_PIVOT_FIELDS

Pivot fields are invalid.

DOWNLOAD_PROBLEM

Problem while downloading the data.

334 EXECUTE_WINWORD

This FM is used to open a winword.doc file from the presentation server.

Example: You have a Word file in your presentation server that you want to work in and then save the data back to the presentation server by clicking a (customizable) button in SAP. You can open the file by passing the file name in the import parameter of this FM and then calling this FM.

[↓] Import

I_FILE

Winword file path. Type: C.

335 FILE_GET_NAME

This FM assigns the physical file name using a logical file name. SAP ERP applications run on various platforms with various file systems. This FM enables you to use platform-independent logical file names in your application programs. Based on definitions maintained in customizing tables for platform-independent file names, the FM converts a logical file name to the corresponding physical file name and path for the hardware platform concerned.

For this conversion to work for different platforms, the definition of a logical file name must include a logical file path, which in turn is converted to different physical file paths, depending on the particular platform. The platform-specific file name returned by the FM is composed of the physical file path for the current platform and the physical file name associated with the logical file name. Placeholders in physical file names and path names are substituted at runtime by the corresponding current values.

Example: Your SAP system can be executed over different operating systems such as Windows and UNIX. You've maintained the logical file details in your ABAP Data Dictionary customizing table. Based on the operating system, you want the path of the physical file for the given operating system. You can fill in the parameters and call this FM. You'll receive the physical file name and its path, which you can use for storing your data. For example, logical filename = Weekly_Sales_Data_File, physical file name = Sales<Param1>, logical path = Sales_path_logical. The physical path of UNIX is /usr/<SYSID>/<FILENAME> and the physical path of Windows is C:\SALES\<Filename>. When you call this FM for UNIX and pass the parameter LOGICAL_FILENAME, export parameter FILE_NAME is returned with the value /usr/SYS12/VALUE_SALES. When you call this FM for Windows and pass the same logical file name, export parameter FILE_NAME is returned with value C:\SALES\VALUE_SALES.

[↓] Import

CLIENT

Logical file name definitions can be client-specific. If you don't want to use the file name definition of the current client, you can specify the client to be used. Type: SY-MANDT.

LOGICAL_FILENAME

You specify the platform-independent logical file name in this parameter. Uppercase letters are required because this is the format in which logical file names are stored in the customizing tables. Type: FILENAME-FILEINTERN.

OPERATING_SYSTEM

Here you can specify the operating system for which the platform-specific file name is to be generated.

The value of this parameter is also used for substituting the reserved word <OPSYS> in the physical file names and path names. The default value is SY-OPSYS. Type: SY-OPSYS.

PARAMETER_1

The reserved word <PARAM_1> in the physical file names and path names is replaced by the contents of the parameter PARAMETER_1. If no value is specified, the reserved word <PARAM_1> is deleted or replaced by blank characters. When you call the FM, you must make sure that the reserved word is given a value.

PARAMETER_2

The reserved word <PARAM_2> in physical file names and path names is replaced by the contents of the parameter PARAMETER_2. If no value is specified, the reserved word <PARAM_2> is deleted or replaced by blank characters. When you call the FM, you must make sure that the reserved word is given a value.

PARAMETER_3

The reserved word <PARAM_3> in physical file names and path names is replaced by the contents of the parameter PARAMETER_3. If no value is specified, the reserved word <PARAM_3> is deleted or replaced by blank characters. When you call the FM, you must make sure that the reserved word is given a value.

USE_PRESENTATION_SERVER

This parameter specifies that the operating system of the presentation server is to be used to determine the file name. As standard, the operating system is identified by the import parameter OPERATING_SYSTEM.

WITH_FILE_EXTENSION

If the value specified here isn't [Space], then the file format of the logical file name is appended as a file name extension to the platform-specific file name.

If no file format is defined for the logical file name, the parameter has no effect.

If the FM is called for a logical file name without a logical path, the parameter isn't evaluated.

USE_BUFFER

If the parameter is a blank space, then entries from control tables for the logical file name and its logical path are re-read at each FM call.
If the parameter isn't [Space], the FM checks whether the table entries required have already been read. This setting improves performance if the FM is called in a loop.

ELEMINATE_BLANKS

This indicator defines whether all spaces are to be removed from the returned file names. If this indicator has the value X, all spaces are removed. If the indicator has the value [Space], only the leading spaces are removed. Spaces within the file name are retained. This can lead to problems, depending on how the file name is used. If you use it in non-Unicode programs, the file names are truncated as of the first space. Type: SY-DATAR.

INCLUDING_DIR

Selection for screen field input. Type: SY-DATAR.

[↑] Export

EMERGENCY_FLAG

If the parameter EMERGENCY_FLAG doesn't have the value [Space], then no physical file path was found for the file. In this case, the path stored in the profile parameter DIR_GLOBAL is used for generating the platform-specific file name. Possible causes:

- The logical file path is unknown.
- The operating system isn't defined.
- The operating system isn't assigned to a group.
- No physical file path is stored for the operating system group.

FILE_FORMAT

The file format defined for the logical file name is returned. You need this value for downloading files to the presentation server. Type: FILENAME-FILEFORMAT.

FILE_NAME

The platform-specific file name and path are returned, completely instantiated. File systems can differ as to the use of mixed-case letters in file names and path names. If mixed-case physical file names and path names have been stored, then a new file may be created in one system, while an existing file is overwritten in another system. Example: The file names *ABCD.C* and *Abcd.c* are different in UNIX; however, they both translate to the same file name in DOS.

[!] Exception

FILE_NOT_FOUND

The logical file name was not found. A possible source of error is that the file name was passed in uppercase and lowercase or a table entry is missing. The table is maintained in the Transaction FILE.

336 FILENAME_GET

This FM is used to select a specific file from the user's computer. The user selects the file from the presentation server and provides the input in the popup, and then that file name is received in the export parameter EV_FILENAME of this FM.

Example: You want to provide F4 help on the file name. You can fill in the parameters and call this FM. The popup appears so that you can enter the file name details.

[↓] Import

IV_FILENAME

Input file name. Type: SDOK_FILNM.

TITLE

Title of the file input popup.

[↑] Export

EV_FILENAME

Returned file name. Type: SDOK_FILNM.

---

337 GUI_DOWNLOAD

This FM transfers an internal table's data from the server to a file on your PC. The GUI_DOWNLOAD module replaces the obsolete modules WS_DOWNLOAD and DOWNLOAD. The file dialog of the DOWNLOAD module is available in the class CL_GUI_FRONTEND_SERVICES.

Example: You want to transfer the internal table data of your report program to Word or Excel or as a comma separated file. Fill in the parameters of this FM and execute it, and your file is downloaded to the presentation server.

[↓] Import

BIN_FILESIZE

Amount of data transferred. This information needs to be specified if the data are transferred in binary format. Type: I.

FILENAME

Name of file into which data should be stored. If no directory path is specified, the file is stored in the current directory; that is, the directory that was used in the last file operation that was completed. Type: STRING.

FILETYPE

Format in which the data should be transferred.

Possible values:

- ASC: Data should be transferred as ASCII text. Data is converted as necessary. This is the default value.
- BIN: Data should be transferred unconverted in binary format.
- DAT: Data should be transported unconverted as ASCII text where the different columns are separated by tabs.
- DBF: Data should be sent in dBASE format in which the data types of the data in the different columns is included in the data sent.
- IBM: Data should be transferred as ASCII text in code page 1103.
- VSS: Data should be sent in ASCII format. Each line has two columns where the first column is a length field for the text in the second column.
- WK1: Data should be transported in Lotus 1-2-3 format.

Type: CHAR10.

APPEND

This parameter indicates whether the data should be appended to an existing file or not.

Possible values:

- [Space]: Data should not be appended to an existing file. If a file with the specified name exists, the data in that file is overwritten.
- X: Data should be appended to the data of an existing file. A new file is created if no file with that name exists.

[Space] is the default value. Type: CHAR01.

WRITE_FIELD_SEPARATOR

Indicates whether a tab character should be inserted between columns in the data transferred. This flag can be used in connection with the ASCII transfer types ASC, DAT, and IBM. With the transfer type DAT, this flag is set implicitly.

Possible values:

- [Space]: No tab character should be inserted between columns. This is the default for all transfer formats except DAT.
- X: Tab characters should be inserted between columns. This is done automatically if the transfer type is DAT.

Type: CHAR01.

HEADER

Hexadecimal string prepended to the data transferred. The length of this string may not exceed 1,023 characters. The default value is 00. Type: XSTRING.

TRUNC_TRAILING_BLANKS

This parameter indicates whether trailing blanks should be removed from the data transferred. This option is only relevant for the ASCII text transfer types ASC, DAT, and IBM.

Possible values:

- [Space]: Trailing blanks should be left untouched.
- X: Trailing blanks should be removed.

The default value is [Space]. Type: CHAR01.

WRITE_LF

Indicates whether a line feed should be included at the end of each table line.

Possible values:

- [Space]: No line feeds should be added at the end of the table line.

- X: A line feed should be appended to each row in the data table.

The default value is X. Type: CHAR01.

COL_SELECT

This parameter indicates whether only specific columns of a table should be added to the data transferred to the client.

This option is only relevant for the transfer formats DAT and DBF.

Possible values:

- [Space]: All columns should be transferred.
- X: Only the columns specified in the selection mask COL_SELECT_MASK should be sent.

The default value is [Space]. Type: CHAR01.

COL_SELECT_MASK

The columns to be transferred to the client can be specified in this parameter. To select a column to be transferred, an X has to be inserted at the corresponding position in the selection mask. Columns not to be added are marked with [Space].

The default value is [Space]. Type: CHAR255.

DAT_MODE

If this flag is set, columns are separated by tabs because they are in the transfer type DAT. Unlike in transfer mode, DAT trailing blanks aren't removed. Conversion exits aren't executed.

Possible values:

- [Space]: If columns should not be sent tab-separated in transfer types different from DAT.
- X: If columns should be sent tab-separated.

The default value is [Space]. Type: CHAR01.

CONFIRM_OVERWRITE

If this option is activated, the user is asked whether an existing file should be overwritten or not. Otherwise, the file will be overwritten without asking the user.

Possible values:

- [Space]: An existing file should be overwritten in any case.
- X: The user should be asked whether to overwrite an existing file or not.

The default value is [Space]. Type: CHAR01.

NO_AUTH_CHECK

This option allows activating/deactivating of the check of the authorization object S_GUI prior to attempting to execute this operation.

Possible values:

- [Space]: Check of authorization object should be deactivated.
- X: Check of authorization object should be done prior to executing operation.

The default value is [Space]. Type: CHAR01.

CODEPAGE

Encoding in which data should be transferred. This is the desired SAP code page. If [Space] is specified instead, the code page in which the SAP GUI is communicating with the server is used. The default value is [Space]. Type: ABAP_ENCODING.

IGNORE_CERR

Flag indicating whether errors in the conversion of the data should be ignored or not.

Possible values:

- ABAP_FALSE: Conversion errors should not be ignored.

- ABAP_TRUE: Conversion errors should be ignored.

The default value is ABAP_TRUE. Type: ABAP_BOOL.

REPLACEMENT

Character to use for characters that can't be converted, that is, they can't be represented in the specified encoding. The default value is #.

Type: ABAP_REPL.

WRITE_BOM

If this parameter is set to ABAP_TRUE, a byte order mark is added to the data transferred to the GUI if the data is encoded in Unicode, that is, UTF-8 or UTF-16 Big Endian/Little Endian.

Possible values:

- ABAP_FALSE: No byte order mark should be prepended to the data.
- ABAP_TRUE: Byte order mark should be added to the data transmitted.

The default value is [Space]. Type: ABAP_BOOL.

TRUNC_TRAILING_BLANKS_EOL

Flag indicating whether the trailing blanks in a line should be removed. This affects only the last column in a line.

Possible values:

- [Space]: The trailing blanks should not be removed.
- X: Trailing blanks should be deleted.

The default value is X. Type: CHAR01.

WK1_N_FORMAT

Specifies how numeric columns should be handled for transfer type WK1.

Possible values:

- [Space] or 0: Standard format should be used.
- 1<x>: Numeric columns should be interpreted as decimal values where 1<x> has to be replaced with the desired number of positions after the decimal point.
- 2<x>: Numeric columns should be interpreted as a currency value with <x> decimal positions.

The default value is [Space].

WK1_N_SIZE

Column width for numeric values for transfer type WK1. The default value is [Space].

WK1_T_FORMAT

Specifies alignment of text in cells containing text for transfer type WK1.

Possible values:

- 0: Text should be left-aligned.
- 1: Text should be right-aligned.
- 2: Text should be center-aligned.

The default value is [Space].

WK1_T_SIZE

Width of text column for WK1 transfer type.

Possible values: [0, 240].

The default value is [Space].

WRITE_LF_AFTER_LAST_LINE

This parameter indicates whether status messages regarding the amount of data transmitted and the code page used should be displayed in the status bar or not.

Possible values:

- [Space]: No status messages should be displayed.
- X: Status message should be displayed.

The default value is ABAP_TRUE. Type: ABAP_BOOL.

SHOW_TRANSFER_STATUS

Flag indicating whether status messages regarding the amount of data transmitted and the code page used should be displayed in the status bar or not.

Possible values:

- [Space]: No status messages should be displayed.
- X: Status message should be displayed.

The default value is ABAP_TRUE. Type: ABAP_BOOL.

VIRUS_SCAN_PROFILE

Virus scan profile. Default: /SCET/GUI_DOWNLOAD. Type: VSCAN_PROFILE.

[↑] Export

FILELENGTH

Number of bytes transferred to the client. Type: I.

Tables

DATA_TAB

Table containing data to be sent to the client.

FIELDNAMES

Field names for the DBF file type.

[!] Exceptions

FILE_WRITE_ERROR

Cannot write to file.

NO_BATCH

Frontend function cannot be executed in background.

GUI_REFUSE_FILETRANSFER

Incorrect frontend.

INVALID_TYPE

Invalid value for parameter FILETYPE.

NO_AUTHORITY

No download authorization.

UNKNOWN_ERROR

Raised if an error occurs during the conversion of data to be transferred to the specified encoding or during the file transfer.

HEADER_NOT_ALLOWED

Header not allowed.

SEPARATOR_NOT_ALLOWED

Separator not allowed.

FILESIZE_NOT_ALLOWED

File size must not be specified.

HEADER_TOO_LONG

The header information is limited to 1,023 bytes at present.

DP_ERROR_CREATE

Cannot create data provider.

DP_ERROR_SEND

Error sending data with DataProvider.

DP_ERROR_WRITE

Error writing data with DataProvider.

UNKNOWN_DP_ERROR

Error calling data provider.

ACCESS_DENIED

Raised if user declines overwrite of existing file.

DP_OUT_OF_MEMORY

Not enough memory in DataProvider.

DISK_FULL

Storage medium full.

DP_TIMEOUT

Timeout of DataProvider.

FILE_NOT_FOUND

Cannot find file.

DATAPROVIDER_EXCEPTION

General exception error in DataProvider.

CONTROL_FLUSH_ERROR

Error in control framework.

---

338 GUI_HAS_ACTIVEX

This FM tests whether ActiveX controls are supported in your GUI or you're using the normal SAP GUI. ActiveX is a software framework created by Microsoft that adapts its earlier Component Object Model (COM) and Object Linking and Embedding (OLE) technologies for content downloaded from a network, particularly from the Internet.

Example: You want to know the frontend of the SAP GUI you're running. Based on the GUI, you need to download the data from the Internet to SAP. You have to first execute this FM to see if the GUI has ActiveX, and then you have to use FM GUI_IS_ITS (FM 340) to see if the GUI is a normal GUI or called from the Internet.

[↑] Export

RETURN

Value X is returned in this parameter if ActiveX Controls are supported.

---

339 GUI_IS_AVAILABLE

This FM checks whether SAP GUI is available for use or SAP GUI is in use.

Example: In a Web Dynpro, you want to know if you've executed the code from SAP GUI or the Internet. You can call this FM. If the return parameter is X, SAP GUI is available, and you've executed the code from ABAP; otherwise, it's from the Internet browser.

[↑] Export

RETURN

Value X is returned if you're running SAP on SAP GUI. [Space] is returned if SAP is running over ITS. Type: C.

---

340 GUI_IS_ITS

This FM tests whether the GUI is a normal Windows GUI or Internet Transaction Server (ITS).

Example: You want to know if the GUI you're using is GUI for Windows or a web UI, so you execute this FM. Based on the GUI, you need to put your logic to download file from the Internet or from the SAP application server.

[↑] Export

RETURN

Value X is returned if you're running SAP on a browser or ITS. [Space] is returned if SAP is running over normal SAP GUI.

---

341 GUI_UPLOAD

This FM uploads the file from the client presentation server to the SAP server either as binary or as text. Numbers and date fields can be interpreted by the system according to the user settings. The GUI_UPLOAD module replaces the obsolete modules WS_UPLOAD and UPLOAD. The file dialog of the download module is available in the class CL_GUI_FRONTEND_SERVICES.

Example: You have an Excel file that contains specific data analysis. You want the data from that file to be publicly available, so you need to upload the file to the SAP server. Microsoft Word or Excel or a comma-separated file format can be uploaded by this FM. You can fill in the parameters of this FM and then execute it. Your file is uploaded to the SAP system.

[↓] Import

FILENAME

Name of file to be uploaded. This path can be either absolute or relative. Environment variables aren't allowed in the path. Type: STRING.

FILETYPE

Format in which the data should be transferred.

Possible values:

- ASC: Data should be transferred as ASCII text. Data is converted as necessary.
- BIN: Data should be transferred unconverted in binary format.
- DAT: Data should be transported unconverted as ASCII text where the different columns are separated by tabs.

The default value is ASC. Type: CHAR10.

HAS_FIELD_SEPARATOR

Flag indicating whether the different fields in the file are separated by tabs.

Possible values:

- [Space]: The fields aren't separated by tabs.
- X: Fields are separated by tabs.

The default value is [Space]. Type: CHAR01.

HEADER_LENGTH

Length of header in BIN transfer format. The default value is 0. Type: I.

READ_BY_LINE

Indicates whether one line in the data table should correspond to one line in the file. The data table may only have one column. Carriage return and line feed characters at the end of a line aren't removed if this option is activated.

Possible values:

- [Space]: This option should not be active.
- X: One line in the data table should correspond to one line in the file.

The default value is X. Type: CHAR01.

DAT_MODE

If this flag is set, columns are separated by tabs as they are in the transfer type DAT. Unlike in transfer mode DAT, trailing blanks aren't removed. Conversion exits aren't executed.

Possible values:

- [Space]: If columns should not be sent tab-separated in transfer types different from DAT.
- X: If columns should be sent tab-separated.

The default value is [Space]. Type: CHAR01.

CODEPAGE

Encoding in which data should be transferred. This is the desired SAP code page. If [Space] is specified instead, the code page in which the SAP GUI is communicating with the server is used. The code page used for the transferred data can be set in the SAP logon for each system entry.

The default value is [Space]. Type: ABAP_ENCODING.

IGNORE_CERR

Parameter indicating whether errors in the conversion of the data should be ignored or not.

Possible values:

- ABAP_FALSE: Conversion errors should not be ignored.
- ABAP_TRUE: Conversion errors should be ignored.

The default value is ABAP_TRUE. Type: ABAP_BOOL.

REPLACEMENT

Character to use for characters that can't be converted, that is, characters which can't be represented in the specified encoding.

The default value is #. Type: ABAP_REPL.

CHECK_BOM

The consistency of the code page and byte order mark are checked. The default value is [Space]. Type: CHAR01.

VIRUS_SCAN_PROFILE

Group of scanners to use for checking the uploaded file (see SAP Notes 797108, 786179, 639486 for details). Type: VSCAN_PROFILE.

NO_AUTH_CHECK

Switch off CHECK FOR ACCESS RIGHTS. The default value is [Space]. Type: CHAR01.

[↑] Export

FILELENGTH

Number of bytes transferred to the client. Type: I.

HEADER

File header in case of binary upload. Type: XSTRING.

Table

DATA_TAB

Table containing data to be sent to the client.

[!] Exceptions

FILE_OPEN_ERROR

File does not exist and cannot be opened.

FILE_READ_ERROR

Error reading file.

NO_BATCH

Frontend function cannot be executed in background.

GUI_REFUSE_FILETRANSFER

Incorrect frontend.

INVALID_TYPE

Invalid value for parameter FILETYPE.

NO_AUTHORITY

No download authorization.

UNKNOWN_ERROR

Raised if an error occurs during the conversion of data to be transferred to the specified encoding or during the file transfer.

BAD_DATA_FORMAT

Cannot interpret data in file.

HEADER_NOT_ALLOWED

Header not allowed.

SEPARATOR_NOT_ALLOWED

Separator not allowed.

HEADER_TOO_LONG

The header information is limited to 1,023 bytes at present.

UNKNOWN_DP_ERROR

Error calling DataProvider.

ACCESS_DENIED

Raised if user declines overwrite of existing file.

DP_OUT_OF_MEMORY

Not enough memory in DataProvider.

DISK_FULL

Storage medium full.

DP_TIMEOUT

Timeout of DataProvider.

342 KCD_EXCEL_OLE_TO_INT_CONVERT

This FM is used to import data from Excel into an SAP internal table. This FM uses Object Linking and Embedding (OLE) technology, which is developed by Microsoft for embedding and linking the documents and other objects. This FM is similar to FM ALSM_EXCEL_TO_INTERNAL_TABLE (FM 332).

Example: You have data in your Excel sheet that you want to analyze and then create an ALV report from the data. For that, you need to get the data from the Excel sheet. You can pass the parameters in this FM and call this FM. You'll get the data imported from Excel to your internal table, which you can use to generate an ALV report.

[↓] Import

FILENAME

Excel file name you want to import the data from. Type: RLGRAP-FILENAME.

I_BEGIN_COL

Column from which you want to import the data. Type: I.

I_BEGIN_ROW

Row from which you want to import the data. Type: I.

I_END_COL

Column until which you want to import the data. Type: I.

I_END_ROW

Row until which you want to import the data. Type: I.

Table

INTERN

Internal table returned with the Excel data. Type: KCDE_INTERN.

[!] Exceptions

INCONSISTENT_PARAMETERS

Exception is raised if the parameters are inconsistent such as the beginning column/row is greater than the ending column/row.

UPLOAD_OLE

Exception while uploading from OLE such as when the file name entered isn't valid.

---

343 KD_GET_FILENAME_ON_F4

This FM is used to locate the file name in F4 help.You need to provide the program name, fieldname, and screen number for which you want F4 help to be displayed. This FM displays a popup in which you can fill in the parameters to select the file from the presentation server if the changing parameter LOCATION_FLAG = P.

Example: You want to give F4 help on the file name. You can fill in the parameters and call this FM. You'll get the popup where you can locate the file.

[↓] Import

PROGRAM_NAME

Module pool program name for the screen field. The default value is SYST-REPID. Type: SY-REPID.

DYNPRO_NUMBER

Screen number where the F4 help is needed. The default value is SYST-DYNNR. Type: SY-DYNNR.

FIELD_NAME

Name of field where the path is to be entered. The default value is [Space].

STATIC

This parameter specifies whether the last-used path can be used again for the next call. If the value of this parameter is [Space], it can be used; otherwise, it cannot. The default value is [Space].

MASK

This parameter isn't used in this FM. Type: C.

FILEOPERATION

Operation to perform on the file. The default value is R. Type: DXFIELDS-FILEOPER.

PATH

Long path name. Type: DXFIELDS_LONGPATH

LOCATION_FLAG

This parameter decides between the application or presentation server. The default value is P. Type: DXFIELDS-LOCATION.

[Δ] Changing

FILE_NAME

Path name selected by user with help of FILEMNGR. Type: RLGRAP-FILENAME.

[!] Exception

MASK_TOO_LONG

The length of MASK must not be > 100.

---

344 MS_EXCEL_OLE_STANDARD_DAT

This FM transfers an ASCII file into Excel format via OLE 2.0 standard formatting. All of the columns of the worksheet must have a column heading.

Example: You want the SAP data to be downloaded into your PC in Excel format from the SAP system. You can fill in the parameters and can call this FM.

[↓] Import

FILE_NAME

Name of download file. Type: RLGRAP-FILENAME.

CREATE_PIVOT

Creating a pivot table. Possible values:

- 0: Don't create a pivot table.
- 1: Create a pivot table in addition to the data sheet.
- 2: Create a pivot table and suppress the data sheet.

DATA_SHEET_NAME

Name of the data sheet. Type: RLGRAP-FILENAME.

PIVOT_SHEET_NAME

Name of the pivot sheet. Type: RLGRAP-FILENAME.

PASSWORD

Password for XLS file.

PASSWORD_OPTION

Regulates the use of the password. Possible values:

- 0: XLS file is saved without a password.
- 1: XLS file with the specified password (in the parameter password) secured.
- 2: It creates its own password. This password is used to open the sheet.

Type: I.

Tables

PIVOT_FIELD_TAB

If the CREATE_PIVOT parameter has a value greater than 0, then data from this table goes to the pivot sheet. Type: EXCELPIVOT.

DATA_TAB

Data from this table goes to the data sheet.

FIELDNAMES

Table that contains column header details.

[!] Exceptions

FILE_NOT_EXIST

Specified file does not exist.

FILENAME_EXPECTED

No file name transferred.

COMMUNICATION_ERROR

SAP GUI receive error.

OLE_OBJECT_METHOD_ERROR

Error while calling OLE method.

OLE_OBJECT_PROPERTY_ERROR

Error while calling OLE attribute.

INVALID_PIVOT_FIELDS

Pivot fields are invalid.

DOWNLOAD_PROBLEM

Problem while downloading the data.

345 RH_START_EXCEL_WITH_DATA

This FM enables you to transfer ASCII data (parameter DATA_TABLE) to a frontend and subsequently to start Excel. The system automatically opens the file with the data (parameter DATA_FILENAME). You can start Excel modally or modelessly (parameter WAIT). If you start Excel modally, you can specify that the file is to be automatically deleted when Excel is closed (DELETE_FILE = X). In addition, when you start Excel, you can simultaneously load a macro. The name of the macro is in the parameter MACRO_NAME.

The FM can't be run in a batch. The runtime takes several seconds from the time that the first FM is called because a control flush must be generated. This FM must not be used in customer programs.

Example: You want to transfer the data from the internal table to Microsoft Excel. You want to make some changes in Excel so you want to open it as well. You can pass the parameters in this FM and execute it. Excel opens up with the data you sent in the DATA_TABLE parameter.

[↓] Import

DATA_FILENAME

Name of data file on frontend. Type: STRING.

DATA_PATH_FLAG

This parameter specifies from where to get the files. The default value is W. Possible values:

- [Space]: No directory, full file name is specified.
- D: Read from desktop directory.
- E: GUI program directory.
- Q: User dialog.
- T: Temp directory.
- W: Current directory for GUI.

Type: HRPATHFLAG.

DATA_ENVIRONMENT

Environment variable for path determination. Type: HRPATHENVM.

DATA_TABLE

Data table for the data sheet of Excel. Type: TABLE.

MACRO_FILENAME

Name of macro to be started. Type: STRING.

MACRO_PATH_FLAG

Flag for DIRECTORY EXTENSION. Possible values:

- [Space]: No directory, full file name is specified.
- D: Read from desktop directory.
- E: GUI program directory.
- Q: User dialog.
- T: Temp directory.
- W: Current directory for the GUI.

The default value is E. Type: HRPATHFLAG.

MACRO_ENVIRONMENT

Environment variable for path determination of macro. Type: HRPATHENVM.

WAIT

If this parameter has the value X, then it starts Excel and waits. The default value is X. Type: FLAG.

DELETE_FILE

General flag for deletion. The default value is X. Type: FLAG.

CODEPAGE

Identifier for character format (UTF-8, UCS-2, etc.). Type: ABAP_ENCOD.

[!] Exceptions

NO_BATCH

Batch call not permitted.

EXCEL_NOT_INSTALLED

Excel isn't installed on the frontend.

INTERNAL_ERROR

An internal error has occurred.

CANCELLED

Action cancelled by user.

DOWNLOAD_ERROR

File could not be created.

NO_AUTHORITY

No download authorization.

FILE_NOT_DELETED

File could not be deleted after call.

346 RS_SEND_MAIL_FOR_SPOOLLIST

This FM is used to generate and send an email to a user from the spool list.

Example: You want to send an email to the user when the scheduled job fails. You can pass the parameters in this FM and call this FM on the failure condition. An email is sent to the user with the failed spool list.

[↓] Import

SPOOLNUMBER

Spool number. The default value is SY-SPONO. Type: SY-SPONO.

MAILNAME

Mail name or details. This is a mandatory parameter. Type: SOOD1-OBJNAM.

MAILTITEL

Title of the email This is a mandatory parameter. Type: SOOD1-OBJDES.

USER

Recipient(s) of the email. The default value is SY-UNAME. Type: SY-UNAME.

Table

TEXT

Body of the email. Type: SOLI.

[!] Exception

ERROR

Error occurred while executing this FM.

347 SAP_CONVERT_TO_XLS_FORMAT

This FM converts SAP internal table data to the Excel format.

Example: You can download SAP data into your PC in Excel format by filling in the parameters and calling this FM.

[↓] Import

I_FIELD_SEPERATOR

Field separator in an internal table. Type: CHAR01.

I_LINE_HEADER

If the value is X for this indicator, the file will have header. Type: CHAR01.

I_FILENAME

Excel file name where data will be saved. This is a mandatory parameter. Type: RLGRAP-FILENAME.

I_APPL_KEEP

The default value is [Space]. Type: CHAR01.

[Δ] Changing

I_TAB_CONVERTED_DATA

This parameter isn't used in this FM. Type: TRUXS_T_TEXT_DATA.

Table

I_TAB_SAP_DATA

Internal table data to be converted. Type: STANDARD TABLE.

[!] Exception

CONVERSION_FAILED

Conversion failed.

---

348 SO_ATTACHMENT_INSERT_API1

This FM is used to create a new attachment for an existing document.

Example: You've created the report from ALV output, and you want to send the document as an email attachment to your supervisor. You can pass the parameters and call this FM to accomplish this task.

[↓] Import

DOCUMENT_ID

ID of a folder entry of the document that is to have the attachment. It's the connection between the document and the folder in which it's contained. Type: SOFOLENTI1-DOC_ID.

ATTACHMENT_DATA

This structure must contain the new attributes of the new attachment. This structure has the following fields:

- OBJ_NAME: Name of the attachment.
- OBJ_DESCR: Title (short description) of the attachment.
- OBJ_LANGU: Language of the attachment.
- ATT_SIZE: Size of the attachment in bytes. In the case of PC documents, the size of the respective file should be entered here. In the case of RAW and SCR documents, the size is calculated from the length of the last line + number of remaining lines multiplied by 255.

Type: SOATTCHGI1.

ATTACHMENT_TYPE

Type of attachment. Apart from folders (FOL) and distribution lists (DLI), all types are allowed. Allowed values:

- SCR: SAP script texts
- RAW: RAW editor texts
- GRA: SAP business graphics
- SHL: Pre-generated search interfaces
- QRY: individual search queries

Type: SOODK-OBJTP.

IV_VSI_PROFILE

Virus scan profile. Type: VSCAN_PROFILE.

[↑] Export

ATTACHMENT_INFO

This structure contains the complete attributes of the newly created attachment. Type: SOATTINFI1.

Tables

ATTACHMENT_HEADER

This table must contain the data dependent on the type of attachment. For example, SAPscript documents contain information on the layout set and style, Excel list viewer contains documents the number of lines and columns, and PC documents contain their original file name.

LINE: Requires the type-dependent information of the attachment by the line. Type: SOLISTI1.

ATTACHMENT_CONTENT

This table must contain the actual contents of the attachment.

LINE: Requires the contents of the attachment by the line. Type: SOLISTI1.

CONTENTS_HEX

Attachment contents in binary format. Type: SOLIX.

ET_VSI_ERROR

Messages are returned. Type: BAPIRET2.

[!] Exceptions

DOCUMENT_NOT_EXIST

The folder entry specified does not exist. An incorrect ID was probably passed or the relevant folder entry deleted.

ATTACHMENT_TYPE_NOT_EXIST

The specified type of attachment does not exist or is not allowed. It is not possible to have folders (FOL) or distribution lists (DLI) as attachments.

OPERATION_NO_AUTHORIZATION

It was not possible to create the attachment. The reason for this may be that the document related to the attachment is another user's private document, or it is a document in a shared folder for which the active user does not have a change authorization. It is also not generally possible to append an attachment to a document that has already been sent.

PARAMETER_ERROR

An invalid combination of parameter values was passed to the FM. An invalid value was probably entered in a field of the structure for the attributes of the attachment.

X_ERROR

Internal error or database inconsistency.

ENQUEUE_ERROR

A lock involved in the create process could not be set. Processing is probably being carried out by another user.

349 SO_ATTACHMENT_READ_API1

This FM returns all information regarding an attachment. This can include the attributes of the attachment (e.g., name, description, creator, date created, priority, expiration date, etc.), and the contents of the attachment and the type-specific information (e.g., the layout set and style used in SAPscript documents).

Example: You want to read an attachment you've received. You can pass the relevant parameters in this FM to read the attachment as well as view all details about it.

[↓] Import

ATTACHMENT_ID

ID of the attachment to be read. It is the connection between the attachment and the document it belongs to. Type: SOATTLSTI1-ATTACH_ID.

[↑] Export

ATTACHMENT_DATA

This structure contains the complete attributes of the newly created attachment. Type: SOATTINFI1.

Tables

ATTACHMENT_HEADER

This table must contain the data dependent on the type of attachment. For example, SAPscript documents contain information on the layout set and style, Excel list viewer documents contain the number of lines and columns, and PC documents contain their original file name.

LINE: Requires the type-dependent information of the attachment by the line. Type: SOLISTI1.

ATTACHMENT_CONTENT

This table must contain the actual contents of the attachment.

LINE: Requires the contents of the attachment by the line. Type: SOLISTI1.

CONTENTS_HEX

Attachment contents in binary format. Type: SOLIX.

[!] Exceptions

ATTACHMENT_NOT_EXIST

The specified attachment does not exist. An incorrect ID was probably passed, or the relevant attachment has been deleted.

OPERATION_NO_AUTHORIZATION

It was not possible to create the attachment. The reason for this may be that the document related to the attachment is another user's private document, or it is a document in a shared folder for which the active user does not have a change authorization. It is also not generally possible to append an attachment to a document that has already been sent.

PARAMETER_ERROR

An invalid combination of parameter values was passed to the FM. An invalid value was probably entered in a field of the structure for the attributes of the attachment.

X_ERROR

Internal error or database inconsistency.

ENQUEUE_ERROR

A lock involved in the create process could not be set. Processing is probably being carried out by another user.

---

350 SO_DLI_INSERT_API1

This FM is used to create a new distribution list.

Example: You want to send the failure notification email to a new set of people. For that, you need to create the distribution list where you can specify the set of mail IDs.

[↓] Import

FOLDER_ID

ID of the folder in which the distribution list is to be inserted. Type: SOFOLDATI1-OBJECT_ID.

DLI_INSERT_DATA

This structure must contain the attributes of the distribution list to be created. Type: SODLICHGI1.

[↑] Export

DLI_INFO

This structure contains the attributes of the newly created distribution list. Type: SODLIDATI1.

Table

DLI_ENTRIES

This table must contain the entries of the distribution list to be created. Type: SODLIENTI1.

[!] Exceptions

DLI_ALREADY_EXIST

The name of the distribution list to be created already exists. If you tried to create a private distribution list, and there is already a distribution list with the same name in the outbox, a private folder, or in the private trash, you'll get this exception.

FOLDER_NOT_EXIST

The folder in which the distribution list is to be created does not exist. An incorrect ID may have been passed, or the relevant folder has been deleted.

OPERATION_NO_AUTHORIZATION

It was not possible to create the attachment. The reason for this may be that the document related to the attachment is another user's private document, or it is a document in a shared folder for which the active user does not have a change authorization. It is also not generally possible to append an attachment to a document that has already been sent.

PARAMETER_ERROR

An invalid combination of parameter values was passed to the FM. An invalid value was probably entered in a field of the structure for the attributes of the attachment.

X_ERROR

Internal error or database inconsistency.

ENQUEUE_ERROR

A lock involved in the create process could not be set. Processing is probably being carried out by another user.

---

351 SO_DLI_READ_API1

This FM returns all information on a distribution list; that is, the attributes of the distribution list (e.g., name, description, creator, date created, sensitivity, etc.) and the entries in the distribution list.

Example: You've created the distribution list for your team. You want to view the IDs that are contained in the distribution list and its details to make sure that no one is missing. You can fill the parameters and call this FM to do so.

[↓] Import

DLI_NAME

Name of the distribution list. The name is unique for the shared folders or for each set of private folders. If both the name and ID of the distribution list are passed, the name is used for determination. The default value is [Space]. Type: SOOBJINFI1-OBJ_NAME.

DLI_ID

Object ID of the distribution list. If both the name and ID of the distribution list are passed, the name is used for determination. The default value is [Space]. Type: SOOBJINFI1-OBJECT_ID.

SHARED_DLI

If this parameter is set to X, the distribution list to be read is a shared distribution list; otherwise, it's assumed that it's a private distribution list. The default value is [Space]. Type: SONV-FLAG.

[↑] Export

DLI_DATA

This structure contains the attributes of the distribution list to read. Type: SODLIDATI1.

Table

DLI_ENTRIES

This table must contain the entries of the distribution list to be created. Type: SODLIENTI1.

[!] Exceptions

DLI_NOT_EXIST

Specified distribution list does not exist.

OPERATION_NO_AUTHORIZATION

It was not possible to create the attachment. The reason for this may be that the document related to the attachment is another user's private document or it's a document in a shared folder for which the active user does not have a change authorization. It is also not generally possible to append an attachment to a document that has already been sent.

PARAMETER_ERROR

An invalid combination of parameter values was passed to the FM. An invalid value was probably entered in a field of the structure for the attributes of the attachment.

X_ERROR

Internal error or database inconsistency.

352 SO_DOCUMENT_INSERT_API1

This FM allows you to create a new document in a specified folder in the SAP system. Documents are created under the folder and can be referred to with an object ID. The folder entries contain the contents and attributes of the document as well as some additional attributes related to the folder entry. These include the sender priority and the expiration date of the entry. Folder entries are the result of resubmissions, links, and sending and creating a new document. Documents are stored in table SOFFCONT1 in binary format.

Example: You want to create a document in the SAP system that's based on the output values of your program. You can do so by passing the parameters and calling this FM.

[↓] Import

FOLDER_ID

ID of the folder in which the document is to be created. Type: SOOBJINFI1-OBJECT_ID.

DOCUMENT_DATA

This structure must contain the attributes of the document to be created. Values in the fields of the attributes belonging to the document affect each future folder entry while entries in the entry-related fields only affect the folder entry generated automatically when the document is created.

Type: SODOCCHGI1.

DOCUMENT_TYPE

Type of document that can be created. Distribution lists (DLI) and folders (FOL) can't be created using this FM. Allowed values:

- SCR: SAP script texts
- RAW: RAW editor texts
- GRA: SAP business graphics
- SHL: Pre-generated search interfaces
- QRY: individual search queries

Type: SOODK-OBJTP.

IV_VSI_PROFILE

Virus scan profile. Type: VSCAN_PROFILE.

[↑] Export

DOCUMENT_INFO

Complete attributes of a document after the document has been created. Type: SOFOLENTI1.

Tables

OBJECT_HEADER

This table must contain the data dependent on the document type. For example, SAPscript documents have information on the layout set and style, Excel list viewer documents contain the number of lines and columns, and PC documents contain their original file names. Type: SOLISTI1.

OBJECT_CONTENT

This table must contain the actual contents of the document. Type: SOLISTI1.

CONTENTS_HEX

Document contents in binary format. Type: SOLIX.

OBJECT_PARA

This table is only used by documents that are to be processed. It must contain the SET/GET parameters that are passed to the processing element. Type: SOPARAI1.

OBJECT_PARB

This table is only used by documents that have been assigned a particular processing type. Depending on this, they have a different meaning. If the processing element is a report or transaction with value transfer to the global memory, the table contents are interpreted as a set of parameters with related values and exported to the memory ID derived from the first line. If the processing element is a function or dialog module, the table is passed to it as the table parameter MSGDIAL. Type: SOPARBI1.

ET_VSI_ERROR

Messages are returned. Type: BAPIRETZ.

[!] Exceptions

FOLDER_NOT_EXIST

The folder in which the distribution list is to be created does not exist. An incorrect ID may have been passed, or the relevant folder has been deleted.

DOCUMENT_TYPE_NOT_EXIST

Document type does not exist.

OPERATION_NO_AUTHORIZATION

It was not possible to create the attachment. The reason for this may be that the document related to the attachment is another user's private document, or it's a document in a shared folder for which the active user does not have a change authorization. It is also not generally possible to append an attachment to a document that has already been sent.

PARAMETER_ERROR

An invalid combination of parameter values was passed to the FM. An invalid value was probably entered in a field of the structure for the attributes of the attachment.

X_ERROR

Internal error or database inconsistency.

ENQUEUE_ERROR

A lock involved in the create process could not be set. Processing is probably being carried out by another user.

---

353 SO_DOCUMENT_READ_API1

This FM provides document information such as attributes of the document (e.g., name, description, creator, date created, priority, expiration date, etc.), transmission information (e.g., sender, date sent, forwarder, date received, send attributes, etc.), contents, type-specific information (e.g., for SAPscript documents, the layout set and style used), and processing information (type of processing, name of the processing element, and specification of transfer parameters).

A distinction needs to be made here between the document and its folder entries. The document itself only exists once in the database. It's used as a template for the folder entries and can be referenced using its object ID. The document can have any number of folder entries. They contain the contents and attributes of the document as well as some additional attributes referring to the folder entry. These include the sender priority and expiration date of the entry. Folder entries are the result of resubmissions, links, and sending and creating a new document.

Example: You need to read the document attached to the purchase order or to the delivery document. You can fill in the parameter and call this FM to read the document.

[↓] Import

DOCUMENT_ID

ID of folder entry to be viewed. It's the connection between the document and the folder in which it's located. Type: SOFOLENTI1-DOC_ID.

FILTER

This structure determines the preselection of required object data.

Values:

- SEND_INFO: If this flag is set (X), the transmission information is also read. The default is X.
- NO_CONTENT: If this flag is set (X), the contents (table OBJECT_CONTENT), the specific header data (table OBJECT_HEADER), and the processing parameters (table OBJECT_PARA and OBJECT_PARB) of the document are also read. The default is [Space].
- REC_LIST: If this flag is set (X), the recipient list (table RECEIVER_LIST) of the document is also read. The default is [Space].

Type: SOFILTERI1.

[↑] Export

DOCUMENT_DATA

Complete attributes of the document after the document has been read. Type: SOFOLENTI1.

Tables

OBJECT_HEADER

This table must contain the data dependent on the document type. For example, SAPscript documents have information on the layout set and style, Excel list viewer documents contain the number of lines and columns, and PC documents contain their original file names. Type: SOLISTI1.

OBJECT_CONTENT

This table must contain the actual contents of the document. Type: SOLISTI1.

OBJECT_PARA

This table is only used by documents that are to be processed. It must contain the SET/GET parameters that are passed to the processing element. Type: SOPARAI1.

OBJECT_PARB

This table is only used by documents that have been assigned a particular processing type. Depending on this, they have a different meaning. If the processing element is a report or transaction with value transfer to the global memory, the table contents are interpreted as a set of parameters with related values and exported to the memory ID derived from the first line. If the processing element is a function or dialog module, the table is passed to it as the table parameter MSGDIAL. Type: SOPARBI1.

ATTACHMENT_LIST

This table contains the essential attributes of the document attachments. Each line stands for an attachment. Type: SOATTLSTI1.

RECEIVER_LIST

This table contains the recipient list of the document. Each line stands for a recipient. Type: SORECLSTI1.

CONTENTS_HEX

Table for binary content. Type: SOLIX.

[!] Exceptions

DOCUMENT_ID_NOT_EXIST

Document does not exist.

OPERATION_NO_AUTHORIZATION

No authorization to view folder entry.

X_ERROR

Internal error or database inconsistency.

---

354 SO_DOCUMENT_SEND_API1

This FM enables you to send a new document, including any existing attachments. The document and the attachments are transferred in the table parameters of this FM. Documents are created when sent and can also be placed in the sender's outbox. This FM also have the ability to provide sender details in email.

Example: You've created a report. You've provided a button in the output screen to send the document as an email. On clicking it, the user needs to provide the email ID to which the document should be sent. You need to fill the import parameters and document details in tables parameter and call this FM to send the document.

[↓] Import

DOCUMENT_DATA

Attributes of a new document. Type: SODOCCHGI1.

PUT_IN_OUTBOX

If this parameter has value X, the newly created document is also placed in the outbox of the active user when it's sent. The default value is [Space]. Type: SONV-FLAG.

SENDER_ADDRESS

Sender's email address. The default value is SY-UNAME. Type: SOEXTRECI1-RECEIVER.

SENDER_ADDRESS_TYPE

Type of sender's address. The default value is B. Type: SOEXTRECI1-ADR_TYP.

COMMIT_WORK

Execute explicit commit works at end of FM is sent if the parameter has the value X. Type: SONV-FLAG.

[↑] Export

SENT_TO_ALL

If this parameter is X, the document was sent to all specified recipients, or, in the case of external forwarding, the corresponding send requests were delivered to the subsystem. If sending or delivery failed in one or more cases, the flag isn't X. Type: SONV-FLAG.

NEW_OBJECT_ID

Object ID of the document created during the send process. Type: SOFOLENTI1-OBJECT_ID.

SENDER_ID

Sender ID is returned in this parameter. Type: SOUDK.

Tables

PACKING_LIST

This table requires information about how the data in the tables OBJECT_HEADER, CONTENTS_BIN, and CONTENTS_TXT are to be distributed to the document and its attachments. Type: SOPCKLSTI1.

OBJECT_HEADER

This table must contain the data dependent on the document type. For example, SAPscript documents have information on the layout set and style, Excel list viewer documents contain the number of lines and columns, and PC documents contain their original file names. Type: SOLISTI1.

CONTENTS_BIN

This parameter is obsolete. Tables CONENTS_HEX is used instead of this parameter. Type: SOLISTI1.

CONTENTS_TXT

This table must contain the summarized content of the objects identified as ASCII objects. Type: SOLISTI1.

CONTENTS_HEX

Binary contents of object and attachments. Type: SOLIX.

OBJECT_PARA

This table is only used by documents that are to be processed. It must contain the SET/GET parameters that are passed to the processing element. Type: SOPARAI1.

OBJECT_PARB

This table is only used by documents that have been assigned a particular processing type. Depending on this, they have a different meaning. If the processing element is a report or transaction with value transfer to the global memory, the table contents are interpreted as a set of parameters with related values and exported to the memory ID derived from the first line. If the processing element is a function or dialog module, the table is passed to it as the table parameter MSGDIAL. Type: SOPARBI1.

RECEIVERS

This table must contain the document recipients. Type: SOMLRECI1.

[!] Exceptions

TOO_MANY_RECEIVERS

Too many recipients, no authorization.

DOCUMENT_NOT_SENT

Document was not sent.

DOCUMENT_TYPE_NOT_EXIST

Document type or attachment type does not exist.

OPERATION_NO_AUTHORIZATION

No authorization to send/create.

PARAMETER_ERROR

Invalid combination of parameter values.

X_ERROR

Internal error or database inconsistency.

ENQUEUE_ERROR

Required locks could not be set.

---

355 SO_NEW_DOCUMENT_ATT_SEND_API1

This FM is a wrapper FM that internally calls the FM SO_DOCUMENT_SEND_API1 (FM 355). All of the parameters in this FM have the same functionality as given and described in FM SO_DOCUMENT_SEND_API1. Note that this FM doesn't pass the sender parameter and its details. To get the sender details, call FM SO_DOCUMENT_SEND_API1. Because this is just a wrapper FM of SO_DOCUMENT_SEND_API1, it's recommended to use SO_DOCUMENT_SEND_API1, which has some additional functionality to pass the sender parameter as well.

Example: See FM SO_DOCUMENT_SEND_API1 (FM 355). End user wants you to create a program that saves the data to ABAP Data Dictionary tables. This program will be executed in background mode. If any error occurs while execution of the program, end user should get an email with the details in the document. You need to call this FM for the same.

[↓] Import

DOCUMENT_DATA

Attributes of new document. Type: SODOCCHGI1.

PUT_IN_OUTBOX

If this parameter has the value X, the newly created document is also placed in the outbox of the active user when it's sent. The default value is [Space]. Type: SONV-FLAG.

COMMIT_WORK

Execute explicit commit works at end of FM is sent if the parameter has the value X. Type: SONV-FLAG.

[↑] Export

SENT_TO_ALL

If this parameter is X, the document was sent to all specified recipients, or, in the case of external forwarding, the corresponding send requests were delivered to the subsystem. If sending or delivery failed in one or more cases, the flag isn't X. Type: SONV-FLAG.

NEW_OBJECT_ID

Object ID of the document created during the send process. Type: SOFOLENTI1-OBJECT_ID.

Tables

PACKING_LIST

Refers to same parameter PACKING_LIST of FM SO_DOCUMENT_SEND_API1 (FM 355). Type: SOPCKLSTI1.

OBJECT_HEADER

This table must contain the data dependent on the document type. For example, SAPscript documents have information on the layout set and style, Excel list viewer documents contain the number of lines and columns, and PC documents contain their original file names. Type: SOLISTI1.

CONTENTS_BIN

This parameter is obsolete. Table CONENTS_HEX is used instead of this parameter. Type: SOLISTI1.

CONTENTS_TXT

This table must contain the summarized content of the objects identified as ASCII objects. Type: SOLISTI1.

CONTENTS_HEX

Binary contents of objects and attachments. Type: SOLIX.

OBJECT_PARA

This table is only used by documents that are to be processed. It must contain the SET/GET parameters that are passed to the processing element. Type: SOPARAI1.

OBJECT_PARB

This table is only used by documents that have been assigned a particular processing type. Depending on this, they have a different meaning. If the processing element is a report or transaction with value transfer to the global memory, the table contents are interpreted as a set of parameters with related values and exported to the memory ID derived from the first line. If the processing element is a function or dialog module, the table is passed to it as the table parameter MSGDIAL. Type: SOPARBI1.

RECEIVERS

This table must contain the document recipients. Type: SOMLRECI1.

[!] Exceptions

TOO_MANY_RECEIVERS

Too many recipients, no authorization.

DOCUMENT_NOT_SENT

Document was not sent.

DOCUMENT_TYPE_NOT_EXIST

Document type or attachment type does not exist.

OPERATION_NO_AUTHORIZATION

No authorization to send/create.

PARAMETER_ERROR

Invalid combination of parameter values.

X_ERROR

Internal error or database inconsistency.

ENQUEUE_ERROR

Required locks could not be set.

---

356 SO_NEW_DOCUMENT_SEND_API1

This FM enables you to send a new document, which has not yet been created, internally and externally. During the send process, the document is created, along with the attributes and content that is to be transferred. This FM also calls FM SO_DOCUMENT_SEND_API1 (FM 355). Fill the table parameters to send the content of the document.

Example: Your program will create a new document, and you want to send the document to your business partner via email. You can fill in the parameters and call this FM to create a new document and then send it to the recipient.

[↓] Import

DOCUMENT_DATA

Attributes of the new document. Type: SODOCCHGI1.

DOCUMENT_TYPE

Document class. All classes are possible except for folders (FOL) and distributions lists (DLI). Allowed values:

- SCR: SAP script texts
- RAW: RAW editor texts
- GRA: SAP business graphics
- SHL: Pregenerated search interfaces
- QRY: individual search queries

The default value is RAW. Type: SOODK-OBJTP.

PUT_IN_OUTBOX

If this parameter has the value X, the newly created document is also placed in the outbox of the active user when it's sent. The default value is [Space]. Type: SONV-FLAG.

COMMIT_WORK

Execute explicit commit works at end of FM is sent if the parameter has the value X. Type: SONV-FLAG.

[↑] Export

SENT_TO_ALL

If this parameter is X, the document was sent to all specified recipients, or, in the case of external forwarding, the corresponding send requests were delivered to the subsystem. If sending or delivery failed in one or more cases, the flag isn't X. Type: SONV-FLAG.

NEW_OBJECT_ID

Object ID of the document created during the send process. Type: SOFOLENTI1-OBJECT_ID.

Tables

OBJECT_HEADER

This table must contain the data dependent on the document type. For example, SAPscript documents have information on the layout set and style, Excel list viewer documents contain the number of lines and columns, and PC documents contain their original file names. Type: SOLISTI1.

OBJECT_CONTENT

Document content. Type: SOLISTI1.

CONTENTS_HEX

Binary contents of objects and attachments. Type: SOLIX.

OBJECT_PARA

This table is only used by documents that are to be processed. It must contain the SET/GET parameters that are passed to the processing element. Type: SOPARAI1.

OBJECT_PARB

This table is only used by documents that have been assigned a particular processing type. Depending on this, they have a different meaning. If the processing element is a report or transaction with value transfer to the global memory, the table contents are interpreted as a set of parameters with related values and exported to the memory ID derived from the first line. If the processing element is a function or dialog module, the table is passed to it as the table parameter MSGDIAL. Type: SOPARBI1.

RECEIVERS

This table must contain the document recipients. Type: SOMLRECI1.

[!] Exceptions

TOO_MANY_RECEIVERS

Too many recipients, no authorization.

DOCUMENT_NOT_SENT

Document was not sent.

DOCUMENT_TYPE_NOT_EXIST

Document type or attachment type does not exist.

OPERATION_NO_AUTHORIZATION

No authorization to send/create.

PARAMETER_ERROR

Invalid combination of parameter values.

X_ERROR

Internal error or database inconsistency.

ENQUEUE_ERROR

Required locks could not be set.

357 SO_OLD_DOCUMENT_SEND_API1

This FM sends a document that already exists internally and externally. This document calls FM SO_OBJECT_SEND internally.

Example: You want to send a document via email to different recipients. You can fill in the parameters of this FM and then call it. On execution of this FM, an email is sent to the recipients.

[↓] Import

DOCUMENT_ID

ID of document to be sent. This is the connection between the document and the folder in which it's located. Type: SOFOLENTI1-DOC_ID.

PUT_IN_OUTBOX

If this parameter has the value X, the newly created document is also placed in the outbox of the active user when it's sent. The default value is [Space]. Type: SONV-FLAG.

[↑] Export

SENT_TO_ALL

If this parameter has the value X, the document can be sent to all specified recipients. If the document can't be sent to at least one participant, the parameter isn't [Space]. Type: SONV-FLAG.

Table

RECEIVERS

This table must contain the recipients of the document. Type: SOMLRECI1.

[!] Exceptions

TOO_MANY_RECEIVERS

Too many recipients, no authorization.

DOCUMENT_NOT_SENT

Document was not sent.

DOCUMENT_TYPE_NOT_EXIST

Document type or attachment type does not exist.

OPERATION_NO_AUTHORIZATION

No authorization to send/create.

PARAMETER_ERROR

Invalid combination of parameter values.

X_ERROR

Internal error or database inconsistency.

ENQUEUE_ERROR

Required locks could not be set.

358 SO_USER_READ_API1

This FM returns all information on an SAPoffice user. The information returned includes various administration information (e.g., user name, creator, date of last access, address number), SAPoffice authorizations of the user, entries on an automatic reply, the IDs of the individual root folders of the user, and so on.

When the FM is called, it's also possible to specify whether the SAPoffice of the respective user should be initialized for the day. If so, the trash is emptied, and the relevant resubmissions are placed in the inbox.

Example: You want to run the program to get the mailbox of employees and send the email with the list of questions to the employees. For this, first you need to get the employees' inbox folder details. You can get those details by passing the relevant parameter in this FM and calling it.

[↓] Import

USER

This structure can be provided with an entry identifying the user to be read. Only one of the fields—USERID, SAPNAME, or OFFICENAME—can be filled. If a value isn't entered, the data of the active user is read. The default value is [Space]. Type: SOUDNAMEI1.

PREPARE_FOR_FOLDER_ACCESS

If this parameter has the value X, initialization of user data is carried out along with reading the user data. Type: SONV-FLAG.

[↑] Export

USER_DATA

This structure contains all read data of a user. Type: SOUDATAI1.

[!] Exceptions

USER_NOT_EXIST

Specified ID or user name does not exist.

PARAMETER_ERROR

Invalid combination of parameter values.

X_ERROR

Internal error or database inconsistency occurred.

359 XXL_FULL_API

Extended Export of Lists (XXL) is a tool for displaying and manipulating list objects from SAP ERP applications. XXL is based on the Microsoft Excel spreadsheet program. The term *list object* is used here because the SAP ERP system transfers not only the list data to XXL but also the corresponding meta information, that is, information on the list structure, the characteristics of the list data, and explanatory texts.

This FM is used as an API to XXL in ABAP/4. It's also used to open and send the internal table data to Excel. Refer to program XXLFTEST for the use of this FM.

Example: You have the internal table data, and you want to open it in Excel with the column headings and other options such as highlighting the key column with a different color. You can pass the relevant parameter of this FM, as well as formatting options in the table parameters, and then call it. You'll see the data of your internal table in Excel.

[↓] Import

DATA_ENDING_AT

Number of the first line to be transferred in DATA. The default value is –1.

DATA_STARTING_AT

Number of the last line to be transferred in DATA. The default value is 1.

FILENAME

File name on the workstation. The default value is [Space]. Type: GXXLT_F-FILE.

HEADER_1

Heading 1. The default value is [Space]. Type: GXXLT_P-TEXT.

HEADER_2

Heading 2. The default value is [Space]. Type: GXXLT_P-TEXT.

NO_DIALOG

If this parameter has the value [Space], the FM requests the FILENAME parameter; the NO_START parameter and dialog box won't appear.

NO_START

If this parameter has the value X, then Excel won't be opened and data will be stored in the system.

N_ATT_COLS

Number of data columns in the list object.

N_HRZ_KEYS

Number of (hierarchical) key lines.

N_VRT_KEYS

Number of (hierarchical) key columns.

SEMA_TYPE

If this parameter has the value X, it convert STR columns.

SO_TITLE

Title of the Excel file. Type: GXXLT_F-SO_TITLE.

Tables

DATA

Contents for key and data columns.

HKEY

Contents for key lines. Type: GXXLT_H.

ONLINE_TEXT

Table with online texts. Type: GXXLT_O.

PRINT_TEXT

Table with print texts. Type: GXXLT_P.

SEMA

Semantic information on DATA columns. Type: GXXLT_S.

VKEY

Headings for key columns. Type: GXXLT_V.

[!] Exceptions

CANCELLED_BY_USER

Cancellation by user (only for NO_DIALOG = [Space]).

DATA_TOO_BIG

Maximum sizes for DATA exceeded.

DIM_MISMATCH_DATA

Nonpresent DATA column is referenced.

DIM_MISMATCH_SEMA

Number of SEMA lines is not equal to N_VRT_KEYS + N_ATT_COLS.

DIM_MISMATCH_VKEY

Number of VKEY lines is not equal to N_VRT_KEYS.

ERROR_IN_HKEY

Error occurred in HKEY parameter.

ERROR_IN_SEMA

Error occurred in SEMA parameter.

FILE_OPEN_ERROR

File FILENAME cannot be opened.

FILE_WRITE_ERROR

File FILENAME cannot be written.

INV_DATA_RANGE

DATA_ENDING or DATA_STARTING is either too large or small. Value is greater than the number of lines in the data table, or it is less than 0.

INV_WINSYS

Wrong window system, DOS windows required.

INV_XXL

Installation at the frontend incorrect.

360 XXL_SIMPLE_API

This FM is similar to XXL_FULL_API (FM 359), which is used for list viewer, but has reduced functions. You can pass less important parameters, so there is less of a drain on the system. This FM is used as an API to XXL in ABAP/4. Refer to program XXLSTEST for the use of this FM.

Example: You have the internal table data and want to open it in Excel, but you don't want the hierarchy of the horizontal keys to be built. You can pass the relevant parameter of this FM and then call it to see the data of your internal table in Excel.

[↓] Import

FILENAME

File name on the workstation. The default value is ULO. Type: GXXLT_F-FILE.

HEADER

Heading for exported list object. The default value is [Space]. Type: GXXLT_P-TEXT.

N_KEY_COLS

Number of (hierarchical) key columns. The default value is 1.

Tables

COL_TEXT

Headings for DATA columns. Type: GXXLT_V.

DATA

Contents for key and data columns.

ONLINE_TEXT

Table with online texts. Type: GXXLT_O.

PRINT_TEXT

Table with print texts. Type: GXXLT_P.

[!] Exceptions

DIM_MISMATCH_DATA

Nonpresent DATA column is referenced.

FILE_OPEN_ERROR

File FILENAME cannot be opened.

FILE_WRITE_ERROR

File FILENAME cannot be written to.

INV_WINSYS

Wrong window system, DOS windows required.

INV_XXL

Installation at the frontend incorrect.

Part VIII
ALE, EDI, and IDoc

Application Link Enabling (ALE) and Electronic Data Interchange (EDI) are used for exchanging business data between different systems. Data are distributed through ALE when you want to communicate from one SAP system to one or more other (mostly SAP) systems. ALE transfers the data in the IDoc format and uses transactional RFC (tRFC) for data transfer. EDI, on the other hand, is generally used if you want to transfer business application documents with external partner systems such as vendors or customers. The SAP system sends EDI messages in IDoc format to an EDI subsystem where they are converted to a universal EDI standard (UN/EDIFACT or ANSI/X12), which enables communication with non-SAP systems. Intermediate documents (IDocs) are the data containers used to transfer data between two systems. IDocs can be outbound or inbound in nature. When the IDoc is sent from the sender system, it's an outbound IDoc, and when the IDoc is received at the receiver system, it's an inbound IDoc. The IDoc structure consists of three records:

- **Control records**
  Contain the header of the IDoc, such as sender and recipient details, IDoc direction and type, and so on.
- **Data records**
  Contain the data to be transferred from one system to another.
- **Status records**
  Contain the IDoc status values with the short text.

This part focuses on the ALE, EDI, and IDoc function modules (FMs) used to send IDoc data from one system to another.

361 ALE_ASYNC_BAPI_GET_RECEIVER

This FM determines the logical receiver systems for an asynchronous BAPI from the ALE distribution model. You can pass a table with logical receiver systems to the FM when the FM is called. If the table is transferred, possible receivers are determined from these receiver systems only if the BAPI isn't defined in the ALE distribution model. If a receiver can't be determined, an empty table for the logical receiver systems is returned.

Example: Using the BAPI ALE interface, the IDoc is triggered by calling the BAPI, so you need to know the receiver of a BAPI. To determine the receiver, the relevant filter object must be assigned. If the filter object isn't known until runtime, you need to get the filter objects by passing the parameter to the FM ALE_BAPI_GET_FILTEROBJECTS (FM 362). After you get the filter objects, you can receive the receiver by calling this FM if the BAPI is asynchronous and the receivers are many. If you have only one unique receiver, you can also pass the parameter to the FM ALE_BAPI_GET_UNIQUE_RECEIVER (FM 363). However, if the BAPI is synchronous, you can call FM ALE_BAPI_GET_UNIQUE_RECIEVER (FM 363) to get the unique receiver.

[↓] Import

OBJECT

BOR object of BAPI. Type: BDI_BAPIOBJECT.

METHOD

BOR method of BAPI. Type: BDI_BAPIMETHOD.

Tables

RECEIVER_INPUT

Default logical receiver systems. Type: BDI_LOGSYS.

RECEIVERS

Logical receiver systems of the BAPI. Type: BDI_LOGSYS.

FILTEROBJECT_VALUES

Filter object types and values. Type: BDI_FOBJ.

[!] Exceptions

ERROR_IN_FILTEROBJECTS

Filter objects either have errors or are incomplete.

ERROR_IN_ALE_CUSTOMIZING

Error in ALE customizing.

362 ALE_BAPI_GET_FILTEROBJECTS

This FM enables the system to determine filter object types of BAPIs from the ALE distribution model at runtime. The filter object types are used to determine the receivers of BAPIs. When the method is called, a table containing logical receiver systems may be transferred to the FM. In this case, the BAPI filter object types are determined from the ALE distribution model for these receiver systems only. If the BAPI isn't maintained in the ALE distribution model, or no filter object types are found in the distribution model, an empty table for the filter object types is returned.

Example: Using the BAPI ALE interface, the IDoc is triggered by calling the BAPI, so you need to know the receiver of a BAPI. To determine the receiver, the relevant filter object must be assigned. If the filter object isn't known until runtime, you need to get the filter objects by passing the relevant parameter to this FM. See also FM ALE_ASYNC_BAPI_GET_RECEIVER (FM 361).

[↓] Import

OBJECT

BOR object of BAPI. Type: BDI_BAPIOBJECT.

METHOD

BOR method of BAPI. Type: BDI_BAPIMETHOD.

Tables

RECEIVER_INPUT

Default logical receiver systems. Type: BDI_LOGSYS.

FILTEROBJECTS

Filter object types and values. Type: BDI_FLTTYP.

[!] Exceptions

ERROR_IN_ALE_CUSTOMIZING

Error in ALE customizing.

363 ALE_BAPI_GET_UNIQUE_RECEIVER

This FM determines one receiver for the synchronous BAPI from the ALE distribution module. The receiver system is determined by the logical system and RFC destination field of the export parameter RECEIVER. If the BAPI isn't maintained in the ALE distribution model, or no receiver has been determined, an empty structure for the receiver system is returned. If more than one receiver system is determined for the BAPI, an exception with an error message is returned.

Example: Using the BAPI ALE interface, the IDoc is triggered by calling the BAPI, so you need to know the receiver of a BAPI. If BAPI is synchronous and unique, you can call this FM to get the unique receiver. See also FM ALE_ASYNC_BAPI_GET_RECEIVER (FM 361).

[↓] Import

OBJECT

BOR object of BAPI. Type: BDI_BAPI-OBJECT.

METHOD

BOR method of BAPI. Type: BDI_BAPI-METHOD.

[↑] Export

RECEIVER

Receiver system of BAPI. Type: BDBAPIDEST.

Tables

FILTEROBJECTS_VALUES

Filter object types and values. Type: BDI_FOBJ.

[!] Exceptions

ERROR_IN_FILTEROBJECTS

Filter objects either have errors or are incomplete.

ERROR_IN_ALE_CUSTOMIZING

Error in ALE customizing.

NOT_UNIQUE_RECEIVER

There are several receiver systems for the BAPI.

NO_RFC_DESTINATION_MAINTAINED

The RFC destination has not been specified for the logical system.

---

364 ALE_IDOCS_CREATE

This FM is an alternative to using FM MASTER_IDOC_DISTRIBUTE (FM 383) for asynchronous BAPIs if you want to generate a receiver. One or more IDocs are generated, depending on whether you have entered one or more recipients. In contrast to FM MASTER_IDOC_DISTRIBUTE (FM 383), no recipient is determined. Also, no linked objects are read from the IDoc.

Example: You've created your custom BAPI according to the requirement. You've created the BAPI ALE interface, and your BAPI is used to create an IDoc. You are sending this IDoc to multiple receiver systems and you are interested in getting the receiver details as well. For such scenarios, you can generate the IDoc by passing the parameters and calling this FM.

[↓] Import

IDOC_CONTROL

Transfer data for the IDoc control record. Type: BDICONTROL.

OBJ_TYPE

Object type for object serialization. Type: SERIAL-OBJ_TYPE.

CHNUM

Channel number for object serialization. Type: SERIAL-CHNUM.

Tables

IDOC_DATA

Data records and entries are changed. Type: EDIDD.

RECEIVERS

Receiver for this IDoc. Type: BDI_LOGSYS.

CREATED_IDOCS

IDoc numbers of the generated IDocs. Type: BDIDOCS.

APPLICATION_OBJECTS

Linked objects from the application. Type: SWOTOBJID.

CREATED_IDOCS_ADDITIONAL

IDoc info for the generated IDocs. Type: EDIDC.

[!] Exceptions

IDOC_INPUT_WAS_INCONSISTENT

Serious error in the transferred data.

---

365 ALE_SYNC_BAPI_GET_RECEIVER

This FM determines the receiver for a synchronous BAPI.

Example: Using the BAPI ALE interface, the IDoc is triggered by calling the BAPI, so you need to know the receiver of a BAPI. You can retrieve the receiver by calling this FM if the BAPI is synchronous and there are many receivers. See also FM ALE_ASYNC_BAPI_GET_RECEIVER (FM 361).

[↓] Import

OBJECT

BOR object of BAPI. Type: BDI_BAPI-OBJECT.

METHOD

BOR method of BAPI. Type: BDI_BAPI-METHOD.

Tables

RECEIVER_INPUT

Default logical receiver systems. Type: BDI_LOGSYS.

RECEIVERS

Logical receiver systems of the BAPI. Type: BDI_LOGSYS.

FILTEROBJECT_VALUES

Filter object types and values. Type: BDI_FOBJ.

[!] Exceptions

ERROR_IN_FILTEROBJECTS

Filter objects either have errors or are incomplete.

ERROR_IN_ALE_CUSTOMIZING

Error in ALE customizing.

NO_RFC_DESTINATION_MAINTAINED

RFC destination has not been specified for the logical system.

---

366 CHANGE_POINTERS_READ

This FM reads the change pointer. Change pointers are used to distribute or send the master data changes with the shared master data (SMD) tool. The SMD tool is connected to the change document interface. If the master data changes are to be distributed, the application writes a change document. The contents of this document are passed to the SMD tool. The tool writes change pointers, reads the application data, and creates the master IDoc. The master IDoc is then passed to the ALE layer, which sends it to all interested systems.

The selection criteria to read the change pointer is made using two time frames—event generation and event activation—that reflect when the change pointer is created or activated. You can also specify whether only processed or only unprocessed change pointers should be read. The default is unprocessed change pointers.

Example: Whenever there is an update in a table, you need to generate an IDoc. You can do so by adding the table fields to the message type and creating the program to trigger an IDoc. Call this FM and pass the message type in the parameter. If the IDoc is successfully triggered, call FM CHANGE_POINTERS_STATUS_WRITE (FM 367) to write the changes to the change pointer status table.

[↓] Import

ACTIVATION_DATE_HIGH

Activation date upper limit. Type: SY-DATUM.

ACTIVATION_DATE_LOW

Activation date lower limit. Type: SY-DATUM.

ACTIVATION_TIME_HIGH

Activation time upper limit. Type: SY-UZEIT.

ACTIVATION_TIME_LOW

Activation time lower limit. Type: SY-UZEIT.

CHANGE_DOCUMENT_OBJECT_CLASS

Object class of change document. Type: CDPOS-OBJECTCLAS.

CREATION_DATE_HIGH

Creation date upper limit. Type: SY-DATUM.

CREATION_DATE_LOW

Creation date lower limit. Type: SY-DATUM.

CREATION_TIME_HIGH

Creation time upper limit. Type: SY-UZEIT.

CREATION_TIME_LOW

Creation time lower limit. Type: SY-UZEIT.

MESSAGE_TYPE

Message type. Type: BDCPS-MESTYPE.

READ_NOT_PROCESSED_POINTERS

If this parameter is set to X, the latest changes of the change pointer are received. The default value is X. Type: BDCPS-PROCESS.

Tables

CHANGE_POINTERS

Change pointers table. Type: BDCP.

MESSAGE_TYPES

ALE message type table. Type: BDMSGTYP.

[!] Exceptions

ERROR_IN_DATE_INTERVAL

Error in date interval.

ERROR_IN_TIME_INTERVAL

Error in time interval.

---

367 CHANGE_POINTERS_STATUS_WRITE

This FM is used to write the status records to the change pointers. For a list of change pointers, a record relating to the message type is written in the change pointer status table. If a record exists, it flags the pointer relating to this message type as processed.

Example: If the IDoc from the previous FM is successfully triggered due to a table update, for instance, call this FM to write the changes to the change pointer status table. See also FM CHANGE_POINTERS_READ (FM 366).

[↓] Import

MESSAGE_TYPE

Message type. Type: EDIDC-MESTYP.

Tables

CHANGE_POINTERS_IDENTS

Change pointer IDs. Type: BDICPIDENT.

---

368 DEQUEUE_ES_EDIDOCE

This FM unlocks the IDoc. This FM is called in the FM EDI_DOCUMENT_DEQUEUE_LATER (FM 373), which is also used to unlock the IDoc. When you've called the MASTER_IDOC_DISTRIBUTE (FM 383), you need to unlock the IDoc by calling this FM.

Example: Your IDoc is in locked status. You can unlock it by passing parameters in this FM and calling it. See FM EDI_DOCUMENT_DEQUEUE_LATER (FM 373).

[↓] Import

MODE_EDIDC

Lock mode for table EDIDC. Type: ENQMODE.

MANDT

Enqueue argument field MANDT. Type: EDIDC-MANDT.

DOCNUM

Enqueue argument field DOCNUM.

X_DOCNUM

Argument DOCNUM with initial values. Type: EDIDC-DOCNUM.

_SCOPE

Scope for the lock.

_SYNCHRON

This parameter identifies whether the locking is synchronous.

_COLLECT

If this parameter is set to X, it collects the unlock request and unlocks it at a later time. Type: DDENQCOLL.

---

369 EDI_CHANGE_CONTROL_RECORD

Use this FM to changes the control record of an IDoc. IDoc control records contain the control or header data, such as receiver and sender details, IDoc type, and so on.

Example: Control records contains the header information of the IDoc. You can change the control record information like recipient address or receiver port of IDoc by calling this FM. See also FM EDI_DOCUMENT_OPEN_FOR_EDIT (FM 374).

[↓] Import

IDOC_CHANGED_CONTROL

Modified IDoc control records are sent into this parameter. Type: EDIDC.

[!] Exceptions

IDOC_NOT_OPEN

IDoc is not opened in EDIT mode.

DIRECTION_CHANGE_NOT_ALLOWED

Direction in the control record cannot be changed.

370 EDI_CHANGE_DATA_SEGMENTS

Use this FM to change the data record of an IDoc. Data records of IDocs contain the actual data that is required to be transferred.

Example: To change the data records of the IDoc you've opened from FM EDI_DOCUMENT_OPEN_FOR_EDIT (FM 374), you need to pass the data records you want to change in the table parameters and call this FM.

Tables

IDOC_CHANGED_DATA_RANGE

IDoc data records are sent into this table. Type: EDIDD.

[!] Exceptions

IDOC_NOT_OPEN

IDoc is not opened.

DATA_RECORD_NOT_EXIST

Data record does not exist.

371 EDI_DOCUMENT_CLOSE_EDIT

This FM closes the IDoc and saves all of the changes.

Example: When you're done with processing, you can close the IDoc from this FM. See FM EDI_DOCUMENT_OPEN_FOR_EDIT (FM 374).

[↓] Import

DOCUMENT_NUMBER

IDoc number. Type: EDIDC-DOCNUM.

DO_COMMIT

If this parameter is set to X, the FM commits the changes.

DO_UPDATE

If this parameter is set to X, the changes are updated to the respective tables.

WRITE_ALL_STATUS

If this parameter is set to X, all of the parameters are written to status table EDIDS. The default value is X.

STATUS_75

IDocs with status 75 are processed only when this flag is set to X because status 75 is a special case. The default value is [Space].

Tables

STATUS_RECORDS

Status records that are added to the IDoc. Type: EDI_DS40.

[!] Exceptions

IDOC_NOT_OPEN

IDoc was not opened in Edit mode.

DIRECTION_CHANGE_NOT_ALLOWED

Error writing to the database.

---

372 EDI_DOCUMENT_CLOSE_PROCESS

This FM closes the IDoc after processing. Also, the changes to the IDoc, such as the new status record and new status value in the control record, are passed on to the database.

Example: When you're done creating status records, you can close the IDoc by passing the IDoc number in the DOCUMENT NUMBER field of this FM.

[↓] Import

DOCUMENT_NUMBER

IDoc number. This IDoc already exists and should have been opened previously in process mode. Type: EDIDC-DOCNUM.

BACKGROUND

Database accesses in the update task. If "B" is entered for this parameter, the database accesses are carried out in the update program. The default value is NO_BACKGROUND. Type: EDI_HELP-BAG_OPTION.

NO_DEQUEUE

Do not unblock IDoc. If a value other than the default value is transferred, the IDoc isn't unblocked, but it's noted in the API that the IDoc still needs to be unblocked. The default value is [Space]. Type: EDI_HELP-UNLOCK.

STATUS_CHECK

Check current status. The default value is [Space]. Type: EDI_HELP-FUNKTION.

QRFC_STATE

qRFC status (values U, Q, T). The default value is U. Type: CHAR1.

[↑] Export

IDOC_CONTROL

The current IDoc control record is returned in this parameter. Type: EDIDC.

[!] Exceptions

DOCUMENT_NOT_OPEN

IDoc is not open for process.

FAILURE_IN_DB_WRITE

Error during database insert or update.

PARAMETER_ERROR

Parameter combination is invalid.

STATUS_SET_MISSING

No status set.

---

373 EDI_DOCUMENT_DEQUEUE_LATER

This FM is used to unlock the IDoc after it's closed. This FM internally calls FM DEQUEUE_ES_EDIDOCE (FM 368), which is used to unlock the IDoc.

Example: You've created the program to send the IDoc at the sender side by calling FM MASTER_IDOC_DISTRIBUTE (FM 383).

Sometimes, your IDoc might get stuck at status 30 (IDoc Ready for Dispatch). You may get this message because your IDoc isn't unlocked. You need to call this FM or FM DEQUEUE_ES_EDIDOCE (FM 368) to unlock the IDoc.

[↓] Import

DOCNUM

IDoc number that is to be unlocked. Type: EDIDC-DOCNUM.

SYNCHRON

Flag: synchronous unlocking? Type: DDENQ_LIKE-SYNCHRON.

[!] Exceptions

IDOC_IS_NOT_TO_DEQUEUE

IDoc number not closed without unlocking.

---

374 EDI_DOCUMENT_OPEN_FOR_EDIT

Use this FM to open an IDoc when you want to change the content of that IDoc. You'll use the following FMs in this process:

- EDI_CHANGE_DATA_SEGMENTS (FM 370)
- EDI_CHANGE_CONTROL_RECORD (FM 369)
- EDI_DOCUMENT_CLOSE_EDIT (FM 371)

Example: When you create an IDoc, you need to change the IDoc data record and control records and save them to the database tables. You can do so by first opening an IDoc for editing with this FM.

[↓] Import

DOCUMENT_NUMBER

IDoc number. An IDoc is to be stored in the database under this number, which should be opened to process the IDoc and make a change to the current status record. Type: EDIDC-DOCNUM.

ALREADY_OPEN

This parameter states whether the IDoc is already open. The default value is N.

[↑] Export

IDOC_CONTROL

The current IDoc control record is returned in this parameter. The control record contains the information about the sender ID, receiver ID, and specifications of IDoc type. Type: EDIDC.

Tables

IDOC_DATA

IDoc data records are returned in this table parameter. Type: EDIDD.

[!] Exceptions

DOCUMENT_FOREIGN_LOCK

IDoc locked.

DOCUMENT_NOT_EXIST

IDoc does not exist.

DOCUMENT_NOT_OPEN

IDoc not open in process mode.

STATUS_IS_UNABLE_FOR_CHANGING

IDoc may not be allowed for editing.

---

375 EDI_DOCUMENT_OPEN_FOR_PROCESS

This FM opens an IDoc for processing. Several IDocs can be processed, imported, or generated in parallel. You'll use the following FMs:

- EDI_STATUS_ADD_BLOCK (FM 377)
- EDI_DOCUMENT_CLOSE_PROCESS (FM 372)

Example: You need to set an IDoc to a certain status via a custom program (note that this FM is also called in standard programs). First, you need to open the IDoc for modification by passing the IDoc number and calling this FM.

[↓] Import

DB_READ_OPTION

With this parameter, you can control whether the data records should be read by the database. The value Y means read from the database; otherwise, do not read. If you don't need the data records, it's favorable for the performance if you override this default parameter. The default value is Y (read from the database). Type: EDI_HELP-DBR_OPTION.

DOCUMENT_NUMBER

IDoc number. An IDoc is to be stored in the database under this number, which should be opened to process the IDoc and make a change to the current status record. Type: EDIDC-DOCNUM.

ENQUEUE_OPTION

When opening an IDoc, a logical lock is relocated. With this parameter, you can control whether the block should be synchronous (value S) or asynchronous (value A). Type: EDI_HELP-ENQ_OPTION.

[↑] Export

IDOC_CONTROL

The current IDoc control record is returned in this parameter. The control record contains the information about the sender ID, receiver ID, and specifications of IDoc type. Type: EDIDC.

[!] Exceptions

DOCUMENT_FOREIGN_LOCK

IDoc locked.

DOCUMENT_NOT_EXIST

IDoc does not exist.

DOCUMENT_NUMBER_INVALID

IDoc number is invalid.

DOCUMENT_IS_ALREADY_OPEN

IDoc already opened in process mode.

---

376 EDI_DOCUMENT_STATUS_SET

This FM is used to change the status of the IDoc programmatically. You can only change the IDoc status when it's in CREATE or PROCESS mode. Status changes are allowed for IDocs that were opened with the FMs EDI_DOCUMENT_OPEN_FOR_EDIT (FM 374) or EDI_DOCUMENT_OPEN_FOR_PROCESS (FM 375). The FM EDI_DOCUMENT_CLOSE_PROCESS (FM 372) should be called to check whether at least one status change has been carried out for the document in Process mode and will save the status changes to the desktop. The following other FMs are involved:

- EDI_DOCUMENT_OPEN_FOR_PROCESS (FM 375)
- EDI_DOCUMENT_CLOSE_PROCESS (FM 372)

Example: You have a requirement that should the IDoc fail based on a specific condition, you need to change the status of IDoc to the Failed status. To do so, first open the IDoc by calling FM EDI_DOCUMENT_OPEN_FOR_PROCESS (FM 375). Next, set the changed status by passing the status to this FM.

[↓] Import

DOCUMENT_NUMBER

IDoc number. Type: EDIDC-DOCNUM.

IDOC_STATUS

Status record. Type: EDI_DS.

[↑] Export

IDOC_CONTROL

The current IDoc control record is returned in this parameter. The control record contains the information about the sender ID, receiver ID, and specifications of IDoc type. Type: EDIDC.

[!] Exceptions

DOCUMENT_NUMBER_INVALID

IDoc number is invalid.

OTHER_FIELDS_INVALID

One field in the status record is invalid.

STATUS_INVALID

Status value is invalid.

377 EDI_STATUS_ADD_BLOCK

This FM is used to create one or more status records after the IDoc is already open. Status changes are allowed for IDocs that are opened. To open an IDoc, you can use FM EDI_DOCUMENT_OPEN_FOR_PROCESS (FM 375).

Example: After the IDoc is open for processing, call this FM by passing the IDoc number in the parameter DOCUMENT_NUMBER. The status record is returned to you.

[↓] Import

DOCUMENT_NUMBER

IDoc number. This IDoc already exists and should have been opened previously in process mode. Type: EDIDC-DOCNUM.

[↑] Export

IDOC_CONTROL

The current IDoc control record is returned in this parameter. The control record contains the information about the sender ID, receiver ID, and specifications of IDoc type. Type: EDIDC.

Tables

IDOC_STATUS_RECORDS

A number of status records are transferred for one IDoc; that is, the status information of the IDoc is transferred. The information covers the IDoc number for which the status is changed, when the status is changed, the cause of the change, and the description of the error if there is an error status. Type: EDI_DS.

[!] Exceptions

DOCUMENT_NUMBER_INVALID

IDoc number is invalid.

IDOC_STATUS_RECORDS_EMPTY

Table IDOC_STATUS_RECORDS is empty.

OTHER_FIELDS_INVALID

One field in status record is invalid.

STATUS_INVALID

Status value is invalid.

378 IDOC_INBOUND_ASYNCHRONOUS

This FM starts inbound processing for port type tRFC. This FM is a successor FM of INBOUND_IDOC_PROCESS, which processes the IDoc record types valid for Release 4.0. If you're processing IDocs prior to Release 4.0, the FM INBOUND_IDOC_PROCESS must be called.

The IDocs are stored by the function block IDOC_INBOUND_WRITE_TO_DB (FM 380) (a `COMMIT` for all IDocs) and then processed by the FM IDOC_START_INBOUND (FM 381).

Example: You have processed an IDoc but it didn't reach the receiver system. You can view a list of issues in tRFC Transaction SM58 with this FM in the FM NAME section of this transaction. The reason for these issues may be that the authorizations aren't appropriate for the user who is sending the IDoc. To resolve these issues, give proper authorization and then reprocess the IDoc.

Tables

IDOC_CONTROL_REC_40

Control record table of the IDoc. Type: EDI_DC40.

IDOC_DATA_REC_40

Data record table of the IDoc. Type: EDI_DD40.

379 IDOC_INBOUND_SINGLE

This FM generates the inbound IDocs.

Example: You need to create an inbound IDoc based on the file in your presentation server. To create an inbound IDoc programmatically, you can pass the parameters to and call this FM.

[↓] Import

PI_IDOC_CONTROL_REC_40

IDoc control record. Type: EDI_DC40.

PI_DO_COMMIT

This parameter controls whether the FM sends a `COMMIT WORK` command. If no `COMMIT` command is sent, and the application carries out a rollback if an error occurs, the IDoc is deleted from the database; that is, there is no longer an IDoc in which the error can be documented. Type: EDI_HELP-DO_COMMIT.

[↑] Export

PE_IDOC_NUMBER

The IDoc number is returned. Type: EDIDC-DOCNUM.

PE_ERROR_PRIOR_TO_APPLICATION

This parameter is set to X if the IDoc doesn't have a status that can be automatically processed further. Type: EDI_HELP-ERROR_FLAG.

Tables

PT_IDOC_DATA_RECORDS_40

IDoc data records are passed in this table. Type: EDI_DD40.

[!] Exceptions

IDOC_NOT_SAVED

IDoc is not saved.

---

380 IDOC_INBOUND_WRITE_TO_DB

This FM is used to create and save the inbound IDoc to the ABAP Data Dictionary database tables.

Example: Sometimes, both the sender and receiver systems of an IDoc are required to be the same. In such cases, IDocs are generally created from reading the data from flat files and then the files are passed to the middleware SAP NetWeaver Process Integration (PI). To create and save the IDoc to the database, you need to pass the parameters and call this FM. When the IDoc is successfully created and you want to process it, you call the FM IDOC_START_INBOUND (FM 381).

[↓] Import

PI_STATUS_MESSAGE

IDoc status record. Type: EDIDS.

PI_DO_HANDLE_ERROR

If this parameter is X, errors are handled. The default value is X. Type: EDIGENERAL-ERRHANDLE.

PI_NO_DEQUEUE

If a value other than [Space] is sent, the IDoc isn't unblocked, but it's noted in the API that the IDoc still needs to be unblocked. The default value is [Space]. Type: EDI_HELP-UNLOCK.

PI_RETURN_DATA_FLAG

If this parameter is set to X, data records are returned in table parameter T_DATA_RECORDS. The default value is X. Type: EDIGENERAL-RET_FLAG.

PI_RFC_MULTI_CP

SAP character set identification. Type: CPCODEPAGE.

[↑] Export

PE_IDOC_NUMBER

IDoc number. Type: EDIDC-DOCNUM.

PE_STATE_OF_PROCESSING

Processing status. Type: SY-SUBRC.

PE_INBOUND_PROCESS_DATA

Process code for inbound processing. Type: TEDE2.

[Δ] Changing

PC_CONTROL_RECORD

IDoc control record. Type: EDIDC.

Tables

T_DATA_RECORDS

IDoc data records. Type: EDIDD.

T_LINKED_OBJECTS

This table returns the IDs of the application objects in the IDoc. Type: SWOTOBJID.

[!] Exceptions

IDOC_NOT_SAVED

IDoc could not be saved.

---

381 IDOC_START_INBOUND

This FM sends an IDoc for inbound processing that has been posted with the use of FM IDOC_INBOUND_WRITE_TO_DB (FM 380).

Example: You need to pass the relevant parameters of FM IDOC_INBOUND_WRITE_TO_DB (FM 380) and call it to create and save. After the IDoc is successfully created, you can process it by passing the relevant parameters of this and calling this FM.

[↓] Import

PI_INBOUND_PROCESS_DATA

Inbound process code. The default value is [Space]. Type: TEDE2.

PI_CALLED_ONLINE

If this parameter is set to X, the workflow to process the task is started in the foreground and not in the background. The default value is [Space]. Type: SWWCOMMIT-DIALOGFLAG.

PI_DO_COMMIT

If this parameter is X, the FM changes are committed to the database. The default value is X. Type: EDIGENERAL-DO_COMMIT.

PI_START_EVENT_ENABLED

If this parameter is X, the IDoc start event can be triggered. The default value is X. Type: EDIGENERAL-STARTEVENT.

PI_ORG_UNIT

If this parameter is set to X, the organizational unit is provided. The default value is [Space]. Type: SWHACTOR.

SUCC_SHOW_FLAG

If this parameter is [Space], a success message is returned; otherwise, an information message is returned when the work item is generated successfully. The default value is [Space]. Type: EDIGENERAL-SUCC_FLAG.

Tables

T_CONTROL_RECORDS

IDoc control records table. Type: EDIDC.

T_DATA_RECORDS

IDoc data record table. Type: EDIDD.

[!] Exceptions

INVALID_DOCUMENT_NUMBER

Invalid IDoc number.

ERROR_BEFORE_CALL_APPLICATION

Error before application was called.

INBOUND_PROCESS_NOT_POSSIBLE

Inbound processing not possible.

OLD_WF_START_FAILED

Process (up to 2.2) cannot be started.

WF_TASK_ERROR

Work item/workflow could not be created.

SERIOUS_INBOUND_ERROR

Some error while inbound processing.

---

382 IDOC_STATUS_WRITE_TO_DATABASE

This FM writes one or more status records of an IDoc to the database.

Example: If you want to get rid of old an IDoc, you can change its status to 68; that is, flagged for deletion. You can do so either by executing the program RC1_IDOC_SET_STATUS or by writing your own program and calling this FM to change the status of the IDoc.

[↓] Import

IDOC_NUMBER

IDoc number. Type: EDIDC-DOCNUM.

IDOC_OPENED_FLAG

If the IDoc is already opened, pass X in this parameter. If this parameter is [Space], the FM opens the IDoc by calling FM EDI_DOCUMENT_OPEN_FOR_PROCESS (FM 375). The default value is [Space]. Type: BD11_FLAG.

NO_DEQUEUE_FLAG

Don't unblock the IDoc. If a value other than the default value is transferred, the IDoc isn't unblocked, but it's noted in the API that the IDoc still needs to be unblocked. The default value is X. Type: BD11_FLAG.

[↑] Export

IDOC_CONTROL

The current IDoc control record is returned in this parameter. The control record contains the information about the sender ID, receiver ID, and specifications of IDoc type. Type: EDIDC.

Tables

IDOC_STATUS

This table contains IDoc status records. Type: BDIDOCSTAT.

[!] Exceptions

IDOC_FOREIGN_LOCK

IDoc is locked.

IDOC_NOT_FOUND

IDoc not found.

IDOC_STATUS_RECORDS_EMPTY

Status record is empty.

IDOC_STATUS_INVALID

Status returned is invalid.

DB_ERROR

Database error.

---

383 MASTER_IDOC_DISTRIBUTE

This FM is the interface from the application to the ALE layer on the outbound side. This FM is used to transfer the IDoc to the ALE layer; that is, to the destination partner system. The application can pass a master IDoc as an internal table using the parameters MASTER_IDOC_CONTROL and MASTER_IDOC_DATA.

This IDoc is then converted into communication IDocs and stored in the ALE layer. IDocs for which no errors occurred are passed to the dispatch control. A `COMMIT WORK` must be dispatched in the calling program; otherwise, the IDocs may not be dispatched.

Example: You have two separate SAP ERP systems. You want to send an IDoc from one system to another. To send an IDoc from the sender system, you call this FM, input the parameters, and this FM puts your IDoc internal table data into the ALE layer.

[↓] Import

MASTER_IDOC_CONTROL

Control record of the master IDoc that you want to send. Type: EDIDC.

OBJ_TYPE

This field specifies the ALE object type. The default value is [Space]. Type: SERIAL-OBJ_TYPE.

CHNUM

This field defines the ALE channel number. The default value is [Space]. Type: SERIAL-CHNUM.

Tables

COMMUNICATION_IDOC_CONTROL

The header records for the communication IDocs created are returned in this parameter. The status field states whether processing was successful. Type: EDIDC.

MASTER_IDOC_DATA

Data records of the master IDoc that you want to send. Type: EDIDD.

[!] Exceptions

ERROR_IN_IDOC_CONTROL

Incorrect entry in IDoc control record.

ERROR_WRITING_IDOC_STATUS

Error when writing IDoc status records.

ERROR_IN_IDOC_DATA

Incorrect entry in IDoc data records.

SENDING_LOGICAL_SYSTEM_UNKNOWN

Own logical system unknown.

Part IX
Users and Authorizations

This part of the book contains the function modules (FMs) related to users who are using SAP and their authorizations to execute transactions. Users are maintained in Transaction SU01 (User Master), and data are saved in user-specific tables, whose names generally start with "USR" such as USR01, USR02, USR03 and so on.

This part contains the FMs that are used to set user details, such as an address; printing the address in Smart Forms; and so on. This part also contains the FM to check whether the user has relevant authorizations to execute any transaction, to execute any ABAP Data Dictionary table or view, or to access any file.

384 ADDR_DIALOG

This FM displays the dialog box in which you can maintain the complete address of a user, vendor, third party, and so on. FM ADDR_DIALOG_PREPARE (FM 385) must be called if you want to parameterize the dialog box, including its title, field selection, exclude functions, and so on.

Example: You want to maintain the address of any person that exists in your SAP system. If you want to parameterize the address dialog box; that is, cause only those specific settings to appear in the address dialog box, you can call FM ADDR_DIALOG_PREPARE (FM 385) first. You can call this FM to open the address dialog box screen and maintain the address without calling FM ADDR_DIALOG_PREPARE (FM 385), which opens the address dialog box with default settings. You can maintain addresses in this dialog box.

[↓] Import

CHECK_ADDRESS

If this parameter is set to X, the address contents are checked. The default value is X. Type: SZAD_FIELD-FLAG.

SUPPRESS_TAXJURCODE_CHECK

The TAX JURISDICTION CODE field is suppressed and isn't handled if this parameter is X. The default value is [Space]. Type: SZAD_FIELD-FLAG.

IV_TIME_DEPENDENT_COMM_DATA

This parameter determines the time validity for communication data. The default value is [Space]. Type: AD_TIMEDEP.

[↑] Export

OK_CODE

Passed for the calling program. Type: SY-UCOMM.

Tables

NUMBER_HANDLE_TAB

Transfer table of addresses to be processed. Type: ADDR1_DIA.

VALUES

Return table with maintained address data. Type: ADDR1_DATA.

[!] Exceptions

ADDRESS_NOT_EXIST

Address does not exist or is invalid.

GROUP_NOT_VALID

Address group does not exist or is invalid.

PARAMETER_ERROR

Incorrect parameter values.

INTERNAL_ERROR

Serious internal error (MESSAGE A...).

385 ADDR_DIALOG_PREPARE

This FM sets parameters for address maintenance or filtering conditions for field selection, which reduces the parameters of address dialog screen. This FM should be called before calling the address maintenance dialog (from FM ADDR_DIALOG [FM 384]) to set up application-specific parameters for the address dialog. These address settings are valid until the end of the LUW or until the FM is called again and overwrites them.

Example: If you want to change the title bar of the address dialog screen, exclude/include any pushbutton, or apply some other condition, you can pass these parameters in this FM, which parameterizes the address dialog screen. Next, when you call FM ADDR_DIALOG (FM 384), it displays the changes you made to parameterize the address dialog box.

[↓] Import

FIELD_SELECTION

Field selection control template. Each character in this parameter string controls an address maintenance screen field group, depending on the character's position and the field group number. Example: the 22nd character controls the field to which "022" is assigned as the 4th modification screen group. Possible Control characters are – (minus, hide the field), (period, display field as optional field), + (plus, display field as required), * (asterisk, display only field), ([Space], standard settings of field) and # (hash, field will be always visible on the standard maintenance screen). Type: ADDR1_FSEL-FISEL.

FIELD_SELECTION_FOR_NATION

Field for the selection control structure. Type: ADDR1_FSEL-FISEL.

USE_FS_FOR_NATION

Checkbox field. Type: SZAD_FIELD-FLAG.

KEYWORDS

Dynamic key word structure. Type: ADDR1_KEYW.

TITLEBAR

Address maintenance screen title bar. Type: SY-TITLE.

CHANGE_DEFAULT_COMM_TYPES

Activate/deactivate communication types. Type: ADDR_COMM.

FRAME_TEXT

Dynamic frame texts. Type: ADDR_FRAME.

DEFAULT_URI_TYPE

Default URI/URL/FTP type. Type: ADURI-URI_TYPE.

SCREEN_VARIANT

Screen (transaction) variant. Type: SHDFV-TCVARIANT.

SHOW_PUSHBOTTONS_AT_TOP

Display toolbar. Only valid for subscreen. Type: SZAD_FIELD-FLAG.

USE_PSEUDO_REQUIRED_FIELDS

Display required entry fields as pseudo-required entry fields. Type: XFELD.

IV_TIME_DEPENDENCE

Time-specific maintenance of communication data. Type: XFELD.

Tables

EXCLUDED_FUNCTIONS

GUI status functions to be excluded.

INCLUDED_FUNCTIONS

Additional functions to include.

ERROR_TABLE

Table with errors, warnings, and information. Type: ADDR_ERROR.

[!] Exceptions

INTERNAL_ERROR

Serious internal error (MESSAGE A...).

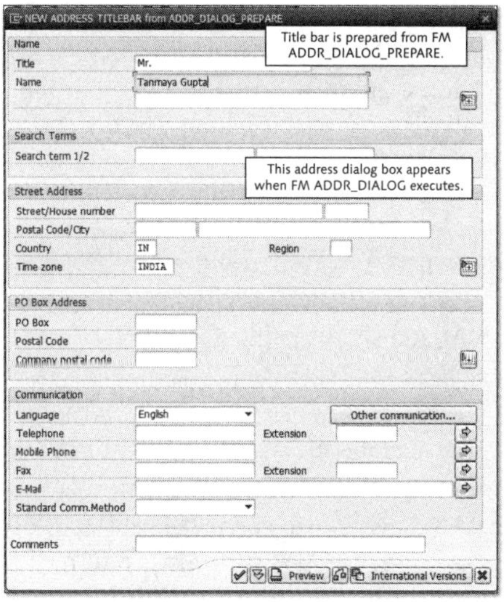

Address Dialog Box

386 ADDR_GET

This FM reads the address of an entity in the SAP system without the dialog box screen. If you want the details of the address and want to build your other program logic, you can call this FM to read the address details.

Example: To get the address for a plant on the selection screen, you can pass the parameter and call this FM.

[↓] Import

ADDRESS_SELECTION

Specification of an address. In the calling program, normally only the address number is recognized by a foreign key relationship (e.g., plant T001W-ADRNR, purchasing organization TVKO-ADRNR). The address number is to be passed to the parameter ADDRNUMBER.

Addresses created but not yet stored in the local memory can be identified and read with the parameter ADDRHANDLE. Only one of the two parameters ADDRNUMBER and ADDRHANDLE has to be filled.

This is a mandatory parameter. Type: ADDR1_SEL.

ADDRESS_GROUP

Address group. Entering an address group can trigger special handling: The system checks whether the data conversion from SADR to ADRC has already occurred for the addresses in the corresponding address group.

In any case, both return parameters ADDR1_VAL and SADR are filled.

If an address group whose addresses have already been converted is passed, ADRC is read, and the structure SADR is returned filled, in addition to ADDR1_VAL.

If no conversion has occurred, table SADR is read, and the SADR structure is returned; structure ADDR1_DATA from SADR is also filled. Type: ADRG-ADDR_GROUP.

READ_SADR_ONLY

Read in table SADR only. Type: SZAD_FIELD-FLAG.

READ_TEXTS

If this parameter is set to X, address texts are also read. Type: SZAD_FIELD-FLAG.

IV_CURRENT_COMM_DATA

This parameter determines the current status of communication data. Type: AD_COMCURR.

[↑] Export

ADDRESS_VALUE

Return data for an address. Type: ADDR1_VAL.

ADDRESS_ADDITIONAL_INFO

Additional communication data information. Type: AD1_FLAGS.

RETURNCODE

Return code. [Space] is success, I is information, W is warning, and E is error. Type: SZAD_FIELD-RETURNCODE.

ADDRESS_TEXT

Texts for key fields in the SY-LANGU language such as language key, country, and so on are returned if the parameter READ_TEXTS is set to X. Type: ADDR1_TEXT.

SADR

Return the old address structure. Type: SADR.

Tables

ADDRESS_GROUPS

Address groups to which the address is assigned. Type: ADAGROUPS.

ERROR_TABLE

Table with errors, warnings, and information. Type: ADDR_ERROR.

VERSIONS

Table with international versions. Type: ADDR_VERS.

[!] Exceptions

PARAMETER_ERROR

Incorrect parameter values.

ADDRESS_NOT_EXIST

Address does not exist.

VERSION_NOT_EXIST

International version of the address does not exist.

INTERNAL_ERROR

Serious internal error (MESSAGE A...).

---

387 ADDR_GET_COMPLETE

This FM provides you with all of the details for the given address number or address handle, including all of the communication types and notes about the address or communication type. All of the valid communication types are defined in table TSAC(T).

Example: To get the complete address of the plant on your selection screen along with the communication details for that address, you can pass the parameter to and call this FM.

[↓] Import

ADDRNUMBER

Address number for which you want the complete information. This parameter is a key of the database table. Type: ADDR1_SEL-ADDRNUMBER.

ADDRHANDLE

Address handle for which you want the complete information. This is a temporary key for the address. Type: ADDR1_SEL-ADDRHANDLE.

ARCHIVE_HANDLE

Read address from archive. Type: SY-TABIX.

IV_CURRENT_COMM_DATA

This parameter determines the current status of communication data The default value is X. Type: AD_COMCURR.

[↑] Export

ADDR1_COMPLETE

The complete address object (including communication) is returned in this parameter. Type: SZADR_ADDR1_COMPLETE.

[!] Exceptions

PARAMETER_ERROR

Incorrect parameter values.

ADDRESS_NOT_EXIST

Address does not exist.

INTERNAL_ERROR

Serious internal error (MESSAGE A...).

WRONG_ACCESS_TO_ARCHIVE

Incorrect access to archived data.

388 ADDR_INSERT

This FM inserts an address without a dialog. This FM creates an address in the local memory of the function group SZA0. These FMs are also used in this process:

- ADDR_NUMBER_GET (FM 391)
- ADDR_MEMORY_SAVE (FM 390)
- ADDR_MEMORY_CLEAR (FM 389)

Example: To create a normal address, you can pass the relevant import parameter into and call this FM, which inserts the address in local memory.

[↓] Import

ADDRESS_DATA

Parameter with address data. Type: ADDR1_DATA.

ADDRESS_GROUP

Address group. Entering an address group can trigger special handling: The system checks whether the data conversion from SADR to ADRC has already occurred for the addresses in the corresponding address group. In any case, both return parameters ADDR1_VAL and SADR are filled.

If an address group whose addresses have already been converted is passed, ADRC is read, and structure SADR is returned filled, in addition to ADDR1_VAL.

If no conversion has occurred, table SADR is read, the SADR structure is returned, and structure ADDR1_DATA from SADR is also filled. Type: ADRG-ADDR_GROUP.

ADDRESS_HANDLE

Address handle for which you want the complete information. This is a temporary key for the address. Type: SZAD_FIELD-HANDLE.

DATE_FROM

Reference date for the address. Type: ADCP-DATE_FROM.

LANGUAGE

Language for the text box. Type: ADRCT-LANGU.

CHECK_EMPTY_ADDRESS

Address input must be checked. Type: SZAD_FIELD-FLAG.

CHECK_ADDRESS

If this parameter is set to X, the address contents are checked. The default value is X. Type: SZAD_FIELD-FLAG.

[↑] Export

ADDRESS_DATA

Parameter with address data. Type: ADDR1_DATA.

RETURNCODE

Return code of the address. if the value of the return code isn't equal to [Space], at least one message is stored in the table ERROR_TABLE. Type: SZAD_FIELD-RETURNCODE.

Tables

ERROR_TABLE

Table with errors, warnings, and information. Type: ADDR_ERROR.

[!] Exceptions

ADDRESS_EXISTS

Address already exists in the handle.

PARAMETER_ERROR

Incorrect parameter values.

INTERNAL_ERROR

Serious internal error (MESSAGE A...).

389 ADDR_MEMORY_CLEAR

This FM deletes the local memory contents. All address data changes made since the first call of any FM or the last local memory save are discarded. This FM should be called after extensive read operations or when returning to the application.

Example: You've modified the address from FM ADDR_INSERT (FM 388), which adds or modifies the address in the local memory. Now you're done with the modification of the address and want to clear the local memory so that it doesn't overflow when you want to modify another address. You can pass the parameter FORCE if you want unsaved data to be cleared as well or just execute the FM without passing any parameters. After this FM is executed, the local memory is cleared (refer to FM ADDR_INSERT [FM 388]).

[↓] Import

FORCE

If this parameter is X, unsaved data are also cleared. Type: SZAD_FIELD-FLAG.

[!] Exceptions

UNSAVED_DATA_EXIST

Missing address number for new address.

INTERNAL_ERROR

Some internal error occurred.

390 ADDR_MEMORY_SAVE

This FM saves all of the address data from the local memory to the database. All address data changes made since the first call of any FM in the function group are then permanent.

Example: You're creating an address from FM ADDR_INSERT (FM 388). To save the address to the respective database table, call this FM. Refer to FM ADDR_INSERT (FM 388).

[↓] Import

EXECUTE_IN_UPDATE_TASK

Database update in the dialog or in the update task. Type: SZAD_FIELD-FLAG.

[!] Exceptions

ADDRESS_NUMBER_MISSING

Missing address number for new address.

PERSON_NUMBER_MISSING

Missing person number for new person.

INTERNAL_ERROR

Serious internal error (MESSAGE A...).

DATABASE_ERROR

Error when writing to the database (dialog).

REFERENCE_MISSING

Reference is missing.

391 ADDR_NUMBER_GET

This FM assigns an address number; that is, converts a temporary address number (address handle) into a permanent address number, which is saved in the database key.

Example: You're creating a new address and need to assign the address number to the address information. This address number is a unique number that identifies the address. You can pass the relevant parameters in this

FM and execute it to get the address number and assign the numbers to the address. Refer to FM ADDR_INSERT (FM 388).

[↓] Import

ADDRESS_HANDLE

Address handle for which you want the complete information. This is a temporary key for the address. Type: SZAD_FIELD-HANDLE.

ADDRESS_REFERENCE

Address reference for the BAS where-used list. The consistency of the where-used list determines the reference integrity for deleting and archiving addresses and is also required for identifying the owner object of an address. Type: ADDR_REF.

PERSONAL_ADDRESS

Personal address indicator. The default value is [Space]. Type: ADRC-PERS_ADDR.

NUMBERRANGE_NUMBER

Number range from which the number is taken. Type: INRI-NRRANGENR.

E071K_WA

Prepopulated structure for the transport entry. Type: E071K.

GENERATE_TRANSPORT_ENTRIES

If this parameter is set to X, entries are generated for the transport request. Type: SZAD_FIELD-FLAG.

OWNER

This parameter indicates that the application object is the owner of the address. Type: SZAD_FIELD-FLAG.

TABLE_NAME

Referencing table for the transport entry. Type: TSADRV-DDIC_TABLE.

FIELD_NAME

Referencing table field for the transport entry. Type: TSADRV-DDIC_FIELD.

OBJKEY

Key of referencing table for the transport entry. Type: ADRCS2-OBJKEY.

[↑] Export

ADDRESS_NUMBER

Number taken from the number range. Type: ADRC-ADDRNUMBER.

RETURNCODE_NUMBERRANGE

Return code when taking the number. Type: INRI-RETURNCODE.

Tables

E071K_TAB

All transport entries that belong to the address. Type: E071K.

[!] Exceptions

ADDRESS_HANDLE_NOT_EXIST

Transferred address handle does not exist.

INTERNAL_ERROR

Serious internal error (MESSAGE A...).

PARAMETER_ERROR

Incorrect parameter values.

---

392 ADDRESS_INTO_PRINTFORM

This FM formats an address for printing in the correct postal form with the required number of lines from the input. This FM also returns a short address; that is, the name and city one-line short address for the data carrier. Many country-specific formats are supported that correspond to the postal regulations of recipient countries.

Example: If your application isn't using the Business Address Services (BAS) for addresses, but you still want to have country-specific addresses displayed in your form, you can do pass the relevant parameters into and call this FM. When you call this FM, you'll receive the country-specific address that can be used for printing the address.

[↓] Import

ADRSWA_IN

Address data/control parameter. This parameter is only used in address type [Space]. Type: ADRS.

ADDRESS_1

Address data transfer structure for type 1. Type: ADRS1.

ADDRESS_2

Address data transfer structure for type 2. Type: ADRS2.

ADDRESS_3

Address data transfer structure for type 3. Type: ADRS3.

ADDRESS_TYPE

There are three address types:

- For address type 1: The normal address (company or organization) structure used in most SAP applications.
- For address type 2: Address of a person.
- For address type 3: Work address, which is usually the address of a contact person in a company.

The default address type [Space] is treated as type 1 and is required for the upward compatibility of the FM.

The interface parameters ADRSWA_IN and ADRSWA_OUT are only used in address type [Space]. Type: SZAD_FIELD-ADDR_TYPE.

ADDRESS_NUMBER

Address number (data is read directly). The address fields (NAME, STREET, CITY, etc.) for formatting can either be passed in the appropriate structure (ADDRESS1/2/3 or ADRSWA_IN) or be read from the BAS database tables, if this address is there, using the address key. To read the database tables, the keys are:

- For address type 1: ADDRESS_NUMBER
- For address type 2: ADDRESS_NUMBER and PERSON_NUMBER
- For address type 3: ADDRESS_NUMBER and PERSON_NUMBER.

Type: ADRC-ADDRNUMBER.

ADDRESS_HANDLE

Address handle for which you want the complete information. This is a temporary key for the address. Type: SZAD_FIELD-HANDLE.

PERSON_NUMBER

Person number (data are read directly). Type: ADRP-PERSNUMBER.

PERSON_HANDLE

Person handle for which you want the complete information. Type: SZAD_FIELD-HANDLE.

SENDER_COUNTRY

Sender country (for type 1, 2, and 3). Type: SZAD_FIELD-SEND_CNTRY.

RECEIVER_LANGUAGE

Recipient language (for type 1, 2, and 3). Type: T002-SPRAS.

NUMBER_OF_LINES

Max. line length for the formatting (for type 1, 2, and 3). The default value is 10. Type: ADRS-ANZZL.

STREET_HAS_PRIORITY

If this parameter is set to X, STREET has priority over PO Box. Type: SZAD_FIELD-STREETPRIO.

LINE_PRIORITY

Line suppression sequence. Type: SZAD_FIELD-PRIORITY.

COUNTRY_NAME_IN_RECEIVER_LANGU

If this parameter is set to X, the country outputs in the recipient language. Type: SZAD_FIELD-FLAG.

LANGUAGE_FOR_COUNTRY_NAME

This parameter specifies the country output language for a foreign address. Otherwise, the sender country language is used. Type: T002-SPRAS.

NO_UPPER_CASE_FOR_CITY

International post guidelines specify that the city and country in foreign addresses are output in uppercase.

If this parameter is set to X, there is no conversion to uppercase. If [Space], country and city are converted to uppercase. Type: SZAD_FIELD-FLAG.

IV_NATION

International version for formatting. Type: ADRC-NATION.

IV_NATION_SPACE

Flag: Use international version space in all cases. Type: T_BOOLE.

IV_PERSON_ABOVE_ORGANIZATION

Print person above organization. Type: T_BOOLE.

IS_BUPA_TIME_DEPENDENCY

Transfer structure for time dependency. Type: ADBUPA_TD.

IV_LANGU_CREA

Language key. Type: T002-SPRAS.

IV_DISPLAY_COUNTRY_IN_SHRTFORM

Display country key in short form. Type: XFELD.

[↑] Export

ADRSWA_OUT

This parameter is only used in address type [Space].

The formatted address is returned in the 10 fields LINE0 to LINE9 in the return structure ADRSWA_OUT.

A one-line short address comprising the name and city is returned in the field LINEK. The format is NAME1/City line (from the formatted address).

A one-line short address for "Payment without documents by data carrier exchange (DCE)" is returned in the field LINED. This usually comprises the postal code, city, and country, and is formatted specific to the country. Type: ADRS.

ADDRESS_PRINTFORM

Formatted address (type 1, 2, and 3). Type: ADRS_PRINT.

ADDRESS_SHORT_FORM

Short form of the formatted address (type 1, 2, and 3). Type: SZAD_FIELD-ADDR_SHORT.

ADDRESS_SHORT_FORM_S

Short form, including street (type 1, 2, and 3). Type: SZAD_FIELD-ADDR_SHORT.

ADDRESS_DATA_CARRIER

Data carrier address (type 1, 2, and 3). Type: SZAD_FIELD-ADDR_DC.

ADDRESS_DATA_CARRIER_0

Second line for the data medium (type 1, 2, and 3). This line is part of the two-line data medium address. The line is the same as the address line, which contains street or PO box. Type: SZAD_FIELD-ADDR_DC.

NUMBER_OF_USED_LINES

Number of lines occupied (type 1, 2, and 3). Type: ADRS-ANZZL.

NAME_IS_EMPTY

Flag: Name fields empty, no formatting. Type: SZAD_FIELD-FLAG.

ADDRESS_NOT_FOUND

This parameter has the value X if the address isn't found in the BAS database table when the address number is entered. Type: SZAD_FIELD-FLAG.

ADDRESS_PRINTFORM_TABLE

Formatted address in table form. The formatted address is returned in an internal table. LINE_TYPE specifies the line type (name, city, street, etc.) per line. Type: SZADR_PRINTFORM_TABLE.

ADDRESS_SHORT_FORM_WO_NAME

Short form street and city (type 1, 2, and 3). Type: SZAD_FIELD-ADDR_SHORT.

EV_NATION

Actual version used if IV_NATION was transmitted. Type: ADRC-NATION.

---

393 ADDRESS_SHOW_PRINTFORM

This FM displays the formatted address in a popup.

Example: To display the print preview of an address according to the postal guidelines of a country, call this FM and fill in the parameters, and the address appears in a popup.

[↓] Import

ADDRESS_1

Address data transfer structure for type 1. Type: ADRS1.

ADDRESS_2

Address data transfer structure for type 2. Type: ADRS2.

ADDRESS_3

Address data transfer structure for type 3. Type: ADRS3.

ADDRESS_TYPE

There are three address types:

- For address type 1: Normal address (company or organization) structure used in most SAP applications.
- For address type 2: Address of a person.
- For address type 3: Work address, which is usually the address of a contact person in a company.

The default address type [Space] is treated as type 1 and is required for the upward compatibility of the FM.

The interface parameters ADRSWA_IN and ADRSWA_OUT are only used in address type [Space]. Type: SZAD_FIELD-ADDR_TYPE.

ADDRESS_NUMBER

Address number (data is read directly). The address fields (NAME, STREET, CITY, etc.) for formatting can either be passed in the appropriate structure (ADDRESS1/2/3 or ADRSWA_IN) or be read from the BAS database tables, if this address is there, using the address key. To read the database tables, the keys are:

- For address type 1: ADDRESS_NUMBER
- For address type 2: ADDRESS_NUMBER and PERSON_NUMBER

- For address type 3: ADDRESS_NUMBER and PERSON_NUMBER.

Type: ADRC-ADDRNUMBER.

ADDRESS_HANDLE

Address handle for which you want the complete information. This is a temporary key for the address. Type: SZAD_FIELD-HANDLE.

PERSON_NUMBER

Person number (data are read directly). Type: ADRP-PERSNUMBER.

PERSON_HANDLE

Person handle for which you want the complete information. Type: SZAD_FIELD-HANDLE.

SENDER_COUNTRY

Sender country (for type 1, 2, and 3). Type: SZAD_FIELD-SEND_CNTRY.

RECEIVER_LANGUAGE

Recipient language (for type 1, 2, and 3). Type: T002-SPRAS.

NUMBER_OF_LINES

Max. line length for the formatting (type 1, 2, and 3). The default value is 10. Type: ADRS-ANZZL.

STREET_HAS_PRIORITY

If this parameter is set to X, street has priority over PO box. Type: SZAD_FIELD-STREETPRIO.

LINE_PRIORITY

Line suppression sequence. Type: SZAD_FIELD-PRIORITY.

COUNTRY_NAME_IN_RECEIVER_LANGU

If this parameter is set to X, country outputs in the recipient language. Type: SZAD_FIELD-FLAG.

LANGUAGE_FOR_COUNTRY_NAME

This parameter specifies the country output language for a foreign address. Otherwise, the sender country language is used. Type: T002-SPRAS.

NO_UPPER_CASE_FOR_CITY

International post guidelines specify that the city and country in foreign addresses are output in uppercase.

If this parameter is set to X, there is no conversion to uppercase. If [Space], country and city are converted to uppercase. Type: SZAD_FIELD-FLAG.

IS_BUPA_TIME_DEPENDENCY

Transfer structure for time dependency. Type: ADBUPA_TD.

[↑] Export

ADDRESS_PRINTFORM

Formatted address (type 1, 2, and 3). Type: ADRS_PRINT.

ADDRESS_SHORT_FORM

Short form of the formatted address (type 1, 2, and 3). Type: SZAD_FIELD-ADDR_SHORT.

ADDRESS_DATA_CARRIER

Data carrier address (type 1, 2, and 3). Type: SZAD_FIELD-ADDR_DC.

NUMBER_OF_USED_LINES

Number of lines occupied (type 1, 2, and 3). Type: ADRS-ANZZL.

NAME_IS_EMPTY

Flag: Name fields empty, no formatting. Type: SZAD_FIELD-FLAG.

ADDRESS_NOT_FOUND

This parameter has the value X if the address isn't found in the BAS database table when the address number is entered. Type: SZAD_FIELD-FLAG.

EV_EXIT_CAUSED_BY_CALLER

Delete list in CALLBACK_USER_COMMAND. Type: CHAR1.

EV_EXIT_CAUSED_BY_USER

How the user left the list. Type: SLIS_EXIT_BY_USER.

394 AUTHORITY_CHECK_DATASET

This FM allows you to check authorizations to access files with the key words OPEN DATASET, READ DATASET, TRANSFER, and DELETE DATASET. The authorization check is performed using the authorization object S_DATASET.

Example: You want to access the file *ABC*. Before opening the file, first you need to check if you're authorized to access the file. You can pass the parameters into and call FM AUTHORITY_CHECK_DATASET. You'll get an exception if you're not authorized to access the file.

[↓] Import

PROGRAM

ABAP program from which authorization needs to be checked. If no program name is specified, the system assumes the current program. Type: AUTHB-PROGRAM.

ACTIVITY

Access type. Possible values are:

- READ: Read file.
- WRITE: Change file.
- READ_WITH_FILTER: Read file with filter function.
- WRITE_WITH_FILTER: Change file with filter function.
- DELETE: Delete file.

Type: C.

FILENAME

File name for authorization. Type: AUTHB-FILENAME.

[!] Exceptions

NO_AUTHORITY

You are not authorized for this access.

ACTIVITY_UNKNOWN

Access type unknown.

---

395 AUTHORITY_CHECK_TCODE

This FM checks whether you're authorized to execute a transaction code. The authorization check is performed using the authorization object S_TCODE. S_TCODE is the authorization object linked with the transaction code authorization.

Example: You want your transaction to be executed only by authorized users. You need to pass the parameters and call this FM in the `INITILIZATION` event so that the transaction can be loaded only for the authorized users. You can also use the ABAP command `AUTORITY-CHECK OBJECT` for the same purpose.

[↓] Import

TCODE

Transaction code. Type: TSTC-TCODE.

[!] Exceptions

OK

Authorized for the transaction.

NOT_OK

Not authorized for the transaction.

---

396 SUSR_USER_AUTH_FOR_OBJ_GET

This FM displays authorizations for a specified object belonging to a user. User authorizations are displayed in an internal table. It's also possible to restrict the authorizations to specified authorization objects.

Example: You want to know whether you have authorization to start Transaction SU01. The user requires authorization for object S_TCODE with the value SU01 in field TCD to be able to do this. You can pass the parameters into and call FM SUSR_USER_AUTH_FOR_OBJ_GET. The value is returned in the table parameter VALUES if you have authorization.

[↓] Import

NEW_BUFFERING

This parameter is used as an indicator for various authorization programs. Type: XUFLAG.

MANDANT

SAP system and client number. Type: SY-MANDT.

USER_NAME

User name in user master record. This is a mandatory parameter. Type: USR02-BNAME.

SEL_OBJECT

Authorization object for restriction. If this parameter is passed, only the permissions to the object SEL_OBJECT are displayed. Type: UST12-OBJCT.

Tables

VALUES

Authorization values are returned in this table. Type: USVALUES.

[!] Exceptions

USER_NAME_NOT_EXIST

User name does not exist.

NOT_AUTHORIZED

No authorization

INTERNAL_ERROR

Internal error.

Part X
Spools and Background Jobs

A spool request is the document for which a print function has been selected and not yet been printed at the printer. The spool system uses the spool request to store the print data temporarily and access it. An output request is created for the spool request document, which outputs the print data of the spool request to a particular output device such as a printer.

SAP background processing automates routine tasks and helps you optimize your organization's SAP computing resources. Background processing tells the SAP system to run programs for you. Background processing lets you move long-running or resource-intensive program runs to times when the system load is low. It also lets you delegate to the system the task of running reports or programs. Background processing isn't tied up with the user interaction and dialog screens.

This part contains the FMs related to spools and jobs such as creating a spool request, finding the spool request, opening the job, submitting the job for background processing, closing the job, and so on.

397 BP_FIND_JOBS_WITH_PROGRAM

This FM finds the background job requests for a given program.

Example: You've created a program that's used to update data into tables in background mode. You want to get the list of all of the job numbers and the details of the jobs for that program. You can pass the program name in the import parameters and call this FM to get a list of jobs for the program.

[↓] Import

ABAP_PROGRAM_NAME

ABAP program name. Type: SY-REPID.

ABAP_VARIANT_NAME

ABAP variant name. Type: RALDB-VARIANT.

EXTERNAL_PROGRAM_NAME

External program name. Type: TBTCSTEP-PROGRAM.

DIALOG

If this parameter has value Y, the dialog box for job selection appears. Type: BTCH0000-CHAR1.

STATUS

Job status. The default value is "*". Type: TBTCJOB-STATUS.

Tables

JOBLIST

Job list with jobs found. Type: TBTCJOB.

[!] Exceptions

NO_JOBS_FOUND

No jobs found.

PROGRAM_SPECIFICATION_MISSING

Program to search for not specified.

INVALID_DIALOG_TYPE

Invalid dialog type.

JOB_FIND_CANCELED

Search canceled by user (dialog box).

398 BP_JOB_DELETE

This FM deletes an aborted background job and its log (if the job has already run). Note that active jobs can't be deleted. Also, a job can't delete itself.

Example: To delete an aborted job and its logs to free up memory space, pass the job number and job name in the importing parameter and call this FM. After the job is deleted, you can go to Transaction SM37 and verify that the job doesn't exist.

[↓] Import

JOBCOUNT

The internal ID number that the background processing system assigned to a job. It's used in addition to the name to uniquely identify the job. This is a mandatory parameter. Type: TBTCJOB-JOBCOUNT.

JOBNAME

Name of the job as assigned by the user. It's required to identify the job. This is a mandatory parameter. Type: TBTCJOB-JOBNAME.

FORCEDMODE

Select this parameter if all errors that occur when deleting table entries or job logs are to be ignored. The default value is [Space]. Type: SY-BATCH.

COMMITMODE

If this parameter is X, COMMIT WORK takes place after deletion. The default value is X. Type: BOOLE-BOOLE.

[!] Exceptions

CANT_DELETE_EVENT_ENTRY

Event scheduling cannot be deleted.

CANT_DELETE_JOB

Job cannot be deleted.

CANT_DELETE_JOBLOG

Job log cannot be deleted.

CANT_DELETE_STEPS

Job steps cannot be deleted.

CANT_DELETE_TIME_ENTRY

Deadline scheduling cannot be deleted.

CANT_DERELEASE_SUCCESSOR

Previous job cannot be modified.

CANT_ENQ_PREDECESSOR

Succeeding job cannot be locked.

CANT_ENQ_SUCCESSOR

Previous job cannot be locked.

CANT_ENQ_TBTCO_ENTRY

Job cannot be locked.

CANT_UPDATE_PREDECESSOR

Previous job cannot be modified.

CANT_UPDATE_SUCCESSOR

Succeeding job cannot be modified.

COMMIT_FAILED

COMMIT WORK failed.

JOBCOUNT_MISSING

Job ID number was not specified.

JOBNAME_MISSING

Job name not specified.

JOB_DOES_NOT_EXIST

Job does not exist.

JOB_IS_ALREADY_RUNNING

Job already active.

NO_DELETE_AUTHORITY

No authorization to delete jobs.

399 BP_JOB_READ

This FM reads the job header table (TBTCO) and associated step list table (TBTCSTEP) of jobs.

Example: You've scheduled a job. Now you want to read that job and display its logs in the program output. You can pass the parameter into and call this FM to get the job details for display.

[↓] Import

JOB_READ_JOBCOUNT

The internal ID number that the background processing system assigned to a job. It's used in addition to the name to uniquely identify the job. This is a mandatory parameter. Type: TBTCJOB-JOBCOUNT.

JOB_READ_JOBNAME

Name of the job as assigned by the user. It's required to identify the job. This is a mandatory parameter. Type: TBTCJOB-JOBNAME.

JOB_READ_OPCODE

This specifies which job data is to be read. Type: BTCH0000-INT4.

JOB_STEP_NUMBER

Job step ID number. Type: TBTCP-STEPCOUNT.

[↑] Export

JOB_READ_JOBHEAD

Job header data of the read job. Type: TBTCJOB.

JOBLOG_ATTRIBUTES

Job log-attribute. Type: BAPIXMJOBLOG.

EPP_ATTRIBUTES

Extended attributes of batch job. Type: BTCJOBEPP.

[Δ] Changing

RET

Special additional error code. Type: I.

Tables

JOB_READ_STEPLIST

Step list of the read job. Type: TBTCSTEP.

SPOOL_ATTRIBUTES

Parameter for spool attributes. Type: BAPIXMSPOOLID.

[!] Exceptions

INVALID_OPCODE

Invalid operation code was discovered.

JOB_DOESNT_EXIST

Job with job name/job count does not exist.

JOB_DOESNT_HAVE_STEPS

No steps were found for the job.

---

400 BP_JOB_SELECT

You can use this FM to find and retrieve a set of jobs that match the selection criteria that you specify.

Example: You want to get the set of jobs based on a specific name (e.g., job name starting with "ZEMP_") for further processing. You can pass the parameters in this FM and call it to get the selected records.

[↓] Import

JOBSELECT_DIALOG

Dialog box. Specify whether the selection criteria for a job are to be specified by the user or provided directly to the FM when it's called. Possible values are:

- BTC_YES: The system displays a selection window. All values that are provided when the FM is called are displayed on the screen.
- BTC_NO: The job selection criteria are provided directly to the FM when it's called.

This is a mandatory parameter. Type: BTCH0000-CHAR1.

JOBSEL_PARAM_IN

Use this structure to pass job selection parameters to the FM. The selection criteria are the same as those available in the background job maintenance function (Transaction SM37). Type: BTCSELECT.

ENDDATE

End date of the job. Type: TBTCO-ENDDATE.

ENDTIME

End time of the job. Type: TBTCO-ENDTIME.

SELECTION

Single-character parameter for selection. Type: CHAR2.

ADK_MODE

If this parameter is X, execution happens in background processing. Type: BTCH0000-CHAR1.

[↑] Export

JOBSEL_PARAM_OUT

Selection parameters for output. Returns the selection criteria used to generate a list of jobs. Type: BTCSELECT.

LOCAL_CLIENT

Reference type CHAR1 for background processing. Type: BTCH0000-CHAR1.

NR_OF_JOBS_FOUND

Number of jobs found is returned in this parameter. Type: I.

[Δ] Changing

ERROR_CODE

Special additional error code. Type: I.

Tables

JOBSELECT_JOBLIST

Selected jobs. This table contains the TBTCJOB entries for the selection jobs. The jobs are displayed to the user in the format that is used in the job overview of the maintenance function for background processing (Transaction SM37). Type: TBTCJOB.

JOBNAME_EXT_SEL

Batch selection range for job name. Type: NJRANGE.

USER NAME_EXT_SEL

Batch selection range for user name. Type: UNRANGE.

[!] Exceptions

INVALID_DIALOG_TYPE

Invalid dialog type.

JOBNAME_MISSING

Job name is missing (wildcards allowed).

NO_JOBS_FOUND

No jobs found that match selection criteria.

SELECTION_CANCELED

User cancels selection.

USER NAME_MISSING

User name missing (wildcards allowed).

401 BP_JOBLOG_READ

This FM reads the job log of the job that you specify in the parameters and stores it in an internal table for further processing.

Example: You want to read the status and the job logs in your program. You can fill in the parameters and call this FM. When you execute this FM, you'll get the job log details for a single job log.

[↓] Import

CLIENT

Client in which the job was scheduled. Type: TBTCJOB-AUTHCKMAN.

JOBCOUNT

The internal ID number that the background processing system assigned to a job. It's used in addition to the name to uniquely identify the job. Type: TBTCJOB-JOBCOUNT.

JOBLOG

Name of the job log in the TemSe database. The TemSe database is an internal SAP database for "temporary, sequential objects," such as log files. The name of the job log must be read from the TBTCJOB table. The job name and number are used to find the correct log. Type: TBTCJOB-JOBLOG.

JOBNAME

Name of the job as assigned by the user. It's required to identify the job. Type: TBTCJOB-JOBNAME.

LINES

Number of lines. Type: BTCINT4.

DIRECTION

Read direction (B = from beginning; E = from end). Type: BTCCHAR1.

Tables

JOBLOGTBL

Table name that was returned by a FM. The table is an internal table that contains the lines of the job log. Type: TBTC5.

[!] Exceptions

CANT_READ_JOBLOG

The system could not access the job log. Possible causes include database problems or a malfunction in the SAP system's TemSe component, in which job logs are held.

JOBCOUNT_MISSING

The job count was not supplied in the call to the FM. The job count is a unique ID number that is assigned to each job by the background processing system.

JOBLOG_DOES_NOT_EXIST

Log not found in TemSe database.

JOBLOG_IS_EMPTY

The job log does not contain any entries. This situation can occur when a job has just started running, and the job log has just been created. If the problem occurs again, a system-wide problem such as network or database failure may have occurred.

JOBLOG_NAME_MISSING

No job log name was supplied with the call to this FM. The job log name can be found in structure TBTCJOB using the job name and ID number.

JOBNAME_MISSING

The job name selected by the user was not provided in the call to the FM. The job name is required for identifying the job.

JOB_DOES_NOT_EXIST

Job already deleted.

---

402 BP_JOBLOG_SHOW

This FM displays the job log of a specified job. A job log contains messages from the background processing system on the processing of a job, with the error messages from the ABAP program and/or the external program running in the job.

Example: You want to display the status and the job logs in the output of your report program. You can fill in the parameters and call this FM. When you execute this FM, you'll be able to see the job log details.

[↓] Import

CLIENT

Client in which the job was scheduled. Type: TBTCJOB-AUTHCKMAN.

JOBCOUNT

The internal ID number that the background processing system assigned to a job. It's used in addition to the name to uniquely identify the job. Type: TBTCJOB-JOBCOUNT.

JOBLOGID

Name of job log in the TemSe database. The TemSe database is an internal SAP database for "temporary, sequential objects," such as log files. The name of the job log must be read from the TBTCJOB table. The job name and number are used to find the correct log. Type: TBTCJOB-JOBLOG.

JOBNAME

Name of the job as assigned by the user. It's required to identify the job. Type: TBTCJOB-JOBNAME.

LINES

Number of lines. Type: BTCINT4.

DIRECTION

Read direction (B = from beginning; E = from end). Type: BTCCHAR1.

[!] Exceptions

ERROR_READING_JOBDATA

The system could not access the job database. Possible causes include database problems or other system problems.

ERROR_READING_JOBLOG_DATA

The system could not access the job log. Possible causes include database problems or malfunctions in the SAP system's TemSe component, in which job logs are held.

JOBCOUNT_MISSING

The job count was not supplied in the call to the FM. The job count is a unique ID number that is assigned to each job by the background processing system.

JOBLOG_DOES_NOT_EXIST

Log not found in TemSe database.

JOBLOG_IS_EMPTY

The job log doesn't contain any entries. This situation can occur when a job has just started running, and the job log has just been created. If the problem occurs again, a systemwide problem such as a network or database failure may have occurred.

JOBLOG_SHOW_CANCELED

Log display canceled.

JOBNAME_MISSING

The job name selected by the user was not provided in the call to the FM. The job name is required for identifying the job.

JOB_DOES_NOT_EXIST

Job already deleted.

NO_JOBLOG_THERE_YET

Job log not yet generated.

NO_SHOW_PRIVILEGE_GIVEN

No authorization to display job log.

---

403 BP_START_DATE_EDITOR

The FM presents the START SPECIFICATIONS screen that is used in the background processing interface (START CONDITION button in Transaction SM36).

Example: You want to schedule the program for background processing. Before scheduling the program for background processing, you want to display the popup for start specification so that you can input the details concerning when you want the background job to be scheduled, such as immediately, at a particular date or time, after some job or program is executed, and so on. You can do so by passing the parameters in this FM and calling it.

[↓] Import

STDT_DIALOG

Dialog box. Specify whether the selection criteria for a job are to be specified by the user or provided directly to the FM when it's called. Possible values are:

- BTC_YES: The system displays a selection window. All values that are provided when the FM is called are displayed on the screen.
- BTC_NO: The job selection criteria are provided directly to the FM when it's called.

This is a mandatory parameter. Type: BTCH0000-CHAR1.

STDT_INPUT

This structure contains any start specifications that are passed to the FM in the call such as start date or period values. If STDT_DIALOG is set to BTC_YES, then the values in this structure are shown to the user as default values that can be overwritten or accepted. This is a mandatory parameter. Type: TBTCSTRT.

STDT_OPCODE

Specify whether the user can only display the job start specifications or can also define them. Possible values are:

- BTC_EDIT_STARTDATE: The user can change the job start specifications for the start date in the editor.
- BTC_SHOW_STARTDATE: The user can't change the start specifications.

Type: BTCH0000-INT4.

[↑] Export

STDT_MODIFY_TYPE

Displays whether the start date data was changed. Value BTC_STDT_MODIFIED specifies that the user has changed the start specifications, whereas value BTC_STDT_NOT_MODIFIED specifies that the user has not changed the start specification value. Type: BTCH0000-INT4.

STDT_OUTPUT

This table contains the start conditions, if STDT_OPCODE was set to BTC_EDIT_STARTDATE. Type: TBTCSTRT.

[!] Exceptions

FCAL_ID_NOT_DEFINED

Factory calendar ID not defined.

INCOMPLETE_LAST_STARTDATE

Incomplete specification of last start date.

INCOMPLETE_STARTDATE

Incomplete start date.

INVALID_DIALOG_TYPE

Invalid dialog type.

INVALID_EVENTID

Event ID specified is invalid.

INVALID_OPCODE

Invalid work mode.

INVALID_OPMODE_NAME

Invalid operation mode name.

INVALID_PERIODBEHAVIOUR

Invalid start date restriction.

INVALID_PREDECESSOR_JOBNAME

Name of predecessor job is invalid.

LAST_STARTDATE_IN_THE_PAST

Last start date is in the past.

NO_PERIOD_DATA_GIVEN

No period specifications.

NO_STARTDATE_GIVEN

No start date specified.

PERIOD_AND_PREDJOB_NO_WAY

Predecessor job cannot be used with period.

PERIOD_TOO_SMALL_FOR_LIMIT

Earlier start and period < one week not allowed.

PREDECESSOR_JOBNAME_NOT_UNIQUE

Name of predecessor job not unique.

STARTDATE_INTERVAL_TOO_LARGE

(Last start date – start date) > one year.

STARTDATE_IN_THE_PAST

Start date is in the past.

STARTDATE_IS_A_HOLIDAY

Start date is not a workday.

STARTDATE_OUT_OF_FCAL_RANGE

Start date is not within factory calendar.

STDT_BEFORE_HOLIDAY_IN_PAST

Workday before public holiday in the past.

UNKNOWN_FCAL_ERROR_OCCURED

Unknown error in factory calendar check.

NO_WORKDAY_NR_GIVEN

Workday not specified.

INVALID_WORKDAY_COUNTDIR

Count direction for workday is invalid.

INVALID_WORKDAY_NR

Workday number is invalid.

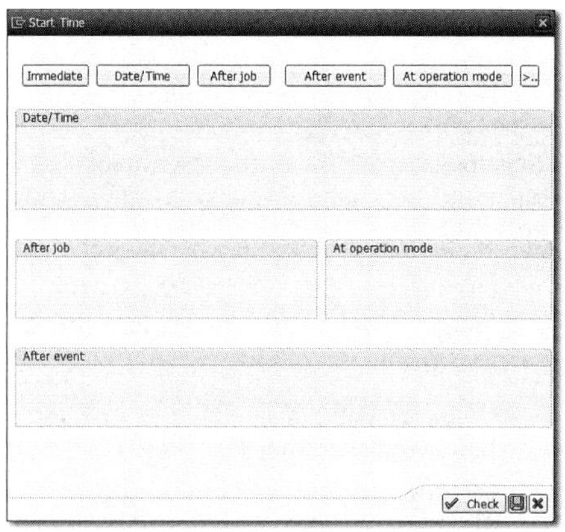

Start Date Popup after Execution of the FM

NOTBEFORE_STDT_MISSING

Earliest start date missing.

WORKDAY_STARTTIME_MISSING

Time for start date "on workday" missing.

NO_EVENTID_GIVEN

No event ID specified.

404 GET_JOB_RUNTIME_INFO

This FM is used for reading a background job's runtime data.

Example: You want to know which jobs are currently running. You can execute this FM to get this information.

[↑] Export

EVENTID

Background processing event. Type: TBTCM-EVENTID.

EVENTPARM

Background event parameters (e.g., JOBNAME/JOBCOUNT). Type: TBTCM-EVENTPARM.

EXTERNAL_PROGRAM_ACTIVE

Indicator for step executed by external program. Type: TBTCM-XPGACTIVE.

JOBCOUNT

Job ID. Type: TBTCM-JOBCOUNT.

JOBNAME

Background job name. Type: TBTCM-JOBNAME.

STEPCOUNT

Job step ID number. Type: TBTCM-STEPCOUNT.

[!] Exceptions

NO_RUNTIME_INFO

If no job is currently running in background, this exception is raised.

405 GET_PRINT_PARAMETERS

This FM is used to define, modify, and display the print and archive parameters. Using this FM is the only correct way to modify a print or archive parameter record. The print parameters are closely related and contain a check total. This check total becomes invalid if any of the print parameters are changed and causes a runtime error when the print parameters are used later.

Example: You have an ALV output and a button for printing in that output. On clicking the button, you want to be able to modify print parameters such as number of copies for print or output printer device where you want to print. You can pass the parameters in this FM and call it to get a popup of print parameters. On clicking CONTINUE, your request is sent for printing.

[↓] Import

ARCHIVE_ID

Archive ID. Type: ARC_PARAMS-ARCHIV_ID.

ARCHIVE_INFO

Additional information for report names. Type: ARC_PARAMS-INFO.

ARCHIVE_MODE

Control of the printing and archiving. This parameter specifies whether the spool request is sent to the printer or the archive. Possible values are 1 (print only), 2 (archive only), or 3 (print and archive both). Type: PRI_PARAMS-ARMOD.

ARCHIVE_TEXT

Info text for the archive request. Type: ARC_PARAMS-ARCTEXT.

AR_OBJECT

Archiving object for document types such as incoming invoices. Type: ARC_PARAMS-AR_OBJECT.

ARCHIVE_REPORT

Archiving report. Type: ARC_PARAMS-REPORT.

AUTHORITY

Default for authorization. Type: PRI_PARAMS-PRBER.

COPIES

Default for number of copies. Type: PRI_PARAMS-PRCOP.

COVER_PAGE

Selection cover sheet output value. Type: PRI_PARAMS-PRBIG.

DATA_SET

Default for name of the spool data. Type: PRI_PARAMS-PRDSN.

DEPARTMENT

Default for department. Type: PRI_PARAMS-PRABT.

DESTINATION

Default for output device. Type: PRI_PARAMS-PDEST.

EXPIRATION

Default for retention period. Type: PRI_PARAMS-PEXPI.

IMMEDIATELY

Default for print immediately. Type: PRI_PARAMS-PRIMM.

IN_ARCHIVE_PARAMETERS

List of all archiving parameters. Type: ARC_PARAMS.

IN_PARAMETERS

List of all default parameters. Type: PRI_PARAMS.

LAYOUT

Default for format. Type: PRI_PARAMS-PAART.

LINE_COUNT

Default for rows. If you specify the value for LINE-COUNT in the report or program statement, this parameter is ignored. Type: PRI_PARAMS-LINCT.

LINE_SIZE

Default for columns. If you specify the value for LINE-SIZE in the report or program statement, this parameter is ignored. If you're printing an online list, this parameter has no effect because the line width has already been set before printing and can't be changed. Type: PRI_PARAMS-LINSZ.

LIST_NAME

Default for spool request name. Type: PRI_PARAMS-PLIST.

LIST_TEXT

Default for spool request text. Type: PRI_PARAMS-PRTXT.

MODE

Mode: BATCH or [Space]. Type: SY-CALLR.

NEW_LIST_ID

Default for new spool request. Type: PRI_PARAMS-PRNEW.

PROTECT_LIST

Can't append any more to generated spool request.

NO_DIALOG

Suppress dialog. If this parameter is set to [Space], the dialog box for print parameters appears where you can change the print parameters. Type: PRI_PARAMS-PRNEW.

RECEIVER

Default for recipient. Type: PRI_PARAMS-PRREC.

RELEASE

Default for keep after printing. Type: PRI_PARAMS-PRREL.

REPORT

REPORT name. Type: SY-REPID.

SAP_COVER_PAGE

Default for the output of the SAP cover sheet. Type: PRI_PARAMS-PRSAP.

HOST_COVER_PAGE

Default for the output of operating system cover sheet. Type: PRI_PARAMS-PRUNX.

PRIORITY

Default for priority. Type: PRI_PARAMS-PRIOT.

SAP_OBJECT

Name of the application for the archiving. Type: ARC_PARAMS-SAP_OBJECT.

TYPE

Default spool request type. Type: PRI_PARAMS-PTYPE.

USER

User name. Type: SY-UNAME.

USE_OLD_LAYOUT

Use old classical list format.

UC_DISPLAY_MODE

Specifies the visual output length of fields. This is important for printing characters with double output length.

- 1: The defined output length applies. For example, a C(5) field occupies five print columns on paper. If the content of this field doesn't fit visually into these five columns, the text is truncated.
- 2: The output length depends on the content. Therefore, a C(5) field with two Asian characters and three European characters has the output length 2 * 2 + 3 * 1 = 7.
- 3: Maximum output length. For example, a C(5) field has the maximum output length of 10.

European characters are connected by a blank character and therefore are of double width; Asian characters are output without blank characters.

DRAFT

Print text only. Type: PRIPAR_EXT-TEXT.

ABAP_LIST

Print ABAP list. Indicates that an ABAP list is to be placed in the spool. Its effect is that the system checks for the document type ALF when archiving.

USE_ARCHIVENAME_DEF

Note %_ARCHIVE-REPORT.

DEFAULT_SPOOL_SIZE

The spool request is 255 characters wide or less—irrespective of the value of LINSZ.

WITH_STRUCTURE

Print with structure information. Additional information is included in the spool request so that the structure recognition can process the print version of the list in the spool display. The data volume of the spool request increases in this mode.

SUPPRESS_SHADING

Suppress colors/shading. During print output, colors or color shadings are suppressed.

PO_FAX_STORE

Print immediately (print parameter). Type: PRI_PARAMS-PRIMM.

NO_FRAMES

No frame characters. During printing, the automatic linking of the frame characters is suppressed.

[↑] Export

OUT_ARCHIVE_PARAMETERS

List of the called up archiving parameters. Type: ARC_PARAMS.

OUT_PARAMETERS

List of the called print parameter. Type: PRI_PARAMS.

VALID

Print parameters in OUT_PARAMTERS are valid. This parameter has the value [Space] in the interactive call if the user canceled the print screen.

VALID_FOR_SPOOL_CREATION

Print parameters in OUT_PARAMETERS are valid for creating a spool request.

[!] Exceptions

ARCHIVE_INFO_NOT_FOUND

Incorrect entries for archive or archive objects.

INVALID_PRINT_PARAMS

No longer used.

INVALID_ARCHIVE_PARAMS

No longer used.

---

406 JOB_CLOSE

This FM closes background job requests with `COMMIT WORK` and passes the background job to the background processing system for execution. Note that after you've closed the job, you can no longer add job steps or change the job. The FM returns an indicator as to whether the job was automatically released or not. A job is automatically released to run only if the user who scheduled the job has *RELE* release authorization for the authorization object operations on background jobs. If a user doesn't have the authorization to release a job, then the job is scheduled with the status Planned. The job can't be started until it's released, regardless of any start specifications made for the job.

Example: You can pass the parameter in this FM to close the job you were working with in FM JOB_OPEN (FM 407).

[↓] Import

AT_OPMODE

Operation mode name. Enter the name of the operation mode for which this job is scheduled. The job can be started (as long as it has been released) when the operation mode becomes active in the SAP system. Type: SPFBA-BANAME.

AT_OPMODE_PERIODIC

Job start is periodic according to the operation mode. Set this parameter to the value X if a job is always to be started when the operation mode specified in the parameter AT_OPMODE is activated. Type: BTCH0000-CHAR1.

CALENDAR_ID

Factory calendar ID for start date restriction. Specify a factory calendar if you're scheduling a job for a particular time or a particular date, and there are restrictions, such as whether the job can be started on nonwork days. The factory calendar is used to specify whether the proposed start date falls on a work-free day or not. Type: TBTCJOB-CALENDARID.

EVENT_ID

Job start after event: event name. This is the name of the event for which a job is to wait. The job can be started if the event is triggered in the background processing system. You can use Transaction SM62 to display the events that have been defined in the system. Type: TBTCJOB-EVENTID.

EVENT_PARAM

Job start after event: event parameter. Enter the event argument for which the job is to wait. If you specify an argument, the job can only be started if both the event specified in the parameter EVENT-ID and this argument are triggered in the background processing system. Type: TBTCJOB-EVENTPARM.

EVENT_PERIODIC

Job start is periodic according to event. Set this parameter to the value X if a job is always to be started and if the event specified in the parameters EVENT_ID and EVENT_PARAM is triggered. Type: BTCH0000-CHAR1.

JOBCOUNT

Job number. The internal ID number that the background processing system assigned to a job. It's used in addition to the name to uniquely identify the job. This is a mandatory parameter. Type: TBTCJOB-JOBCOUNT.

JOBNAME

Name of the job as assigned by the user. It's required to identify the job. This is a mandatory parameter. Type: TBTCJOB-JOBNAME.

LASTSTRTDT

The last possible start date for the job. This specification is optional and applies only for jobs that were scheduled for a particular time or a particular date. Type: TBTCJOB-LASTSTRTDT.

LASTSTRTTM

The last possible start time for the job. This specification is optional and applies only to jobs that were scheduled for a particular time or a particular date. Type: TBTCJOB-LASTSTRTTM.

PRDDAYS

Day portion of job repetition period. Specify a number of days in the repetition period of a job that is to be automatically repeated. Type: TBTCJOB-PRDDAYS.

PRDHOURS

Specify a number of hours in the repetition period of a job that is to be automatically repeated. Type: TBTCJOB-PRDHOURS.

PRDMINS

Specify a number of minutes in the repetition period of a job that is to be automatically repeated. Type: TBTCJOB-PRDMINS.

PRDMONTHS

Specify a number of months in the repetition period of a job that is to be automatically repeated. Type: TBTCJOB-PRDMONTHS.

PRDWEEKS

Specify a number of weeks in the repetition period of a job that is to be automatically repeated. Type: TBTCJOB-PRDWEEKS.

PREDJOB_CHECKSTAT

Set this parameter to X so that the job is only started after the predecessor job specified in PRED_JOBNAME has been successfully completed. If this parameter isn't set, the job starts when the predecessor job is completed, regardless of whether the job was successfully executed or not. Type: TBTCSTRT-CHECKSTAT.

PRED_JOBCOUNT

ID number of the predecessor job. The ID number is assigned to a job by the background processing system and is required, together with the job name, to identify the job. Type: TBTCJOB-JOBCOUNT.

PRED_JOBNAME

Job name of the predecessor job. The job name is a unique parameter along with the job ID, which is required to identify the job. Type: TBTCJOB-JOBNAME.

SDLSTRTDT

Start date of the background job. Type: TBTCJOB-SDLSTRTDT.

SDLSTRTTM

Time of start date of the background job. Type: TBTCJOB-SDLSTRTTM.

STARTDATE_RESTRICTION

Restrictions for the start date. Type: TBTCJOB-PRDBEHAV.

STRTIMMED

Immediate execution of the background job. Set this parameter to X, if the job is to be scheduled with immediate start. Type: BTCH0000-CHAR1.

TARGETSYSTEM

Target system of the job. You can enter the name of the host system on which this job is to be processed. This is useful, for example, if a job is to use a resource that is available only on a particular host.

START_ON_WORKDAY_NOT_BEFORE

Earliest start date for start on workday. Type: TBTCSTRT-NOTBEFORE.

START_ON_WORKDAY_NR

Number of workday on which the job is to start. Type: TBTCSTRT-WDAYNO.

WORKDAY_COUNT_DIRECTION

Direction of workday count. Type: TBTCSTRT-WDAYCDIR.

RECIPIENT_OBJ

The mail recipient of the spool lists. Type: SWOTOBJID.

TARGETSERVER

Target application server for the job. Application server is a more specific definition than target system and should therefore be used. Type: BTCTGTSRVR-SRVNAME.

DONT_RELEASE

Job not released in spite of start condition. Type: BTCH0000-CHAR1.

TARGETGROUP

Target application server group for the job. Type: BPSRVGRP.

DIRECT_START

X: Start without converting to time-based job. Type: BTCH0000-CHAR1.

[↑] Export

JOB_WAS_RELEASED

This parameter is X if the job is released. Type: BTCH0000-CHAR1.

[Δ] Changing

RET

Special additional error code. Type: I.

[!] Exceptions

CANT_START_IMMEDIATE

Cannot start immediately.

INVALID_STARTDATE

Start condition is invalid.

JOBNAME_MISSING

Job name missing (wildcards allowed).

JOB_CLOSE_FAILED

Error during JOB_CLOSE, see SYSLOG.

JOB_NOSTEPS

Job specified does not contain any steps.

JOB_NOTEX

Specified job does not exist.

LOCK_FAILED

Lock attempt failed.

INVALID_TARGET

Target server or group is invalid.

407 JOB_OPEN

This FM opens job scheduling without the dialog screen. You can also create a background job. The FM returns the unique ID number, which is required along with the job name to identify the job. The other FMs you'll use in this process are the following:

- JOB_SUBMIT (FM 408)
- JOB_CLOSE (FM 406)

Example: You need to copy a group that involves a large amount of data in 20 database tables. This copying activity doesn't require any user interaction or dialog screens during the copy. You want to schedule the job for background processing. First, you need to open the job by passing the relevant parameters and calling this FM.

[↓] Import

DELANFREP

Do not use. Type: TBTCJOB-DELANFREP.

JOBGROUP

Group to which the job is to be assigned. This parameter isn't used. Type: TBTCJOB-JOBGROUP.

JOBNAME

Job name. Type: TBTCJOB-JOBNAME.

SDLSTRTDT

Do not use. Type: TBTCJOB-SDLSTRTDT.

SDLSTRTTM

Do not use. Type: TBTCJOB-SDLSTRTTM.

JOBCLASS

Job classification. Type: TBTCJOB-JOBCLASS.

CHECK_JOBCLASS

Reference type CHAR1 for background processing. Type: BTCH0000_CHAR1.

[↑] Export

JOBCOUNT

This parameter contains the ID number of the background job. This number is returned when the job is opened. Use the number together with the job name to identify the job in other calls from background FMs. Type: TBTCJOB-JOBCOUNT.

[Δ] Changing

RET

Special additional error code. Type: I.

[!] Exceptions

CANT_CREATE_JOB

Job cannot be created, see SYSLOG.

INVALID_JOB_DATA

Job contains invalid job data, see SYSLOG.

JOBNAME_MISSING

Job name not specified.

408 JOB_SUBMIT

This FM Inserts a background task in the background request. Use this FM to add a job step to a background job that you've opened with JOB_OPEN (FM 407). A job step is an independent unit of work in a job, the execution of an ABAP/4 or external program. Each job step can have its own authorizations user and printer/optical archiving specifications.

Example: After using FM JOB_OPEN (FM 407), you need to submit the job steps to the job. This can be done by calling this FM. You then need to close the job for background processing. To do so, see FM JOB_CLOSE (FM 406).

[↓] Import

ARCPARAMS

Archiving parameters. The arguments determine whether and how the output of the job step is used in the optical archiving system. Type: ARC_PARAMS.

AUTHCKNAM

Background user name for authorization check. This parameter is valid only for SAP programs. No authorization checking can be done for external programs. This is a mandatory parameter. Type: TBTCJOB-AUTHCKNAM.

COMMANDNAME

Name of the command. Type: SXPGCOLIST-NAME.

OPERATINGSYSTEM

Name of the operating system. Type: SY-OPSYS.

EXTPGM_NAME

Name of external program. Specify the path name of the program that is to be executed by the background processing system. All file types that can be executed are permitted. Type: TBTCSTEP-PROGRAM.

EXTPGM_PARAM

Parameters of external program. Type: TBTCSTEP-PARAMETER.

EXTPGM_SET_TRACE_ON

Switch on flow trace for the control program. Set this parameter to the value X to switch to the third and most detailed form of the SAP program flow tracing. Type: BTCH0000-CHAR1.

EXTPGM_STDERR_IN_JOBLOG

Log standard error output. Set this parameter to the value X if you want to add the standard error of the external program to the background processing job log. The default value is X. Type: BTCH0000-CHAR1.

EXTPGM_STDOUT_IN_JOBLOG

Log standard output of external program. Set this parameter to the value X if you want to add the standard output of the external programs to the background processing job log. The default value is X. Type: BTCH0000-CHAR1.

EXTPGM_SYSTEM

Target host for executing the external program. Name of the host system on which an external program is to be run. You must supply a value for this parameter. Type: TBTCSTEP-XPGTGTSYS.

EXTPGM_RFCDEST

RFC destination. Type: TBTCSTEP-XPGRFCDEST.

EXTPGM_WAIT_FOR_TERMINATION

Wait for termination of external program. Type: BTCH0000-CHAR1.

JOBCOUNT

ID number of the background job. Type: TBTCJOB-JOBCOUNT.

JOBNAME

Job name. Type: TBTCJOB-JOBNAME.

LANGUAGE

Language. The default value is SY-LANGU. Type: SY-LANGU.

PRIPARAMS

Print parameters. Use this structure to pass arguments for the spool system to the background processing system. The arguments determine how and whether output from this job step is handled by the SAP spool system. Type: PRI_PARAMS.

REPORT

Report to run in the background job. Specify the name of the ABAP program that is to be executed in this job step. This parameter can't be used if an external program is to be executed in this job step. Type: SY-REPID.

VARIANT

Variant for the specified report. This is the name of the variant for the ABAP program specified in the REPORT parameter. Type: RALDB-VARIANT.

[↑] Export

STEP_NUMBER

Job step ID number. Type: TBTCJOB-STEPCOUNT.

[!] Exceptions

BAD_PRIPARAMS

Incorrect print/archiving parameters.

BAD_XPGFLAGS

Incorrect control flags for external program.

INVALID_JOBDATA

Error in job data, see SYSLOG.

JOBNAME_MISSING

Job name not specified.

JOB_NOTEX

Specified job does not exist.

JOB_SUBMIT_FAILED

Job submit failed, see SYSLOG.

LOCK_FAILED

Lock is not set.

PROGRAM_MISSING	PROG_ABAP_AND_EXTPG_SET
Program to execute is not specified.	ABAP and external program specified.

409 LOAD_PRINT_PARAMETERS

This FM displays a set of print parameters for a specified key. This key is obtained from a previous call of FM STORE_PRINT_PARAMETRS (FM 423). You'll use the following FMs sequentially in this process:

- STORE_PRINT_PARAMETERS (FM 423)
- LOAD_PRINT_PARAMETERS (FM 409)
- SET_PRINT_PARAMETERS (FM 421)
- STORE_PRINT_PARAMETERS (FM 423)

Example: You want to add an SAP cover page every time you print. You therefore need to change the printer settings for your user ID so that when the printout is done, it is according to your settings. First, you need to get the default printer settings for your user name by calling this FM. In order to execute this FM, you need a key for print parameter set. You will get this key by calling the FM STORE_PRINT_PARAMETERS (FM 423). Based on the key, you need to call this FM and get the default printer settings. Next, you need to perform the modification and pass those modifications in the FM SET_PRINT_PARAMETERS (FM 421) to set the settings. Lastly, you need to save and store the print settings you made via FM STORE_PRINT_PARAMETERS (FM 423).

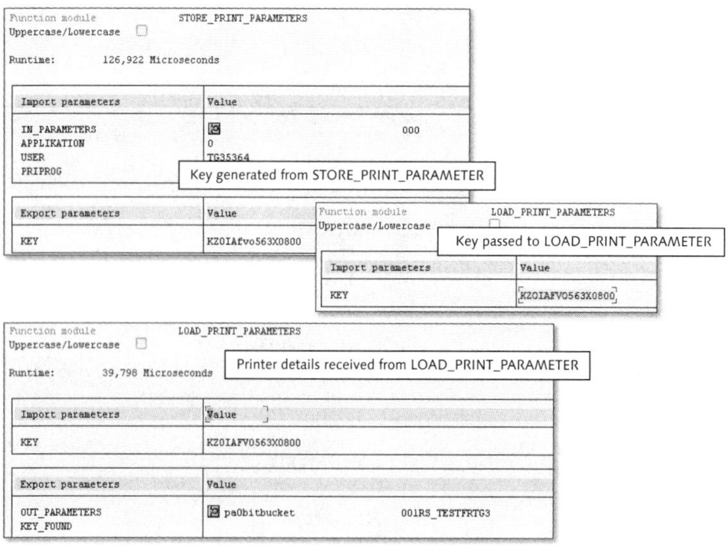

Load and Save Print Parameters

[↓] Import

KEY

Key for print parameter set. This is a mandatory parameter. Type: CHAR16.

[↑] Export

OUT_PARAMETERS

Print parameter from table TPRI_PAR. Type: PRI_PARAMS.

KEY_FOUND

The key entered is found. Type: C.

[!] Exceptions

ERROR_OCCURED

An error occurred.

410 PRINT_SELECTIONS

This FM is used to print the selection screen data on the report output.

Example: You're scheduling multiple jobs of a program Z_PROG. Now you want to create a program that will determine if the job is already scheduled for the screen variant by the user. For this, you need the selection screen information for the already-scheduled job. You can pass the program name and variant name in this FM to get the selection screen data. You can also print this data in your output.

[↓] Import

MODE

This parameter isn't used in the FM. However, this parameter is marked as M.

RNAME

Report program name. Type: TRDIR-NAME.

RVARIANTE

Report variant name. Type: RALDB-VARIANT.

Tables

INFOTAB

Output information of the selection screen.

411 RSPO_DISPLAY_SPOOLJOB

This FM displays the details for a specified spool job. Note that FM RSPO_RETURN_ABAP_SPOOLJOB (FM 415) also performs the same operation. The only difference is that the data is returned in table BUFFER in FM RSPO_RETURN_ABAP_SPOOLJOB (FM 415), and you can see the spool data by clicking the DISPLAY OUTPUT LIST OF FUNCTION MODULE button in this FM.

Example: You've displayed a list of spool jobs you've created in an ALV output (similar to Transaction SP01). Now you want to be able to select one spool, click DISPLAY, and see the spool's details. This can be done by providing the relevant parameters in this FM and calling it. Note that parameters FIRST_LINE and LAST_LINE don't work correctly. You need to implement SAP Note 1299738 for these parameters to work correctly.

[↓] Import

RQIDENT

Spool request number. This is a mandatory parameter. Type: TSP01-RQIDENT.

FIRST_LINE

First row to be displayed. Type: I.

LAST_LINE

Last line to be displayed. The value 0 means the spool details will be displayed until the end of spool. Type: I.

[!] Exceptions

NO_SUCH_JOB

No spool request number found with RQIDENT.

JOB_CONTAINS_NO_DATA

Order is empty.

SELECTION_EMPTY

Selected area is empty.

NO_PERMISSION

Permission is not enough.

CAN_NOT_ACCESS

Type cannot be determined.

READ_ERROR

Data cannot be read.

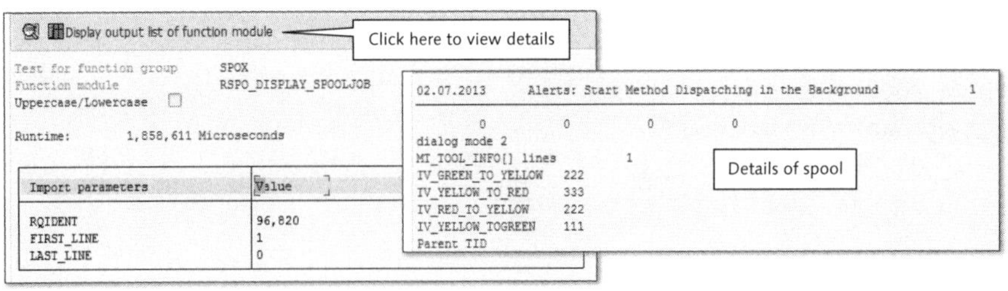

Details of Spool Generated

412 RSPO_DOWNLOAD_SPOOLJOB

This FM downloads the spool data to the presentation server's local file. You should specifically use this FM to download the spool with ABAP list details, not for the spool with images.

Example: You want to download the spool you've created in your system's local file. You can pass the path, file name, and other details to this FM. You'll receive the spool in your file after you execute this FM.

[↓] Import

ID

Number of the spool request. Type: TSP01-RQIDENT.

FNAME

File name without the path. Type: RLGRAP-FILENAME.

ENCODING

Identifier for the characters (UTF-8, UCS-2, etc.). Type: ABAP_ENCOD.

FULLPATH

File name with the path. Type: STRING.

[Δ] Changing

DIRECTORY

This parameter specifies the directory where the file is downloaded. Type: STRING.

[!] Exceptions

INVALID_DOCTYPE

Invalid document type.

DOWNLOAD_FAILED

Download failed.

CANNOT_RETURN

Spool ID is not found.

413 RSPO_FIND_SPOOL_REQUESTS

The FM transfers the selection criteria you specify in the import parameters to the matching spool request and returns the spool data in the internal table SPOOLREQUESTS.

Example: You've created a program that is making multiple spool requests on every execution. You want to get the details of the spool requests you ran for further processing. You can do so by passing your user name and other relevant parameters in this FM. On execution, you can get the details of the spool request you ran.

[↓] Import

ALLCLIENTS

An X should be passed if the request is to be searched on all clients. Omit the parameter or value [Space] to search only in your own clients.

AUTHORITY

Name of the function or transaction that has performed the authorization. Type: RSTSTYPE-AUTHCHECKD.

DATATYPE

The type that should have the spooled data. Generic specification is possible. Omit the parameter or value "*" to disable this selection criterion. Type: RSTSTYPE-TYPE.

HAS_OUTPUT_REQUESTS

Use X to search only spool requests with output orders. Use the value [Space] to search only spool requests without issuing orders. Omit parameters or the "*" value to disable this selection criterion. Type: RSPOTYPE-HASOUTREQ.

RQ0NAME

The first part of the three-spool job name (DSN). Generic specification is possible. Examples: LIST1S or * LIST. Omit the parameter or value "*" to disable this selection criterion. Type: RSPOTYPE-RQ0NAME.

RQ1NAME

The second part of the three-spool job name (suffix 1). Generic specification is possible. Examples: LP01 or LT *. Omit the parameter or value "*" to disable this selection criterion. Type: RSPOTYPE-RQ1NAME.

RQ2NAME

The third part of the three-spool job name (suffix-2). Generic specification is possible. Examples: RSLG0000_MIT or rslg *. Omit the parameter or value "*" to disable this selection criterion. Type: RSPOTYPE-RQ2NAME.

RQDEST

The name of the output device to be output to the spool request. Generic specification is possible. Omit the parameter or value "*" to disable this selection criterion. Type: RSPOTYPE-DEVICE.

RQIDENT

Spool request number. If this parameter is omitted or 0 is passed, then the selection criteria are disabled. Type: RSPOTYPE-RQIDENT.

RQOWNER

Name of the user who generated the spool request. Generic specification is possible. If you pass [Space], no values will be returned. If you pass *, all the value of all the users will be displayed. Type: RSPOTYPE-OWNER.

Tables

SPOOLREQUESTS

Table with appropriate spool requests. Type: RSPORQ.

[!] Exceptions

NO_PERMISSION

No authorization for so far-reaching search.

414 RSPO_OUTPUT_SPOOL_REQUEST

This FM creates an output request for the spool request.

Example: To print the spool to the printer directly on executing your program, you can pass the parameter in this FM and call it. On execution of the FM, the output request is created for your spool request.

[↓] Import

COPIES

Number of copies. The default value is 0. Type: TSP02-PJCOPIES.

DEVICE

Output device. The default value is "*". Type: TSP03-PADEST.

DIVISION

Department name (on the title page). The default value is "*". Type: TSP02-PJDIVISION.

ENDPAGE

Last page to be printed. The default value is 0. Type: TSP02-PJENDPAGE.

PRIO

Priority for printing. The default value is 0. Type: TSP02-PJPRIO.

RECEIVER

Recipient name on the title page. The default value is "*". Type: TSP02-PJRECEIVER.

REQEST_TITLE

Title name on the title page. The default value is "*". Type: TSP01-RQTITLE.

SPOOL_REQUEST_ID

Spool request number. This is a mandatory parameter. Type: TSP01-RQIDENT.

STARTPAGE

First page to be printed. The default value is 0. Type: TSP02-PJSTRTPAGE.

TELELAN

Country code for the fax number. The default value is "*". Type: ITCPP-TDTELELAND.

TELENUM

Fax number. The default value is "*". Type: TSKPA-TELNR.

POSNAME

Long name of printers for the host spooler. Type: TSP02-PJPOSNAME.

ACTTIME

System log time stamp. Type: TSP02-PJACTTIME.

Tables

ATTRIBUTES

Attributes for the object description. Type: RSPOATTR.

[!] Exceptions

ARCHIVE_DEST_INVALID

No archive device at archive destination.

ARCHIVE_DEST_NOT_FOUND

Archive destination does not exist.

ARCHIVE_DEST_NO_RIGHT

No authorization to use archive destination.

CANNOT_ARCHIVE

Cannot archive spool request at present.

CHANGE_ARCHDEST_NO_RIGHT

No authorization to change archive device.

CHANGE_COPIES_NO_RIGHT

No authorization to change number of copies.

CHANGE_DEST_NO_RIGHT

No authorization to change output device.

CHANGE_DEVTYPE_NO_RIGHT

No authorization to change device type.

CHANGE_PRIO_NO_RIGHT

No authorization to change priority.

CHANGE_TELENUM_NO_RIGHT

No authorization to change number/country for fax.

CHANGE_TITLE_NO_RIGHT

No authorization to change title, recipient, or department.

DEST_INVALID

Output device not allowed.

DEST_NOT_FOUND

Output device does not exist.

DEST_NO_RIGHT

No authorization to use output device.

INTERNAL_PROBLEM

Problem in FM.

INVALID_FAX_ATTRIBUTE

Incorrect country, number, or device for telefax.

INVALID_PARAMETERS

Parameters entered aren't valid.

NON_OWNER_NO_RIGHT

No authorization to process foreign spool request.

NO_LAYOUT

Format and device don't match.

NO_SPOOL_REQUEST

Spool request doesn't exist (any longer).

OUT_AGAIN_NO_RIGHT

No authorization to output again.

SPOOLER_PROBLEM

Problem has occurred in spooler.

415 RSPO_RETURN_ABAP_SPOOLJOB

This FM provides an ABAP spool list as an internal table in ASCII format.

Example: You are working on a program that's creating different print documents and spool request. An end user doesn't have authorization of SP01, so he requested you to provide a display of selected spool requests in the output report program. Use this FM to display the output of the spool request. You'll receive the spool data in the BUFFER table. Refer to FM RSPO_DISPLAY_SPOOLJOB (FM 411). Note that parameters FIRST_LINE and LAST_LINE don't work correctly. You need to implement SAP Note 1299738 for these parameters to work correctly.

[↓] Import

RQIDENT

Spool request number. This is a mandatory parameter. Type: TSP01-RQIDENT.

FIRST_LINE

First row to be displayed. Type: I.

LAST_LINE

Last line to be displayed. The value 0 means spool details will be displayed until the end of Spool. Type: I.

PAGES

Pages of the spool. Type: C.

Tables

BUFFER

You'll receive the data of the spool in this table.

[!] Exceptions

NO_SUCH_JOB

No spool request number found with RQIDENT.

NOT_ABAP_LIST

Spool order is not an ABAP list.

JOB_CONTAINS_NO_DATA

Order is empty.

SELECTION_EMPTY

Selected area is empty.

NO_PERMISSION

Permission is not enough.

CAN_NOT_ACCESS

Type cannot be determined.

READ_ERROR

Data cannot be read.

416 RSPO_RID_SPOOLREQ_LIST

This FM displays a list of spool requests with their IDs. The spool requests are displayed using ALV in the list.

Example: You've created a program that is creating many spool requests together. You want to display a list of all of the spool requests you've created in the output. You can do so by passing the spool ID to the FM and executing it. The spool details are returned.

[↓] Import

ID_LIST

List of spool request numbers. Type: SP01R_ID_LIST.

SUMMARY

Summary. This parameter doesn't have any significance in the program. Type: C.

[!] Exceptions

ERROR

An error occurred.

417 RSTS_CLOSE

This FM closes the TemSeobject that is opened for editing using FM RSTS_OPEN_RLC.

Example: After you're done with the changes, you need to close the TemSe object by calling this FM.

[↓] Import

FBHANDLE

Handle for open object. This parameter should be specified if FBHANDLE is received from FM RSTS_OPEN_RLC. Type: RSTSTYPE-FBHANDLE.

[!] Exceptions

FB_CALL_HANDLE

Error with handles at the CALL interface, for example, if both the FM and the CALL interface are working with the TemSe.

FB_ERROR

Error calling the FM or in the FM itself.

FB_RSTS_OTHER

Other errors reported by the TemSe C system.

418 RSTS_GET_ATTRIBUTES

TemSe is a store for temporary sequential data; that is, objects that aren't normally permanently held in the system are stored in the TemSe system. The spool system uses TemSe to store output data temporarily. You can set whether the spool data is to be stored by TemSe in the SAP database or in the file system. This FM gets the attributes of an entire TemSe object. The other FMs that used in this process are the following:

- RSTS_OPEN_RLC (FM 419)
- RSTS_READ (FM 420)

Example: You want to read the TemSe object spool request. First, you need to get the attributes in which the log is stored, that is, the TemSe object details and its attributes. You can pass the relevant parameters in the import parameters of this FM to the next FM.

[↓] Import

AUTHORITY

Name of the function or transaction that has already checked the necessary permissions. Type: RSTSTYPE-AUTHCHECKD.

CLIENT

The client in whose area the TemSe object has been saved. Type: RSTSTYPE-CLIENT.

NAME

Name of the TemSe object. This is a mandatory parameter. Type: RSTSTYPE-NAME.

PART

Number of the part of the TemSe object. By passing a 0, the attributes of all of the parts are required.

The default value is 0, that is, together all parts. Type: TST01-DPART.

[↑] Export

CHARCO

The character set in which the data is stored. Type: RSTSTYPE-CHARCO.

CREATER

Producer or owner of the object. Type: RSTSTYPE-CREATER.

CREDATE

Creation date. If there are multiple parts, the oldest creation date is stored in parameter CREDATE, and the most recent creation date of a part is stored in parameter MAX_CREDATE. Type: RSTSTYPE-CREDATE.

DELDATE

Expiration date. From this point of cleanups, TemSe administration may delete this object without prompting. The TemSe management internally stores the time. If there are multiple parts, the oldest expiration date is in DELDATE, and the latest expiry date is written in MAX_DELDATE. Type: RSTSTYPE-DELDATE.

MAX_CREDATE

Latest creation date of a part. Refer to parameter CREDATE. Type: RSTSTYPE-CREDATE.

MAX_DELDATE

Latest expiry date of a part. Refer Parameter DELDATE. Type: RSTSTYPE-DELDATE.

NON_UNIQ

If the object has only a part, or if all of the key attributes are the same, [Space] is returned. Type: RSTSTYPE-SEL_OK.

NOOF_PARTS

Number of parts that make up the object. Most TemSe objects consist of only one part. But if there are multiple items, then they can have different attributes. Input parameter PART controls whether the attributes of all or part of a specific part are to be determined. Type: TST01-DNOPARTS.

RECTYP

Record type. Type: RSTSTYPE-RECTYP.

SIZE

Total size of the object in bytes. Type: TST01-DSIZE.

STOTYP

Memory type. Type: TST01-DSTOTYP.

TYPE

Type of TemSe object such as TEXT. Type: RSTSTYPE-TYPE.

OBJTYPE

Type of TemSe object. This is a new parameter. Type: RSTSTYPE-TYPE.

LOCAT

Storage location (SID/value) for the TemSe object. Type: TST01-DLOCAT.

LANG

Language key. Type: TST01-DLANG.

[!] Exceptions

FB_ERROR

Error in/on the function block.

FB_RSTS_OTHER

Others reported multiple errors of TemSe system.

NO_OBJECT

Object not found/not available.

NO_PERMISSION

No permission for this operation.

419 RSTS_OPEN_RLC

This FM opens a TemSe object to read the line-by-line conversion. TemSe objects save the data in sequential files. To edit, the objects must be opened, and when editing is finished, they must be closed. The other FMs that are included in this process are the following:

- RSTS_READ (FM 420)
- RSTS_CLOSE (FM 417)

Example: To open the TemSe object for reading the spool request, pass the relevant parameters in this FM.

[↓] Import

ALLINE

When ALLINE is set to X, all rows are read. A column is added at the beginning that includes the row type for each row read. If ALLINE isn't set to X, the system reads all normal rows (with data) and ignores rows with only row type information (but no data). The default value is [Space]. Type: RSTSTYPE-ALLINE.

AUTHORITY

Name of the function or transaction that has already checked the necessary permissions. Type: RSTSTYPE-AUTHCHECKD.

CHARCO

This specifies the character set to use for the data output. Usually, the system character set is used. The default value is "*". Type: RSTSTYPE-CHARCO.

CLIENT

Client in which the TemSe object is saved. This is normally the current client. The default value is SY-MANDT. Type: RSTSTYPE-CLIENT.

NAME

Name of the TemSe object. This is a mandatory parameter. Type: RSTSTYPE-NAME.

OWN_FBHANDLE

This parameter is specified if FBHANDLE is used. The default value is [Space]. Type: RSTSTYPE-WFBHANDLE.

PROM

Processing mode. Possible values are:

- I: Current internal session.
- E: All internal sessions with this current external session.
- W: Work process.

The default value is E. Type: RSTSTYPE-PROM.

RECTYP

Record type of the object. The default value is "*". Type: RSTSTYPE-RECTYP.

OBJTYPE

Data type. The default value is TEXT. Type: RSTSTYPE-TYPE.

ENQLOCK

Lock mode of the object. Possible values are:

- N: Don't use the enqueue lock (this is compatible with the previous implementation).
- E: Set an exclusive enqueue lock.
- S: Set a shared enqueue lock.

The lock logic assumes that all users use the locks. The RSTS_CLOSE call resets the lock. The default value is N. Type: RSTSTYPE-PROM.

[↑] Export

FBHANDLE

Handle for open object. Type: RSTSTYPE-FBHANDLE.

[!] Exceptions

FB_CALL_HANDLE

Error with handles at the CALL interface.

FB_ERROR

FM error.

FB_RSTS_NOCONV

Required conversion not possible.

FB_RSTS_OTHER

Other errors reported by TemSe C system.

NO_OBJECT

Required TemSe object does not exist.

NO_PERMISSION

No authorization for this operation.

ENQ_FOREIGN_LOCK

Object locked by another user.

ENQ_SYS_FAIL

Enqueue system is failed.

ENQ_OTHER

Some other problem with enqueue.

---

420 RSTS_READ

This FM reads data from TemSe object that was opened from the FM RSTS_OPEN_RLC or any other FM. Reading can be based on lines or block.

Example: To read the TemSe object data that you've opened, pass the relevant parameters in the FM RSTS_READ. You'll get the data in the table DATATAB, which you can use to process the internal table.

[↓] Import

BUFFLG

Maximum storage length per line. This parameter exists for certain special cases. In general, the line length is determined by the width of the internal tables passed.

FBHANDLE

Handle for open object. Type: RSTSTYPE-FBHANDLE.

MAXIX

Maximum number of lines to be read. This is a protection to avoid, for example, 100MB being read at once. The default setting is a maximum of 1,000 lines. You can enter here the maximum number of lines that can be appended to an internal table. As a special case, zero (0) means that an unlimited number can be read.

PARTS1BY1

This FM normally tries to read a complete TemSe object at once. If the TemSe object consists of several parts, they are all opened separately in this FM, read, and then closed again. You can set a protection here to prevent, for example, too many TemSe objects from being opened in a database transaction. Type: RSTSTYPE-PARTS1BY1.

SHOWLG

If this parameter is set to X, the length of the line is put in front of each line. Type: RSTSTYPE-SHOWLG.

HFLAG

Indicator for hexadecimal display – for internal use only. Type: RSTSTYPE-SHOWLG.

[↑] Export

ALLDATA

If this parameter is set to X, it specifies that all data is processed. Type: RSTSTYPE-ALLDATA.

Tables

DATATAB

An internal table in which the data are written. If the table isn't empty, entries are appended.

[!] Exceptions

FB_CALL_HANDLE

Error with handles at the CALL interface.

FB_ERROR

Error in the FM or in its call.

FB_RSTS_NOCONV

Desired conversion was not possible. Details can be obtained with the FM RSTS_LAST_ERROR (this FM returns the short analysis of last error received from the TemSe object).

FB_RSTS_OTHER

Other errors, which are reported by the TemSe C system. Details can be obtained with the FM RSTS_LAST_ERROR.

421 SET_PRINT_PARAMETERS

You can use this FM to set spool print parameters and archive parameters. This FM sets the print parameter when starting a report with [Ctrl]+[P] (this FM should be called before the print parameter screen, i.e., during `INITIALIZATION` or `AT SELECTION-SCREEN` event) or when printing an online list (this FM should be called before the end of the program).

Example: You have a PRINT button on the selection screen. You want to change the printer name based on the user input so that the printout goes to that printer on the execution of the program. You can pass the printer name at event `AT SELCTION-SCREEN` and call this FM. This FM sets the printer settings.

[↓] Import

ARCHIVE_ID

Default for the archiving ID. Type: ARC_PARAMS-ARCHIV_ID.

ARCHIVE_INFO

Additional information for report names. Type: ARC_PARAMS-INFO.

ARCHIVE_MODE

Control of the printing and archiving. Type: PRI_PARAMS-ARMOD.

ARCHIVE_TEXT

Info text for the archive request. Type: ARC_PARAMS-ARCTEXT.

AR_OBJECT

Archiving object. Type: ARC_PARAMS-AR_OBJECT.

AUTHORITY

Default for authorization. Type: PRI_PARAMS-PRBER.

COPIES

Default for number of copies. Type: PRI_PARAMS-PRCOP.

COVER_PAGE

Selection cover sheet output value. Type: PRI_PARAMS-PRBIG.

DATA_SET

Default for name of the spool data. Type: PRI_PARAMS-PRDSN.

DEPARTMENT

Default for department. Type: PRI_PARAMS-PRABT.

DESTINATION

Default for output device. Type: PRI_PARAMS-PDEST.

EXPIRATION

Default for retention period. Type: PRI_PARAMS-PEXPI.

IMMEDIATELY

Default for print immediately. Type: PRI_PARAMS-PRIMM.

IN_ARCHIVE_PARAMETERS

List of all archiving parameters. Type: ARC_PARAMS.

IN_PARAMETERS

List of all default parameters. Type: PRI_PARAMS.

LAYOUT

Default for format. Type: PRI_PARAMS-PAART.

LINE_COUNT

Default for rows. If you specify the value for LINE-COUNT in the report or program statement, this parameter is ignored. Type: PRI_PARAMS-LINCT.

LINE_SIZE

Default for columns. If you specify the value for LINE-SIZE in the report or program statement, this parameter is ignored. If you're printing an online list, this parameter has no effect because the line width has already been set before printing and can't be changed. Type: PRI_PARAMS-LINSZ.

LIST_NAME

Default for spool request name. Type: PRI_PARAMS-PLIST.

LIST_TEXT

Default for spool request text. Type: PRI_PARAMS-PRTXT.

NEW_LIST_ID

Default for new spool request. Type: PRI_PARAMS-PRNEW.

RECEIVER

Default for recipient. Type: PRI_PARAMS-PRREC.

RELEASE

Default for retention period. Type: PRI_PARAMS-PRREL.

SAP_COVER_PAGE

Default for the output of the SAP cover sheet. Type: PRI_PARAMS-PRSAP.

HOST_COVER_PAGE

Default for the host cover page. Type: PRI_PARAMS-PRUNX.

PRIORITY

Type: PRI_PARAMS-PRIOT.

SAP_OBJECT

Name of the application for the archiving. Type: ARC_PARAMS-SAP_OBJECT.

TYPE

Default for spool request type. Type: PRI_PARAMS-PTYPE.

FOOT_LINE

Footer line on/off. Type: PRI_PARAMS-FOOTL.

422 SHOW_JOBSTATE

This FM returns the status of a job when you enter the job name and job number. It returns one of the six possible statuses of the job in the export parameter.

Example: You've scheduled some jobs and need to display the job statuses in your ALV display. You can pass the job number and job name in the importing parameter and call this FM to get the job status. Job status can be Aborted, Finished, Preliminary, Ready, Running, or Scheduled.

[↓] Import

JOBCOUNT

ID number of job. Type: TBTCO-JOBCOUNT.

JOBNAME

Job name. Type: TBTCO-JOBNAME.

[↑] Export

ABORTED

Status Canceled. Type: TBTCV-ABORT.

FINISHED

Status Finished. Type: TBTCV-FIN.

PRELIMINARY

Status Scheduled Temporarily. Type: TBTCV-PRELIM.

READY

Status "Ready for Execution". Type: TBTCV-READY.

RUNNING

Status Active. Type: TBTCV-RUN.

SCHEDULED

Status Scheduling Released. Type: TBTCV-SCHED.

[!] Exceptions

JOBCOUNT_MISSING

No job ID number specified.

JOBNAME_MISSING

No job name specified.

JOB_NOTEX

Specified job does not exist.

423 STORE_PRINT_PARAMETERS

This FM stores the contents of a variable of type PRI_PARAMS (which is the structure containing most of the fields of SAP print parameter storage table TPRI_PAR) in the ABAP Data Dictionary.

This FM returns a key, with which the contents of this structure can be accessed again. The key is in three parts: The first two parts identify and report the user who created these print parameters, and the third part is a sequential number. This key can be accessed by FM LOAD_PRINT_PARAMETERS (FM 409).

Example: You want to print the document in duplex format. Hence, you want to change the printer settings for your user ID so the printout appears according to your settings. You need to save and store the print settings you made. You can pass the parameters and call this FM to store the print parameters. See also FM LOAD_PRINT_PARAMETERS (FM 409).

[↓] Import

IN_PARAMETERS

Importing parameters of structure PRI_PARAMS. Type: PRI_PARAMS.

APPLIKATION

Application name. Type: SYPRAPPL.

USER

User name. Type: SY-UNAME.

PRIPROG

Report name. Type: SY-REPID.

[↑] Export

KEY

Generated key value. Type: CHAR16.

[!] Exceptions

ERROR_OCCURED

Some error occurred.

Part XI
Messages and Popups

This part of the book provides the function modules (FMs) you'll use to communicate with end users via messages or popups. Messages are texts that are generally used to provide any information, warning, success, or error details to the user. Messages are stored in Table T100 and are maintained by Transaction SE91. This transaction maintains the messages by defining a message class that stores the message, message number in the message class, message language, and message text (72characters of the actual message). The types of messages include success, error, warning, abort, exit, or information.

Popups are the dialog box screens that are displayed to the user. A popup screen is a smaller screen whose size is specified by coordinates. Popup screens are also sometimes used to give messages to users or to get confirmation from the user.

FMs in this part include creating a message, sending and collecting the messages to and from the message handler, and taking user confirmation in the popup screen. Popup screens are also used to show the time and date, display help, and so on.

424 COPO_POPUP_TO_DISPLAY_TEXTLIST

This FM displays the text lists (i.e., texts in a table) in a popup dialog box. You can also use this FM to show the F1 help information of the selection screen field.

Example: You want to create a popup with multiple lines of long texts. You can fill in the table TEXT_TABLE with the text lines you want to display in the popup, pass the other importing parameters, and then call this FM to get the popup with the text lines.

[↓] Import

TASK

Processing type. Possible values:

- Display: This is used only for displaying the desired text.
- Decide: Transfer, change, or cancel can take place with this option.

The default value is DISPLAY. Type: CLIKE.

TITEL

Dialog box title. Type: CLIKE.

[↑] Export

FUNCTION

Function code when exiting. Returned values:

- ACCE: When TRANSFER is clicked.
- MODI: When CANCEL is clicked.
- BACK: When the CANCEL button is clicked.

Type: SY-UCOMM.

Table

TEXT_TABLE

Table of texts to be displayed in the popup. Type: TLINE.

425 ERGO_TEXT_SHOW

This FM displays the F1 help documentation data. You can maintain the documentation from Transaction SE61 or from the DOCUMENTATION button. Data is saved in table DOKHL; that is, the documentation header table.

Example: You want to display the F1 help documentation that is maintained for the data element, which is referred by the selected option field. You can pass the parameter in this FM and then call it to get the documentation in the popup.

[↓] Import

TEXTNAME

Name of the general text to be displayed. Value comes from the field OBJECT of Table DOKHL. Type: DOKHL-OBJECT.

ID

Documentation ID. Value comes from the field ID of Table DOKHL. Type: DOKHL-ID.

LANGU

Language in which the text is required. Type: SY-LANGU.

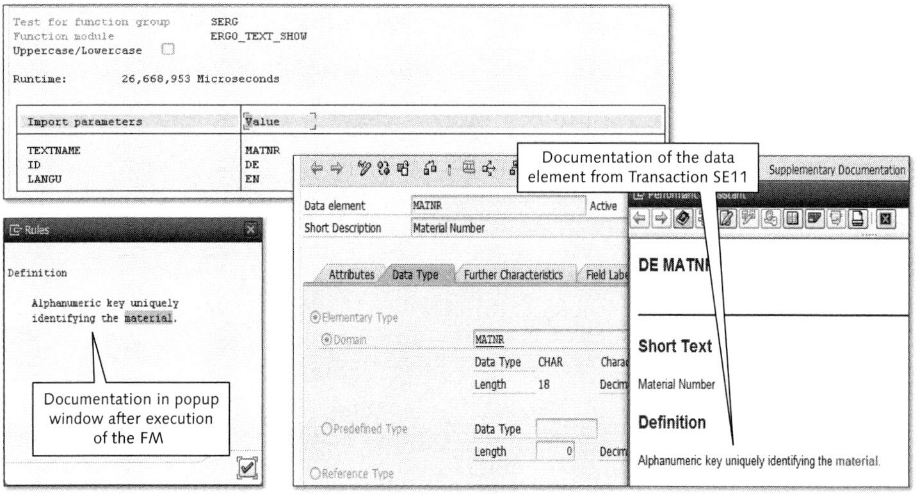

Documentation Displayed from This Function Module

---

426 F4_CLOCK

This FM displays a digital clock in the F4 help on time fields.

Example: If you have a parameter of type `sy-uzeit`, you can access the automatic F4 help. This FM is called to display the F4 help. You can also call this FM programmatically if you want to display the F4 help for time.

[↓] Import

START_TIME

The time you want to display in the F4 pop-up. The default value is SY-UZEIT. Type: SY-UZEIT.

DISPLAY

If this parameter is X, the popup is display only.

[↑] Export

SELECTED_TIME

Time returned from the F4 popup. Type: SY-UZEIT.

---

427 F4_FILENAME

This FM provides the F4 help for the file name or file manager support to locate a file in a directory.

Example: You have a parameter in the selection screen that defines the file name, and you want to upload the data to the file name. This FM displays the popup to select the file name and, when selected, the file path is returned.

[↓] Import

PROGRAM_NAME

Module pool program name for the screen field. The default value is SYST-CPROG. Type: SY-REPID.

DYNPRO_NUMBER

Dynpro number where F4 help is needed. The default value is SYST-DYNNR. Type: SY-DYNNR.

FIELD_NAME

Name of field where the path is to be entered. The default value is [Space]. Type: DYNPREAD-FIELDNAME.

[↑] Export

FILE_NAME

Path name selected by the user with the help of the file manager. Type: IBIPPARMS-PATH.

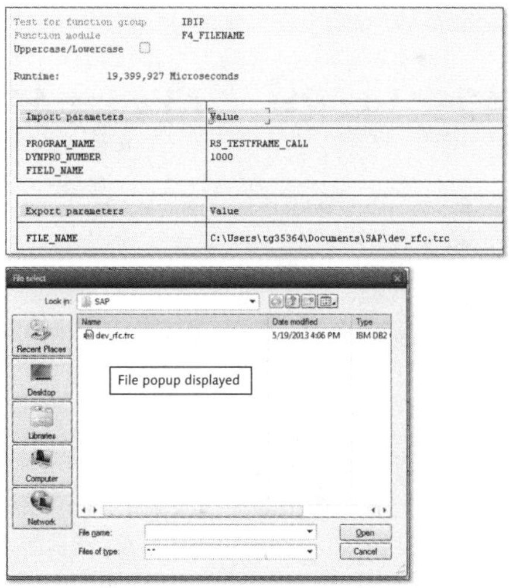

File Popup from This Function Module

---

428 F4_FILENAME_SERVER

This FM provides the F4 help for the files present in the directories of an application server.

Example: You have a parameter on your selections screen where you need to give the file name of the application server to save the data. You want to display the F4 help for this field. You can pass the relevant parameters in this FM and then call it.

[↓] Import

PFAD

Path of the file. Type: IBIPPARMS-PATH.

[↑] Export

FILE_NAME

File name returned. Type: IBIPPARMS-PATH.

[!] Exception

NO_FILE_ON_SERVER

File is not present on the server.

---

429 FITRV_CALCULATOR

This FM displays the pocket calculator in a popup.

Example: You want an end user to perform a calculation and see the result on the input field of screen. You can call this FM to display the calculator. When the calculation is done, you can pass the export parameter's returned value to the screen field.

[↓] Import

INPUT_VALUE

Value to be displayed when called.

CURRENCY

Currency for rounding the result. Type: WAERS.

START_COLUMN

Column for window positioning.

START_ROW

Line for window positioning.

[↑] Export

OUTPUT_VALUE

Calculated value returned.

[!] Exceptions

INVALID_INPUT

Input is not valid.

CALCULATION_CANCELED

Calculation is canceled by the user.

430 FORMAT_MESSAGE

This FM formats the message. This FM is similar to RPY_MESSAGE_COMPOSE (FM 458), but this FM doesn't return long text.

Example: See FM RPY_MESSAGE_COMPOSE (FM 458). You have a message "Order &1 is not valid for the given date &2' in message class." You want to write the message in the output of your program based on the value of Order 123 and Date 10/31/2013. You can pass the parameters in this FM and execute this FM to get the message "Order 123 is not valid for the given Date 10/31/2013."

[↓] Import

ID

Message class. Type: SY-MSGID.

LANG

Message language. Type: T100-SPRSL.

NO

Message number. Type: SY-MSGNO.

V1

Variable 1 parameter for the message. Type: SY-MSGV1.

V2

Variable 2 parameter for the message. Type: SY-MSGV2.

V3

Variable 3 parameter for the message. Type: SY-MSGV3.

V4

Variable 4 parameter for the message. Type: SY-MSGV4.

[↑] Export

MSG

Message text returned.

[!] Exception

NOT_FOUND

Message is not found.

431 MD_POPUP_SHOW_INTERNAL_TABLE

This FM displays the internal table data of your program in the returned popup screen. You can select the row of the internal table data that you want to process further from the popup screen. You can select one material, say DELIVERY, and click CONTINUE.

Example: You have a SELECT OPTION in the selections screen. On entering the range of values in the SELECT OPTION, you want to display the range of data selected in the popup and select one bit of data to use it for further processing. You can pass the parameter in this FM and then call this FM to get the data in the popup. You can choose the data and write further logic based on it as well.

[↓] Import

TITLE

Title of the window.

[↑] Export

INDEX

Index of the selected row. Type: SY-TABIX.

Tables

VALUES

Total number of lines of an internal table.

COLUMNS

Columns of the internal table. Type: HELP_VALUE.

[!] Exception

LEAVE

FM popup is canceled.

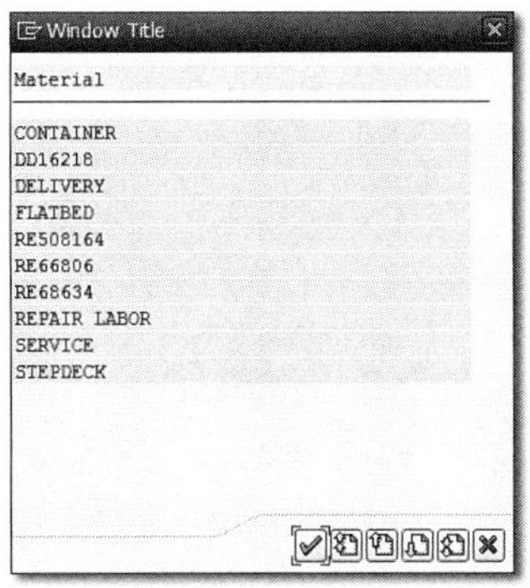

Popup Screen with Internal Table Data

432 MESSAGES_ACTIVE

This FM is used to check whether the message handler for collecting messages is active or not. If the message handler isn't active, you should call the FM MESSAGE_INITIALIZE (FM 435) to activate it.

This FM along with other FMs like MESSAGE_INITIALIZE (FM 435), MESSAGE_STORE (FM

438), etc., are used during transaction messages or error messages from various subroutines, external performs, and function blocks (not from dialog modules) to collect and display together when needed. You'll use the following FMs in this process:

- MESSAGES_INITIALIZE (FM 435)
- MESSAGE_STORE (FM 438)
- MESSAGES_STOP (FM 437)
- MESSAGES_SHOW (FM 436)
- MESSAGES_GIVE (FM 434)
- MESSAGES_COUNT (FM 433)

Example: You want to store a message for your program in memory and display all of the messages in the log when the log button is clicked. This process of displaying messages is followed in many standard SAP transactions such as Transaction VL09 (Reverse Goods Movement), which doesn't stop the processing of the further logic and program when one error has occurred and doesn't display the messages one by one. Instead, you can click on a MESSAGE LOG button to display all of the messages. First, you need to check if the message handler for storing the messages is active. See FM MESSAGES_INITIALIZE (FM 435).

[↑] Export

MAX_SEVERITY

SY-SUBC value of the message. The following values are possible:

- 2: Success message.
- 4: Information message.
- 8: Warning message.
- 12: Error message.
- 16: Abort message.

Type: SY-SUBRC.

ZEILE

Default message line if it's set.

[!] Exception

NOT_ACTIVE

Collection of messages is not active. You should call FM MESSAGES_INITIALIZE to activate the message handler.

433 MESSAGES_COUNT

The function block determines the number of collected messages in a line interval. Import parameters LINE_FROM and LINE_TO define the range from which the message is to be counted.

Example: If you want to determine the number of messages you've collected in the placeholder "&1" of the message and want to display a message to end user like "You have &1 messages that need your attention," you can get the number of messages by calling this FM.

[↓] Import

ONLY_FOR_ACTUAL_LINE

This parameter is set to X if the messages are to be returned only for the actual line. Type: SY-BATCH.

LINE_FROM

This parameter allows restricting the messages to be displayed. It displays only those messages whose line numbers are greater than or equal to LINE_FROM.

LINE_TO

This parameter allows restricting the messages to be displayed. It displays only those messages whose line numbers are less than or equal to LINE_TO.

SEVERITY_FROM

Messages from the degree of severity. You can provide the SY-SUBRC of the messages from which you want to display. If you want to display only error and abort messages you can set this parameter as 12 (12 is the severity of the error message, and 16 is the severity of abort message). Type: SY-SUBRC.

SEVERITY_TO

Messages up to the degree of severity. Type: SY-SUBRC.

[↑] Export

COUNT

Number of collected messages. Type: SY-TABIX.

MAX_SEVERITY

Maximum degree of severity of the errors occurred. See parameter MAX_SEVERITY in FM MESSAGES_ACTIVE (FM 432). Type: SY-SUBRC.

E_CHECK_ON_COMMIT

This parameter is returned to X if it was already set by some other FM of the same function group, which signifies that messages are validated before commit. Type: SY-BATCH.

E_HANDLER_ACTIVE

This parameter is set to X if the message handler is active. Type: BOOLE_D.

[!] Exceptions

INCONSISTENT_RANGE

LINE_TO is shorter than LINE_FROM.

INCONSISTENT_RANGE_SEVERITY

SEVERITY_TO is smaller than SEVERITY_FROM.

434 MESSAGES_GIVE

If the messages that you've collected should be issued by a different program, use this FM as an alternative to MESSAGES_SHOW (FM 436).

This FM receives the internal table that contains all previously collected messages, work areas, message numbers, message types, and other variables in addition to the processed message.

Example: You've stored the message in the message handler and created another report program that checks the messages at a particular time. Because this is a different program from the program used to generate the message, you can pass the relevant parameters in this FM and call it to display the messages in that report program and get the messages in the internal table.

[↓] Import

I_ZEILE

If this parameter is set, messages are returned to the default line. If this parameter isn't set, all messages are returned. Type: MSGZEILE.

I_INCL_TITLE

This parameter defines the title. Type: C.

Table

T_MESG

Table of collected messages returned. Type: TSMESG.

435 MESSAGES_INITIALIZE

This FM deletes all of the previously collected messages so that you can collect the future messages from FM MESSAGE_STORE (FM 438) in the same message handler. This FM is also used to initialize the handler and activate it.

Example: If the message handler isn't active, you need to initialize the handler by calling this FM.

[↓] Import

COLLECT_AND_SEND

If this parameter is set to X, the FM collects the messages and sends them to output. The default value is [Space].

RESET

If this parameter is set to X, it resets the collected messages. The default value is X.

LINE_FROM

This parameter allows restricting the messages to be reset. Only those messages whose line numbers are greater than or equal to LINE_FROM are reset. The default value is [Space].

LINE_TO

This parameter allows restricting the messages to be displayed. Only those messages whose line numbers are less than or equal to LINE_TO are reset. The default value is [Space].

I_STORE_DUPLICATES

If this parameter is X, the FM saves identical messages more than once. The default value is X. Type: BOOLE_D.

I_NO_DUPLICATE_COUNT

Number of messages up to which no duplicates are saved. This parameter is active only if I_STORE_DUPLICATES is active. When the specified number of messages is reached, duplicate messages are saved more than once. The default value is 500. Type: SYTABIX.

I_IDENTIFICATION

Identification from first call. When the FM MESSAGES_INITIALIZE (FM 435) is called for the first time, the caller is given an identification (see the exporting parameter E_IDENTIFICATION). When this FM is called for the second time (and later), the identification must be specified if the parameter RESET = X. This prevents the called functions from deleting messages from the message handler without being checked by the caller first.

An exception to this is messages in lines that start with "$". These lines can be used to collect messages locally in deeper level functions and to remove them if necessary. The default value is X. Type: SY-UZEIT.

CHECK_ON_COMMIT

This parameter isn't used. Type: SY-BATCH.

I_ALLOW_FOREIGN_RESET

When this FM is called for the first time, the parameter I_ALLOW_FOREIGN_RESET can be used to decide whether the message handler can be reset (without specifying the valid identification) in future calls.

If not, message BL195 ("Message handler call error: incorrect I_IDENTIFICATION value") notifies you that a reset isn't allowed. The default value is X. Type: BOOLE-BOOLE.

I_RESET_LINE

If this parameter is set to X, the FM resets LINE. The default value is X. Type: BOOLE-BOOLE.

[↑] Export

E_IDENTIFICATION

When this FM is used to call the message handler the first time, the caller is given an identification. Only programs that know this identification can perform all message handler functions. Type: SY-UZEIT.

[!] Exceptions

LOG_NOT_ACTIVE

Each message handler log is registered with the message dispatcher. If the log has been deleted (importing parameter RESET = X), the log must be deregistered with the dispatcher. Only active logs can be deregistered.

The log specified by I_IDENTIFICATION isn't currently active, which means that at least one further inactive log has been registered in the program call hierarchy; alternatively, I_IDENTIFICATION has an incorrect value (e.g., it's initial).

WRONG_IDENTIFICATION

The log is only reset when the correct identification is passed.

436 MESSAGES_SHOW

This FM displays the dialog box in which the previously collected messages are displayed. In this popup, you can scroll down and select the branches over the long text of the message. If it has been working with the LINE parameter in the FM MESSAGE_STORE (FM 438), the display can be restricted to a range of rows. Also, it's possible to suppress the display of the line number by using LINE_FROM and LINE_TO parameters.

Example: Call this FM to display the messages which you have collected in Transaction VL09 (Reverse Goods Movement). If the messages are already stored, you can directly call this FM.

[↓] Import

CORRECTIONS_OPTION

If CORRECTIONS_OPTION equals X and the message type equals W, it's a warning. There is an extra pushbutton for the user to decide if the process should be stopped. If the user uses the extra pushbutton, the exception is raised as CORRECTIONS_WANTED.
Type: C.

CORRECTIONS_FUNC_TEXT

Text for function key CORRECTIONS_WANTED. Type: ANY.

LINE_FROM

This parameter allows restricting the messages to be displayed. It displays only those messages whose line number is greater than or equal to LINE_FROM. Type: ANY.

LINE_TO

This parameter allows restricting the messages to be displayed. It displays only those messages whose line number is less than or equal to LINE_TO. Type: ANY.

OBJECT

This parameter is incorporated into the title bar of the message window to be displayed. This title has the form "Message to show OBJECT".

If OBJECT isn't specified, the default "line item" in the title will be displayed. Type: ANY.

SEND_IF_ONE

Message is sent directly if the number = 1. Type: BOOLE_D.

BATCH_LIST_TYPE

This parameter defines whether the messages should be printed in the job log or spool request. Possible values:

- J: Messages are printed in the job log only.
- L: Messages are printed in the spool list only.
- B: Messages are printed both in the job log and in the spool list.

Type: C.

SHOW_LINNO

This parameter allows you to hide a column of line numbers when displaying the collected messages. Type: BOOLE_D.

SHOW_LINNO_TEXT

Column header for the row. Type: C.

SHOW_LINNO_TEXT_LEN

Column width for the line in the display.

I_USE_GRID

If this parameter is specified, the ALV grid is used for display; otherwise, the ALV list is used. Type: C.

I_AMODAL_WINDOW

If this parameter is X, the dialog box will be displayed. Type: BOOLE-BOOLE.

[↑] Export

CORRECTIONS_WANTED

Refer to parameter CORRECTIONS_OPTION of this FM for details.

E_EXIT_COMMAND

Popup function code details for enter, cancel, and exit are returned in this parameter. Type: BAL_S_EXCM.

[!] Exceptions

INCONSISTENT_RANGE

LINE_TO is shorter than LINE_FROM.

NO_MESSAGES

No messages in the desired interval.

---

437 MESSAGES_STOP

This FM stops collecting messages and provides information on the severity of the error messages.

Example: When you're done storing all of the messages you need, you can stop the message handler by calling this FM.

[↓] Import

I_RESET_IDENTIFICATION

If this parameter is set to X, it resets the identification (see parameter E_IDENTIFICATION of FM MESSAGES_INITIALIZE (FM 435) to learn more about identification). Type: BOOLE_D.

I_IDENTIFICATION

Identification for reset messages. Type: SY-UZEIT.

I_RESET_MESSAGES

If this parameter is set to X, it reset collected messages. Type: BOOLE_D.

[!] Exceptions

A_MESSAGE

Maximum occurring error level: termination.

E_MESSAGE

Maximum occurring error level: E.

W_MESSAGE

Maximum occurring error level: W.

I_MESSAGE

Maximum occurring error level: I.

S_MESSAGE

Maximum occurring error level: S.

DEACTIVATED_BY_MD

Log deactivated by message dispatcher. The message log of the SMSG message handler has been deactivated by the message dispatcher. A different log is currently active. The SMSG message log cannot be deactivated while this other log is active.

---

438 MESSAGE_STORE

This FM builds a message similar to the ABAP statement `MESSAGE` and stores it in the global memory of the function group. FM MESSAGES_INITIALIZE (FM 435) must be called before calling this FM; otherwise, the exception NOT_ACTIVE is triggered.

Example: This FM builds the message that you want to store. When you get the message, you can pass the parameter to build it and store it in the log by passing the relevant parameters into and calling this FM.

[↓] Import

ARBGB

Message ID. Type: SMESG-ARBGB.

EXCEPTION_IF_NOT_ACTIVE

If this parameter is set to X, the exception is triggered as NOT_ACTIVE. Otherwise, the message is sent by the block MESSAGE_STORE itself.

MSGTY

Type of message (I, S, W, E, A). Type: SMESG-MSGTY.

MSGV1

First variable parameter of the message.

MSGV2

Second variable parameter of the message.

MSGV3

Third variable parameter of the message.

MSGV4

Fourth variable parameter of the message.

TXTNR

Message number.

ZEILE

Reference line. The parameter can be used for sorting and grouping successively handed messages. The subsequent output of the collected messages can be restricted to a range of row values. If such sorting or grouping isn't desired, this parameter can be omitted.

[↑] Export

ACT_SEVERITY

Severity level of the current message. Possible values:

- 02: Message type S.
- 04: Message type I.
- 08: Message type W.
- 12: Message type E.
- 16: Message type A.

Type: SY-SUBRC.

MAX_SEVERITY

Maximum level of severity. Possible values:

- 02: Message type S.
- 04: Message type I.
- 08: Message type W.
- 12: Message type E.
- 16: Message type A.

Type: SY-SUBRC.

[!] Exceptions

MESSAGE_TYPE_NOT_VALID

Type of message is not valid; that is, not I, S, W, E, or A.

NOT_ACTIVE

Collection of messages not activated (with the help of the function block MESSAGES_INITIALIZE). Therefore, the given message cannot be saved.

439 POPUP_CONTINUE_YES_NO

This FM provides a dialog box in which you can enter a question and the response alternatives YES and NO (one is preselected), where the user must confirm a processing step. The dialog box appears on the screen with a fixed size.

TITEL is displayed in the title bar; TEXTLINE1 and TEXTLINE2 are displayed in the second or third line of the dialog box. The title is displayed in the center and must not be longer than 35 characters. Lines that are longer than 35 characters are truncated.

Example: When someone is making changes to data and has unsaved data, you want to trigger a popup message before they navigate away from the screen. If, for instance, the BACK button is clicked, you can configure this FM to trigger a popup message asking a custom message "Do you want to save?" with options YES and NO.

[↓] Import

DEFAULTOPTION

This parameter determines the default settings of the selection alternatives:

- J or Y: Yes. This is the default value.
- N: No.

TEXTLINE1

First line of the dialog box. The line can be a maximum of 35 characters. Lines longer than 35 characters are truncated. This is a mandatory parameter.

TEXTLINE2

Second line of the dialog box. The line can be a maximum of 35 characters. Lines longer than 35 characters are truncated.

TITEL

Title of the dialog box. This is a mandatory parameter.

START_COLUMN

Start column of the dialog box. The dialog box starts from this column. The default value is 25. Type: SY-CUCOL.

START_ROW

Start line of the dialog box. The dialog box starts from this line. The default value is 6. Type: SY-CUROW.

[↑] Export

ANSWER

This parameter defines the end user's ANSWER decision. ANSWER can contain the following values:

- A: User selects CANCEL.
- J: User has confirmed the processing step.
- N: User has withdrawn the processing step.

440 POPUP_DISPLAY_TEXT

This FM displays the text of the document created by document maintenance Transaction SE61 in the popup dialog box. If the document doesn't exist in the transferred language, the English version of the text is displayed. If the English version is also not found, the German version is shown.

Example: You can use this FM to configure the system so that when the user presses F1 over a certain field, the system displays the document maintained for it in the modal dialog box.

[↓] Import

LANGUAGE

Language in which the text should be displayed. The default value is SY-LANGU. Type: SY-LANGU.

POPUP_TITLE

Text of the title line. This is a mandatory parameter.

START_COLUMN

Start column of the dialog box. The default value is 10. Type: SY-CUCOL.

START_ROW

Start line of the dialog box. The default value is 3. Type: SY-CUROW.

TEXT_OBJECT

Name of the text document. This is a mandatory parameter. Type: DOKHL-OBJECT.

HELP_MODAL

If this parameter is X, a modal dialog box appears; otherwise, it's modeless. The default value is X.

[↑] Export

CANCELLED

This parameter specifies whether the text display was ended by the pushbutton CANCEL or the pushbutton CONTINUE.

The parameter can have the following values:

- X: Display ended with CANCEL.
- Space: Display ended with CONTINUE.

[!] Exception

TEXT_NOT_FOUND

This exception is raised if the document is not found. The exception is raised with MESSAGE E... RAISING. If the exception is captured in the calling program, the error message is captured in the following form:

```
MESSAGE ID SY-MSGID TYPE SY-MSGTY
NUMBER SY-MSGNO WITH SY-MSGV1 SY-
MSGV2.
```

The field SY-MSGV1 contains the document name; the field SY-MSGV2 contains the language key.

441 POPUP_DISPLAY_TEXT_USER_BUTTON

This FM has the same functionality as FM POPUP_DISPLAY_TEXT (FM 440), but you can define your own buttons in the popup.

Example: If you want your own buttons for further processing such as COPY, DELETE, and so on, you can call this FM.

[↓] Import

LANGUAGE

Language in which the text should be displayed. The default value is SY-LANGU. Type: SY-LANGU.

POPUP_TITLE

Text of the title line. This is a mandatory parameter.

START_COLUMN

Start column of the dialog box. The default value is 5. Type: SY-CUCOL.

START_ROW

Start line of the dialog box. The default value is 5. Type: SY-CUROW.

DOCU_CLASS

Document class ID. The valid document classes can be found in table TTXID. Type: DOKIL-ID.

TEXT_OBJECT

Name of the text document. This is a mandatory parameter. Type: DOKHL-OBJECT.

DISPLAY_OK_BUTTON

OK button display switch. If this parameter is X, the OK button appears.

DISPLAY_CANCEL_BUTTON

CANCEL button display switch. If this parameter is X, the CANCEL button appears.

[↑] Export

ANSWER

OK button code. Type: RSNEWLENG-FCODE.

Tables

USER_BUTTONS

Pushbutton transfer table. Following are the details of the table fields:

- TEXT: Pushbutton TEXT.
- ICON_ID: Icon on the pushbutton.
- ICON_TEXT: Pushbutton icon text. If there is no ICON_ID, the text is ignored.
- QUICKINFO: Pushbutton quick info.

Type: SMP_DYNTXT.

PARAMETER

This table passes the parameters that are passed to the document to be displayed. Type: SPAR.

[!] Exceptions

ERROR_IN_TEXT

Parameters not restricted correctly in text.

TEXT_NOT_FOUND

Document not found.

442 POPUP_DISPLAY_TEXT_WITH_PARAMS

This FM shows a dialog box to the system user in which the text of a dynamic message is displayed. The parameters in the document are replaced (usually the placeholder is &) as specified in the internal table. If the document doesn't exist in the transferred language, the English version of the text is displayed. If the English version is also not found, the German version is shown.

Example: If your document has placeholders for parameters, then you can pass those parameters in the parameters table of this FM and then call it. There can be up to four placeholder in one text. For example, with the following placeholder text, "User &1 isn't authorized to run the program," when you try to run the program and you're not authorized to do so, you'll get the message "User Laura isn't authorized to run the program." If you want to provide the user-defined buttons along with parameters, you can pass those values in FM POPUP_DISPLAY_TEXT_USER_BUTTON (FM 441) and call that FM.

[↓] Import

LANGUAGE

Language in which text should be displayed. The default value is SY-LANGU. Type: SY-LANGU.

POPUP_TITLE

Text of the title line. This is a mandatory parameter.

START_COLUMN

Start column of the dialog box. The default value is 10. Type: SY-CUCOL.

START_ROW

Start line of the dialog box. The default value is 3. Type: SY-CUROW.

TEXT_OBJECT

Name of the text document. This is a mandatory parameter. Type: DOKHL-OBJECT.

HELP_MODAL

If this parameter is X, a modal dialog box appears; otherwise, it's modeless. The default value is X.

[↑] Export

CANCELLED

This parameter specifies whether the text display was ended by the pushbutton CANCEL or the pushbutton CONTINUE.

The parameter can have the following values:

- X: Display ended with Cancel.
- [Space]: Display ended with CONTINUE.

Table

PARAMETER

This table passes the parameters that are passed to the document to be displayed. Type: SPAR.

[!] Exceptions

ERROR_IN_TEXT

Parameters not restricted correctly in text.

TEXT_NOT_FOUND

This exception is raised if the document is not found. The exception is raised with `MESSAGE E... RAISING`. If the exception is captured in the calling program, the error message is captured in the following form:

```
MESSAGE ID SY-MSGID TYPE SY-MSGTY
NUMBER SY-MSGNO WITH SY-MSGV1 SY-
MSGV2.
```

The field SY-MSGV1 contains the document name; the field SY-MSGV2 contains the language key.

---

443 POPUP_FOR_INTERACTION

You can use this FM to provide an end user with a custom message.

Example: You want to give a long message to the user such as "Are you sure you want to delete the records. Do you also want to delete the records from the related table Tab2 and Tab3?" You have buttons labeled YES DELETE ALL TABLES DATA, DELETE ONLY MAIN TABLE DATA, and NO, DO NOT DELETE THE DATA. You can provide this type of messages and buttons by passing the parameters in this FM and calling this FM.

[↓] Import

HEADLINE

Heading of the popup.

TEXT1

First text line for the popup.

TEXT2

Second text line for the popup.

TEXT3

Third text line for the popup.

TEXT4

Fourth text line for the popup.

TEXT5

Fifth text line for the popup.

TEXT6

Sixth text line for the popup.

TICON

Icon for the message in the dialog box.

BUTTON_1

Title of text Button 1.

BUTTON_2

Title of text Button 2.

BUTTON_3

Title of text Button 3.

[↑] Export

BUTTON_PRESSED

This parameter returns the function code of the button clicked by the user.

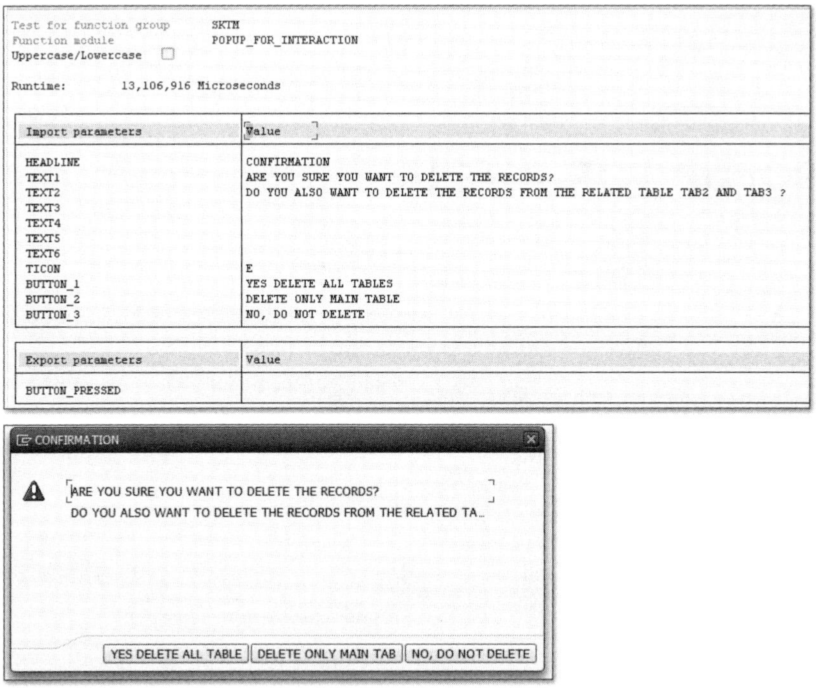

User Defines Message Displayed from This Popup

444 POPUP_GET_VALUES

This FM sends a dialog box to the user and includes the fields specified in the FIELDS interface table. The fields specified in the FIELDS table must be defined in the ABAP Data Dictionary. If no text is specified for a field, the ABAP Data Dictionary key word is displayed. The key words and fields can be displayed normally or can be highlighted. You must also define whether the fields are ready for input and whether they are mandatory. Defaults can be defined for the fields. After the display, the values entered by the user in the date and time fields are checked. Other FMs that you can use in this process include the following:

- POPUP_GET_VALUES_DB_CHECKED (FM 445)
- POPUP_GET_VALUES_USER_BUTTONS (FM 447)
- POPUP_GET_VALUES_USER_CHECKED (FM 448)

Example: You want to display a popup that contains fields such as NAME, AGE, and ADDRESS. The user can fill in the details, and then the values should be returned to be saved in the table. You can pass these fields in this FM and then call this FM. On execution of this FM, the end user gets the popup with these fields. The user can type in the values and click CONTINUE. The values are received in the table FIELDS, which can be used for saving it to the database. Refer to program RSSPO410 for an example.

[↓] Import

NO_VALUE_CHECK

Deactivates the data type check. With this parameter, you can deactivate the check on the data types (deactivate = X). If this switch is set, it applies to all screen fields. In this case, there is no formatting of date and time fields. Date and time fields are output in the internal format; that is, 19960926 instead of 26.09.1996. Format, value, and plausibility checks must be carried out by the caller. F4 help isn't supported. The default value is [Space].

POPUP_TITLE

Text of the title line. This is a mandatory parameter.

START_COLUMN

Start column of the dialog box.

START_ROW

Start line of the dialog box.

[↑] Export

RETURNCODE

The user action is returned in this parameter. It can take the following values:

- [Space]: The user wants to perform the action. FIELDS contains the changed values.
- A: The user has cancelled the action. FIELDS contains the original values.

Tables

FIELDS

The table contains the fields, values, and attributes for the layout of the dialog box. The fields appear in the sequence that they have in this FIELDS table. One line must be entered for each field that is to appear in the dialog box. This table has the following important fields:

- TABNAME (table name): The table must be active in the ABAP Data Dictionary.
- FIELDNAME (name of the field to be displayed): The field must belong to the table TABNAME. If a field is to appear twice, the second field must be passed on with * before the table name (e.g., *T001).
- VALUE (field value): VALUE can be assigned a value when the FM is called up. This value is displayed. If the user doesn't cancel the action, VALUE contains the value entered by the user.

The values must be converted to character format with MOVE not with ASSIGN.

[!] Exceptions

ERROR_IN_FIELDS

This exception is raised if fields were transferred incorrectly. It's raised with MESSAGE E... RAISING. If the exception is captured when calling, the error message with the following statement can be created:

```
MESSAGE ID SY-MSGID TYPE SY-MSGTY
NUMBER SY-MSGNO WITH SY-MSGV1
SY-MSGV2 SY-MSGV3 SY-MSGV4.
```

445 POPUP_GET_VALUES_DB_CHECKED

This FM sends a dialog box and displays the fields listed in the interface table FIELDS for entering or displaying values in that field. The fields specified in the table FIELDS must be defined in the ABAP Data Dictionary and belong to a database table or view. The entries of the user are checked according to the relational operator specified in each case against these database tables or views (EQ, NE, GT, GE, LT, LE, or NC = do not check). This FM provides the same functionality as FM POPUP_GET_VALUES (FM 444) with an additional check for database tables.

Example: If you want to check the field entered in the FIELDS table with the ABAP Data Dictionary database table, you can pass the parameters and call this FM.

[↓] Import

CHECK_EXISTENCE

This indicator specifies whether appropriate table lines must exist (e.g., to open another object) or must be missing (e.g., to insert an object). Accordingly, an error message is sent when the check against the database is performed, and the dialog box is processed once more, or processing is terminated. The indicator must have one of the following values:

- [Space]: When table lines exist, an error message is sent, for example, when an object is to be inserted. No data may exist for the specified object.
- X: When table lines are missing, an error message is sent, for example, when another object is to be opened. Data must exist for this object.

POPUP_TITLE

Text of the title line. This is a mandatory parameter.

START_COLUMN

Start column of the dialog box.

START_ROW

Start line of the dialog box.

[↑] Export

RETURNCODE

The user action is returned in this parameter. It can take the following values:

- [Space]: The user wants to perform the action. FIELDS contains the changed values.
- A: The user has cancelled the action. FIELDS contains the original values.

Table

FIELDS

The table contains the fields, values, and attributes for the layout of the dialog box. The fields appear in the sequence that they have in this FIELDS table. One line must be entered for each field that is to appear in the dialog box. Following are the important fields of this table:

- TABNAME (table name): The table must be active in the ABAP Data Dictionary.
- FIELDNAME (name of the field to be displayed): The field must belong to the table TABNAME. If a field is to appear twice, the second field must be passed on with * before the table name (e.g., *T001).
- VALUE (field value): VALUE can be assigned a value when the FM is called up. This value is displayed. If the user doesn't cancel the action, VALUE contains the value entered by the user.

[!] Exception

ERROR_IN_FIELDS

This exception is raised if fields were transferred incorrectly. It's raised with `MESSAGE E... RAISING`. If the exception is captured when calling, the error message with the following statement can be created:

```
MESSAGE ID SY-MSGID TYPE SY-MSGTY
NUMBER SY-MSGNO WITH SY-MSGV1
SY-MSGV2 SY-MSGV3 SY-MSGV4
```

446 POPUP_GET_VALUES_SET_MAX_FIELD

This FM sets the maximum number of fields in the popup window to be displayed. If more than the maximum number of fields is specified, the dialog box is sent with scroll bars.

Example: You're processing an internal table that has named fields. You want users to see a popup where they input text in all of these fields. With this FM, you'll first identify the number of fields in the internal table and set the popup for that number of fields. You can pass the number of lines of the internal table in this FM and then set number of fields for the popup. Next, you can call the FM POPUP_GET_VALUES (FM 444) for user input on these fields.

[↓] Import

NUMBER_OF_FIELDS

Maximum number of fields in a dialog box. Up to 30 fields can be specified.

[!] Exception

OUT_OF_RANGE

NUMBER_OF_FIELDS is not between 1 and 30.

---

447 POPUP_GET_VALUES_USER_BUTTONS

The FM sends a dialog box and provides the fields listed in the interface table FIELDS for input or for display similar to FM POPUP_GET_VALUES (FM 444). The fields specified in the FIELDS table must be defined in the ABAP Data Dictionary. Apart from the functionality offered by POPUP_GET_VALUES (FM 444), this FM also offers the option to add more user-defined pushbuttons. The user inputs are passed to an external subroutine (user exit), which must be specified in the interface of this function block. The interface of the subroutine must include the table FIELDS, OK code of buttons, an error structure, and a flag parameter indicating whether the dialog should be ended after exiting the subroutine. The external subroutine evaluates the OK codes and performs the appropriate functions. The error structure is filled in case of error with an error message. It's also possible to specify user exits for the F1 and F4 help when standard help for F1 and F4 won't be displayed.

Example: You want to have more user-specific pushbuttons such as CREATE and CHANGE. You can pass the parameters in this FM instead of calling FM POPUP_GET_VALUES (FM 444). You can also use it to provide the same features as POPUP_GET_VALUES, with additional user-specific buttons like COPY, PASTE, etc. To enter the fields on the popup and to validate the user input, you can pass the parameters in the FM POPUP_GET_VALUES_USER_CHECKED (FM 448) and execute it. User input is checked against the logic written in the FORM exit. Refer to program RSSPO450 in SAP ERP for examples and usage information.

[↓] Import

F1_FORMNAME

The parameter must contain the name of the external subroutine for F1 help if you want user-specific F1 help to be displayed instead of standard F1 help. The form routine must be defined as follows:

```
FORM <FORMNAME> USING TABNAME
FIELDNAME.
* Display F1 help for field TABNAME-
FIELDNAME
ENDFORM.
```

F1_PROGRAMNAME

The parameter must contain the name of the program or module pool where the form routine F1_FORMNAME is defined if you want user-specific F1 help. Type: TRDIR-NAME.

F4_FORMNAME

Name of the external subroutine for the F4 help, if you don't want to display the standard F4 help. Define the form routine as follows:

```
FORM <FORMNAME> USING TABNAME
FIELDNAME DISPLAY
CHANGING RETURNCODE VALUE.
ENDFORM.
```

Display F4 help for the field TABNAME-FIELDNAME

- If DISPLAY is passed with X, the values can only be displayed and not selected (e.g., for the F4 help of fields not ready for input).
- RETURNCODE specifies whether the user has canceled the action (A) or selected a field ([Space]), or whether the associated field on the dialog box has been updated (F).
- Pass the selected value in the VALUE field.

F4_PROGRAMNAME

Name of the program or module pool in which the form routine F4_FORMNAME is defined unless the standard F4 help is to be displayed. Type: TRDIR-NAME.

FORMNAME

Name of the external subroutine for processing. This is a mandatory parameter.

PROGRAMNAME

Name of the module pool from FORMNAME. This is a mandatory parameter. Type: TRDIR-NAME.

POPUP_TITLE

Text of the title line. This is a mandatory parameter.

OK_PUSHBUTTONTEXT

Standard pushbutton text. This is a mandatory parameter. Type: SVALBUTTON-BUTTONTEXT.

ICON_OK_PUSH

Icon for the standard pushbutton text. Type: ICON-NAME.

QUICKINFO_OK_PUSH

Quick info for the standard pushbutton text. Type: SMP_DYNTXT-TEXT.

FIRST_PUSHBUTTON

The field must contain the text of the first additional pushbutton. Up to 10 characters are available. If no text is passed on, no second pushbutton is displayed. Type: SVALBUTTON-BUTTONTEXT.

ICON_BUTTON_1

Icon for the first additional pushbutton text. Type: ICON-NAME.

QUICKINFO_BUTTON_1

Quick info for the first additional pushbutton text. Type: SMP_DYNTXT-TEXT.

SECOND_PUSHBUTTON

The field must contain the text of the second additional pushbutton. Up to 10 characters are available. If no text is passed on, no second pushbutton is displayed. Type: SVALBUTTON-BUTTONTEXT.

ICON_BUTTON_2

Icon for the second additional pushbutton text. Type: ICON-NAME.

QUICKINFO_BUTTON_2

Quick info for the second additional pushbutton text. Type: SMP_DYNTXT-TEXT.

START_COLUMN

Start column of the dialog box.

START_ROW

Start line of the dialog box.

NO_CHECK_FOR_FIXED_VALUES

This parameter controls whether the input values are to be checked against any existing constants or not. Values:

- [Space]: Checks are made against any existing constants.
- X: No checks are made against constants.

[↑] Export

RETURNCODE

User response. Refer to the RETURNCODE parameter in FM POPUP_GET_VALUES (FM 444).

Table

FIELDS

Table fields, values, and attributes. Refer to the FIELDS parameter in FM POPUP_GET_VALUES (FM 444). Type: SVAL.

[!] Exception

ERROR_IN_FIELDS

Fields were transferred incorrectly. Refer to parameter ERROR_IN_FIELDS in FM POPUP_GET_VALUES (FM 444).

---

448 POPUP_GET_VALUES_USER_CHECKED

This FM has the same functionality as POPUP_GET_VALUES (FM 444) with an addition that when the user provides the input on the fields of the popup, the values can be validated and checked with the ABAP Data Dictionary table field values. Table parameter FIELDS has the fields TABNAME, FIELDNAME, and VALUE. The VALUE is validated against the ABAP Data Dictionary field of the table, which is specified in the field TABNAME and FIELDNAME. You need to provide an external subroutine whose interface must consist of the table FIELDS and an error structure. The external subroutine analyzes the data, processes it, and sends the error message if an error occurs. See program RSSPO430 for details on using this FM in SAP ERP.

Example: You want to display a popup that contains fields such as NAME, AGE, and ADDRESS. When a user enters a value for NAME in the popup, you want that value to appear in the respective field of your Z_USER table. You can pass these fields in this FM and then call this FM. On execution of this FM, you'll get the popup with these fields. Users can type in the values and click CONTINUE. The values are received in the table FIELDS, which can be used for saving them to the database.

[↓] Import

FORMNAME

Names of external subroutines (user exits) that are called at event PAI, after value entry. Define the form routine as follows:

```
FORM <FORMNAME> TABLES  FIELDS
STRUCTURE SVAL
                    CHANGING ERROR
STRUCTURE SVALE.
ENDFORM.
```

POPUP_TITLE

Text of the title line.

PROGRAMNAME

Name of the module pool of the external subroutine. Type: TRDIR-NAME.

START_COLUMN

Start column of the dialog box.

START_ROW

Start line of the dialog box.

NO_CHECK_FOR_FIXED_VALUES

This parameter controls whether the input values are to be checked against any existing constants or not. Values:

- Space: Checks are made against any existing constants.
- X: No checks are made against constants.

[↑] Export

RETURNCODE

User response. Refer to the RETURNCODE parameter in FM POPUP_GET_VALUES (FM 444).

Table

FIELDS

Table fields, values, and attributes. Refer to the FIELDS parameter in FM POPUP_GET_VALUES. Type: SVAL.

[!] Exceptions

ERROR_IN_FIELDS

Fields were transferred incorrectly. Refer to the ERROR_IN_FIELDS parameter in FM POPUP_GET_VALUES.

449 POPUP_GET_VALUES_USER_HELP

This FM provides the same functionality as POPUP_GET_VALUES (FM 444). In addition, this FM provides the option to specify the F1 help and F4 help for the fields provided in the FIELDS table. An external subroutine is specified as a user exit to provide the F1 and F4 help.

Example: You want to display NAME, ADDRESS, and USER ID fields in the popup. You want to give the F4 help for the USER ID and NAME field from user master table USR01. You can provide these details in the form and can pass the form name and other parameters in this FM. The popup is returned with fields that have F1 and F4 options enabled.

[↓] Import

F1_FORMNAME

The parameter must contain the name of the external subroutine for F1 help if you want user-specific F1 help to be displayed instead of standard F1 help. The form routine must be defined as follows:

```
FORM <FORMNAME> USING TABNAME
FIELDNAME.
* Display F1 help for field TABNAME-
FIELDNAME
ENDFORM.
```

F1_PROGRAMNAME

The parameter must contain the name of the program or module pool where the form routine F1_FORMNAME is defined if you want user-specific F1 help. Type: TRDIR-NAME.

F4_FORMNAME

Name of the external subroutine for the F4 help, if you don't want to display the standard F4 help. Define the form routine as follows:

```
FORM <FORMNAME> USING TABNAME
FIELDNAME DISPLAY
CHANGING RETURNCODE VALUE.
ENDFORM.
```

- Display F4 help for the field TABNAME-FIELDNAME
- If DISPLAY is passed with X, the values can only be displayed and not selected (e.g., for the F4 help of fields not ready for input).
- RETURNCODE specifies whether the user has canceled the action (A) or selected a field ([Space]), or whether the associated field on the dialog box has been updated (F).
- Pass the selected value in the VALUE field.

F4_PROGRAMNAME

Name of the program or module pool in which the form routine F4_FORMNAME is defined unless the standard F4 help is to be displayed. Type: TRDIR-NAME.

FORMNAME

Name of the external subroutine for processing. This is a mandatory parameter.

POPUP_TITLE

Text of the title line. This is a mandatory parameter.

PROGRAMNAME

Name of the module pool from FORMNAME. This is a mandatory parameter. Type: TRDIR-NAME.

START_COLUMN

Start column of the dialog box.

START_ROW

Start line of the dialog box.

NO_CHECK_FOR_FIXED_VALUES

This parameter controls whether the input values are to be checked against any existing constants or not. Values:

- [Space]: Checks are made against any existing constants.
- X: No checks are made against constants.

[↑] Export

RETURNCODE

User response. Refer to the RETURNCODE parameter in FM POPUP_GET_VALUES (FM 444).

Tables

FIELDS

Table fields, values, and attributes. Refer to the FIELDS parameter in FM POPUP_GET_VALUES (FM 444). Type: SVAL.

[!] Exception

ERROR_IN_FIELDS

Fields were transferred incorrectly. Refer to the ERROR_IN_FIELDS parameter in FM POPUP_GET_VALUES (FM 444).

---

450 POPUP_NO_LIST

This FM generates a dialog box with the message that no list is formatted. This dialog box is generated when no page is created; that is, SY-PAGNO is 0. When the user confirms the dialog box, the system continues processing. If you use the NEW-PAGE statement, SY-PAGNO is never equal to 0, and the dialog box doesn't appear after NEW-PAGE even when the page is blank.

Example: You have a selection screen criteria. Based on the selection criteria, you're fetching the data from a table, processing the data, and displaying the output list. You want to give the popup message with "No List Generated" if no list is generated and then exit for further processing of the logic. You can pass the parameter and call this FM to display the message as "No List Generated."

[↓] Import

ON_EMPTY_LIST

Value X controls whether the message "No List Generated" should appear on a list that contains only the headers but is otherwise blank. Value [Space] suppresses these empty lists and popups won't appear.

FORCE

If this parameter has a value X, the message "No List Generated" in each case on the list is displayed.

---

451 POPUP_TO_CONFIRM

You can use this FM to configure a message for any situation that poses a question to the user and performs a certain action based on the user's response. Text elements should be used for questions so that it can be easily translated. This FM has replaced these other previously used POPUP_TO_* FMs:

- POPUP_TO_CONFIRM_LOSS_OF_DATA
- POPUP_TO_CONFIRM_STEP
- POPUP_TO_CONFIRM_WITH_MESSAGE
- POPUP_TO_CONFIRM_WITH_VALUE
- POPUP_TO_CONFIRM_DATA_LOSS
- POPUP_TO_DECIDE
- POPUP_TO_DECIDE_WITH_MESSAGE

Example: A user clicked the BACK button, resulting in unsaved data on the output processing. You can configure this FM to give the message, "Do you want to save the data?" with YES and NO buttons. If the user selects YES, the data are saved; otherwise, the data are not saved.

[↓] Import

TITLEBAR

Title of dialog box.

DIAGNOSE_OBJECT

The document passed must be created with the document maintenance Transaction SE61 as a text in dialog class document. No check is made that all parameters used in the text document are passed in the internal table PARAMETERS. Missing parameters aren't displayed. You should use numbered texts to make the parameter values translatable. Type: DOKHL-OBJECT.

TEXT_QUESTION

Question text in the dialog box. Maximum characters for text questions is 400. This is a mandatory parameter.

TEXT_BUTTON_1

Text on the first pushbutton. Maximum length is 8 characters. Default: Ja.

ICON_BUTTON_1

Icon on the first pushbutton. Type: ICON-NAME.

TEXT_BUTTON_2

Text on the second pushbutton. Maximum length is 8 characters. Default: Nein.

ICON_BUTTON_2

Icon on the second pushbutton. Type: ICON-NAME.

DEFAULT_BUTTON

Cursor position on the pushbutton. 1 is for the left pushbutton; 2 is for the right pushbutton.

DISPLAY_CANCEL_BUTTON

Button for displaying the CANCEL pushbutton. Values:

- [Space]: CANCEL pushbutton isn't displayed.
- X: CANCEL pushbutton is displayed.

USERDEFINED_F1_HELP

The help text must be created and maintained in Transaction SE61 as dialog in text. The document name is passed. If there is no document for the document name passed, an error message is output. Type: DOKHL-OBJECT.

START_COLUMN

Column in which the popup begins. Type: SY-CUCOL.

START_ROW

Line in which the popup begins. Type: SY-CUROW.

POPUP_TYPE

The POPUP_TYPE specifies which icon appears to the left of the text. Possible values:

- ICON_MESSAGE_QUESTION (default)
- ICON_MESSAGE_INFORMATION
- ICON_MESSAGE_WARNING
- ICON_MESSAGE_ERROR
- ICON_MESSAGE_CRITICAL
- NO_ICON

Type: ICON-NAME.

IV_QUICKINFO_BUTTON_1

Quick info on the first pushbutton. Maximum length is 30 characters. Type: TEXT132.

IV_QUICKINFO_BUTTON_2

Quick info on the second pushbutton. Maximum length is 30 characters. Type: TEXT132.

[↑] Export

ANSWER

Return values after the user clicks the pushbutton. Possible values:

- 1: Left pushbutton.
- 2: Next pushbutton.
- A: CANCEL pushbutton.

Table

PARAMETER

The internal table contains the parameters that are to be put in the parameter for DIAGNOSE_OBJECT. Type: SPAR.

[!] Exception

TEXT_NOT_FOUND

Diagnosis text not found.

452 POPUP_TO_DECIDE_LIST

This FM displays a dialog box that prompts the user to choose an option from a list of options. These options are in the form of radio buttons or checkboxes. You can specify up to 25 options.

Example: You have a parameter for tables. You want to display the F4 help and want to list all of the tables with the checkboxes so that users can select those tables. Based on the selection, you should change the data and save those data to these tables. You need to fill in the table parameter T_SPOPLi with the options you want to display as checkboxes and call this FM. This FM displays the popup with the table selections.

[↓] Import

CURSORLINE

This parameter specifies the cursor or proposal radio button line. Type: SY-LILLI.

MARK_FLAG

If the indicator is empty, the selection list is preceded by radio buttons. If the indicator is filled with X, checkboxes appear in the list. This is necessary if the user is allowed to select multiple options.

MARK_MAX

If the list is displayed with checkboxes and if this parameter is X, it's possible to set the maximum number of checkboxes.

START_COL

Column in which the dialog box begins. Type: SY-CUCOL.

START_ROW

Line in which the popup starts. Type: SY-CUROW.

TEXTLINE1

First text line. In the fields TEXTLINE1 to TEXTLINE3, explanatory texts or questions can be given that appear in the popup at the start. Each line can accommodate 35 characters. This is a mandatory parameter.

TEXTLINE2

Second text line.

TEXTLINE3

Third text line.

TITEL

Popup header line. This is a mandatory parameter.

DISPLAY_ONLY

If this parameter is set to X, popup fields are display only and the user can't select the fields.

[↑] Export

ANSWER

User response is returned in this parameter.

Table

T_SPOPLI

This table contains the selection rows that are displayed in the popup. If a user selects any row, field SELFLAG of this table is returned with X. Type: SPOPLI.

[!] Exceptions

NOT_ENOUGH_ANSWERS

Not enough responses proposed.

TOO_MUCH_ANSWERS

Too many responses (>15) proposed.

TOO_MUCH_MARKS

Too many selections.

---

453 POPUP_TO_DISPLAY_TEXT

This FM displays the popup to show any text to the user. You can display up to two text lines with a maximum of 70 characters each. There is no option for Yes or No; the user is only presented with a Continue button. If you need additional lines of text, refer to FM POPUP_TO_INFORM (FM 454).

Example: You want to provide information to the user stating that the user has entered a list of valid and invalid materials in the selection screen's select option. When the user clicks on the Invalid Materials button you've provided, you want all of the invalid materials to be shown to the user. You can concatenate all of the invalid material in two lines and pass those lines as parameters TEXTLINE1 and TEXTLINE2 and then call this FM. On execution, a popup with the invalid materials as specified in the text lines is displayed.

[↓] Import

TITEL

Title of the dialog box.

TEXTLINE1

Text line 1. This is a mandatory parameter.

TEXTLINE2

Text line 2.

Popup Displayed from This Function Module

START_COLUMN

Start column for the dialog box. Type: SY-CUCOL.

START_ROW

Start row for the dialog box. Type: SY-CUROW.

454 POPUP_TO_INFORM

The FM returns a dialog box with a diagnosis message and response options. In this FM, you can include up to four lines of text as opposed to only two lines in FM POPUP_TO_DISPLAY_TEXT (FM 440). The user must confirm a processing step. The dialog box is presented on the screen in a fixed size.

Example: You want to inform the user that clicking the DELETE button not only deletes this record, but also deletes all of the other foreign key records from foreign key tables. You can fill in the parameters and call this FM to give the user this information.

[↓] Import

TITEL

Title of the dialog box. Maximum length is 35 characters.

TXT1

Text line 1. This is a mandatory parameter.

TXT2

Text line 2. This is a mandatory parameter.

TXT3

Text line 3. Type: SY-CUCOL.

TXT4

Text line 4. Type: SY-CUROW.

455 POPUP_TO_SELECT_MONTH

This FM provides the popup to the user to choose the month.

Example: You want users to be able to select the specific month from a screen field to specify the starting month of a project. You can pass the parameters and call this FM to get the popup with MONTH and YEAR selections. You can select the value that will be returned in the export parameter SELECTED_MONTH.

[↓] Import

ACTUAL_MONTH

Month in YYYYMM format. Type: ISELLIST-MONTH.

FACTORY_CALENDAR

Factory calendar to be considered in the popup. Type: TFACD-IDENT.

HOLIDAY_CALENDAR

Public holiday calendar to be considered in the popup. Type: THOCI-IDENT.

LANGUAGE

Language in which the text of the popup should display. Type: SY-LANGU.

START_COLUMN

Start column for the popup. Type: SY-SCOLS.

START_ROW

Start row for the popup. Type: SY-SROWS.

[↑] Export

SELECTED_MONTH

This parameter returns the selected month value. Type: ISELLIST-MONTH.

RETURN_CODE

This parameter returns the SY-SUBRC code. Type: SY-SUBRC.

[!] Exceptions

FACTORY_CALENDAR_NOT_FOUND

Factory calendar specified is not found.

HOLIDAY_CALENDAR_NOT_FOUND

Holiday calendar specified is not found.

MONTH_NOT_FOUND

Month entered is not found.

---

456 POPUP_WITH_3_BUTTONS_TO_CHOOSE

This FM displays a custom dialog box that allows the user to make alternative decisions with defined buttons.

Example: You want to display the dialog box message, which can have three user inputs; that is, there is one header table and two children tables. You want to copy the data on clicking the COPY button. You want to provide the message, "Do you want to copy?" along with COPY ALL, COPY HEADER ONLY, CANCEL buttons. You can pass the parameters and can call this FM. User responses are returned in the returning parameter ANSWER.

[↓] Import

DEFAULTOPTION

Cursor position on the first or second alternative.

DIAGNOSETEXT1

First line of the diagnostic text.

DIAGNOSETEXT2

Second line of the diagnostic text.

DIAGNOSETEXT3

Third line of the diagnostic text.

TEXTLINE1

First line of the dialog box.

TEXTLINE2

Second line of the dialog box.

TEXTLINE3

Third line of the dialog box.

TEXT_OPTION1

Text of the first alternative.

TEXT_OPTION2

Text of the second alternative.

TEXT_OPTION3

Text of the third alternative.

TITEL

Title bar of the dialog box.

[↑] Export

ANSWER

Selected user's response.

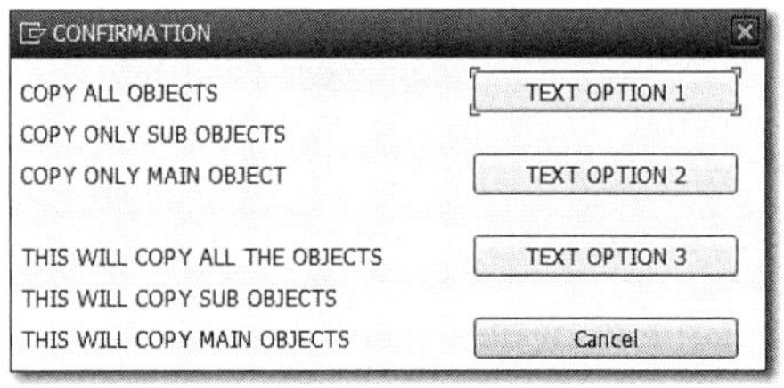

Popup with Three Options

---

457 POPUP_WITH_TABLE_DISPLAY

This FM displays the internal table data in the dialog box popup.

Example: You have an internal table from your processing logic. You want to see the internal table in the popup and select an entry from it. Based on the entry you've selected, you want to delete or copy it. You can pass the parameters in this FM and display the internal table data in a popup. Next, you can select a record you want to process. You'll get the response back in export parameter CHOISE.

[↓] Import

ENDPOS_COL

Ending column position of the popup. Type: INT4.

ENDPOS_ROW

Ending row position of the popup. Type: INT4.

STARTPOS_COL

Starting column position of the popup. Type: INT4.

STARTPOS_ROW

Starting row position of the popup. Type: INT4.

TITLETEXT

Text in the title bar of the popup. Type: CHAR80.

[↑] Export

CHOISE

Number of the table entry. Type: SY-TABIX.

Table

VALUETAB

Table with possible values.

[!] Exception

BREAK_OFF

Some exception occurred.

---

458 RPY_MESSAGE_COMPOSE

This FM is used to format any type of message.

Example: You want to display a message that is returned from the message class (from table T100) so that you can process that message accordingly. You can pass the parameter to this FM and call this FM to display the message in the language you want to display. You can also use the FMs FORMAT_MESSAGE (FM 430) or FM WRITE_MESSGAGE (FM 462) for the same purpose.

[↓] Import

LANGUAGE

Language key for the message to display. Type: T100-SPRSL.

MESSAGE_ID

Message class. Type: SY-MSGID.

MESSAGE_NUMBER

Message number. Type: SY-MSGNO.

MESSAGE_VAR1

Variable 1 parameter for the message. Type: SY-MSGV1.

MESSAGE_VAR2

Variable 2 parameter for the message. Type: SY-MSGV2.

MESSAGE_VAR3

Variable 3 parameter for the message. Type: SY-MSGV3.

MESSAGE_VAR4

Variable 4 parameter for the message. Type: SY-MSGV4.

[↑] Export

MESSAGE_TEXT

Message text returned. Type: SY-LISEL.

Table

LONGTEXT

Long text of the message returned. Type: TLINE.

[!] Exception

MESSAGE_NOT_FOUND

Message is not found.

---

459 SO_EXPRESS_FLAG_SET

You can use this FM to select a group of users to receive a message immediately or when they next login. A popup message is sent to the users.

Example: The SAPoffice Express feature sends a document with the attribute EXPRESS. The recipients get a popup where they are informed of the receipt of the document (including the title and author). You can view the document directly from the popup, jump to the entrance, or continue the previous processing. You can pass the parameters in this FM and call this FM for the same feature.

[↓] Import

CLIENT

Client in which the express message is to be triggered.

The parameter is optional. If it isn't set, the express message is triggered in the current client. If a client other than the current client is specified, the recipients have to be entered with the SAP user name. Type: SY-MANDT.

TEXT_INFO

Text that is to be displayed.

The parameter consists of a structure with the following fields:

- msgid: Message class.
- msgno: Message number.
- msgv1: Variable text section 1.
- msgv2: Variable text section 2.
- msgv3: Variable text section 3.
- msgv4: Variable text section 4.

You have to maintain the texts in table T100, which is defined for messages. Type: SOTXTINFO.

PROCESS_PARAM

Transfer of identification and method of the object to be executed. The parameter consists of a structure with the fields:

- logsys: Logical system.
- objtype: Object type.
- objkey: Object key.
- method: Method.

Type: SOPROCPAR.

PROCDIRECT

Indicator denoting whether the object can be executed directly from the dialog box of the express message. If this indicator is set to X, the EXECUTE symbol is displayed in the dialog box. The prerequisite for this is that the structure parameter PROCESS_PARAM should be maintained. Type: SOEP-PROCDIRECT.

SINGLENTRY

Indicator denoting whether a collective message is issued or whether the express messages are displayed individually if several express messages with the same text have accumulated for one user. Type: SOEP-SINGLENTRY.

INBOX

Indicator denoting whether the express message is to contain a link to the Business Workplace inbox. The link is started with the pushbutton. Type: SOEP-INBOX.

POPUP_TITLE

Title of the express dialog box. Type: SOEP-TITLE.

PERIOD_OF_VALIDITY

Validity period of the express document. Type: SOVALTIME.

CALL_AT_ONCE

Sets whether the task handler should be notified immediately. If this indicator isn't set, the task handler is notified only at COMMIT WORK. Type: SONV-FLAG.

[↑] Export

SENT_TO_ALL

Indicator denoting whether the sending was successful.

Possible values:

- X: All recipients have received the express message.
- Space: At least one recipient hasn't received the express message.

Type: SONV-FLAG.

Table

REC_TAB

Internal table for transfer of a list of recipients. Type: SOOS7.

[!] Exceptions

NO_RECEIVER_EXIST

No recipient could be determined.

OFFICE_NAME_ERROR

The caller of the FM is not assigned a Business Workplace.

460 TERM_CONTROL_EDIT

This FM creates a text editor in a popup. You can edit the text editor lines, and values are returned to the table parameter TEXTLINES.

Example: You want to display the text control. You can pass the parameter and call this FM to get the popup for the text edit. You can also call the FM TXW_TEXTNOTE_EDIT (FM 461) for the same purpose.

[↓] Import

TITEL

Title for the text edit.

LANGU

Language. Type: SY-LANGU.

Table

TEXTLINES

Text lines returned.

[!] Exception

USER_CANCELLED

User has clicked the CANCEL button.

461 TXW_TEXTNOTE_EDIT

This FM displays the TEXTNOTE editor control.

Example: See FM TERM_CONTROL_EDIT (FM 460). There, you can give the title bar name to the TEXTNOTE editor and the language. In this FM, you can't change the title bar or language, but there is an option to open the TEXTEDIT in change or display mode.

Example, you have a Text control which contains the long text description of the parameter screen field in your program. You want to display this information in text control. Use this FM for the same.

[↓] Import

EDIT_MODE

If this parameter is X, it opens the TEXTNODE control in editing mode. The default value is X. Type: C.

Table

T_TXWNOTE

Note with plain text. Type: TXW_NOTE.

---

462 WRITE_MESSAGE

This FM writes the message from table T100 into the message field.

Example: See FM RPY_MESSAGE_COMPOSE (FM 458). That FM only displays the message text, whereas this FM can display message length and error information as well. For example, you have a message "Order &1 is not valid for the given date &2" in the message class. You want to write the message in the output of your program based on the value of Order 123 and Date 10/31/2013. You can pass the parameters in this FM and execute this FM to get the message "Order 123 is not valid for the given Date 10/31/2013."

[↓] Import

MSGID

Message ID or message class. This is a mandatory parameter. Type: SY-MSGID.

MSGNO

Message number. This is a mandatory parameter. Type: SY-MSGNO.

MSGTY

Message type. This is a mandatory parameter. Type: SY-MSGTY.

V1

Variable 1 parameter for the message. Type: SY-MSGV1.

V2

Variable 2 parameter for the message. Type: SY-MSGV2.

V3

Variable 3 parameter for the message. Type: SY-MSGV3.

V4

Variable 4 parameter for the message. Type: SY-MSGV4.

[↑] Export

ERROR

This indicator is set to X if the message returned is an abort or error message.

MESSG

Message details returned in this structure. Type: MESSAGE.

MSGLN

Length of message returned.

Part XII
Transport Management

The Transport Management System plays an important role in the development and customizing environment. All of the development is locked under the task of the transport in the development system, which is moved to the quality system and finally to the production system. The Transport Management System administers and controls the new development, and records the version of the changes made for a particular program. This part of the book focuses on the function modules (FMs) used in the Transport Management System.

463 TR_APPEND_LOG

This FM is used for log output in XPRAs. XPRA (EXecution of PRogram After import) is added to a transport request using statement `R3TR XPRA rename`. If the XPRA report is called with this FM during the upgrade, the FM adds the messages, which is passed with table parameter XMSG to a file. If the XPRA report is called with Transaction SE38, the FM writes the messages to memory. To display the messages from memory on the screen, you're strongly advised to call the FM TR_FLUSH_LOG (FM 467) in the XPRA report after this FM is called for the last time. The FM TR_FLUSH_LOG reads the data from memory and then displays it.

Example: You're upgrading from the SAP ERP 4.6 release to SAP 6.0 release, and you need to transfer old user roles and authorizations to the new maintenance transaction after the upgrade. You may need to create the XPRA program and want to write the transport log of your report to the log file. Execute this FM to add messages to the transport log.

[↓] Import

OPEN_FILE

Don't use this parameter. Type: SPROT_I-XFELD.

CLOSE_FILE

Don't use this parameter. Type: SPROT_I-XFELD.

CONDENSE

Lines are condensed. Type: SPROT_I-XFELD.

MASTER_LANGU

Alternative language. Type: T100-SPRSL.

ACCEPT_NOT_INIT

Don't use this parameter. Type: SPROT_I-XFELD.

IV_SUPPRESS_STATISTICS

End time stamp not output. Type: SPROT_I-XFELD.

Table

XMSG

Table with messages. Type: SPROT_U.

[!] Exceptions

FILE_NOT_FOUND

Access to file not possible.

WRONG_CALL

Incorrect call.

464 TR_APPEND_TO_COMM_OBJS_KEYS

This FM appends the object to the transport task and locks it. Objects are appended to the table WT_E071 and the corresponding keys in the table WT_E071K.

Example: You've created a table maintenance program for a customizing table (table whose entries are transported to different SAP clients through customizing transports) and want to maintain an entry for the table. When you create an entry, it should be added to the customizing transport so that it can be transported by using this FM.

[↓] Import

WI_SIMULATION

If this parameter is set to X, all of the functions except database update are performed. This function should be used for testing purposes. Type: TRPARI-W_SIMULAT.

WI_SUPPRESS_KEY_CHECK

This parameter deactivates the key check syntax for the table parameter WT_E071K. Type: TRPARI-W_NO_CHECK.

WI_TRKORR

Task in which the object is to be added. Type: E070-TRKORR.

IT_E071K_STR

Table key with string field. Type: E071K_STRTYP.

Tables

WT_E071

Table of objects to be added. Type: E071.

WT_E071K

Table of keys to be added. Type: E071K.

[!] Exceptions

KEY_CHAR_IN_NON_CHAR_FIELD

Character in a non-character field in a task key.

KEY_CHECK_KEYSYNTAX_ERROR

Syntax error in the table E071K entry.

KEY_INTTAB_TABLE

Table is the internal table for a task key.

KEY_LONGER_FIELD_BUT_NO_GENERC

Table key longer than E071K table key but not generic.

KEY_MISSING_KEY_MASTER_FIELDS

Empty master field(s) in a task key.

KEY_MISSING_KEY_TABLEKEY

Field table key empty in a task key.

KEY_NON_CHAR_BUT_NO_GENERIC

Noncharacter key but no generic in task key.

KEY_NO_KEY_FIELDS

Table: no key fields for a task key.

KEY_STRING_LONGER_CHAR_KEY

Table string longer than character key. This exception is not raised from this FM.

KEY_TABLE_HAS_NO_FIELDS

Table has no fields for a task key.

KEY_TABLE_NOT_ACTIV

Table is not active for a task key.

KEY_UNALLOWED_KEY_FUNCTION

Invalid function in a task key.

KEY_UNALLOWED_KEY_OBJECT

Invalid object in the key of a task key.

KEY_UNALLOWED_KEY_OBJNAME

Invalid object name in the key of a task key.

KEY_UNALLOWED_KEY_PGMID

Invalid program ID in the key of a task key.

KEY_WITHOUT_HEADER

E071K entry without E071 header entry.

OB_CHECK_OBJ_ERROR

Syntax error in E071 entry (see message table).

OB_DEVCLASS_NO_EXIST

Development class does not exist.

OB_EMPTY_KEY

Select fields in LOCK_KEY are empty (system error).

OB_GENERIC_OBJECTNAME

Generic object name only in command files.

OB_ILL_DELIVERY_TRANSPORT

Invalid transport to delivery system.

OB_ILL_LOCK

Lock in task of another system.

OB_ILL_PARTS_TRANSPORT

Single elements in original request.

OB_ILL_SOURCE_SYSTEM

Consolidation request only from integration system.

OB_ILL_SYSTEM_OBJECT

Invalid system object.

OB_ILL_TARGET

Consolidation request not in consolidation system.

OB_INTTAB_TABLE

Table is an internal table.

OB_LOCAL_OBJECT

Local private object.

OB_LOCKED_BY_OTHER

Locked by another task.

OB_MODIF_ONLY_IN_MODIF_ORDER

Modifications only in tasks.

OB_NAME_TOO_LONG

Invalid object name length.

OB_NO_APPEND_OF_CORR_ENTRY

Invalid manual E071 correction entries.

OB_NO_APPEND_OF_C_MEMBER

C-member is not appended.

OB_NO_CONSOLIDATION_TRANSPORT

Invalid transport in consolidation layer.

OB_NO_ORIGINAL

Original request from non-original system.

OB_NO_SHARED_REPAIRS

No link of repairs possible.

OB_NO_SYSTEMNAME

System name cannot be determined or invalid length.

OB_NO_SYSTEMTYPE

System type cannot be determined.

OB_NO_TADIR

Missing TADIR entry, no lock necessary.

OB_NO_TADIR_NOT_LOCKABLE

Missing TADIR entry, lock necessary.

OB_PRIVAT_OBJECT

Private object (in consolidation request).

OB_REPAIR_ONLY_IN_REPAIR_ORDER

Repairs only allowed in repair requests.

OB_RESERVED_NAME

Object in reserved name range.

OB_SYNTAX_ERROR

Incorrect object syntax.

OB_TABLE_HAS_NO_FIELDS

Table has no fields (does not exist).

OB_TABLE_NOT_ACTIV

Table is not activated.

TR_ENQUEUE_FAILED

No short-term lock possible.

TR_ERRORS_IN_ERROR_TABLE

Object error, exact error in WT_TRMESS_INT.

TR_ILL_KORRNUM

Task does not correspond to <sys>K<nnnnnn>.

TR_LOCKMOD_FAILED

TLOCK modification not possible.

TR_LOCK_ENQUEUE_FAILED

Error in competing TLOCK access.

TR_NOT_OWNER

Task has another owner.

TR_NO_SYSTEMNAME

System name cannot be determined or invalid length.

TR_NO_SYSTEMTYPE

System type cannot be determined.

TR_ORDER_NOT_EXIST

Task does not exist.

TR_ORDER_RELEASED

Task already released.

TR_ORDER_UPDATE_ERROR

Error saving the task header E070.

TR_WRONG_ORDER_TYPE

Incorrect task type.

OB_INVALID_TARGET_SYSTEM

No application import in Basis system.

TR_NO_AUTHORIZATION

No authorization for this function.

OB_WRONG_TABLETYP

Wrong table type entered.

OB_WRONG_CATEGORY

Request and object categories conflict.

OB_SYSTEM_ERROR

System error in determined request/task.

OB_UNLOCAL_OBJEKT_IN_LOCAL_ORD

Transportable object not in local request.

TR_WRONG_CLIENT

Task editing in invalid client.

OB_WRONG_CLIENT

Object editing in invalid clients.

KEY_WRONG_CLIENT

Key editing in invalid client.

OB_WRONG_TABLETYP

Identifier <-> object: table <-> view intermixed.

KEY_WRONG_CLIENT

Key editing in invalid client.

465 TR_CHECK_TYPE

This FM analyzes the class type of an E071 entry (i.e., PGMID, OBJECT, OBJ_NAME) for syntactic correctness, transportability, existence of a corresponding TADIR entry (with corresponding key entry), and lock ability to request a lock (with corresponding lock key entry). The analysis covers all object types defined in the Change and Transport System (CTS).

Example: You have a bunch of transport which you want to move from development to quality box. Business has requested you to create a report program which can check the correctness and transportability of the transport. You can use this FM in your report program to check the same.

[↓] Import

WI_E071

Input key. Type: E071.

IV_TRANSLATE_OBJNAME

If the value is X, it translates the object name to uppercase. Type: TRBOOLEAN.

[↑] Export

PE_RESULT

Values returned in this parameter signify if this FM has analyzed the class type for syntactic correctness, transportability, TADIR entry, and lock ability. Following value are returned.

- Space: Doesn't analyze anything.
- S: Analyze syntactic correctness.
- R: Analyze syntactic correctness and transportability.
- T: Analyze syntactic correctness, transportability, and TADIR table entry.
- L: Analyze syntactic correctness, transportability, TADIR, and TLOCK entry.
- P: Illegal name selection or a system object.

Type: TRPARI-S_CHECKED.

WE_E071

Output string E071. Type: E071.

WE_LOCK_KEY

Structure of the request lock is displayed in this parameter. Type: TLOCK_INT.

WE_TADIR

Object directory the key created. Type: TADIR.

WE_COMPLETE_ONLY

If this parameter is X, only complete objects can be transported. Type: TRPARI-W_COMPLETE.

WE_CATEGORY

Category of the object:

- SYST: Repository object.
- CUST: Client-dependent customizing object.
- CUSY: Client-independent customizing object.
- APPL: Application object.

Type: E070-KORRDEV.

WE_CLI_DEP

If this parameter is X, the object is client dependent. Type: TRPARI-W_CLI_DEP.

WE_TABLE_TYP

Table type. Possible values:

- POOL: Pooled table.
- CLUSTER: Cluster table.
- TRANSP: Transparent table.

- VIEW: Database view.
- INTTAB: Internal table.

Type: DD02L-TABCLASS.

WE_TABLE_CONTFLAG

Table delivery class. Type: DD02L-CONTFLAG.

PE_NAME_TOO_LONG

Object name too long and exceeded the maximum permissible length. Type: TRPARI-S_CHECKED.

EV_NO_OBJECT_NAME

If this parameter is set to X, no object name is transferred. Type: TRPARI-S_CHECKED.

EV_R3TRTEXT_SYNTAXERROR

Value X in this parameter specifies the syntax error in an R3TR TEXT object. Type: TRPARI-S_CHECKED.

EV_LANGTYP

Language dependency of object:

- I: Object with no significance for language.
- F: Object without language-dependent components.
- M: Object partly language-dependent.
- L: Object consists only of language-dependent components.

Type: TRPARI-LANGTYP.

EV_LANGTYPA

Type of language dependency of an object.

- M: Purely language-dependent object in master language.
- E: Purely language-dependent object in first level language.
- L: Purely language-dependent object in second level language.

Type: TRPARI-LANGTYPA.

EV_LOGO_OBJ

If X is the value, the object is a TLOGO object. Type: TRPARI-S_EXISTS.

EV_CURR_SETTING

If this parameter is X, it specifies whether the object belongs to current settings. Type: OBJH-CURSETTING.

EV_NAME_TOO_LONG_FOR_OLD

Object name is too long (according to rules prior to 4.0). Type: TRPARI-S_CHECKED.

EV_NAME_WITH_NONSTANDARD_CHARS

Object name contains invalid characters. Type: TRPARI-S_CHECKED.

EV_NAMESPACE

Namespace. Type: TRNSPACE-NAMESPACE.

EV_INVALID_PREFIX_SYNTAX

Prefix of the object name is incorrect. Type: TRPARI-S_CHECKED.

EV_PREFIX_NOT_ALLOWED

Object name prefix not valid for object type. Type: TRPARI-S_CHECKED.

EV_CASE_SENSITIVE

If this parameter is [Space], data are translated to uppercase. Type: TRPARI-S_CHECKED.

EV_OBJLEN

Maximum name length for the object type. Type: TLOCK-LEN.

EV_INVALID_LANGUAGE_FIELD

If X, the LANGUAGE field is filled incorrectly. Type: TRPARI-S_CHECKED.

---

466 TR_DEVCLASS_GET

This FM returns the specified package details. A package can be considered as a folder where all of the objects reside.

Example: You want to display all of the package information for your program. If the change record parameter of the package field KORRFLAG is X, you want to delete the entry from the repository object directory (TADIR) table. You can pass the relevant parameters in this FM and call this FM to get the package information. Based on the value you get in field KORRFLAG of export parameter ES_TDEVC, you can delete the repository object.

[↓] Import

IV_DEVCLASS

Package. Type: TDEVC-DEVCLASS.

IV_LANGU

Language for the short text. Type: TDEVCT-SPRAS.

[↑] Export

ES_TDEVC

Read package. Type: TDEVC.

[!] Exception

DEVCLASS_NOT_FOUND

Package does not exist.

---

467 TR_FLUSH_LOG

This FM reads the messages that the FM TR_APPEND_LOG (FM 463) has written to memory from the memory and displays them on the screen. We recommend calling this FM after calling FM TR_APPEND_LOG (FM 463).

Example: You're upgrading from the SAP 4.6 release to the SAP 6.0 release and found some errors which are logged in memory. To display the log messages from memory to screen after the upgrade, you can call this FM.

468 TR_INSERT_NEW_COMM

Use this FM to create a new transport request or task in the Transport Organizer. Based on the value in the parameter WI_TRFUNCTION, one of the request types from the Import parameters is created.

The transport/task number is automatically generated by the system. The number has the format *<3-character system name>K<6-character number range number>*. For SAP, these 6 characters go from 000000 to 899999; for the customer, these 6 characters go from 900000 to 999999.

Example: This FM generally is called in the background whenever you create a new transport or task from Transport Management. When you're changing the new program that isn't yet locked, the FM generates a popup dialog box where you need to enter the transport details. If you click New Request, a screen for creation of a new task comes up. On saving, the task is created. In the program, this FM is called to create a new task.

[↓] Import

WI_KURZTEXT

Short text for task/request. Type: E07T-AS4TEXT.

WI_TRFUNCTION

Request type that can be created in the Transport Organizer. The following request types can be created:

- S: Correction (task).
- R: Repair (task).
- X Unclassified task.
- K: Transportable change request.
- L: Local change request.
- A: Unclassified request.
- W: Customizing request (client-dependent customizing objects only).
- T: Transport of copies.
- C: Relocation without development class change.
- O: Relocation with development class change.
- E: Relocation of a complete development class.

Type: E070-TRFUNCTION.

WI_STRKORR

In this parameter, tasks and requests that can be linked together are entered. Type: E070-STRKORR.

WI_CATEGORY

This parameter assigns the request category to the request/task. Possible values:

- SYST: System objects.
- CUST: Client-dependent customizing objects.

Type: E070-KORRDEV.

IV_TARSYSTEM

Target system for request. Type: E070-TARSYSTEM.

WI_CLIENT

Export client of request/task. Type: E070C-CLIENT.

WI_TARGET_CLIENT

Don't use. Type: E070C-TARCLIENT.

IV_TARDEVCL

Target package of the request (type O only). Type: E070M-TARDEVCL.

IV_DEVCLASS

Package name. This parameter is for type E category only. Type: E070M-DEVCLASS.

IV_TARLAYER

Target transport layer of the request. This parameter is for type E category only. Type: E070M-TARLAYER.

WI_PROTECT

Request is created with protected status. Type: TRPARI-S_LOCKFLAG.

IV_SIMULATION

Value X in this parameter determines the simulation mode; that is, no update will occur. Type: TRPARI-W_SIMULAT.

[↑] Export

WE_TRKORR

Task number/request number. Type: E070-TRKORR.

WE_E070

Output details of table E070. Type: E070.

WE_E07T

Output details of table E07T. Type: E07T.

WE_E070C

Output details of table E070C. Type: E070C.

ES_E070M

Output details of table E070M. Type: E070M.

[!] Exceptions

CLIENT_RANGE_FULL

Customer request number range full, that is, exceeded the maximum limit of sysK999999.

E070L_INSERT_ERROR

Database error when inserting a new request/task in ABAP Data Dictionary table E070L.

E070L_UPDATE_ERROR

Database error when updating a new request/task in ABAP Data Dictionary table E070L.

E070_INSERT_ERROR

Database error when inserting a new request/task in ABAP Data Dictionary table E070.

E07T_INSERT_ERROR

Database error when inserting a new request/task in ABAP Data Dictionary table E07T.

E070C_INSERT_ERROR

Database error when inserting a new request/task in ABAP Data Dictionary table E070C.

E070M_INSERT_ERROR

Database error when inserting a new request/task in ABAP Data Dictionary table E070M.

NO_SYSTEMNAME

System name cannot be determined or invalid length.

NO_SYSTEMTYPE

System type cannot be determined.

SAP_RANGE_FULL

SAP request number range full; that is, exceeded the maximum limit of sysK899999.

UNALLOWED_TRFUNCTION

Unknown request type/task type.

UNALLOWED_USER

Invalid user name; that is, contains '*.

ORDER_NOT_FOUND

Request/task does not exist.

INVALID_TARGETSYSTEM

Invalid target system.

INVALID_TARGET_DEVCLASS

Invalid target package.

INVALID_DEVCLASS

Invalid package.

INVALID_TARGET_LAYER

Invalid target transport layer.

INVALID_STATUS

Request/task not in status DOCU.

NOT_AN_ORDER

Input parameter request is not a request.

ORDER_LOCK_FAILED

Lock on request rejected.

NO_AUTHORIZATION

No authorization to execute function.

WRONG_CLIENT

Different clients for given object request/task.

WRONG_CATEGORY

Different category of object request/task entered.

FILE_ACCESS_ERROR

Error when accessing file.

INTERNAL_ERROR

Internal error.

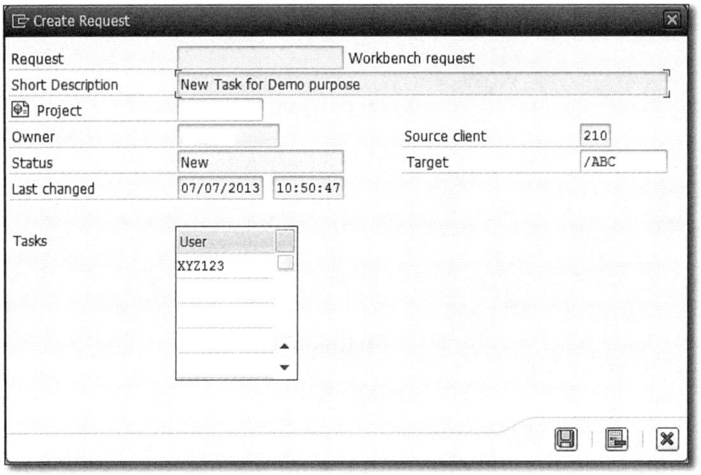

Popup to Create New Transport

469 TR_OBJECT_INSERT

This FM is used to insert an object and key into a transport. This FM performs the same functionality that FM TR_OBJECTS_INSERT (FM 472) performs, but this FM is used for a single object whereas FM TR_OBJECTS_INSERT is used for many objects. This FM must be called before the object is saved and also locks the object if required.

Example: You've made some entries in a table. You want to add this object and its entries in your transport. First, you need to call FM TR_OBJECT_CHECK (FM 470) to check if you're allowed to maintain the object. Next, you need to call this FM to insert the object details to the transport.

[↓] Import

WI_ORDER

Default transport request number. Type: E070-TRKORR.

WI_KO200

Input for object to be edited. This is a mandatory parameter. Type: KO200.

IV_NO_STANDARD_EDITOR

If this parameter is set to X, a special editor is called. Type: TRPARI-FLAG.

IV_NO_SHOW_OPTION

If this parameter is set to X, the cause of the error won't be displayed in the dialog box, if any error occurs. Type: TRPARI-FLAG.

IV_EXTERNALPS

External project system details, if any, are given in this parameter. Type: CTSPROJECT-EXTERNALPS.

IV_EXTERNALID

External project ID details. Type: CTSPROJECT-EXTERNALID.

IV_NO_PS

If this parameter is X, it specifies that the call of this FM isn't made from an external project system. Type: TRPARI-FLAG.

IV_OLD_CALL

Input for object to be edited. This is a mandatory parameter. Type: TRBOOLEAN.

IT_OBJ_ENTRY

Interface structure for objects with keys in CTS. Type: CTS_OBJ_ENTRY.

[↑] Export

WE_ORDER

Selected request (transport request). Type: E070-TRKORR.

WE_TASK

Selected request (correction/repair). Type: E070-TRKORR.

WE_KO200

Output for object to be edited. Type: KO200.

ES_TADIR

Output object directory entry. Type: TADIR.

Table

WT_E071K

Input table for object keys to be edited. Type: E071K.

[!] Exceptions

CANCEL_EDIT_OTHER_ERROR

Cancel.

SHOW_ONLY_OTHER_ERROR

Cancel, user wants to go to display mode.

470 TR_OBJECT_CHECK

This FM is used to check objects and keys. You can use this for only one object at a time.

Example: You've made some entries in a table. You want to add this object and its entries in your transport. Call this FM to check if you're allowed to maintain the object.

[↓] Import

WI_KO200

Input for the object to be edited. This is a mandatory parameter. Type: KO200.

IV_NO_STANDARD_EDITOR

If this parameter is set to X, a special editor is called. Type: TRPARI-FLAG.

IV_NO_SHOW_OPTION

If this parameter is set to X, the cause of error isn't displayed in the dialog box, if any error occurs. Type: TRPARI-FLAG.

IV_EXTERNALPS

External project system details, if any, are given in this parameter. CTSPROJECT-EXTERNALPS.

IV_EXTERNALID

External project ID details. Type: CTSPROJECT-EXTERNALID.

IV_NO_PS

If this parameter is X, it specifies that the call of this FM isn't made from an external project system. Type: TRPARI-FLAG.

IT_OBJ_ENTRY

Interface structure for objects with keys in CTS. Type: CTS_OBJ_ENTRY.

[↑] Export

WE_ORDER

Transport request and its details are returned in this parameter. Type: E070-TRKORR.

WE_TASK

Task and its details are returned in this parameter. Type: E070-TRKORR.

WE_KO200

Details of the object to be edited are returned in this parameter. Type: KO200.

WE_OBJECT_APPENDABLE

This parameter has the following values:

- X: Objects must be recorded. Before changing the object, you must call FM TR_OBJECTS_INSERT (FM 472).
- [Space]: You don't have to record changes. You don't have to call FM TR_OBJECTS_INSERT.

Type: TRPARI-S_CHECKED.

ES_TADIR

Outputs object directory details. Type: TADIR.

Table

WT_E071K

Input table for object keys to be edited. Type: E071K.

[!] Exceptions

CANCEL_EDIT_OTHER_ERROR

Cancel.

SHOW_ONLY_OTHER_ERROR

Cancel, user wants to go to display mode.

471 TR_OBJECTS_CHECK

This FM must be called when you start maintaining, that is, creating, changing, or deleting an object, to check if the user is allowed to edit the specified object. If you're changing the object for the first time, and the object is already present in any transport, a new task number is generated in the same transport in which the object already exists, and the objects is placed in this task. If you create a new object, a dialog box for the object directory entry appears. One or more objects can be copied in the table WT_KO200. The fields PGMID, OBJECT, and OBJ_NAME need to be filled for each object. The objects can be either lockable or nonlockable. This FM is a part of the interface for the Transport Organizer object tools.

Example: You've made some entries in many tables. You want to add this object and its entries in your transport. You need to call this to check if you're allowed to maintain the object.

[↓] Import

IV_NO_STANDARD_EDITOR

If this parameter is set to X, a special editor is called. Type: TRPARI-FLAG.

IV_NO_SHOW_OPTION

If this parameter is set to X, the cause of the error isn't displayed in the dialog box, if any error occurs. Type: TRPARI-FLAG.

IV_EXTERNALPS

External project system details, if any, are given in this parameter. Type: CTSPROJECT-EXTERNALPS.

IV_EXTERNALID

External project ID details. Type: CTSPROJECT-EXTERNALID.

IV_NO_PS

If this parameter is X, it specifies that the call of this FM isn't made from an external project system. Type: TRPARI-FLAG.

IT_OBJ_ENTRY

Interface structure for objects with keys in CTS. Type: CTS_OBJ_ENTRY.

[↑] Export

WE_ORDER

Transport request and its details are returned in this parameter. Type: E070-TRKORR.

WE_TASK

Task and its details are returned in this parameter. Type: E070-TRKORR.

WE_OBJECT_APPENDABLE

This parameter has the following values:

- X: Objects must be recorded. Before changing the object, you must call FM TR_OBJECTS_INSERT (FM 472).
- [Space]: You don't have to record changes. You don't have to call FM TR_OBJECTS_INSERT (FM 472).

Type: TRPARI-S_CHECKED.

Tables

WT_KO200

Input table for objects to be edited. Type: KO200.

WT_E071K

Input table for object keys to be edited. Type: E071K.

TT_TADIR

Output table for object directory entries. Type: TADIR.

[!] Exceptions

CANCEL_EDIT_OTHER_ERROR

Cancel.

SHOW_ONLY_OTHER_ERROR

Cancel, user wants to go to display mode.

---

472 TR_OBJECTS_INSERT

This FM is used as an interface that allows you to insert objects and keys into the transport request. This FM must be called before the objects are saved. Before calling this FM, FM TR_OBJECTS_CHECK (FM 471) must be called for the objects you want to maintain. If you're editing a single object, you can use the alternative FM TR_OBJECT_CHECK (FM 470). This FM is a part of the interface for the Transport Organizer object tools.

Example: You've made some entries in a table. You want to add those entries in your transport. First, you need to call the FM TR_OBJECTS_CHECK to check if you're allowed to maintain the object. Next, you need to call the FM TR_OBJECTS_INSERT to insert the object details to the transport. If you want to add only one object, you can use FM TR_OBJECT_CHECK for checking and TR_OBJECT_INSERT (FM 469) for inserting.

[↓] Import

WI_ORDER

Default transport request number. Type: E070-TRKORR.

IV_NO_STANDARD_EDITOR

If this parameter is set to X, a special editor is called. Type: TRPARI-FLAG.

IV_NO_SHOW_OPTION

If this parameter is set to X, the cause of the error isn't displayed in the dialog box, if any error occurs. Type: TRPARI-FLAG.

IV_EXTERNALPS

External project system details, if any, are given in this parameter. Type: CTSPROJECT-EXTERNALPS.

IV_EXTERNALID

External Project ID details. Type: CTSPROJECT-EXTERNALID.

IV_NO_PS

If this parameter is X, it specifies that the call of this FM isn't made from an external project system. Type: TRPARI-FLAG.

IT_OBJ_ENTRY

Interface structure for objects with keys in CTS. Type: CTS_OBJ_ENTRY.

[↑] Export

WE_ORDER

Selected transport request details come in this parameter. Type: E070-TRKORR.

WE_TASK

Selected task details come in this parameter. Type: E070-TRKORR.

Tables

WT_KO200

One or more objects can be copied in this table. The fields PGMID, OBJECT, and OBJ_NAME need to be filled for each object. The objects can be lockable or nonlockable. Type: KO200.

WT_E071K

Input table of edited object keys. Type: E071K.

TT_TADIR

Output table for object directory entries. Type: TADIR.

[!] Exceptions

CANCEL_EDIT_OTHER_ERROR

Cancel.

SHOW_ONLY_OTHER_ERROR

Cancel, user wants to go to display mode.

473 TR_READ_COMM

This FM reads all of the transport request data. Request header information is stored in tables E070 and E070C, request text information is stored in table E07T, request objects information is stored in table E071, keys for transport objects are stored in table E071K, and corresponding namespaces for keys are stored in table E071KF. If you want to read table E071 and the request has the status Released, then table E071 is read in addition to table E071C. This table contains information on the import. The value of the lock flag is determined from table E071C. If no entry is found here, the value is taken from table E071.

Example: You need to create a report in which the transport number is provided on the selection screen field, and you want to read and display its header, object, and key information in the output screen for analysis of which transport contains which objects. You can pass the relevant parameters into and call this FM.

[↓] Import

WI_TRKORR

Request/task whose details are to be read. Type: E070-TRKORR.

WI_DIALOG

If this parameter is set to X, information messages, if any, are sent to the user. Type: TRPARI-W_DIALOG.

WI_LANGU

Language in which the request or task is maintained. Type: SY-LANGU.

WI_SEL_E070

If this parameter is set to X, table E070 is read. Type: TRPARI-W_E070.

WI_SEL_E071

If this parameter is set to X, table E0701 is read. Type: TRPARI-W_E071.

WI_SEL_E071K

If this parameter is set to X, table E071K is read. Type: TRPARI-W_E071K.

IV_SEL_E071KF

If this parameter is set to X, table E071KF is read. Type: TRPARI-W_E071KF.

WI_SEL_E07T

If this parameter is set to X, table E07T is read. Type: TRPARI-W_E07T.

WI_SEL_E070C

If this parameter is set to X, table E070C is read. Type: TRPARI-W_E070C.

IV_SEL_E070M

If this parameter is set to X, table E070M is read. Type: TRPARI-W_E070M.

IV_SEL_E070A

If this parameter is set to X, table E070A is read. Type: TRPARI-FLAG.

[↑] Export

WE_E070

This parameter returns table E070 details. Type: E070.

WE_E07T

This parameter returns table E07T details. Type: E07T.

WE_E070C

This parameter returns table E070C details. Type: E070C.

ES_E070M

This parameter returns table E070M details. Type: E070M.

WE_E07T_DOESNT_EXIST

Value X specifies that table E07T doesn't exist on the database. Type: TRPARI-W_E07T.

WE_E070C_DOESNT_EXIST

Value X specifies that table E070C doesn't exist on the database. Type: TRPARI-W_E070C.

EV_E070M_DOESNT_EXIST

Value X specifies that table E070M doesn't exist on the database. Type: TRPARI-W_E070M.

Tables

WT_E071

This parameter returns table E071 details. Type: E071.

WT_E071K

This parameter returns table E071K details. Type: E071K.

ET_E071KF

This parameter returns table E071KF details. Type: E071KF.

ET_E070A

This parameter returns table E070A details. Type: E070A.

[!] Exceptions

NOT_EXIST_E070

Request/task does not exist.

NO_AUTHORIZATION

No authorization to read the request/task.

474 TR_SYS_PARAMS

This FM determines the global system parameters such as system name, system type, system editing, client modification, client-independent object modification, and the client role.

Example: You can't transport the customizing objects of your system. You need to check the system settings by using this FM. You can then edit the settings if needed.

[↑] Export

SYSTEMEDIT

System editing can be set so that everything is editable if the value is [Space], or nothing is editable if the value is N. Type: TADIR-EDTFLAG.

SYSTEMNAME

System name in which you logged in. Type: SY-SYSID.

SYSTEMTYPE

System type: (SAP or CUSTOMER). Type: SY-SYSID.

SYSTEM_CLIENT_EDIT

Change option for client-dependent customizing objects. Possible values:

- [Space]: No automatic recording of changes for transport.
- 1: Changes are recorded in the transport request.
- 2. Customizing objects in the client can't be changed.
- 3: Customizing objects can be changed as requested but can't be transported.

Refer to field CCCORACTIV in client table T000. Type: T000-CCCORACTIV.

SYS_CLIINDDEP_EDIT

Change option for repository objects in the logon client. Possible values:

[Space]: Changes to repository and cross client customizing is allowed

- 1: No changes to cross-client Customizing objects is allowed
- 2: No changes to repository objects
- 3: No changes to repository and cross-client customizing objects

Refer to field CCNOCLIIND in client table T000. Type: T000-CCNOCLIIND.

SYSTEM_CLIENT_ROLE

Client role specifies the role of the client. Possible values:

- P: Productive client.
- T: Test client.
- C: Customizing client.
- D: Demo client.
- E: Training client.
- R: SAP reference client.

Type: T000-CCCATEGORY.

EV_SFW_BCSET_REC

Recording client for switch BC sets. Type: T000-CCORIGCONT.

EV_C_SYSTEM

Obsolete: Always returned empty. Type: TRPARI-S_CHECKED.

[!] Exceptions

NO_SYSTEMNAME

System name cannot be determined or invalid length.

NO_SYSTEMTYPE

System type cannot be determined.

---

475 TR_TADIR_INTERFACE

This FM is used to create, change, delete, or read the object directory entry from the TADIR table. This FM acts as an interface for the object directory (TADIR) of the ABAP Workbench. You can only make modifications to the object directory with this FM. Internal checks are made that guarantee the consistency of the data. You need to provide the PGMID, OBJECT, and OBJ_NAME fields of TADIR to identify the object.

Example: Whenever an entry in the object directory is created, read, or changed, you can call this FM. You can see the call to the FM in standard transport reports. When you create a new transport from the Transport Organizer, the entry is created in the object directory by calling this FM. If you want to create an entry programmatically, you can do so by passing the relevant parameters into and calling this FM.

[↓] Import

WI_DELETE_TADIR_ENTRY

If this parameter is X, it deletes the object directory entry. This parameter should be set to X if you want to delete an entry in the object directory. Type: TRPARI-S_CHECKED.

WI_REMOVE_REPAIR_FLAG

If this parameter is X, it resets the repair flag. Type: TRPARI-S_CHECKED.

WI_SET_REPAIR_FLAG

If this parameter is X, it sets the repair flag. Type: TRPARI-S_CHECKED.

WI_TEST_MODUS

If this parameter is X, it enables the simulation mode, but no update occurs. If you want to update the database, you should set it as [Space]. The default value is X. Type: TRPARI-S_CHECKED.

WI_TADIR_PGMID

Input for the PGMID field of the TADIR table. Type: TADIR-PGMID.

WI_TADIR_OBJECT

Input for the OBJECT field of the TADIR table. Type: TADIR-OBJECT.

WI_TADIR_OBJ_NAME

Input for the OBJ_NAME field of the TADIR table. Type: TADIR-OBJ_NAME.

WI_TADIR_KORRNUM

Input for the KORRNUM field of the TADIR table. Type: TADIR-KORRNUM.

WI_TADIR_SRCSYSTEM

Input for the SRCSYSTEM field of the TADIR table. Type: TADIR-SRCSYSTEM.

WI_TADIR_AUTHOR

Input for the AUTHOR field of the TADIR table. This parameter is mandatory when inserting the entry to the object directory. Type: TADIR-AUTHOR.

WI_TADIR_DEVCLASS

Input for the DEVCLASS field of the TADIR table. This parameter is mandatory when you're inserting an entry in the object directory or changing the package of the object. Each time you change a package, you must transport the change into all systems in your system group. This isn't currently done automatically. Transport the change by inserting the entry LIMU ADIR R3TRPROGZREPORT in a transportable change request, in this example, for the report program ZREPORT. Type: TADIR-DEVCLASS.

WI_TADIR_MASTERLANG

Input for the MASTERLANG field of the TADIR table. Type: TADIR-MASTERLANG.

WI_TADIR_CPROJECT

Input for the CPROJECT field of the TADIR table. This is for internal use only. Type: TADIR-CPROJECT.

WI_TADIR_VERSID

Input for the VERSID field of the TADIR table. This is for internal use only. Type: TADIR-VERSID.

WI_REMOVE_GENFLAG

This parameter deletes the generation flag of the TADIR table. This parameter should be set to X if you want to remove the generation flag for an object. Type: TRPARI-S_CHECKED.

WI_SET_GENFLAG

This parameter sets the generation flag of the TADIR table. This parameter should be set when you want to flag an object as generated. Possible values:

- X: The object is generated and is retained during the upgrade.
- T: The object is generated but can be deleted during the upgrade.
- [Space]: The generated flag isn't changed.

Type: TADIR-GENFLAG.

WI_READ_ONLY

Read object directory entry. This parameter must be set to X if you want to read the entry in the object directory. Type: TRPARI-S_CHECKED.

IV_SET_EDTFLAG

Editing lock for standard editors. Value X in this parameter prevents the object from being edited with the standard object editors. Type: TADIR-EDTFLAG.

IV_TRANSL_TECH_TEXT

Translate technical texts into development language. Type: TADIR-TRANSLTTXT.

IV_DELFLAG

This parameter sets the deletion flag from the TADIR table. Type: TADIR-DELFLAG.

IV_NO_PAK_CHECK

Exception indicator for package check. Type: TADIR-PAKNOCHECK.

IV_OBJ_STABILITY

Release status of a development object. Type: TADIR-OBJSTABLTY.

WI_TADIR_CHECK_DATE

SCC: Check date. Type: TADIR.

[↑] Export

NEW_GTADIR_ENTRY

New entry in the TADIR table. Type: GTADIR.

NEW_TADIR_ENTRY

Modified TADIR entry. This parameter is returned when the object directory entry is read or changed. Type: TADIR.

[!] Exceptions

TADIR_ENTRY_NOT_EXISTING

Object directory entry does not exist.

TADIR_ENTRY_ILL_TYPE

Transferred TADIR key not compatible with E071.

NO_SYSTEMNAME

System name not found.

NO_SYSTEMTYPE

System type not defined.

ORIGINAL_SYSTEM_CONFLICT

Object already exists in another system.

OBJECT_RESERVED_FOR_DEVCLASS

Object reserved for name range.

OBJECT_EXISTS_GLOBAL

Object exists globally.

OBJECT_EXISTS_LOCAL

Object exists locally.

OBJECT_IS_DISTRIBUTED

Object is distributed.

OBJ_SPECIFICATION_NOT_UNIQUE

Object specification for import is not sufficient.

NO_AUTHORIZATION_TO_DELETE

No permission to delete.

DEVCLASS_NOT_EXISTING

Package unknown.

SIMULTANIOUS_SET_REMOVE_REPAIR

Repair flag set/reset simultaneously.

ORDER_MISSING

Repair request was not transferred.

NO_MODIFICATION_OF_HEAD_SYST

Modification of HEAD-SYST entry not allowed.

PGMID_OBJECT_NOT_ALLOWED

PGMID entry not permitted.

MASTERLANGUAGE_NOT_SPECIFIED

Master language not specified.

DEVCLASS_NOT_SPECIFIED

Package not specified.

SPECIFY_OWNER_UNIQUE

Owner is not unique.

LOC_PRIV_OBJS_NO_REPAIR

No repair to local-private objects.

GTADIR_NOT_REACHED

The GTADIR cannot be accessed.

OBJECT_LOCKED_FOR_ORDER

Object is locked for an order.

CHANGE_OF_CLASS_NOT_ALLOWED

Change of the class is not allowed.

NO_CHANGE_FROM_SAP_TO_TMP

Do not switch SAP objects to customer development class.

---

476 TR_TADIR_POPUP_ENTRY_E071

This FM displays the object directory entry (TADIR) for the object sent in the parameters WI_E071_PGMID, I_E071_OBJECT, and WI_E071_OBJ_NAME.

Example: To display the object directory entry programmatically for your object changes so that you can lock the entry in the object directory, pass the parameter, and call this FM.

[↓] Import

WI_E071_PGMID

E071-PGMID of requested TADIR entry. This is a mandatory parameter. Type: E071-PGMID.

WI_E071_OBJECT

E071-OBJECT of requested TADIR entry. This is a mandatory parameter. Type: E071-OBJECT.

WI_E071_OBJ_NAME

E071-OBJ_NAME of requested TADIR entry. This is a mandatory parameter. Type: E071-OBJ_NAME.

WI_E071_GENNUM

E071-GENNUM of the requested TADIR entry. Type: E071-GENNUM.

WI_MESSAGE_ENTER_DEVCLASS

This parameter creates an "Enter development class" message on the dialog box. Type: TRPARI-W_MSG_DEVC.

WI_READ_ONLY

If this parameter is set to X, TADIR is in display mode. Type: TRPARI-S_READONLY.

WI_TADIR_DEVCLASS

TADIR_DEVCLASS (default for creating new TADIR). Type: TADIR-DEVCLASS.

WI_TADIR_GENFLAG

TADIR-GENFLAG parameter. Possible values:

- [Space]: Not generated.
- X: Generated and retained during the upgrade.
- T: Generated but can be deleted during the upgrade.

Type: TADIR-GENFLAG.

WI_TADIR_MASTERLANG

TADIR-MASTERLANG parameter of the TADIR table. Language of the package. Type: TADIR-MASTERLANG.

IV_SUPPRESS_MOD_SHOW

Suppress display-change function on the dialog box. Type: TRPARI-FLAG.

[↑] Export

WE_TADIR

New object directory entry is returned in this parameter. Type: TADIR.

ES_TDEVC

New package entry is returned in this parameter. Type: TDEVC.

EV_DELETED

This parameter is set to X, if the object entry is deleted successfully. Type: TRPARI-FLAG.

[!] Exceptions

DISPLAY_MODE

Only display object.

EXIT

Function aborted.

GLOBAL_TADIR_INSERT_ERROR

Error inserting into global TADIR.

NO_REPAIR_SELECTED

No repair request selected.

NO_SYSTEMNAME

System name cannot be determined or invalid length.

NO_SYSTEMTYPE

System type cannot be determined.

NO_TADIR_TYPE

Object type cannot be included in TADIR.

RESERVED_NAME

Violation of reserved TRESN names.

TADIR_ENQUEUE_FAILED

TADIR locked by short-term lock.

DEVCLASS_NOT_FOUND

Package not in TDEVC.

TADIR_NOT_EXIST

TADIR does not exist, deletion rejected.

OBJECT_EXISTS

Object exists, deletion rejected.

INTERNAL_ERROR

Internal program error.

OBJECT_APPEND_ERROR

Error when locking and saving the E071 entries.

TADIR_MODIFY_ERROR

Error when saving TADIR entry.

OBJECT_LOCKED

Object locked by external request lock.

NO_OBJECT_AUTHORITY

No authorization to execute the function.

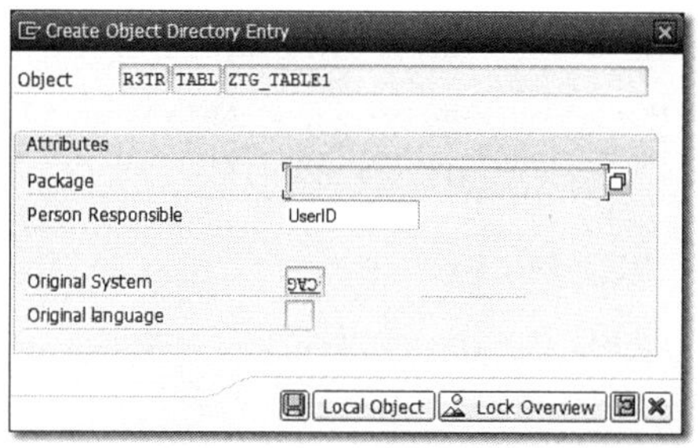

Object Directory Entry on Execution of This Function Module

Part XIII
Application Log

Application logs record the progress of the execution of an application by logging application events. This process is similar to using a system log for logging system event information. The advantage of using an application log is that it's a system-wide standardized process and is convenient to analyze.

Several application logs can be written at the same time by an application. Logs should be deleted in intervals—maybe weekly—to avoid high-volume database load. An example of using an application log is in delivery processing where the negative value of the dangerous goods needs to be checked and written to the application log. Transaction SLG0 creates the application log, SLG1 analyze the application log, and Transaction SLG2 deletes the application log.

There are many function modules (FMs) you can use with application logs for reading, saving application logs, and more. FMs starting with APPL_LOG_* have been around since release 3.0 of SAP ERP. These are the old FMs that internally call the new BAL_* FMs. The new BAL_* FMs—available since release 4.6—are more powerful and useful.

477 APPL_LOG_DELETE

This FM deletes all of the protocols of the application log and frees the log memory so it can be used again. Authorization object S_APPL_LOG is checked when this FM is executed.

Example: You're done with your log processing and want to delete all of the messages of the log. You can pass the parameters, such as object name, subobject name, external number, and other parameters, and call this FM to delete the messages of the application log.

[↓] Import

OBJECT

Object name. This is a mandatory parameter. Type: BALHDR-OBJECT.

SUBOBJECT

Subobject name. Type: BALHDR-SUBOBJECT.

EXTERNAL_NUMBER

External ID of application log. Type: BALHDR-EXTNUMBER.

DATE_TO

Delete-by date. This is a mandatory parameter. Type: BALHDR-ALDATE.

TIME_TO

Delete-by time. This is a mandatory parameter. Type: BALHDR-ALTIME.

LOG_CLASS

Problem class. Type: BALM-PROBCLASS.

I_WITH_COMMIT_WORK

Database will commit after 100 log entries. Type: BOOLEAN.

I_PACKAGE_SIZE

Package size for DB_COMMIT. Type: I.

[↑] Export

NUMBER_OF_DELETED_LOGS

Number of logs deleted. Type: SY-TABIX.

NUMBER_OF_NON_DELETED_LOGS

Number of records that could not be deleted because the expiration date had not been reached. Type: SY-TABIX.

[!] Exceptions

NO_AUTHORITY

No authorization for deleting logs. Authority object S_APPL_LOG is checked.

478 APPL_LOG_DISPLAY

This FM is used to display the selected application log specified in the import parameters.

Example: You can display the messages in the application log, which is saved in the database, by using this FM. Refer to program RSSLG150.

[↓] Import

OBJECT

Object name. Type: BALHDR-OBJECT.

SUBOBJECT

Subobject name. Type: BALHDR-SUBOBJECT.

EXTERNAL_NUMBER

External number. Type: BALHDR-EXTNUMBER.

OBJECT_ATTRIBUTE

This parameter defines the display properties of OBJECT. Possible values: 0 = ready for input, 1 = display, 2 = hide.

SUBOBJECT_ATTRIBUTE

This parameter defines the display properties of SUBOBJECT _ATTRIBUTE. Possible values: 0 = ready for input, 1 = display, 2 = hide.

EXTERNAL_NUMBER_ATTRIBUTE

This parameter defines the display properties of EXTERNAL_NUMBER_ATTRIBUTE. Possible values: 0 = ready for input, 1 = display, 2 = hide.

DATE_FROM

Date: Lower limit. Type: BALHDR-ALDATE.

TIME_FROM

Time: Lower limit. Type: BALHDR-ALTIME.

DATE_TO

Date: Upper limit. Type: BALHDR-ALDATE.

TIME_TO

Time: Upper limit. Type: BALHDR-ALTIME.

TITLE_SELECTION_SCREEN

Selection screen title.

TITLE_LIST_SCREEN

Log display title.

COLUMN_SELECTION

Selection of columns in the header list. Type: BALDISP.

SUPPRESS_SELECTION_DIALOG

Suppress selection screen.

COLUMN_SELECTION_MSG_JUMP

This parameter isn't used. Type: BALDISP2-MSG_JUMP.

EXTERNAL_NUMBER_DISPLAY_LENGTH

Display length. Type: I.

I_S_DISPLAY_PROFILE

This parameter defines how the log should appear. Type: BAL_S_PROF.

I_VARIANT_REPORT

Variant of the report to be displayed. Type: SY-REPID.

I_SRT_BY_TIMSTMP

Sort logs by time stamp (X) or log number ([Space]). Type: BOOLEAN.

[↑] Export

NUMBER_OF_PROTOCOLS

Number of logs read. Type: SY-DBCNT.

[!] Exceptions

NO_AUTHORITY

No authorization for deleting logs. Authority object S_APPL_LOG is checked.

---

479 APPL_LOG_DISPLAY_INTERN

This FM analyzes and displays the logs in local memory.

Example: You've collected the logs only during the program execution, and you only want to analyze and display the logs without writing them to the database. You can pass the parameters and call this FM to do so, so you don't overload the database.

[↓] Import

OBJECT

Object name. Type: BALHDR-OBJECT.

SUBOBJECT

Subobject name. Type: BALHDR-SUBOBJECT.

TITLE_LIST_SCREEN

Log display title.

COLUMN_SELECTION

Selection of columns in the header list. Type: BALDISP.

COLUMN_SELECTION_MSG_JUMP

This parameter isn't used. Type: BALDISP2-MSG_JUMP.

EXTERNAL_NUMBER_DISPLAY_LENGTH

Display length. Type: I.

I_S_DISPLAY_PROFILE

This parameter defines how the log should appear. Type: BAL_S_PROF.

[↑] Export

NUMBER_OF_PROTOCOLS

Number of logs read. Type: SY-DBCNT.

[!] Exceptions

OBJECT_NOT_FOUND

Object not found.

SUBOBJECT_NOT_FOUND

Subobject not found.

---

480 APPL_LOG_INIT

This FM initializes the local memory; that is, this FM deletes all of the existing entries for the specified object and subobject in the local memory.

Example: You want to enter new messages to the log. You need to check and delete already existing messages for your log object. You can pass the parameter in this FM and can call this FM to delete all of the entries of the object and subobjects.

[↓] Import

OBJECT

Object name. Type: BALHDR-OBJECT.

SUBOBJECT

Subobject name. Type: BALHDR-SUBOBJECT.

LOG_HANDLE

Log handle. Type: BALHDR-LOG_HANDLE.

[!] Exceptions

OBJECT_NOT_FOUND

Object not found.

SUBOBJECT_NOT_FOUND

Subobject not found.

---

481 APPL_LOG_INIT_MESSAGES

This FM initializes the log in local memory; that is, deletes the messages from local memory for the specified object and subobject. The header data isn't initialized, so you can write a new log for the current object or subobject with the same header data.

Example: If you want to delete only selected messages from the local memory, use this FM.

[↓] Import

OBJECT

Object name. Type: BALHDR-OBJECT.

SUBOBJECT

Subobject name. Type: BALHDR-SUBOBJECT.

CONTINUE_PROTOCOL

This parameter specifies whether the current log is continued or a new log is created when the log header or message is written for the next time. The default value is [Space], which means a new log should be created.

LOG_HANDLE

Log handle. Type: BALHDR-LOG_HANDLE.

[!] Exceptions

OBJECT_NOT_FOUND

Object not found.

SUBOBJECT_NOT_FOUND

Subobject not found.

---

482 APPL_LOG_READ_DB

This FM reads the log from database and loads it into the database's memory.

Example: You want to display the application log and read it from the database. After calling it, this FM reads the data from the database, so if you want to see the application after some time—maybe two to three days—you should save it to the memory to see it. You can do the processing and call FM APPL_LOG_DISPLAY (FM 478) to show the application.

[↓] Import

OBJECT

Object name. Type: BALHDR-OBJECT.

SUBOBJECT

Subobject name. Type: BALHDR-SUBOBJECT.

EXTERNAL_NUMBER

External number. Type: BALHDR-EXTNUMBER.

DATE_FROM

Read-from date. Type: BALHDR-ALDATE.

DATE_TO

Read-by date. Type: BALHDR-ALDATE.

TIME_FROM

Read-from time. Type: BALHDR-ALTIME.

TIME_TO

Read-by time. Type: BALHDR-ALTIME.

LOG_CLASS

Problem class. Type: BALM-PROBCLASS.

PROGRAM_NAME

Program name. Type: BALHDR-ALPROG.

TRANSACTION_CODE

Transaction name. Type: BALHDR-ALTCODE.

USER_ID

User name. Type: BALHDR-ALUSER.

MODE

Operating mode, that is, batch, batch input, or dialog mode. Type: BALHDR-ALMODE.

PUT_INTO_MEMORY

If this parameter is [Space], then it refreshes the logs that are loaded from memory. Type: BOOLEAN.

[↑] Export

NUMBER_OF_LOGS

Number of logs read.

Tables

HEADER_DATA

This table contains the log header data. Type: BALHDR.

HEADER_PARAMETERS

This table contains the log parameters for the read logs. Type: BALHDRP.

MESSAGES

This table contains the messages for the read logs. Type: BALM.

MESSAGE_PARAMETERS

This table contains the messages parameters for the read logs. Type: BALMP.

CONTEXTS

Message context information. Type: BALC.

T_EXCEPTIONS

Exceptions in log. Type: BAL_S_EXCEPTION.

---

483 APPL_LOG_READ_INTERN

This FM reads all log data for the specified object and subobject from the local memory. The data is transferred to the calling program in four internal tables (log header data, log parameters, messages, message parameters). The number of read logs is also returned.

Example: You've created an application for a delivery process, and you want to check the logs for all of the dangerous goods. If the logs are already present in the local memory, then you aren't required to read the logs from the database from this FM; instead, you can read it from the local memory. You can pass the relevant parameters in this FM, which reads the log from the local memory.

[↓] Import

OBJECT

Object name. Type: BALHDR-OBJECT.

SUBOBJECT

Subobject name. Type: BALHDR-SUBOBJECT.

LOG_HANDLE

Log handle. Type: BALHDR-LOG_HANDLE.

LOG_CLASS

Log class. Type: BALHDR-PROBCLASS.

LANGUAGE

Message language. Type: SY-LANGU.

[↑] Export

NUMBER_OF_LOGS

Number of logs read.

Tables

HEADER_DATA

This table contains the log header data. Type: BALHDR.

HEADER_PARAMETERS

This table contains the log parameters for the read logs. Type: BALHDRP.

MESSAGES

This table contains the messages for the read logs. Type: BALM.

MESSAGE_PARAMETERS

This table contains the messages parameters for the read logs. Type: BALMP.

CONTEXTS

Message context information. Type: BALC.

MESSAGE_PREPARED

Formatted log messages. Type: T_PREPARED_MESSAGES.

T_EXCEPTIONS

This table contains the exceptions for the logs that have been read. Type: BAL_S_EXCEPTION.

T_PREPARED_EXC

This table contains the formatted exceptions in the log. This table will be filled if table parameter T_PREPARED_EXC and parameter LANGUAGE are passed in the FM. Type: BAL_S_PREPARED_EXC.

[!] Exceptions

OBJECT_NOT_FOUND

Object not found.

SUBOBJECT_NOT_FOUND

Subobject not found.

FUNCTION_NOT_COMPLETED

Formatting unsuccessful.

MESSAGE_NOT_FOUND

Log message not found.

PARAMETER_MISSING

Input parameter missing.

---

484 APPL_LOG_SET_OBJECT

This FM checks whether the specified subobject exists for an object. If no subobject has been specified, the module checks whether there are subobjects for the object or not. A missing object or subobject triggers the exceptions OBJECT_NOT_FOUND or SUBOBJECT_NOT_FOUND, respectively. If the check is successful, the FM resets the object, subobject, and parameter NO_LOG in the local memory.

Example: You want to collect the application log messages to the output screen of your program. You need to give the object and subobject details along with the parameter NO_LOG and call this FM. Next, when you call FM APPL_LOG_WRITE_MESSAGES (FM 489) the message is written to the local memory.

[↓] Import

OBJECT

Object name. Type: BALHDR-OBJECT.

SUBOBJECT

Subobject name. Type: BALHDR-SUBOBJECT.

LOG_HANDLE

Log handle. Type: BALHDR-LOG_HANDLE.

NO_LOG

This parameter determines if the messages should be collected to the local memory or displayed on the screen.

[!] Exceptions

OBJECT_NOT_FOUND

Object not found.

SUBOBJECT_NOT_FOUND

Subobject not found.

---

485 APPL_LOG_WRITE_DB

This FM writes the application log data from the local memory to the database.

Example: You've already opened and read a log in your local memory, and now you want to save the log to the database. Pass the object name, subobject name, and the log handle in this FM, and your log will be written to the database.

[↓] Import

OBJECT

Object name. Type: BALHDR-OBJECT.

SUBOBJECT

Subobject name. Type: BALHDR-SUBOBJECT.

LOG_HANDLE

Log handle. Type: BALHDR-LOG_HANDLE.

UPDATE_TASK

If this parameter is set to X, the FM is called in the update task. Type: BOOLEAN.

[↑] Export

OBJECT_WITH_LOGNUMBER

In this table, the log number for which the log was written to the database is transferred to the calling program, for every object, subobject, and external number. The log number is only transferred when the log is written for the first time, not when extending. Type: BALNRI.

[!] Exceptions

OBJECT_NOT_FOUND

Object not found.

SUBOBJECT_NOT_FOUND

Subobject not found.

INTERNAL_ERROR

Internal error.

486 APPL_LOG_WRITE_HEADER

This FM writes the log header data specified in the HEADER structure to the local memory. When the same object and subobject are called repeatedly, the data is overwritten.

Example: You want to write the log header details to memory and get back the log handle for your log header. This log handle is required for you to pass in other application log FMs, such as APPL_LOG_WRITE_DB (FM 485), and to write the logs to the database.

[↓] Import

HEADER

Log header data. Type: BALHDRI.

LOG_HANDLE

Handle of log to be changed. Type: BALLOGHNDL.

[↑] Export

UPDATE_OR_INSERT

If the same object is called, this parameter determines whether the data should be updated (if the value is U) or inserted (if the value is I).

E_LOG_HANDLE

Handle of created/changed log. Type: BALLOGHNDL.

[!] Exceptions

OBJECT_NOT_FOUND

Object not found.

SUBOBJECT_NOT_FOUND

Subobject not found.

ERROR

Log create/change error.

---

487 APPL_LOG_WRITE_LOG_PARAMETERS

This FM writes log parameters and corresponding values into the local memory. When multiple calls of this FM are made with the same object or subobject, the existing parameters are deleted, and the new parameters are added. If no object/subobject is specified, the last object and subobjects are used.

Example: You want to write the specified log parameters to memory; for example, you can add "Order 1000123 has been deleted" in the application log. You can pass the text and parameter 1000123 in the respective table parameter along with the import parameter and call this FM to write the log parameter to the local memory.

[↓] Import

OBJECT

Object name. Type: BALHDR-OBJECT.

SUBOBJECT

Subobject name. Type: BALHDR-SUBOBJECT.

LOG_HANDLE

Log handle. Type: BALHDR-LOG_HANDLE.

Table

PARAMETERS

Parameters and corresponding values are passed in this parameter. If the parameter values are text, text symbols should be used so that they can be translated to other languages. Type: SPAR.

[!] Exceptions

OBJECT_NOT_FOUND

Object not found.

SUBOBJECT_NOT_FOUND

Subobject not found.

NO_HEADER

No log header data entered.

NO_TEXT_AND_NO_USER_EXIT

Neither text module nor user exit specified.

---

488 APPL_LOG_WRITE_MESSAGE_PARAMS

This FM writes a message with parameters in local memory and works similar to the FM APPL_LOG_WRITE_SINGLE_MESSAGE (FM 489).

Example: You want to write the message to the application log. If your message contains parameters, you can pass those in and call this FM. Your message is written in local memory.

[↓] Import

OBJECT

Object name. Type: BALHDR-OBJECT.

SUBOBJECT

Subobject name. Type: BALHDR-SUBOBJECT.

LOG_HANDLE

Log handle. Type: BALHDR-LOG_HANDLE.

MESSAGE

Message to be written. Type: BALMI.

UPDATE_OR_INSERT

If the same object is called, this parameter determines whether the data should be updated (if the value is U) or inserted (if the value is I).

[↑] Export

E_MSG_HANDLE

Message handle returned. Type: BALMSGHNDL.

Table

PARAMETERS

Parameters and corresponding values are passed in this parameter. If the value of the parameter is text, text symbols should be used so that it can be translated to other languages. Type: SPAR.

[!] Exceptions

OBJECT_NOT_FOUND

Object not found.

SUBOBJECT_NOT_FOUND

Subobject not found.

NO_TEXT_AND_NO_USER_EXIT

Neither text module nor user exit specified.

---

489 APPL_LOG_WRITE_MESSAGES

This FM writes multiple messages without parameters in the local memory. The functionality of the FM is similar to FM APPL_LOG_WRITE_SINGLE_MESSAGE (FM 490).

Example: You can call this FM if you want to write multiple messages to local memory and your message doesn't contain parameters.

[↓] Import

OBJECT

Object name. Type: BALHDR-OBJECT.

SUBOBJECT

Subobject name. Type: BALHDR-SUBOBJECT.

LOG_HANDLE

Log handle. Type: BALHDR-LOG_HANDLE.

UPDATE_OR_INSERT

If the same object is called, this parameter determines whether the data should be updated (if the value is U) or inserted (if the value is I).

Table

MESSAGES

Internal table with messages. This table contains the message type, message class, message number, message parameters, standard text (displayed when the log is analyzed), and user exit (form routine can be specified that is called instead of the text display). Type: BALMI.

[!] Exceptions

OBJECT_NOT_FOUND

Object not found.

SUBOBJECT_NOT_FOUND

Subobject not found.

---

490 APPL_LOG_WRITE_SINGLE_MESSAGE

This FM writes a single message that doesn't have parameters in local memory.

Example: If you want to write a single message to memory, you can pass the parameter in this FM.

[↓] Import

OBJECT

This parameter contains the name of the object to which the data in the local memory should be associated. Type: BALHDR-OBJECT.

SUBOBJECT

This parameter contains the name of the subobject to which the data in the local memory should be associated. Type: BALHDR-SUBOBJECT.

LOG_HANDLE

Log handle. Type: BALHDR-LOG_HANDLE.

MESSAGE

Messages to be written. This table contains the message type, message class, message number, message parameters, standard text (displayed when the log is analyzed), user exit (form routine can be specified that is called instead of the text display). Type: BALMI.

UPDATE_OR_INSERT

This flag determines whether repetition of the same message causes the message to be updated or repeatedly entered. Value I specifies that the same message is entered repeatedly (possibly with different standard text or user exit). Value U specifies that existing entries are updated. If there are already several entries for this message, the last one is updated.

[↑] Export

E_MSG_HANDLE

Message handle returned. Type: BALMSGHNDL.

[!] Exceptions

OBJECT_NOT_FOUND

Object not found.

SUBOBJECT_NOT_FOUND

Subobject not found.

---

491 BAL_DB_DELETE

This FM deletes the log from the database. You can pass the logs to be deleted by passing any one of three parameters (depending on the information you have): I_T_LOG_HANDLE, I_T_LOGNUMBER, or I_T_LOGS_TO_DELETE.

Example: You've created the application log for your application (say a library system), and you've logged the message of the defaulters who haven't returned their books on time. The log entry needs to be deleted when the student returns the book. This FM deletes the application log entries.

[↓] Import

I_T_LOGS_TO_DELETE

Table with log handles. This method is useful if you've kept the LOG_HANDLE in your application. Type: BALHDR_T.

I_T_LOG_HANDLE

Table with log numbers. You can use this table if you have the log number LOGNUMBER as reference to the log in the application table and not the LOG_HANDLE. Type: BAL_T_LOGH.

I_T_LOGNUMBER

Table with log headers. This table is returned by the FM BAL_DB_SEARCH, which you use when you have no reference to the log in your application table, and the link is established via the field EXTNUMBER in the log header. In this case, specify the application log object or subobject and the external number in the BAL_DB_SEARCH filter. Type: BAL_T_LOGN.

I_CLIENT

Client field for deleting entries in alternative with I_T_LOGNUMBER. If this parameter isn't specified, entries from the current client are deleted. Type: SY-MANDT.

I_IN_UPDATE_TASK

If this parameter is set to X, this FM is called in the update task. Type: BOOLEAN.

I_WITH_COMMIT_WORK

If this parameter is set to X, the FM commits the deleted entries to the database based on the parameter I_PACKAGE_SIZE. Type: BOOLEAN.

I_PACKAGE_SIZE

Package size for database to commit. Type: I.

[!] Exceptions

NO_LOGS_SPECIFIED

No logs specified.

492 BAL_DB_LOAD

This FM loads log entries from the database. Depending on your information, you can load the logs based on one of three parameters: I_T_LOG_HANDLE, I_T_LOGNUMBER, or I_T_LOGS_TO_DELETE.

Example: You want to display the created logs for the book *Function Modules in ABAP* from your library system log application. You can first search the logs in the database with this FM, and then pass the relevant parameters into and call this FM to load the logs. Refer to program SBAL_DEMO_05 in SAP ERP.

[↓] Import

I_T_LOGS_TO_DELETE

Table with log handles. This method is useful if you've kept the LOG_HANDLE in your application. This option searches the logs in all of the clients. Type: BALHDR_T.

I_T_LOG_HANDLE

Table with log numbers. You can use this table if you have the log number LOGNUMBER as reference to the log in the application table and not the LOG_HANDLE. If the I_CLIENT parameter is specified, the log is searched in that client; otherwise, the log is searched in the current client. Type: BAL_T_LOGH.

I_T_LOGNUMBER

Table with log headers. This table is returned by the FM BAL_DB_SEARCH, which you use when you have no reference to the log in your application table, and the link is established via the field EXTNUMBER in the log header. In this case, specify the application log object or subobject and the external number in the BAL_DB_SEARCH filter. Logs are searched on the client received from the table field MANDANT of export parameter E_T_LOG_HEADER from FM BAL_DB_SEARCH. Type: BAL_T_LOGN.

I_CLIENT

Client field for deleting entries in alternative with I_T_LOGNUMBER. If this parameter isn't specified, entries from the current client are deleted. Type: SY-MANDT.

I_DO_NOT_LOAD_MESSAGES

If this parameter is set to X, only log header details are loaded in memory. Type: BOOLEAN.

I_EXCEPTION_IF_ALREADY_LOADED

If this parameter is set to X, no logs are loaded to the memory, and exception LOG_ALREADY_LAODED is raised. If this parameter is [Space], the FM ignores the log if it's already in memory, and all other logs are loaded. Type: BOOLEAN.

[↑] Export

E_T_LOG_HANDLE

Table of handles of logs read. Type: BAL_T_LOGH.

E_T_MSG_HANDLE

Table of handles of messages read. Type: BAL_T_MSGH.

[!] Exceptions

NO_LOGS_SPECIFIED

No logs specified.

LOG_NOT_FOUND

Log not found.

LOG_ALREADY_LOADED

Log is already loaded.

493 BAL_DB_SAVE

This FM saves the log data from the memory to the database. All of the data or a subset of the data specified by the set of log handles can be saved. Log handle tables can be received from the FM BAL_GLB_SEARCH_LOG (FM 507), which searches for logs in memory.

Example: You've created log files with the names of the students who have issued the book *Function Modules in ABAP*, and you want to save the logs in the respective database table. You can pass the parameters and call this FM to save log details to the database.

[↓] Import

I_CLIENT

Client in which the new log is to be saved. Type: SY-MANDT.

I_IN_UPDATE_TASK

If this parameter is set to X, logs are saved in UPDATE TASK. Type: BOOLEAN.

I_SAVE_ALL

If this parameter is set to X, all logs in memory are saved. Type: BOOLEAN.

I_T_LOG_HANDLE

If parameter I_SAVE_ALL is [Space], all of the log handles specified in this parameter are saved to memory. Type: BAL_T_LOGH.

[↑] Export

E_NEW_LOGNUMBERS

When logs are saved, a new log number is issued. At runtime, this field has a temporary value such as $001. Type: BAL_T_LGNM.

[!] Exceptions

LOG_NOT_FOUND

Log not found.

SAVE_NOT_ALLOWED

Cannot save.

NUMBERING_ERROR

Number assignment error.

494 BAL_DB_SEARCH

This FM finds logs in the database, depending on your input parameters. You can pass the log header filter criteria in parameter I_S_LOG_FILTER, and the sorted table of log headers in structure E_T_LOG_HEADER is returned.

Example: You've maintained the log for the library system in the database table. You want to search who has issued the *Function Modules in ABAP* book. Pass the parameter in this FM to search the logs on the database. Next you should load the logs from the database via FM BAL_DB_LOAD (FM 492). Then display the logs to the database via FM BAL_DSP_LOG_DISPLAY (FM 495).

[↓] Import

I_CLIENT

Client to be searched. Type: SY-MANDT.

I_S_LOG_FILTER

Log header data filter. This structure should be passed to avoid the full table scan for log search. LOGNUMBER, LOG_HANDLE, EXTNUMBER, OBJECT, and SUBOBJECT are passed for the identification of the log. Type: BAL_S_LFIL.

[↑] Export

E_T_LOG_HEADER

Table of log header data is returned in this parameter. Type: BALHDR_T.

[!] Exceptions

LOG_NOT_FOUND

No log found.

NO_FILTER_CRITERIA

Filter criteria missing.

495 BAL_DSP_LOG_DISPLAY

This FM displays the application logs. Logs can also be displayed and checked from Transaction SLG1, which calls this FM.

Example: You want to display the application log in the output screen via Transaction SLG1, rather than in the tree format. First you need to set the desired profile in the import parameters.

[↓] Import

I_S_DISPLAY_PROFILE

Display profile. This parameter specifies how the log should appear. The application log provides predefined display profiles, which you can get with FMs BAL_DSP_

PROFILE_*. You can also define your own display profile. If no display profile is specified, the standard display profile of Transaction SLG1 is chosen. Type: BAL_S_PROF.

I_T_LOG_HANDLE

Restrict display by log handle. FM BAL_LOG_CREATE is used to create the log handle. Type: BAL_T_LOGH.

I_T_MSG_HANDLE

Restrict display by message handle. The message handle consists of the log handle and an internally assigned sequence number. Type: BAL_T_MSGH.

I_S_LOG_FILTER

Restrict display by log filter. Type: BAL_S_LFIL.

I_S_MSG_FILTER

Restrict display by message filter. Type: BAL_S_MFIL.

I_T_LOG_CONTEXT_FILTER

Restrict display by log context filter. Type: BAL_T_CFIL.

I_T_MSG_CONTEXT_FILTER

Restrict display by message context filter. Type: BAL_T_CFIL.

I_AMODAL

If this parameter is set to X, the log is displayed in a new session. Type: BOOLEAN.

I_SRT_BY_TIMSTMP

If this parameter is set to X, logs are sorted by time stamp, log number, and message number. If this parameter is set to [Space], logs are sorted by log number and message number. Type: BOOLEAN.

[↑] Export

E_S_EXIT_COMMAND

This parameter returns the function code of the button clicked by the user. It can be BACK, EXIT, CANCEL, CONTINUE, or EXTERNAL (callback routine is called and exit flag is set). Type: BAL_S_EXCM.

[!] Exceptions

PROFILE_INCONSISTENT

Inconsistent display profile.

INTERNAL_ERROR

Internal data formatting error.

NO_DATA_AVAILABLE

No data to be displayed found.

NO_AUTHORITY

No display authorization.

496 BAL_DSP_LOG_PRINT

This FM prints the application log.

Example: You want to print the application log for the library system application you've created. You can pass the relevant parameters in and call this FM.

[↓] Import

I_S_PRINT_OPTIONS

Print options. Type: SLIS_PRINT_ALV.

I_S_LIST_APPEND

Output of a list. If this parameter is set to X, several lists generated within one report are put in a single spool request. Value Space specifies that one spool request is created for each list generated in the report. Type: BOOLEAN.

I_S_DISPLAY_PROFILE

Display profile. This parameter specifies how the log should appear. The application log provides predefined display profiles, which you can get with FM BAL_DSP_PROFILE_*. You can also define your own display profile. If no display profile is specified, the standard display profile of Transaction SLG1 is chosen. Type: BAL_S_PROF.

I_T_LOG_HANDLE

Restrict display by log handle. FM BAL_LOG_CREATE is used to create the log handle. Type: BAL_T_LOGH.

I_T_MSG_HANDLE

Restrict display by message handle. Message handle consists of the log handle and an internally assigned sequence number. Type: BAL_T_MSGH.

I_S_LOG_FILTER

Restrict display by log filter. Type: BAL_S_LFIL.

I_S_MSG_FILTER

Restrict display by message filter. Type: BAL_S_MFIL.

I_T_LOG_CONTEXT_FILTER

Restrict display by log context filter. Type: BAL_T_CFIL.

I_T_MSG_CONTEXT_FILTER

Restrict display by message context filter. Type: BAL_T_CFIL.

I_SRT_BY_TIMSTMP

If this parameter is set to X, logs are sorted by time stamp, log number, and message number. If this parameter is set to Space, logs are sorted by log number and message number. Type: BOOLEAN.

[!] Exceptions

PROFILE_INCONSISTENT

Inconsistent display profile.

INTERNAL_ERROR

Internal data formatting error.

NO_DATA_AVAILABLE	NO_AUTHORITY
No data to be printed found.	No authorization.

497 BAL_DSP_PROFILE_DETLEVEL_GET

This FM returns the display profile that the navigation tree creates from the message texts. The field DETLEVEL of the structure BAL_S_MSG of FM BAL_LOG_MSG_ADD (used to create a message) specifies the level in the tree.

Example: If you want the logs to be displayed in tree layout to see parent-child relations or log messages at header and item data, call this FM.

[↑] Export

E_S_DISPLAY_PROFILE

Display Profile returned. Type: BAL_S_PROF.

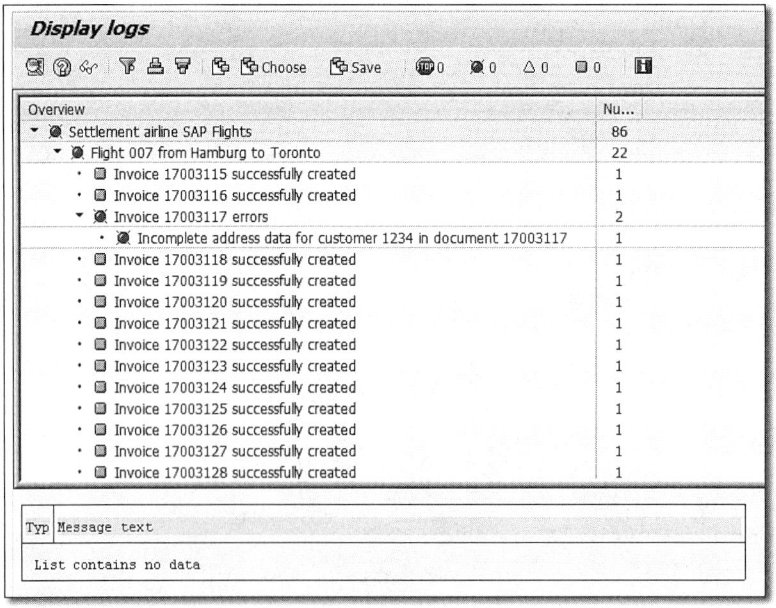

Profile with Tree Navigation

498 BAL_DSP_PROFILE_NO_TREE_GET

This FM returns the display profile, which displays the message in full screen. The navigation tree isn't created in this case

Example: If you want the layout with full screen display and no navigation tree, you can generate the profile from this FM. You can refer to Program SBAL_DEMO_04_NO_TREE for details in SAP ERP.

[↑] Export

E_S_DISPLAY_PROFILE

Display profile returned. Type: BAL_S_PROF.

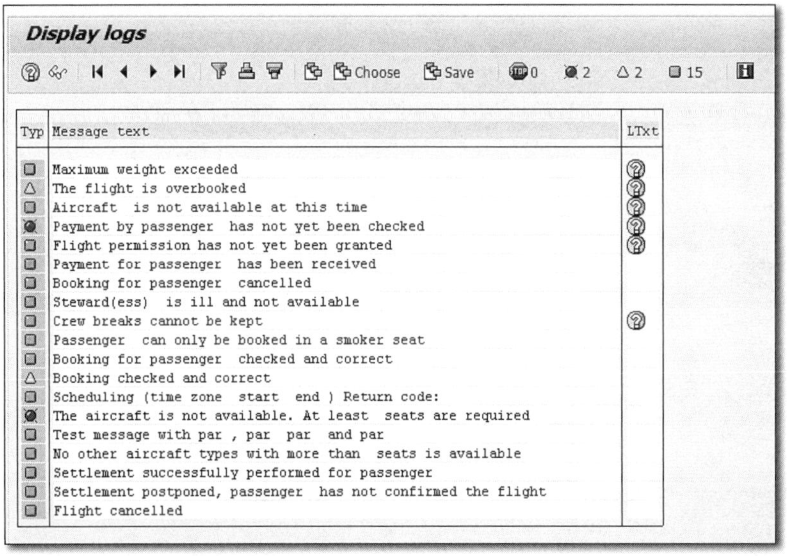

Profile with No Navigation Tree Display

499 BAL_DSP_PROFILE_POPUP_GET

This FM displays the list of application log messages in a popup.

Example: You want to display the application log messages in a popup for the *Function Modules in ABAP* book from your library system application. You can select the book and call this FM to set the profile to display the log in a popup window. Next, you need to call the FM BAL_DSP_LOG_DISPLAY (FM 495) to display the application log. You can refer to program SBAL_DEMO_04_POPUP.

[↓] Import

START_COL

Start column of the popup.
Type: BALCOORD.

START_ROW

Start row of the popup. Type: BALCOORD.

END_COL

End column of the popup.
Type: BALCOORD.

END_ROW

End row of the popup. Type: BALCOORD.

[↑] Export

E_S_DISPLAY_PROFILE

Display profile. Type: BAL_S_PROF.

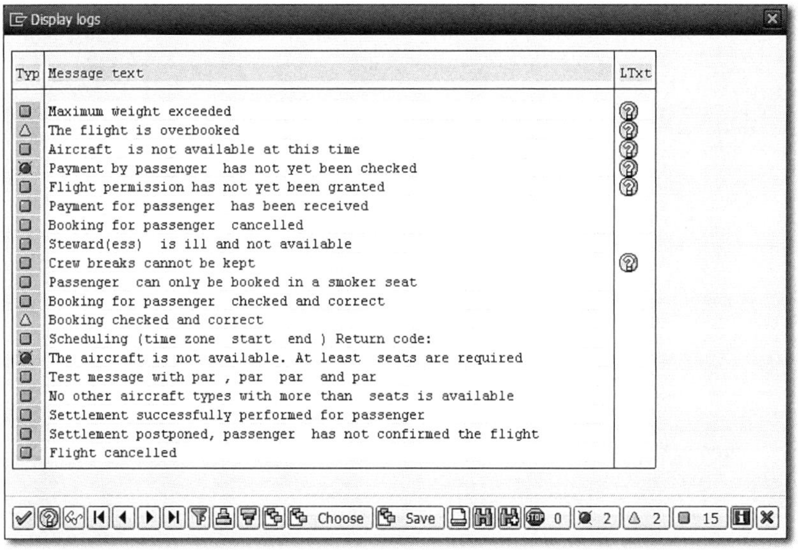

Profile in Popup

500 BAL_DSP_PROFILE_SINGLE_LOG_GET

This FM returns the display profile when only one log entry is to be displayed. In this format, the navigation tree level 1 contains the log header, and level 2 contains the problem class. The message text is displayed in the message list.

Example: You've selected the *Function Module in ABAP* book, which contains only one log entry from your library system application. You want to display only this single entry. You can pass the relevant parameters in this FM and call this FM to set the profile for single log entry. Next, you need to call FM BAL_DSP_LOG_DISPLAY (FM 495) to display the same. Refer to program SBAL_DEMO_04_SINGLE_LOG.

[↑] Export

E_S_DISPLAY_PROFILE

Display profile. Type: BAL_S_PROF.

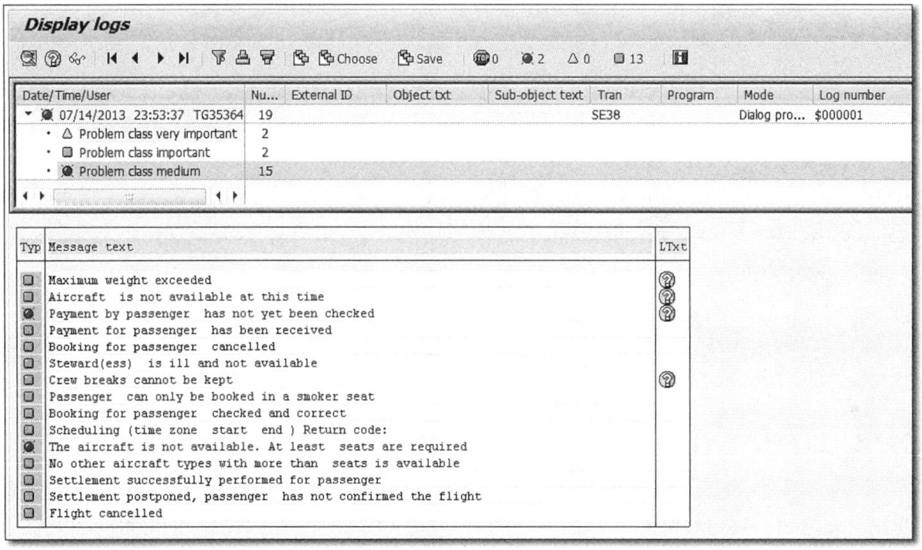

Profile with Single Log Display

501 BAL_DSP_PROFILE_STANDARD_GET

This FM returns the profile that is returned in the standard application log Transaction SLG1, which displays several logs at once. In this format, navigation tree level 1 contains the log header and its data, and level 2 is classified by problem class. The message text is displayed in the message list. The message isn't displayed until the user selects a log.

Example: You want to display the log of your application in standard format as in Transaction SLG1. You can pass the relevant parameter in this FM to set the profile for setting the standard format log display. Next, you need to call FM BAL_DSP_LOG_DISPLAY (FM 495) to display the log. You can refer to program SBAL_DEMO_04_STANDARD.

[↑] Export

E_S_DISPLAY_PROFILE

Display profile. Type: BAL_S_PROF.

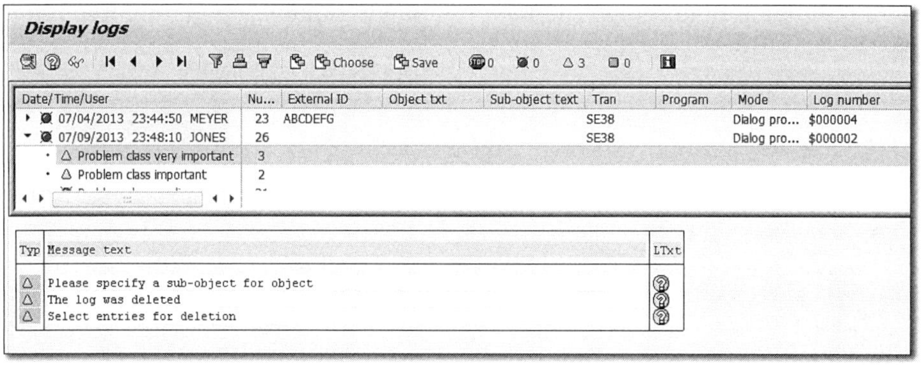

Standard Profile Layout

---

502 BAL_GLB_AUTHORIZATION_GET

This FM is used to assign an authorization to a user to perform critical functions such as configuration (by FM BAL_GLB_CONFIG_SET [FM 505]) and initialization (by FM BAL_GLB_MEMORY_REFRESH [FM 506]) of application logs. This FM gives the authorization to the parent program (i.e., the first program executed that requires authorization) by allocating the unique key returned from the exporting parameter of this FM. Other critical functions can only be performed by specifying this authorization key in their authorization object parameter I_AUTHORIZATION. If I_AUTHORIZATION isn't specified or has an incorrect value, the action is refused. This FM can't be repeated, so you can't get the second key for execution of the same task. First you need to reset the key with FM BAL_GLB_AUTHORIZATION_RESET (FM 503).

The following FMs require authorizations:

- BAL_GLB_AUTHORIZATION_GET (FM 504)
- BAL_GLB_AUTHORIZATION_RESET (FM 503)
- BAL_GLB_CONFIG_SET (FM 505)
- BAL_GLB_MEMORY_REFRESH (FM 506)

Example: You want to collect the error messages and display only those messages. For this, you need to maintain the configuration by calling FM BAL_GLB_CONFIG_SET (FM 505). You should be the authorized user to execute this FM. Authorization can be set to your user name by passing the parameter in this FM and calling this FM. The authorization parameter that is returned from this FM should be used as an import parameter in FM BAL_GLB_CONFIG__SET.

[↑] Export

E_AUTHORIZATION

Authorization ID returned. Type: BALAUTH.

[!] Exception

AUTHORIZATION_ALREADY_DONE

Authorization already assigned.

503 BAL_GLB_AUTHORIZATION_RESET

This FM resets the authorization key, which is received from FM BAL_GLB_AUTHRIZATION_GET (FM 502).

Example: When you are done with changing the configuration of the application log, you should call this FM to reset the authorization key for application log.

[↓] Import

I_AUTHORIZATION

Authorization. Type: BALAUTH.

[!] Exception

NOT_AUTHORIZED

Authorization not correct.

504 BAL_GLB_CONFIG_GET

This FM returns the existing configuration of the application log.

Example: You want to change the configuration of the application log. In order to change the configuration, you need to have the current configuration. You need to pass the parameters in this FM to get the current configuration.

[↑] Export

I_S_CONFIGURATION

This table identifies which messages are to be collected and which messages are to be displayed as soon as they are received. Following are the main fields of this structure:

- COLLECT-INACTIVE: This parameter deactivates the message collection if the value is X.
- COLLECT-MSG_FILTER and COLLECT-CON_FILTER: These parameters specify which messages are to be collected by the application log. These parameters have no effect if COLLECT-INACTIVE = X.
- DISPLAY-INACTIVE: If this parameter is set to X, it prevents any message from being displayed when sent to the application log. If this parameter is [Space], messages are displayed with the callback routine specified in parameter DISPLAY-CALLBACK.
- DISPLAY-MSG_FILTER and DISPLAY-CON_FILTER: These are similar to COLLECT-MSG_FILTER and COLLECT-CON_FILTER.
- DISPLAY-CALLBACK: Display callback routine used for messages stored in table T100.

Parameter DISPLAY-INACTIVE should be [Space]. Type: BAL_S_CONF.

505 BAL_GLB_CONFIG_SET

This FM is used to configure the application log at the start of the transaction.

Example: To collect and display error and warning messages and discard any other messages, pass the parameters in this FM and call it. You should have proper authorization to execute this FM. You can call FM BAL_GLB_AUTHORIZATION_GET (FM 502) to get the authorization ID. You can also pass the parameter in the FM BAL_GLB_CONFIG_GET (FM 504) to get the current configuration. You can modify that configuration and set your own. When you're done with the configuration, you can reset the authorization ID by passing the parameter in FM BAL_GLB_AUTHORIZATION_RESET (FM 503).

[↓] Import

I_AUTHORIZATION

Authorization ID. Type: BALAUTH.

I_S_CONFIGURATION

Configuration data. This table identifies which messages are to be collected and which messages are to be displayed as soon as they are received. Following are the main fields of this structure:

- COLLECT-INACTIVE: This parameter deactivates the message collection if the value is X.
- COLLECT-MSG_FILTER and COLLECT-CON_FILTER: These parameters specify which messages are to be collected by the application log. These parameters have no effect if COLLECT-INACTIVE = X.
- DISPLAY-INACTIVE: If this parameter is set to X, it prevents any message from being displayed when sent to the application log. If this parameter is [Space], messages are displayed with the callback routine specified in parameter DISPLAY-CALLBACK.
- DISPLAY-MSG_FILTER and DISPLAY-CON_FILTER: These are similar to

COLLECT-MSG_FILTER and COLLECT-CON_FILTER.

- DISPLAY-CALLBACK: Display callback routine used for messages stored in table T100.

Parameter DISPLAY-INACTIVE should be [Space]. Type: BAL_S_CONF.

[!] Exceptions

NOT_AUTHORIZED

Authorization not correct.

---

506 BAL_GLB_MEMORY_REFRESH

This FM refreshes the entire application log memory or a subset of the application log memory. This FM should be used carefully because it can delete all of the logs from the main memory of the application log.

Example: You want to refresh your application log data from memory. First, you need to identify your data and get the log handle via FM BAL_GLB_SEARCH_LOG (FM 507), which returns the log handle table. Then, pass in this FM to refresh the memory.

[↓] Import

I_AUTHORIZATION

Authorization. Refer to FM BAL_GLB_AUTHORIZATION_GET. Type: BALAUTH.

I_REFRESH_ALL

If this parameter is set to X, all of the application log memory is refreshed. Subsets of memory are refreshed if this parameter is [Space]. Logs that are passed in the parameter I_T_LOGS_TO_BE_REFRESHED are also refreshed. The default value is X. Type: BOOLEAN.

I_T_LOGS_TO_BE_REFRESHED

Table of log handles. If the parameter I_REFRESH_ALL is [Space], then log entries present in this table are refreshed. Type: BAL_T_LOGH.

[!] Exception

NOT_AUTHORIZED

Authorization not correct.

---

507 BAL_GLB_SEARCH_LOG

This FM searches the log in memory. If the application program has no handle to access a log in the application log, this FM is used to search a log in the memory.

Example: To search the log handle from the header information of the log, pass the relevant parameters and call this FM.

[↓] Import

I_S_LOG_FILTER

Log header filter criteria. Type: BAL_S_LFIL.

I_T_LOG_CONTEXT_FILTER

Log header context filter criteria. Type: BAL_T_CFIL.

I_T_LOG_HANDLE

Log handle table. Type: BAL_T_LOGH.

[↑] Export

E_T_LOG_HANDLE

Parameter returned with the log details. Type: BAL_T_LOGH.

[!] Exception

LOG_NOT_FOUND

Log not found.

508 BAL_GLB_SEARCH_MSG

This FM searches the message in memory. If the application program has no handle to access an application log message, it can get one by calling this FM. If many filter criteria are passed together, messages are searched with the logical AND of these filters.

Example: You want to display all of the error messages that have a high critical problem class. First, you need to fill the ranges for the message filter and call this FM, which returns the error messages of the problem class and the message handle. Next, you need to read the message. You should pass the handle received from this FM to FM BAL_LOG_MSG_READ (FM 515). Your message details are returned in the export parameter E_TXT_MSG.

[↓] Import

I_S_LOG_FILTER

Log header filter criteria. Type: BAL_S_LFIL.

I_T_LOG_CONTEXT_FILTER

Log header context filter criteria. Type: BAL_T_CFIL.

I_T_LOG_HANDLE

Log handle table. Type: BAL_T_LOGH.

I_S_MSG_FILTER

Message data filter criteria. Type: BAL_S_MFIL.

I_T_MSG_CONTEXT_FILTER

Message context filter criteria. Type: BAL_T_CFIL.

I_T_MSG_HANDLE

Message handle table. Type: BAL_T_MSGH.

[↑] Export

E_T_LOG_HANDLE

Log handle table returned. Type: BAL_T_LOGH.

E_T_MSG_HANDLE

Message handle table returned. Type: BAL_T_MSGH.

[!] Exception

MSG_NOT_FOUND

No notifications found.

509 BAL_LOG_CREATE

This FM opens the application log whose header data is specified in the importing parameter I_S_LOG_HEADER and returns the log handle that uniquely identifies the log.

Example: You need to add a message to the application log. First you need to create a log handler. You can pass the parameter in this FM and execute this FM to create the log handler.

[↓] Import

I_S_LOG

Log header data structure. Type: BAL_S_LOG.

[↑] Export

E_LOG_HANDLE

Log handle information is returned in this parameter. The log header is a GUID (Globally Unique Identifier), which uniquely identifies a log. The log can be accessed with this handle. Type: BALLOGHNDL.

[!] Exceptions

LOG_HEADER_INCONSISTENT

Log header is inconsistent.

510 BAL_LOG_HDR_READ

This FM is used to read the header data of the log. Log header data is returned in the export parameter E_S_LOG.

Example: You want to display the log. You need to first search the log handle by passing the parameter in FM BAL_LOG_SEARCH_LOG. You'll receive the handle that you need to pass in this FM. This FM returns the header data of the log handle. Refer to program SBAL_DEMO_03 for detailed use of this FM in SAP ERP.

[↓] Import

I_LOG_HANDLE

Log handle information is returned in this parameter. The log header is a GUID (Globally Unique Identifier), which uniquely identifies a log. The log can be accessed with this handle. This is a mandatory parameter. Type: BALLOGHNDL.

I_LANGU

Language. Type: SYLANGU.

[↑] Export

E_S_LOG

Log header data. Refer to parameter E_S_LOG of FM BAL_LOG_CREATE for details. Type: BAL_S_LOG.

E_EXISTS_ON_DB

Log already exists in the database. Type: BOOLEAN.

E_CREATED_IN_CLIENT

Client in which the log was created. Type: SY-MANDT.

E_SAVED_IN_CLIENT

Client in which the log was saved. Type: SY-MANDT.

E_IS_MODIFIED

Log changed. Type: BOOLEAN.

E_LOGNUMBER

Internal log number. If the log isn't yet saved, the log number is a temporary number beginning with $. Type: BALOGNR.

E_STATISTICS

Statistics (number of A, E, W, I, and S messages) of log data. Type: BAL_S_SCNT.

E_TXT_OBJECT

Application log object text. Type: C.

E_TXT_SUBOBJECT

Application log subobject text. Type: C.

E_TXT_ALTCODE

Transaction code text. Type: C.

E_TXT_ALMODE

Operating mode text. Type: C.

E_TXT_ALSTATE

Status text. Type: C.

E_TXT_PROBCLASS

Problem class text. Type: C.

E_TXT_DEL_BEFORE

"Delete Before Expiry" flag text. Type: C.

E_WARNING_TEXT_NOT_FOUND

One or more texts not found. Type: BOOLEAN.

---

511 BAL_LOG_MSG_ADD

This FM is used to add the message to the application log. The message to be added is passed in the importing parameter I_S_MSG. The message handle, which uniquely identifies the message, is returned in the parameter E_S_MSG_HANDLE.

Example: You've got an error message and want to add the message to the application log you've created. You can pass the log handle and message details in this FM and execute it. This FM adds the message to the application log.

[↓] Import

I_LOG_HANDLE

Log handle information is returned in this parameter. The log header is a GUID (Globally Unique Identifier), which uniquely identifies a log. The log can be accessed with this handle. This is a mandatory parameter. Type: BALLOGHNDL.

I_S_MSG

Notification data. This parameter has the following fields:

MSGTY, MSGID, MSGNO, MSGV1, MSGV2, MSGV3, MSGV4

- T100 message data: The field's message type, message class, and message number are required. The fields for the four message variables, MSGV1 to MSGV4, are optional.
- PROBCLASS, DETLEVEL, ALSORT, TIME_STMP: These are T100 message attributes such as problem class (i.e., very serious), level of detail (lies between 1 and 9), sort criterion (field ALSORT), and time stamp (TIME_STMP).
- MSG_COUNT: If a message is cumulated, the cumulated value can be put in this parameter. This parameter's value increments when FM BAl_LOG_MSG_CUMULATE is called and adds messages.

Type: BAL_S_MSG.

[↑] Export

E_S_MSG_HANDLE

Message handle. Message handle comprises the log handle (LOG_HANDLE) to which the message belongs and an internally assigned sequential number (MSGNUMBER). Type: BALMSGHNDL.

E_MSG_WAS_LOGGED

Message collected. If this parameter is returned as X, messages are collected. Type: BOOLEAN.

E_MSG_WAS_DISPLAYED

Message output. If this parameter is returned as X, messages are displayed immediately. Type: BOOLEAN.

[!] Exceptions

LOG_NOT_FOUND

Log not found.

MSG_INCONSISTENT

Message inconsistent.

LOG_IS_FULL

Message number 999999 reached. Log is full.

---

512 BAL_LOG_MSG_ADD_FREE_TEXT

This FM inserts the free text to the application log with the log handler.

Example: When you're adding a message to the application log, you want to add additional information with the message you've received. You can pass the parameter in this FM and execute to add free text.

[↓] Import

I_LOG_HANDLE

Log handle information is returned in this parameter. The log header is a GUID (Globally Unique Identifier), which uniquely identifies a log. The log can be accessed with this handle. This is a mandatory parameter. Type: BALLOGHNDL.

I_MSGTY

Message type (A, E, W, I, S). This is a mandatory parameter. Type: SYMSGTY.

I_PROBCLASS

Problem class (1, 2, 3, 4). The default value is 4. Type: BALPROBCL.

I_TEXT

Message data. This is a mandatory parameter. Type: C.

I_S_CONTEXT

Context information for free text message. Type: BAL_S_CONT.

I_S_PARAMS

Parameter set for free text message. Type: BAL_S_PARM.

[↑] Export

E_S_MSG_HANDLE

Message handle. The message handle comprises the log handle (LOG_HANDLE) to which the message belongs and an internally assigned sequential number (MSGNUMBER). Type: BALMSGHNDL.

E_MSG_WAS_LOGGED

Message collected. If this parameter is returned as X, messages are collected. Type: BOOLEAN.

E_MSG_WAS_DISPLAYED

Message output. If this parameter is returned as X, messages are displayed immediately. Type: BOOLEAN.

[!] Exceptions

LOG_NOT_FOUND

Log not found.

MSG_INCONSISTENT

Message inconsistent.

LOG_IS_FULL

Message number 999999 reached. Log is full.

513 BAL_LOG_MSG_DELETE

This FM deletes a message from the application log.

Example: If you think that a message is no longer relevant to the application log, execute this FM to delete your log message.

[↓] Import

I_S_MSG_HANDLE

Message handle. The message handle comprises the log handle (LOG_HANDLE) to which the message belongs and an internally assigned sequential number (MSGNUMBER). Type: BALMSGHNDL.

[!] Exceptions

MSG_NOT_FOUND

Message not found.

LOG_NOT_FOUND

Log not found.

514 BAL_LOG_MSG_DELETE_ALL

This FM deletes all messages of the log handle specified with I_LOG_HANDLE from main memory. This FM deletes all of the log messages but not the log header. If you want to delete the log header along with the log messages, use FM BAL_LOG_REFRESH (FM 516).

Example: You want to delete all of the messages of the application log so that you can create new messages in that log. Upon execution, this FM deletes all of the messages of the log. If you want to delete the application log also, you can pass the log handler in FM BAL_LOG_REFRESH and execute it. It deletes the log handle along with the messages within it.

[↓] Import

I_LOG_HANDLE

Log handle information is returned in this parameter. The log header is a GUID (Globally Unique Identifier), which uniquely identifies a log. A log can be accessed with this handle. This is a mandatory parameter. Type: BALLOGHNDL.

[!] Exceptions

LOG_NOT_FOUND

Log not found.

---

515 BAL_LOG_MSG_READ

This FM reads messages and the text of the application log. You need to pass the message handle to read the message of that handle.

Example: You've created the messages in the log and want to display message log. First you need to read the messages. You can pass the parameter in this FM and call this FM to read the logs. You can call the FM BAL_DSP_LOG_DISPLAY (FM 495) to display the application log.

[↓] Import

I_S_MSG_HANDLE

Message handle. The message handle comprises the log handle (LOG_HANDLE) to which the message belongs and an internally assigned sequential number (MSGNUMBER). Type: BALMSGHNDL.

I_LANGU

Language. Type: SYLANGU.

[↑] Export

E_S_MSG

Message data, For fields of this structure, refer to parameter I_S_MSG of FM BAL_LOG_MSG_ADD. Type: BAL_S_MSG.

E_EXISTS_ON_DB

Value X is returned if the message exists in the database. Type: BOOLEAN.

E_TXT_MSGTY

Message type. Type: C.

E_TXT_MSGID

Application area or message class. Type: C.

E_TXT_DETLEVEL

Level of detail text. Type: C.

E_TXT_PROBCLASS

Problem class text. Type: C.

E_TXT_MSG

Message text (formatted). Type: C.

E_WARNING_TEXT_NOT_FOUND

At least one text not found. Type: BOOLEAN.

[!] Exceptions

LOG_NOT_FOUND

Log not found.

MSG_NOT_FOUND

Message not found.

---

516 BAL_LOG_REFRESH

This FM deletes the log from memory. Log header and messages are deleted from the application log memory, but the data already saved in the database isn't affected. If you want to delete only the log message and not the log header, use FM BAL_LOG_MSG_DELETE_ALL (FM 514).

Example: You want to delete the entire log from the memory (i.e., the header information of log and messages of the application log) so that you can create new messages in that log. Use this FM for the same purpose.

[↓] Import

I_LOG_HANDLE

Log handle information is returned in this parameter. The log header is a GUID (Globally Unique Identifier), which uniquely identifies a log. A log can be accessed with this handle. This is a mandatory parameter. Type: BALLOGHNDL.

[!] Exceptions

LOG_NOT_FOUND

Log not found.

Part XIV
BAPI

Business Application Programming Interfaces (BAPIs) define the API methods for business objects in SAP. These business object types and their BAPIs are described and stored in the Business Object Repository (BOR). BAPIs are always RFC-enabled, so they can also be called from external application systems and other programs. BAPIs are also related to SAP components such as Sales and Distribution (SD), Materials Management (MM), Production Planning (PP), Quality Management (QM), Warehouse Management (WM), and so on. This part contains all of the commonly used BAPIs.

SAP has released more than 3,000 BAPIs. In this part, more than 150 commonly used BAPIs are covered. Other BAPIs aren't commonly used.

517 BAPI_ACC_ACTIVITY_ALLOC_POST

This FM posts a controlling activity allocation document. Activity allocation involves measuring, recording, and allocating business services performed. Activity allocation occurs when business transactions are confirmed and when activity quantities are posted to accounts.

Example: Service cost center Machine Repair performs 1 hour of activity Repair for your machine M1, which is under cost object CO1. Total cost involved in this process is say $10. Service cost center Machine Repair should be credited with $10, and cost center C1 should be debited with $10. The activity allocation document should be posted by this FM.

[↓] Import

DOC_HEADER

This structure contains the header information of the document to be posted. The CO_AREA field must contain the controlling area in which you want to post. For an external number range, the DOC_NO field is intended for the document number. The filling of the DOCDATE (document date) and POSTGDATE (posting date) fields is optional. With empty fields, the current date is added. The OBJ_KEY field allows you to enter the number and organization of a reference document (10 characters). The OBJ_TYPE field is for the object type of the reference document for the sending system.

You should only fill the LOGSYSTEM field if this document belongs to a non-SAP system. You can enter a document header text in the DOC_HDR_TX field. If you know a suitable screen variant, you can store it in the VARIANT field. Filling the VERSION field is only advisable if you're posting on a business process because if Activity-Based Costing is active as a parallel accounting system, you must enter a delta version. Otherwise, you can post only in version 0, which you can also set as the default. The USERNAME field is a required field. Type: BAPIDOCHDRP.

IGNORE_WARNINGS

If this parameter is set to X, warnings that occurred during document checks are ignored. Type: BAPIIW-IGNWARN.

[↑] Export

DOC_NO

Document number of the posted document is received in this field. Type: BAPIDOCHDRP-DOC_NO.

Tables

DOC_ITEMS

Line item data of the document to be posted such as account assignment objects, amount to be posted, unit of measure, and so on. Type: BAPIAAITM.

RETURN

Collected messages are stored in this structure. Type: BAPIRET2.

CRITERIA

You use this table to transfer the characteristics of a receiver profitability segment to the BAPI. The table must have the structure

BAPIACKECR. The name of the characteristic must be entered in the FIELDNAME field, and the value must be entered in the CHARACTER field. To assign the characteristics to a document item, the ITEMNO_ACC field must be filled as well. The value of this field must match the value of the field of the same name in the corresponding line of the DOC_ITEMS table. Type: BAPIACKECR.

CUSTOMER_FIELDS

You can use this table to transfer to the BAPI (COBL structure) the customer fields (customer include CI_COBL) that were added to the account assignment block. In addition to the customer fields, a reference to the document row used in the ITEMNO_ACC field must be transferred. The table must, therefore, have the BAPICOBL_CI structure. Type: BAPIEXTC.

---

518 BAPI_ACC_BILLING_POST

This FM posts the billing document in Financial Accounting (FI). Billing document is a general term for invoices, credit memos, debit memos, cancellation documents, and so on. The billing document is made up of a header and a number of items.

Example: You've generated some of your revenue by selling goods on credit. Now you should post the credit-based revenues in a billing document and send it to accounting, which you do by executing this FM with the specific parameters.

[↓] Import

DOCUMENTHEADER

This parameter contains the document header data. The following fields must be filled out:

- OBJ_TYPE: Sender ID.
- OBJ_KEY: Reference number of source document.
- OBJ_SYS: Logical system of source document.
- COMP_CODE: Company code.
- PSTING_DATE: Posting date.
- USERNAME: User name.

The field COMPO_ACC must remain blank in the case of a general posting. If an SAP application component is entered in this field, posting only takes place in that application component. Type: BAPIACHE01.

CUSTOMERCPD

One-time customer. Type: BAPIACPA00.

[↑] Export

OBJ_TYPE

Reference procedure. Type: BAPIACHE01-OBJ_TYPE.

OBJ_KEY

Reference key. Type: BAPIACHE01-OBJ_KEY.

OBJ_SYS

Reference system. Type: BAPIACHE01-OBJ_SYS.

Tables

ACCOUNTRECEIVABLE

Parameter for transferring customer line items. If the CUSTOMER number isn't given, only the General Ledger is updated. In this case, an entry must be made in the field GL_ACCOUNT (for GL accounting).

The currency fields aren't contained in this parameter. They must be transferred separately in the parameter CURRENCY AMOUNT for the various currency types and currencies. The system creates a link to the relevant line item via the line item number ITEMNO_ACC. Type: BAPIACAR01.

ACCOUNTGL

Parameter for transferring GL account line items. The G/L_ACCOUNT field must be filled in this table. You can also pass other fields such as company code, posting date, document type, account document number, fiscal year, and fiscal period. If this information isn't provided in this table, it's taken from the document header. Type: BAPIACGL01.

ACCOUNTTAX

Tax line items. Field GL_ACCOUNT and TAX_CODE must be filled in this table. Type: BAPIACTX01.

CRITERIA

CO-PA characteristics. You can use this parameter to transfer the Profitability Analysis (CO-PA) characteristics to an account assignment. The system creates a link to the relevant line item via the line item number ITEMNO_ACC, which for this reason must be unique for each line item. Type: BAPIACKECR.

VALUEFIELD

This parameter can be used to transfer the value fields for the profitability analysis (CO-PA) for an accounting assignment. The system creates a link to the relevant line item via the line item number ITEMNO_ACC, which therefore must be unique for each line item. Type: BAPIACKEVA.

CURRENCYAMOUNT

This parameter describes the currency fields for a particular line item. The parameter is a required field for the method POST. If the parameter isn't maintained for the method CHECK, only a general account assignment check is carried out. Individual line items or several documents can be checked at once in this way. Field CURRENCY and AMT_DOCCUR must be provided in this table. Field CURR_TYPE can have following values:

- 00: Transaction currency.
- 10: Company code currency.
- 20: Controlling area currency.
- 30: Group currency.
- 40: Hard currency.
- 50: Index-based currency.
- 60: Global company currency.

Type: BAPIACCR01.

RETURN

This table contains the return message from the FM call. Type: BAPIRET2.

SALESORDER

This table transfers the sales order line item. Field OBJ_KEY and OBJ_POS must be filled. The currency field isn't contained in this parameter and must be passed from parameter SALESAMOUNT. The system creates the link to the relevant line item via the line item number OBJ_POS and the consecutive identification number OBJ_ID. Type: BAPIACSO00.

SALESAMOUNT

This field defines the currency fields for the sales order or customer quotation. Currency type CURR_TYPE field can have value 00 for transaction currency or 10 for company code currency. The following fields must be passed:

- OBJ_KEY: Reference key.
- OBJ_POS: Reference line item.
- ORVALTYPE: Payment or cost.
- CURRENCY: Currency key.
- AMT_ORIGIN: Currency amount.

Type: BAPIACCRSO.

EXTENSION1

Table for customer enhancement. Container used to transfer data for calling up user exits when processing update BAPIs in accounting. The user exit must be activated with Transaction CMOD (extension ACBAPI01) for this to be possible.

In addition to the container (structure BAPIEXTC), the document header (structure ACCHD) and line items (structure ACCIT) from the accounting document are also transferred in the exit. The accounting document header and line items can be supplemented or modified in this way.

The container doesn't have a structure, which means that the user can transfer any data in character format. If line items are to be supplemented, the relevant line item (ITEMNO_ACC), which is also contained in other parameters, must be transferred, as well as the data. Type: BAPIEXTC.

519 BAPI_ACC_DOCUMENT_CHECK

This FM checks whether you're able to post a document to accounting. The check depends on one of the following possible business transactions.

- Postings affect the GL (RFBU).
- For billing in SD, accounting is supplied with the relevant billing data (SD00).
- Accounting can use the data of a logistics system that results from an invoice receipt (RMRP).
- Goods movements are triggered by operations in PP and in SD or by inventory postings. Within logistics, they lead to a change in the warehouse stock of Inventory Management. This results in a posting in accounting. This is why accounting is supplied with the relevant data from logistics (RMWA).

Example: Raw materials are purchased in accordance with targets. The invoice receipt is posted in the target system. The data from the raw material purchase should be first checked by an end user and then forwarded to the accounting system. This check can be performed by passing the parameters in this FM. Posting can be done by passing the parameter in FM BAPI_ACC_DOCUMENT_POST (FM 520).

[↓] Import

DOCUMENTHEADER

This parameter contains the document header data. The following fields must be filled out:

- OBJ_TYPE: Sender ID.
- OBJ_KEY: Reference number of source document.
- OBJ_SYS: Logical system of source document.
- COMP_CODE: Company code.
- PSTING_DATE: Posting date.
- USERNAME: User name.

The field COMPO_ACC must remain blank in the case of a general posting. If an SAP application component is entered in this field, posting only takes place in that application component. Type: BAPIACHE01.

CUSTOMERCPD

One-time customer. Type: BAPIACPA00.

CONTRACTHEADER

Additional contract accounts receivable and payable header line. By filling in this field, the sender calls up FI-CA as open item accounting, instead of accounts receivable accounting. Type: BAPIACCAHD.

Tables

ACCOUNTGL

Parameter for transferring GL account line items. The GL_ACCOUNT field must be filled in this table. You can also pass other fields such as company code, posting date, document type, account document number, fiscal year, and fiscal period. If this information isn't provided in this table, it's taken from the document header. The currency fields aren't contained in this parameter. They must be transferred separately in the parameter CURRENCY AMOUNT for the various currency types and currencies. The system creates a link to the relevant line item via the line item number ITEMNO_ACC. Type: BAPIACGL01.

ACCOUNTRECEIVABLE

Parameter for transferring customer line items. If the CUSTOMER number isn't given, only the GL is updated. In this case, an entry must be made in the field GL_ACCOUNT (for GL accounting).

The currency fields aren't contained in this parameter. They must be transferred separately in the parameter CURRENCY AMOUNT for the various currency types and currencies. The system creates a link to the relevant line item via the line item number ITEMNO_ACC. Type: BAPIACAR01.

ACCOUNTPAYABLE

Parameter for transferring vendor line items. If the vendor account number VENDOR_NO isn't given, only the GL is updated. In this case, an entry must be made in the field GL_ACCOUNT (for GL accounting). The currency fields aren't contained in this parameter. They must be transferred separately in the parameter CURRENCY AMOUNT for the various currency types and currencies. The system creates a link to the relevant line item via the line item number ITEMNO_ACC. Type: BAPIACAP09.

ACCOUNTTAX

Tax line items. Fields GL_ACCOUNT and TAX_CODE must be filled in this table. Type: BAPIACTX01.

CURRENCYAMOUNT

This parameter describes the currency fields for a particular line item. The parameter is a required field for the method POST. If the parameter isn't maintained for the method CHECK, only a general account assignment check is carried out. Individual line items or several documents can be checked at once in this way. Fields CURRENCY and AMT_DOCCUR must be provided in this table. Field CURR_TYPE can have the following values:

- 00: Transaction currency.
- 10: Company code currency.
- 20: Controlling area currency.
- 30: Group currency.
- 40: Hard currency.
- 50: Index-based currency.
- 60: Global company currency.

Type: BAPIACCR01.

CRITERIA

CO-PA characteristics. You can use this parameter to transfer the Controlling–Profitability Analysis (CO-PA) characteristics to an account assignment. The system creates a link to the relevant line item via the line item number ITEMNO_ACC, which for this reason must be unique for each line item. Type: BAPIACKECR.

VALUEFIELD

This parameter can be used to transfer the value fields for CO-PA for an accounting assignment. The system creates a link to the relevant line item via the line item number ITEMNO_ACC, which therefore must be unique for each line item. Type: BAPIACKEVA.

EXTENSION1

Table for customer enhancement. This exit can be used to edit the accounting document before it's transferred to the accounting components for the update.

For this, BTE technology (Business Transaction Events) is used. Type: BAPIEXTC.

RETURN

This table contains the return message from the FM call. Type: BAPIRET2.

PAYMENTCARD

Payment card information. In this parameter, you can transfer data for credit card payments. Type: BAPIACPC09.

CONTRACTITEM

In this parameter, fields are transferred that are concerned with an accounting document at the item level not including the contract accounts receivable and payable (FI-CA). Type: BAPIACCAIT.

EXTENSION2

This enhancement structure can be used to transfer additional fields. There are four fields available for each field (VALUEPART1 – 4) that each contain 240 characters. In the STRUCTURE field, a defined ABAP Data Dictionary structure can be transferred, on the basis of which the field contents can be read.

The accounting document can be supplemented and changed in an implemented BAdI before the active accounting component is called up. Type: BAPIPAREX.

REALESTATE

In this parameter, account assignments can be transferred for Real Estate Management.

The current line represents an addition to the accounting document line items (ACCOUNTGL, ACCOUNTRECEIVABLE, ACCOUNTPAYABLE). The line in parameter REALESTATE must be linked with the document line number using the line number (ITEMNO_ACC). Type: BAPIACRE09.

ACCOUNTWT

Parameter for transferring withholding tax information. Fields ITEMNO_ACC (item number), WT_TYPE (withholding tax type) and WT_CODE (withholding tax code) should be passed. Type: BAPIACWT09.

520 BAPI_ACC_DOCUMENT_POST

After you've checked to see whether a document can post to accounting, you use this FM to actually create a posting in accounting for those business transactions.

Example: Raw materials are purchased in accordance with targets. The invoice receipt should be posted in the target system. Posting can be done by passing the parameter in this FM.

[↓] Import

DOCUMENTHEADER

Refer to the same parameter in FM BAPI_ACC_DOCUMENT_CHECK (FM 519). Type: BAPIACHE09.

CUSTOMERCPD

Refer to the same parameter in FM BAPI_ACC_DOCUMENT_CHECK (FM 519). Type: BAPIACPA09.

CONTRACTHEADER

Refer to the same parameter in FM BAPI_ACC_DOCUMENT_CHECK (FM 519). Type: BAPIACCAHD.

[↑] Export

OBJ_TYPE

Reference procedure. When a business transaction is updated in accounting, it may be that several documents are updated at the same time (FI document, CO document, etc.). If the business transaction isn't initiated in accounting, a source document must exist outside accounting. The information from this document must be transferred when the FI/CO interface is called up. The source information is updated in all accounting documents and is made up of the following fields:

▶ AWTYP (OBJ_TYPE): Reference transaction sender ID.

▶ AWKEY (OBJ_KEY): Reference key, that is, source document number.

The reference key is made up of the following:

▶ AWREF reference document number (10 digits): Source document number.

▶ AWORG reference organizational unit (10 digits): Number range ID (if required).

▶ AWSYS (OBJ_SYS) Logical system: Logical system of the source document.

When the document is being processed, the system checks whether a reference document number (first part of the reference key) has been transferred.

Type: BAPIACHE09-OBJ_TYPE.

OBJ_KEY

Reference key. Refer to parameter OBJ_TYPE. Type: BAPIACHE09-OBJ_KEY.

OBJ_SYS

Logical system of source document. Refer to parameter OBJ_TYPE. Type: BAPIACHE09-OBJ_SYS.

Tables

ACCOUNTGL

Refer to the same parameter in FM BAPI_ACC_DOCUMENT_CHECK (FM 519). Type: BAPIACGL09.

ACCOUNTRECEIVABLE

Refer to the same parameter in FM BAPI_ACC_DOCUMENT_CHECK (FM 519). Type: BAPIACAR09.

ACCOUNTPAYABLE

Refer to the same parameter in FM BAPI_ACC_DOCUMENT_CHECK (FM 519). Type: BAPIACAP09.

ACCOUNTTAX

Refer to the same parameter in FM BAPI_ACC_DOCUMENT_CHECK (FM 519). Type: BAPIACTX09.

CURRENCYAMOUNT

Refer to the same parameter in FM BAPI_ACC_DOCUMENT_CHECK (FM 519). Type: BAPIACCR09.

CRITERIA

Refer to the same parameter in FM BAPI_ACC_DOCUMENT_CHECK (FM 519). Type: BAPIACKEC9.

VALUEFIELD

Refer to the same parameter in FM BAPI_ACC_DOCUMENT_CHECK (FM 519). Type: BAPIACKEV9.

EXTENSION1

Refer to the same parameter in FM BAPI_ACC_DOCUMENT_CHECK (FM 519). Type: BAPIACEXTC.

RETURN

Refer to the same parameter in FM BAPI_ACC_DOCUMENT_CHECK (FM 519). Type: BAPIRET2.

PAYMENTCARD

Refer to the same parameter in FM BAPI_ACC_DOCUMENT_CHECK (FM 519). Type: BAPIACPC09.

CONTRACTITEM

Refer to the same parameter in FM BAPI_ACC_DOCUMENT_CHECK (FM 519). Type: BAPIACCAIT.

EXTENSION2

Refer to the same parameter in FM BAPI_ACC_DOCUMENT_CHECK (FM 519). Type: BAPIPAREX.

REALESTATE

Refer to the same parameter in FM BAPI_ACC_DOCUMENT_CHECK (FM 519). Type: BAPIACRE09.

ACCOUNTWT

Refer to the same parameter in FM BAPI_ACC_DOCUMENT_CHECK (FM 519). Type: BAPIACWT09.

---

521 BAPI_ACC_DOCUMENT_REV_POST

You use this FM to reverse an external document in accounting (i.e., the document is negatively posted).

Example: You've posted an accounting document. Now you want to reverse the posting. You can pass the parameters in this FM and can call this FM for reversal.

[↓] Import

REVERSAL

Parameter for transferring a reversal document. The reversal document for the source document isn't transferred; the reversal documents are generated in the individual components from the documents to be reversed. This is only possible if the corresponding documents are available, meaning that line item update is active in all components, and no documents have been archived. If this isn't the case, an appropriate error message is issued in the return parameter. If this happens, the document can only be reset by means of a corresponding inverse posting by the sender. Type: BAPIACREV.

BUS_ACT

To reverse the document, the same business transaction used for posting should be used. Type: BAPIACHE09-BUS_ACT.

[↑] Export

OBJ_TYPE

Reference procedure. When a business transaction is updated in accounting, it may be that several documents are updated at the same time (FI document, CO document, etc.). If the business transaction isn't initiated in accounting, a source document must exist outside accounting. The information from this document must be transferred when the FI/CO interface is called up. The source information is updated in all accounting documents, which have the following fields:

- AWTYP (OBJ_TYPE): Reference transaction sender ID.
- AWKEY (OBJ_KEY): Reference key, that is, source document number.
- AWSYS (OBJ_SYS): Logical system of source document.

When the document is being processed, the system checks whether a reference document number (first part of the reference key) has been transferred. Type: BAPIACHE09-OBJ_TYPE.

OBJ_KEY

Reference key. Refer to parameter OBJ_TYPE. Type: BAPIACHE09-OBJ_KEY.

OBJ_SYS

Logical system of source document. Refer to parameter OBJ_TYPE. Type: BAPIACHE09-OBJ_SYS.

Table

RETURN

This table contains the return messages from the method call. The first message contains the general status. Message tables have messages in sequences such as the following:

- S RW 605: Document posted successfully.
- S RW 614: Document checked with errors.
- E RW 609: Document has errors.

Type: BAPIRET2.

522 BAPI_ACC_GL_POSTING_CHECK

You can use this FM to validate business transactions in accounting. These transactions mainly affect the GL and don't update in subledgers that are relevant to the balance sheet. Other FMs you may use in this process to post/reverse post are listed here:

- BAPI_ACC_GL_POSTING_POST (FM 523)
- API_ACC_GL_POSTING_REV_POST (FM 524)

Example: Foreign currency revaluation for receivable or payable goods is caused by considerable exchange rate fluctuations. These fluctuations aren't updated in accounts payable or accounts receivable ledgers. You can pass the parameter in this FM and can call this FM to validate the posting. See FM BAPI_ACC_GL_POSTING_POST (FM 523) for the next step.

[↓] Import

DOCUMENTHEADER

Document header. Refer to the same parameter in FM BAPI_ACC_DOCUMENT_CHECK (FM 519). Type: BAPIACHE08.

Tables

ACCOUNTGL

GL account line items. Refer to the same parameter in FM BAPI_ACC_DOCUMENT_CHECK (FM 519). Type: BAPIACGL08.

CURRENCYAMOUNT

Line item currency fields. Refer to the same parameter in FM BAPI_ACC_DOCUMENT_CHECK (FM 519). Type: BAPIACCR08.

RETURN

This table contains the return message from the FM call. Type: BAPIRET2.

EXTENSION1

Table for customer enhancement. Refer to the same parameter in FM BAPI_ACC_DOCUMENT_CHECK (FM 519). Type: BAPIEXTC.

---

523 BAPI_ACC_GL_POSTING_POST

This FM is used to update the business transaction in accounting. This FM finds the information that's been affected in the GL but hasn't updated the subledgers and is relevant to the balance sheet. This object can't be used to update business transactions that are normally posted to one of the subledgers that affects the balance sheet; for this, you need to consider the business object account billing and its FMs.

Example: After you've validated the posting from the previous FM, you need to call this FM to post the document and update the ledgers.

[↓] Import

DOCUMENTHEADER

Document header. Refer to the same parameter in FM BAPI_ACC_DOCUMENT_CHECK (FM 519). Type: BAPIACHE08.

[↑] Export

OBJ_TYPE

Reference procedure. When a business transaction is updated in accounting, it may be that several documents are updated at the same time (FI document, CO document, etc.). If the business transaction isn't initiated in accounting, a source document must exist outside accounting. The information from this document must be transferred when the FI/CO interface is called up. The source information is updated in all accounting documents and is made up of the following fields:

- AWTYP (OBJ_TYPE) Reference transaction sender ID.
- AWKEY (OBJ_KEY) Reference key; that is, source document number.

The reference key is made up of the following:

- AWREF Reference document number (10 digits): Source document number.
- AWORG Reference organizational unit (10 digits): Number range ID (if required).
- AWSYS (OBJ_SYS): Logical system of source document.

When the document is being processed, the system checks whether a reference document number (first part of the reference key) has been transferred. Type: BAPIACHE02-OBJ_TYPE.

OBJ_KEY

Reference key. Refer to parameter OBJ_TYPE for more details. Type: BAPIACHE02-OBJ_KEY.

OBJ_SYS

Logical system of source document. Refer to parameter OBJ_TYPE for more details. Type: BAPIACHE02-OBJ_SYS.

Tables

ACCOUNTGL

GL account line items. Refer to the same parameter in FM BAPI_ACC_DOCUMENT_CHECK (FM 519). Type: BAPIACGL08.

CURRENCYAMOUNT

Line item currency fields. Refer to the same parameter in FM BAPI_ACC_DOCUMENT_CHECK (FM 519). Type: BAPIACCR08.

RETURN

Return Parameter. Refer to the same parameter in FM BAPI_ACC_DOCUMENT_CHECK (FM 519). Type: BAPIRET2.

EXTENSION1

Table for customer enhancement. Refer to the same parameter in FM BAPI_ACC_DOCUMENT_CHECK (FM 519). Type: BAPIEXTC.

524 BAPI_ACC_GL_POSTING_REV_POST

This FM reverses the GL accounting posting.

Example: You want to reverse a document because it's incorrectly posted. You need to pass the parameters in this FM and call this FM.

[↓] Import

REVERSAL

This parameter is used for transferring reversal documents. The reversal document is generated in the individual components from the documents to be reversed. This is only possible if the corresponding documents are available; that is, line item update is active in all components, and no documents have been archived. If this isn't the case, an appropriate error message is issued in the return parameter. If this happens, the document can only be reset by means of a corresponding inverse posting by the sender. Type: BAPIACREV.

[↑] Export

OBJ_TYPE

Reference transaction. Type: BAPIACREV-OBJ_TYPE.

OBJ_KEY

Reference key. Type: BAPIACREV-OBJ_KEY.

OBJ_SYS

Reference system. Type: BAPIACREV-OBJ_SYS.

Table

RETURN

This table contains the return message from the FM call. Type: BAPIRET2.

525 BAPI_ACC_INVOICE_RECEIPT_CHECK

This FM checks if the invoice receipts can be posted in accounting. Data received from the invoice receipts in a logistics system can be transferred to accounting.

Example: Raw materials are bought on credit, and the invoice receipt is posted in a logistics system. The data from the purchase of the raw materials is passed on to accounting. First you need to check if you can post the invoice receipt with this FM. If the check is valid, you can post the invoice by calling the FM BAPI_ACC_INVOICE_RECEIPT_POST (FM 526).

[↓] Import

DOCUMENTHEADER

This parameter contains the document header data. The following fields must be filled out:

- OBJ_TYPE: Sender ID.
- OBJ_KEY: Reference number of source document.
- OBJ_SYS: Logical system of source document.
- COMP_CODE: Company code.

- PSTING_DATE: Posting date.
- USERNAME: User name.

The field COMPO_ACC must remain blank in the case of a general posting. If an SAP application component is entered in this field, posting only takes place in that application component. Type: BAPIACHE03.

CUSTOMERCPD

One-time customer details for a customer posting. One-time customer signifies any customer with whom you don't do business regularly. Type: BAPIACPA00.

Tables

ACCOUNTPAYABLE

Parameter for transferring vendor line items. If the vendor account number VENDOR_NO isn't given, only the GL is updated. In this case, an entry must be made in the field GL_ACCOUNT (for GL accounting). The currency fields aren't contained in this parameter. They must be transferred separately in the parameter CURRENCY AMOUNT for the various currency types and currencies. The system creates a link to the relevant line item via the line item number ITEMNO_ACC. Type: BAPIACAP09.

ACCOUNTGL

Parameter for transferring GL account line items. The GL_ACCOUNT field must be filled in this table. You can also pass other fields such as company code, posting date, document type, account document number, fiscal year, and fiscal period. If this information isn't provided in this table, it's taken from the document header. The currency fields aren't contained in this parameter. They must be transferred separately in the parameter CURRENCY AMOUNT for the various currency types and currencies. The system creates a link to the relevant line item via the line item number ITEMNO_ACC. Type: BAPIACGL09.

ACCOUNTTAX

Tax Line Items. Field GL_ACCOUNT and TAX_CODE must be filled in this table. Type: BAPIACTX09.

CURRENCYAMOUNT

This parameter describes the currency fields for a particular line item. The parameter is a required field for the method POST. If the parameter isn't maintained for the method CHECK, only a general account assignment check is carried out. Individual line items or several documents can be checked at once in this way. Field CURRENCY and AMT_DOCCUR must be provided in this table. Field CURR_TYPE can have the following values:

- 00: Transaction currency.
- 10: Company code currency.
- 20: Controlling area currency.
- 30: Group currency.
- 40: Hard currency.
- 50: Index-based currency.
- 60: Global company currency.

Type: BAPIACCR-09.

PURCHASEORDER

This table is used for transferring purchase order information. The OBJ_KEY and OBJ_POS fields needs to be passed. The currency field must be passed in parameter PURCHASEAMOUNT. The link to the line item is created via line item number ITEMNO_ACC and the consecutive identification number OBJ_ID. This identification number OBJ_ID can be used to subdivide the line item. A payment plan with several due dates at the line item level or a purchase order item that is assigned to several accounts can be represented in this way. The division of the amounts is the responsibility of the BAPI user. Type: BAPIACPO00.

PURCHASEAMOUNT

Purchase order or purchase requisition currency fields. The original amount, that is, the purchase order amount, can be copied in the field AMT_ORIGIN. The open amount, that is, the commitment, must be transferred in the field AMT_OPEN. The amount as reduced by follow-on documents must be transferred in the field AMT_ACTUAL. Possible values:

- 00: Transaction currency.
- 10: Company code currency.

Type: BAPIACCRPO.

RETURN

This table contains the return message from the FM call. Type: BAPIRET2.

CRITERIA

CO-PA characteristics. You can use this parameter to transfer the Controlling–Profitability Analysis (CO-PA) characteristics to an account assignment. The system creates a link to the relevant line item via the line item number ITEMNO_ACC, which for this reason must be unique for each line item. Type: BAPIACKEC9.

VALUEFIELD

This parameter can be used to transfer the value fields for CO-PA for an accounting assignment. The system creates a link to the relevant line item via the line item number ITEMNO_ACC, which therefore must be unique for each line item. Type: BAPIACKEV9.

EXTENSION1

This table is used to transfer data for calling the user exit when processing the update BAPI in accounting. The user exit must be activated with Transaction CMOD. The document header (structure ACCHD) and line items (structure ACCIT) from the accounting document are also transferred in the exit. The accounting document header and line items can be supplemented or modified in this way. The container doesn't have a structure, which means that the user can transfer any data in character format. If line items are to be supplemented, the relevant line item (ITEMNO_ACC), which is also contained in other parameters, must be transferred, as well as the data. Type: BAPIACEXTC.

526 BAPI_ACC_INVOICE_RECEIPT_POST

Data received from the invoice receipts in a logistics system can be transferred to accounting. This FM posts the invoice receipts in the accounting system.

Example: In continuation of the example in the preceding FM, after you check the invoice receipt, you'll post it to accounting. You need to pass the relevant parameters in this FM and call this FM.

[↓] Import

DOCUMENTHEADER

Document header. Refer to the same parameter in FM BAPI_ACC_DOCUMENT_CHECK (FM 519). Type: BAPIACHE03.

CUSTOMERCPD

One-time customer details for customer posting. One-time customer signifies any customer with whom you don't do business regularly. Type: BAPIACPA00.

[↑] Export

OBJ_TYPE

Reference transaction. Refer to the same parameter of FM BAPI_ACC_DOCUMENT_REV_POST (FM 521). Type: BAPIACHE03-OBJ_TYPE.

OBJ_KEY

Reference key. Refer to OBJ_KEY of FM BAPI_ACC_DOCUMENT_REV_POST (FM 521). Type: BAPIACHE03-OBJ_KEY.

OBJ_SYS

Reference system. Refer to OBJ_KEY of FM BAPI_ACC_DOCUMENT_REV_POST (FM 521). Type: BAPIACHE03-OBJ_SYS.

Tables

ACCOUNTPAYABLE

Refer to the same parameter in FM BAPI_ACC_DOCUMENT_CHECK (FM 519). Type: BAPIACAP09.

ACCOUNTGL

Refer to the same parameter in FM BAPI_ACC_DOCUMENT_CHECK (FM 519). Type: BAPIACGL09.

ACCOUNTTAX

Refer to the same parameter in FM BAPI_ACC_DOCUMENT_CHECK (FM 519). Type: BAPIACTX09.

CURRENCYAMOUNT

Refer to the same parameter in FM BAPI_ACC_DOCUMENT_CHECK (FM 519). Type: BAPIACCR09.

PURCHASEORDER

This table is used for transferring purchase order information. The OBJ_KEY and OBJ_POS fields need to be passed. The currency field must be passed in parameter PURCHASEAMOUNT. The link to the line item is created via line item number ITEMNO_ACC and the consecutive identification number OBJ_ID. This identification number OBJ_ID can be used to subdivide the line item. A payment plan with several due dates at the line item level or a purchase order item that is assigned to several accounts can be represented in this way. The division of the amounts is the responsibility of the BAPI user. Type: BAPIACPO00.

PURCHASEAMOUNT

Purchase order or purchase requisition currency fields. The original amount; that is, the purchase order amount, can be copied in the field AMT_ORIGIN. The open amount, that is, the commitment, must be transferred in the field AMT_OPEN. The amount as reduced by follow-on documents must be transferred in the field AMT_ACTUAL. Possible values:

- 00: Transaction currency.
- 10: Company code currency.

Type: BAPIACCRPO.

RETURN

Refer to the same parameter in FM BAPI_ACC_DOCUMENT_CHECK. Type: BAPIRET2.

CRITERIA

Refer to the same parameter in FM BAPI_ACC_DOCUMENT_CHECK (FM 519). Type: BAPIACKEC9.

VALUEFIELD

Refer to the same parameter in FM BAPI_ACC_DOCUMENT_CHECK (FM 519). Type: BAPIACKEV9.

EXTENSION1

This table is used to transfer data for calling the user exit when processing an update BAPI in accounting. The user exit must be activated with Transaction CMOD. The document header (structure ACCHD) and line items (structure ACCIT) from the accounting document are also transferred in the exit. The accounting document header and line items can be supplemented or modified in this way. The container doesn't have a structure, which means that the user can transfer any data in character format. If line items are to be supplemented, the relevant line item (ITEMNO_ACC), which is also contained in other parameters, must be transferred, as well as the data. Type: BAPIACEXTC.

---

527 BAPI_ACC_MANUAL_ALLOC_POST

This FM posts and reverses the manual cost allocation to accounting.

Example: You have two cost centers for your material. You've consumed material but, by mistake, $100 of that material was allocated to cost center C2 instead of C1. Now you need to manually allocate the cost to C1 and reverse the cost from C2; that is, you'll post the negative cost of $100 in C2. You can pass the relevant parameters in this FM and can call this FM to post the cost change to accounting.

[↓] Import

DOC_HEADER

Document header data. Type: BAPIDOCHDRU12P.

IGNORE_WARNINGS

This parameter is used to determine if the warnings issued during document checks are to be ignored. Type: BAPIIW-IGNWARN.

[↑] Export

DOC_NO

Document number of the posting document returned. Type: BAPIDOCHDRU12P-DOC_NO.

Tables

DOC_ITEMS

Item data of the document to be posted such as account assignment objects or costs to be posted. Type: BAPIMAITM.

RETURN

This table contains the messages returned from the FM. Type: BAPIRET2.

CUSTOMER_FIELDS

You can use this table to transfer the customer fields (customer include CI_COBL) that were added to the account assignment block (COBL structure) to the BAPI. In addition to the customer fields, you need to transfer the document row used in the ITEMNO_ACC field. The table must, therefore, have the BAPICOBL_CI structure. Type: BAPIEXTC.

---

528 BAPI_ALM_CONF_CREATE

This FM is used to maintain time confirmations for operations from maintenance or service orders. After executing the method, you must execute the FM BAPI_TRANSACTION_COMMIT (FM 644), so that the data is written to the database.

Example: An end user has performed a service order operation at work center WRKCN1, and he wants to maintain the confirmation that the work has been done. You've created a program for the end user to maintain confirmation by using this FM.

[↓] Import

POST_WRONG_ENTRIES

This parameter is used to control if the confirmations to be entered in the event of exceptions (incorrect data/locked objects) should be taken from the SAP system and entered in the pool of incorrect confirmations. The following types of POSTWRONGENTRIES parameters are possible:

- [Space]: Confirmations for which no exceptions occur are posted. All other confirmations are rejected and aren't transferred to the error pool.
- 1: Confirmations for which no exceptions occur are posted. All confirmations for which incorrect data is determined are transferred to the error pool. Confirmations that can't be posted due to a lock situation aren't processed. That is, such confirmations aren't transferred to the error pool.
- 2: Confirmations for which no exceptions occur are posted. All confirmations for which incorrect data or lock situations are determined are transferred to the error pool.

Type: BAPI_CORU_PARAM-INS_ERR.

TESTRUN

This parameter is used to control if the transferred confirmation data is only checked for correctness. In this mode, objects such as order/message and reservation aren't locked, and the databases aren't updated. To activate the TESTRUN mode, set the value X. Type: BAPI_CORU_PARAM-TESTRUN.

[↑] Export

RETURN

Return parameter for the error information. Type: BAPIRET2.

Tables

TIMETICKETS

The data for the time confirmation are transferred to this table. Type: BAPI_ALM_TIME-CONFIRMATION.

DETAIL_RETURN

For each confirmation, the DETAILRETURN table gives information on whether the confirmation was entered successfully, and whether the confirmation can't be entered due to a lock conflict or which error occurred. If the confirmation could be entered successfully, the key for the confirmation created is logged in the DETAILRETURN-CONF_NO and DETAILRETURN-CONF_CNT fields. If there is a lock conflict, the DETAILRETURN-FLGLOCKED indicator is set. If incorrect confirmation data is determined, the relevant error message is issued in the corresponding fields of the DETAILRETURN table. Type: BAPI_ALM_RETURN.

---

529 BAPI_ALM_NOTIF_CREATE

This FM is used to create a new maintenance or service notification. You can create certain fields from the notification header as well as items, causes, tasks, actions, and long texts for this. You can also maintain partner data. Other FMs you'll use for this process are listed here:

- BAPI_ALM_NOTIF_SAVE (FM 530)
- BAPI_TRANSACTION_COMMIT (FM 644)

Example: You want to create a new service notification for the damage of a certain part. You can pass the details to the parameter and call this FM to create the service notification.

[↓] Import

EXTERNAL_NUMBER

Notification number. Type: BAPI2080_NOTHDRE-NOTIF_NO.

NOTIF_TYPE

Notification type. Type: BAPI2080-NOTIF_TYPE.

NOTIFHEADER

Notification header information. Type: BAPI2080_NOTHDRI.

TASK_DETERMINATION

Execution indicator parameter to indicate functions. Type: BAPIFLAG.

SENDER

Logical system of sender. Type: BAPI_SENDER.

ORDERID

Order number assigned to the notification. Type: BAPI2080_NOTHDRE-ORDERID.

IV_DONT_CHK_MANDATORY_PARTNER

Don't check mandatory partner for notification. Type: BAPIFLAG-BAPIFLAG.

[↑] Export

NOTIFHEADER_EXPORT

BAPI service notification header. Type: BAPI2080_NOTHDRE.

Tables

NOTITEM

Notification item for creation. Type: BAPI2080_NOTITEMI.

NOTIFCAUS

Notification cause for creation. Type: BAPI2080_NOTCAUSI.

NOTIFACTV

Notification activities for creation. Type: BAPI2080_NOTACTVI.

NOTIFTASK

Notification task for creation. Type: BAPI2080_NOTTASKI.

NOTIFPARTNR

Partner for creation/change. Type: BAPI2080_NOTPARTNRI.

LONGTEXTS

Notification long text. Type: BAPI2080_NOTFULLTXTI.

KEY_RELATIONSHIPS

Key relationships within the notification data. Type: BAPI2080_NOTKEYE.

RETURN

Return parameters. This table contains the list of errors that occurred during processing. Type: BAPIRET2.

530 BAPI_ALM_NOTIF_SAVE

This FM saves the changes made in maintenance or service notifications to the system.

Example: You want to create a new service notification for the damage of a certain part. You've created the notification from the previous FM. Now, you can save the notification by calling this FM.

[↓] Import

NUMBER

Notification number to be processed. Type: BAPI2080_NOTHDRE-NOTIF_NO.

[↑] Export

NOTIFHEADER

Notification header data returned. Type: BAPI2080_NOTHDRE.

Tables

RETURN

This table contains the list of messages that occurred during the processing of the FM. Type: BAPIRET2.

531 BAPI_ALM_ORDER_GET_DETAIL

This FM reads the detail data for the maintenance or service order.

Example: You want to get the details of the order so that you can change it. You can pass the relevant parameters in this FM, which gets the order details.

[↓] Import

NUMBER

Order number to be read. Type: BAPI2080_NOTHDRE-NOTIF_NO.

[↑] Export

ES_HEADER

Order header data of an order. Type: BAPI_ALM_ORDER_HEADER_E.

ES_SRVDATA

Service-specific data of an order. Type: BAPI_ALM_ORDER_SRVDAT_E.

ES_REFORDER_ITEM

Header component for the refurbishment order. Type: BAPI_REFORDER_ITEM.

Tables

ET_PARTNER

Partner data table. Type: BAPI_ALM_ORDER_PARTNER.

ET_OPERATIONS

Operations table. Type: BAPI_ALM_ORDER_OPERATION_E.

ET_COMPONENTS

Order components table. Type: BAPI_ALM_ORDER_COMPONENT_E.

ET_RELATIONS

Order relationships table. Type: BAPI_ALM_ORDER_RELATION_EXPORT.

ET_SRULES

Settlement rule table. Type: BAPI_ALM_ORDER_SRULE_E.

ET_OLIST

Object list table. Type: BAPI_ALM_ORDER_OBJECTLIST.

ET_OPROL

Relation operation object list table. Type: BAPI_ALM_OLIST_RELATION.

ET_TEXTS

Text header for ALM order BAPIs. Type: BAPI_ALM_TEXT.

ET_TEXT_LINES

Text Lines for ALM BAPIs. Type: BAPI_ALM_TEXT_LINES.

ET_PRTS

Structure for production resources/tools. Type: BAPI_ALM_ORDER_PRT_E.

ET_COSTS_SUM

Structure for order costs sum. Type: BAPI_ALM_ORDER_COSTS_SUM_E.

ET_COSTS_DETAILS

Structure for order costs detail. Type: BAPI_ALM_ORDER_COSTS_DETAIL_E.

RETURN

Return table with messages. Type: BAPIRET2.

EXTENSION_IN

Reference structure for BAPI parameters EXTENSIONIN. Type: BAPIPAREX.

EXTENSION_OUT

Reference Structure for BAPI parameter EXTENSIONOUT. Type: BAPIPAREX.

ET_REFORDER_SERNO_OLIST

BAPI: Structure for serial number object list. Type: BAPI_REFORDER_SERNO_OLIST.

---

532 BAPI_ALM_ORDER_MAINTAIN

This FM is used to change maintenance or service orders and their subobjects. The following objects of an order can be processed: order header, partner, user status, order operations, relationships, components, long text for order header, operations and components, settlement rules object list, object list links, and production resources/tools. Objects that are to be processed are passed in the tables parameter. Also the subobjects of an existing order can be changed without changing the header.

Example: You want to change the user status of the service order. You can pass the relevant parameters in this FM and can call this FM to change the user status.

[↓] Import

IV_MMSRV_EXTERNAL_MAINTENANCE

Data for order item is maintained externally. Type: BAPI_FLAG.

Tables

IT_METHODS

The methods table specifies which methods the BAPI executes. The following fields of the method are required:

- REFNUMBER: A reference number for a data table line.
- OBJECTTYPE: An object type, which specifies which data table is referenced.
- METHOD: A method which specifies what should be done to the object.
- OBJECTKEY: A key as a reference to the corresponding higher-level object.

Type: BAPI_ALM_ORDER_METHOD.

IT_HEADER

Header of the order. Type: BAPI_ALM_ORDER_HEADERS_I.

IT_HEADER_UP

Update structure for ALM order headers. If the structure is left empty, all of the field values are updated. If the value is passed as X in the fields, then respective fields are updated. Type: BAPI_ALM_ORDER_HEADERS_UP.

IT_HEADER_SRV

ALM order BAPIs: service-specific data, external structure. Service data of the order must only be specified for service orders. The same index as for IT_HEADER is used. Type: BAPI_ALM_ORDER_SRVDAT_E.

IT_HEADER_SRV_UP

ALM order BAPIs: service-specific data, external structure. If the structure is left empty, all of the field values are updated. If the value is passed as X in the fields, then respective fields are updated. Type: BAPI_ALM_ORDER_SRVDAT_UP.

IT_USERSTATUS

Structure for ALM order, user status change. Type: BAPI_ALM_ORDER_USRSTAT.

IT_PARTNER

ALM order BAPIs: partner data for several orders. Type: BAPI_ALM_ORDER_PARTN_MUL.

IT_PARTNER_UP

ALM order BAPIs: partner data for several orders update table. Type: BAPI_ALM_ORDER_PARTN_MUL_UP.

IT_OPERATION

Input structure for PM/CS BAPI operations. Type: BAPI_ALM_ORDER_OPERATION.

IT_OPERATION_UP

Update structure for PM/CS BAPI operations. Type: BAPI_ALM_ORDER_OPERATION_UP.

IT_RELATION

BAPI structure for ALM order relationships. Type: BAPI_ALM_ORDER_RELATION.

IT_RELATION_UP

BAPI structure for ALM order relationships. Type: BAPI_ALM_ORDER_RELATION_UP.

IT_COMPONENT

PM/CS BAPI order components. Type: BAPI_ALM_ORDER_COMPONENT.

IT_COMPONENT_UP

PM/CS BAPI order components (update structure). Type: BAPI_ALM_ORDER_COMPONENT_UP.

IT_OBJECTLIST

Object list: BAPI structure. Type: BAPI_ALM_ORDER_OBJECTLIST.

IT_OBJECTLIST_UP

Update structure for object list PM/CS. Type: BAPI_ALM_ORDER_OLIST_UPL.

IT_OLIST_RELATION

Relationship of object list entry to operations. Type: BAPI_ALM_OLIST_RELATION.

IT_TEXT

Text header for ALM order BAPIs. Type: BAPI_ALM_TEXT.

IT_TEXT_LINES

SAPscript: text lines. Type: BAPI_ALM_TEXT_LINES.

IT_SRULE

PM/CS BAPI settlement rule. Type: BAPI_ALM_ORDER_SRULE.

IT_SRULE_UP

PM/CS BAPI settlement rule (update structure). Type: BAPI_ALM_ORDER_SRULE_UP.

IT_TASKLISTS

Parameter for integrating task lists. Type: BAPI_ALM_ORDER_TASKLISTS_I.

EXTENSION_IN

Reference structure for BAPI parameters EXTENSIONIN/EXTENSIONOUT. Type: BAPIPAREX.

RETURN

Return parameter(s). Type: BAPIRET2.

ET_NUMBERS

Return structure with transferred and issued order numbers. Type: BAPI_ALM_NUMBERS.

IT_REFORDER_ITEM

BAPI: header component for refurbishment order (import). Type: BAPI_REFORDER_ITEM_I.

IT_REFORDER_ITEM_UP

BAPI: update structure for refurbishment order item. Type: BAPI_REFORDER_ITEM_UP.

IT_REFORDER_SERNO_OLIST_INS

BAPI: import structure for serial number object list. Type: BAPI_REFORDER_SERNO_OLIST_I.

IT_REFORDER_SERNO_OLIST_DEL

BAPI: import structure for serial number object list. Type: BAPI_REFORDER_SERNO_OLIST_I.

IT_PRT

ALM order BAPIs: input structure for production resources/tools. Type: BAPI_ALM_ORDER_PRT_I.

IT_PRT_UP

ALM order BAPIs: changes for production resources/tools. Type: BAPI_ALM_ORDER_PRT_UP.

533 BAPI_AR_ACC_GETKEYDATEBALANCE

This FM returns the customer's total account balance at a given key date. The total balance includes both standard and special GL transactions. The balance is displayed per currency transaction but can be further broken down per special GL transaction.

Example: You want to know the current account balance of your customer for a given date. If you provide X in the BALANCESPGLI parameter, it will return the value per special ledger in the table parameter KEYBALANCE.

[↓] Import

COMPANYCODE

Company code. Type: BAPI3007_1-COMP_CODE.

CUSTOMER

Customer. Type: BAPI3007_1-CUSTOMER.

KEYDATE

Key date. Type: BAPI3007-KEY_DATE.

BALANCESPGLI

Balance per special GL indicator. Type: BAPI3007-BAL_SGLIND.

NOTEDITEMS

Noted items requested.
Type: BAPI3007-NTDITMS_RQ.

[↑] Export

RETURN

Return code. Type: BAPIRETURN.

Tables

KEYBALANCE

Balance at key date. Type: BAPI3007_3.

---

534 BAPI_AR_ACC_GETOPENITEMS

This FM displays the open items of the customer account at a given date. You can also display the noted items.

Example: You want to display the open items of a customer so that you can clear them. You can pass the parameters in this FM and execute this to get the open items.

[↓] Import

COMPANYCODE

Company code. Type: BAPI3007_1-COMP_CODE.

CUSTOMER

Customer. Type: BAPI3007_1-CUSTOMER.

KEYDATE

Key date. Type: BAPI3007-KEY_DATE.

NOTEDITEMS

Noted items requested.
Type: BAPI3007-NTDITMS_RQ.

SECINDEX

Secondary indexes required.
Type: BAPI3007-SINDEX_RQ.

[↑] Export

RETURN

Return code. Type: BAPIRETURN.

Tables

LINEITEMS

Line items. Type: BAPI3007_2.

---

535 BAPI_AR_ACC_GETPERIODBALANCES

This FM provides the customer's current balance for the current fiscal year along with the transactions and sales for that period. The annual values for special GL transactions are also displayed in the table.

Example: To create a report showing get all of your customer balances in the output for a fiscal year, pass the customer number in this FM and execute it.

[↓] Import

COMPANYCODE

Company code. Type: BAPI3007_1-COMP_CODE.

CUSTOMER

Customer. Type: BAPI3007_1-CUSTOMER.

[↑] Export

ACTUAL_BALANCE

Actual balance of the customer.
Type: BAPI3007_9.

RETURN

Return code. Type: BAPIRETURN.

Tables

DEBITOR_BALANCES

Account balance and sales per period. Type: BAPI3007_7.

DEBITOR_SPECIAL_BALANCES

Special GL balances. Type: BAPI3007_A.

---

536 BAPI_ASSET_RETIREMENT_POST

This FM is used to post the asset retirement in Asset Accounting (AA); the resulting document updates the posting in Financial Accounting (FI). The document contains the same posting as a document that is posted manually by using the nonintegrated acquisition posting transaction in AA. FM BAPI_TRANSACTION_COMMIT (FM 644) is required to commit the work after posting.

Example: Asset retirement is a process of removing an asset or a part of an asset from an asset portfolio. You want to remove the laptop that is older than five years from your organization's asset portfolio. After you retire the asset, you want to update your financial accounts as well. When you execute this FM, it posts the asset retirement in AA and updates FI as well.

[↓] Import

ORIGINDOCREFERENCE

Reference to the original document. You must pass this parameter if the accounting transaction has already been the cause of posting in another system. Fields OBJ_TYPE, OBJ_KEY, and OBJ_SYS should be passed in this parameter. Type: BAPIFAPO_DOC_REF.

GENERALPOSTINGDATA

Generally necessary posting specifications. Fields USERNAME, PSTNG_DATE, COMP_CODE, ASSETMAINO, and ASSETSUBNO should be passed in this parameter. Type: BAPIFAPO_GEN_INFO.

RETIREMENTDATA

Specifications about asset retirement. If none of the fields are transferred, the system post the complete retirement and determines the value date based on the customizing settings for the transaction. Type: BAPIFAPO_RET.

ACCOUNTASSIGNMENTS

Possible additional account assignments for the document. Type: BAPIFAPO_ACC_ASS.

FURTHERPOSTINGDATA

Parameter for transferring additional information for creating the document. Type: BAPIFAPO_ADD_INFO.

[↑] Export

DOCUMENTREFERENCE

This parameter is only relevant if the ORIGINDOCREFERENCE parameter was not transferred when the BAPI was called. The system then uses this parameter to provide all documents that were created in FI/CO that offer unique, updated reference information for identifying the accounting transaction. Type: BAPIFAPO_DOC_REF.

RETURN

Result messages. Type: BAPIRET2.

---

537 BAPI_BILLINGDOC_CANCEL1

This FM is used to cancel a customer's billing document.

Example: You have a billing document but you want to cancel it because it's no longer valid. You can pass the relevant parameters in this FM and can call this FM to cancel the billing document.

[↓] Import

BILLINGDOCUMENT

Billing document number to be canceled. Type: BAPIVBRKSUCCESS-BILL_DOC.

TESTRUN

Whether this method should be used in the simulation run (when TESTRUN is X) or in update run (when TESTRUN is [Space]). Type: BAPIVBRKTESTRUN-TESTRUN.

NO_COMMIT

If this parameter is set to X, changes won't be committed to the database. Type: BAPI_NCOMT.

BILLINGDATE

Billing date. Type: BF_DATM1EB.

Tables

RETURN

Table of error messages entered. Type: BAPIRETURN1.

SUCCESS

Table of successfully processed documents. Type: BAPIVBRKSUCCESS.

---

538 BAPI_BILLINGDOC_CREATEMULTIPLE

This FM is used to create a billing document.

Example: You want to create a billing document based on the sales order. You can fill in table BILLINGDATAIN along with other relevant parameters to create a billing document.

[↓] Import

CREATORDATAIN

This parameter contains the name of the person who created the object and the date on which the record was created. Type: BAPICREATORDATA.

TESTRUN

Value X determines the simulation run, and value [Space] determines the update run of the transaction. Type: BAPIVBRKTESTRUN-TESTRUN.

POSTING

This parameter indicates if the transfer of payment is carried out directly. Type: POSTING_TYPE_CT.

Tables

BILLINGDATAIN

Table for item data to be processed. Type: BAPIVBRK.

CONDITIONDATAIN

Table for conditions to be processed. Type: BAPIKOMV.

CCARDDATAIN

Table for method of payment to be processed. Type: BAPICCARD_VF.

TEXTDATAIN

Communication structure texts for the billing interface. Type: BAPIKOMFKTX.

ERRORS

Information on incorrect processing of the preceding items. Type: BAPIVBRKERRORS.

RETURN

Table for processing errors. Type: BAPIRET1.

SUCCESS

Table for successfully processed items. Type: BAPIVBRKSUCCESS.

NFMETALLITMS

BAPI communication structure. Type: /NFM/BAPIDOCITM.

---

539 BAPI_BUPA_ADDRESS_CHANGE

This FM is used to change the address of the business partners. The system checks that the business partner and the address exist, and then checks the new data for consistency and customizing settings. Any errors that occur are logged and issued in the RETURN table. If any of the checks run across errors, the address changes aren't carried out. Depending on the business partner category of the business partner you've entered, either the address of a person or that of an organization is changed.

Example: You want to change the address of your business partner. You can pass the parameters in this FM and can execute this FM to do so.

[↓] Import

BUSINESSPARTNER

Business partner number. Type: BAPIBUS1006_HEAD-BPARTNER.

ADDRESSGUID

Address GUID uniquely identifies an address over all systems. Type: BUT020-GUID.

ADDRESSDATA

Address data of the business partner. Type: BAPIBUS1006_ADDRESS.

ADDRESSDATA_X

Change information on ADDRESSDATE. Type: BAPIBUS1006_ADDRESS_X.

DUPLICATE_MESSAGE_TYPE

You can use this parameter to control whether the message type with the duplicate check should be an E message or a W message, in other words, an error message or a warning. With E messages, further processing is canceled. Type: BAPIBUS1006_HEAD-CONTROLDUPLICATEMESSAGE.

ACCEPT_ERROR

You use this parameter to control whether a message from the regional structure check should be category E or W. Enter the value X in this parameter if a possible message of category W should be issued. Type: BAPI4001_1-ACCEPT_ERR.

Tables

BAPIADTEL

Telephone numbers. Type: BAPIADTEL.

BAPIADFAX

Telefax numbers. Type: BAPIADFAX.

BAPIADTTX

Teletex numbers. Type: BAPIADTTX.

BAPIADTLX

Telex numbers. Type: BAPIADTLX.

BAPIADSMTP

Email addresses. Type: BAPIADSMTP.

BAPIADRML

RML addresses. Type: BAPIADRML.

BAPIADX400

X400 addresses. Type: BAPIADX400.

BAPIADRFC

RFC addresses. Type: BAPIADRFC.

BAPIADPRT

Printer addresses. Type: BAPIADPRT.

BAPIADSSF

SSF addresses. Type: BAPIADSSF.

BAPIADURI

URI addresses. Type: BAPIADURI.

BAPIADPAG

Pager numbers. Type: BAPIADPAG.

BAPIAD_REM

Address notes. Type: BAPIAD_REM.

BAPICOMREM

Notes on communication types. Type: BAPICOMREM.

ADDRESSUSAGE

Address usages. Type: BAPIBUS1006_ADDRESSUSAGE.

BAPIADVERSORG

International address version for organization/group. Type: BAPIAD1VD.

BAPIADVERSPERS

International address version for persons. Type: BAPIAD2VD.

BAPIADUSE

BAPI structure for communication usages (BAS). Type: BAPIADUSE.

BAPIADTEL_X

Change information on BAPIADTEL. Type: BAPIADTELX.

BAPIADFAX_X

Change information on BAPIADFAX. Type: BAPIADFAXX.

BAPIADTTX_X

Change information on BAPIADTTX. Type: BAPIADTTXX.

BAPIADTLX_X

Change information on BAPIADTLX. Type: BAPIADTLXX.

BAPIADSMT_X

Change information on BAPIADSMTP. Type: BAPIADSMTX.

BAPIADRML_X

Change information on BAPIADRML. Type: BAPIADRMLX.

BAPIADX40_X

Change information on BAPIADX400. Type: BAPIADX40X.

BAPIADRFC_X

Change information on BAPIADRFC. Type: BAPIADRFCX.

BAPIADPRT_X

Change information on BAPIADPRT. Type: BAPIADPRTX.

BAPIADSSF_X

Change information on BAPIADSSF. Type: BAPIADSSFX.

BAPIADURI_X

Change information on BAPIADURI. Type: BAPIADURIX.

BAPIADPAG_X

Change information on BAPIADPAG. Type: BAPIADPAGX.

BAPIAD_RE_X

Change information on BAPIAD_REM. Type: BAPIAD_REX.

BAPICOMRE_X

Change information on BAPICOMREM. Type: BAPICOMREX.

ADDRESSUSAGE_X

Change information on ADDRESSUSAGE. Type: BAPIBUS1006_ADDRESSUSAGE_X.

BAPIADVERSORG_X

Change information for international address version for organizations/groups. Type: BAPIAD1VDX.

BAPIADVERSPERS_X

Change information for international address version for persons. Type: BAPIAD2VDX.

BAPIADUSE_X

Change reference structure for BAPIADUSE. Type: BAPIADUSEX.

RETURN

Messages. Type: BAPIRET2.

ADDRESSDUPLICATES

The addresses that the system has recognized as being the same or similar are stated in this table. The field ADDRNUMBER contains the address number of the addresses recognized as being the same or similar. If the addresses are personal addresses, the field PERSNUMBER is filled with the relevant personal number. Depending on whether the tool for the duplicate check offers the possibility, the field PERCENTAGE contains the probability of the given address record being a duplicate. Type: BAPIBUS1006_ADDRESS_DUPLICATES.

540 BAPI_BUPA_ADDRESS_GETDETAIL

This FM is used to read the address and communication data of a business partner address.

Example: You're displaying the details of your business partner in the report of all your business partners. You need to display the address of your business partner. You can pass the relevant parameters in this FM and can call this FM to display the address and other communication details of your business partner, and then copy and paste as needed.

[↓] Import

BUSINESSPARTNER

Business partner number. Type: BAPIBUS1006_HEAD-BPARTNER.

ADDRESSGUID

Address GUID. This parameter uniquely identifies an address for all systems. Type: BUT020-GUID.

VALID_DATE

If no ADDRESSGUID is entered, the standard address is returned that is valid on this validity date. Type: BAPIBUS1006_VALIDITY-VALID_DATE.

RESET_BUFFER

If this parameter is set to X, it resets the buffer. Type: BOOLE-BOOLE.

[↑] Export

ADDRESSDATA

Address data of the business partner is returned in this parameter. Type: BAPIBUS1006_ADDRESS.

Tables

BAPIADTEL

Telephone numbers. Type: BAPIADTEL.

BAPIADFAX

Telefax numbers. Type: BAPIADFAX.

BAPIADTTX

Teletex numbers. Type: BAPIADTTX.

BAPIADTLX

Telex numbers. Type: BAPIADTLX.

BAPIADSMTP

Email addresses. Type: BAPIADSMTP.

BAPIADRML

RML addresses. Type: BAPIADRML.

BAPIADX400

X400 addresses. Type: BAPIADX400.

BAPIADRFC

RFC addresses. Type: BAPIADRFC.

BAPIADPRT

Printer addresses. Type: BAPIADPRT.

BAPIADSSF

SSF addresses. Type: BAPIADSSF.

BAPIADURI

URI addresses. Type: BAPIADURI.

BAPIADPAG

Pager numbers. Type: BAPIADPAG.

BAPIAD_REM

Address notes. Type: BAPIAD_REM.

BAPICOMREM

Notes on communication types. Type: BAPICOMREM.

ADDRESSUSAGE

Address usages. Type: BAPIBUS1006_ADDRESSUSAGE.

BAPIADVERSORG

International address version for organization/group. Type: BAPIAD1VD.

BAPIADVERSPERS

International address version for persons. Type: BAPIAD2VD.

BAPIADUSE

BAPI structure for communication usages (BAS). Type: BAPIADUSE.

RETURN

Error messages are returned in this table. Type: BAPIRET2.

541 BAPI_BUPA_ADDRESS_REMOVE

This FM deletes the business partner's address along with related communication data. If no address GUID is provided, the system determines and deletes the standard address of the business partner. If several addresses exist for this business partner, the standard address isn't deleted as long as other addresses continue to exist. If you want to delete the standard address of the business partner and other addresses still exist, you have to flag one of the other addresses as being the standard address by using the FM BAPI_BUPA_ADDRESS_CHANGE (FM 539). You must change the previous address in such a way so that it loses its status as a standard address. When an address is deleted, the relevant international address versions are also deleted.

Example: One of the offices of your business partner is closed, so you want to remove that address of the business partner from your records. You can pass the relevant parameters in this FM and can call this FM for the removal of the address.

[↓] Import

BUSINESSPARTNER

Business partner number. Type: BAPIBUS1006_HEAD-BPARTNER.

ADDRESSGUID

Address GUID. This parameter uniquely identifies an address for all systems. Type: BUT020-GUID.

Tables

RETURN

Error messages are returned in this table. Type: BAPIRET2.

542 BAPI_BUPA_CENTRAL_CHANGE

This FM changes the central or general data of the business partner such as name, address, bank details, and so on. You can also change the address-independent communication data such as telephone, fax, email, printer details, and so on of the business partner.

Example: Your business partner, Partner1, has changed his telephone number, and you want to change it in your SAP system. Pass the parameter in this FM and call it to make the change.

[↓] Import

BUSINESSPARTNER

Business partner number. Type: BAPIBUS1006_HEAD-BPARTNER.

CENTRALDATA

Central data. The structure contains general data, which is independent of the partner category, for the business partner. Type: BAPIBUS1006_CENTRAL.

CENTRALDATAPERSON

This structure contains general data for the business partner of the category PERSON. Type: BAPIBUS1006_CENTRAL_PERSON.

CENTRALDATAORGANIZATION

Central data, organization. Type: BAPIBUS1006_CENTRAL_ORGAN.

CENTRALDATAGROUP

Central data, group. Type: BAPIBUS1006_CENTRAL_GROUP.

CENTRALDATA_X

Change information on CENTRALDATA. Type: BAPIBUS1006_CENTRAL_X.

CENTRALDATAPERSON_X

Change information on CENTRALDATAPERSON. Type: BAPIBUS1006_CENTRAL_PERSON_X.

CENTRALDATAORGANIZATION_X

Change information on CENTRALDATAORGANIZATION. Type: BAPIBUS1006_CENTRAL_ORGAN_X.

CENTRALDATAGROUP_X

Change information on CENTRALDATAGROUP. Type: BAPIBUS1006_CENTRAL_GROUP_X.

VALID_DATE

Initial date for business partner data. Type: BAPIBUS1006_VALIDITY-VALID_DATE.

DUPLICATE_MESSAGE_TYPE

Message type control. Type: BAPIBUS1006_HEAD-CONTROLDUPLICATEMESSAGE.

DUPLICATE_CHECK_ADDRESS

SAP BP: BAPI structure for address data. Type: BAPIBUS1006_ADDRESS.

Tables

TELEFONDATANONADDRESS

Telephone numbers (address-independent). Type: BAPIADTEL.

FAXDATANONADDRESS

Fax numbers (address-independent). Type: BAPIADFAX.

TELETEXDATANONADDRESS

Teletex numbers (address-independent). Type: BAPIADTTX.

TELEXDATANONADDRESS

Telex numbers (address-independent). Type: BAPIADTLX.

E_MAILDATANONADDRESS

Email addresses (address-independent). Type: BAPIADSMTP.

RMLADDRESSDATANONADDRESS

RML addresses (address-independent). Type: BAPIADRML.

X400ADDRESSDATANONADDRESS

X.400 addresses (address-independent). Type: BAPIADX400.

RFCADDRESSDATANONADDRESS

RFC addresses (address-independent). Type: BAPIADRFC.

PRTADDRESSDATANONADDRESS

PRT addresses (address-independent). Type: BAPIADPRT.

SSFADDRESSDATANONADDRESS

SSF addresses (address-independent). Type: BAPIADSSF.

URIADDRESSDATANONADDRESS

URI addresses (address-independent). Type: BAPIADURI.

PAGADDRESSDATANONADDRESS

Pager numbers (address-independent). Type: BAPIADPAG.

COMMUNICATIONNOTESNONADDRESS

Notes on communication types (address-independent). Type: BAPICOMREM.

COMMUNICATIONUSAGENONADDRESS

Communication usages (address-independent (BAS). Type: BAPIADUSE.

TELEFONDATANONADDRESSX

Telephone numbers (address-independent) update bar. Type: BAPIADTELX.

FAXDATANONADDRESSX

Fax numbers (address-independent) update bar. Type: BAPIADFAXX.

TELETEXDATANONADDRESSX

Teletex numbers (address-independent) update bar. Type: BAPIADTTXX.

TELEXDATANONADDRESSX

Telex numbers (address-independent) update bar. Type: BAPIADTLXX.

E_MAILDATANONADDRESSX

Email addresses (address-independent) update bar. Type: BAPIADSMTX.

RMLADDRESSDATANONADDRESSX

RML addresses (address-independent) update bar. Type: BAPIADRMLX.

X400ADDRESSDATANONADDRESSX

X.400 addresses (address-independent) update bar. Type: BAPIADX40X.

RFCADDRESSDATANONADDRESSX

RFC addresses (address-independent) update bar. Type: BAPIADRFCX.

PRTADDRESSDATANONADDRESSX

PRT addresses (address-independent) update bar. Type: BAPIADPRTX.

SSFADDRESSDATANONADDRESSX

SSF addresses (address-independent) update bar. Type: BAPIADSSFX.

URIADDRESSDATANONADDRESSX

URI addresses (address-independent) update bar. Type: BAPIADURIX.

PAGADDRESSDATANONADDRESSX

Pager numbers (address-independent) update bar. Type: BAPIADPAGX.

COMMUNICATIONNOTESNON-ADDRESSX

Notes on communications types (address-independent) update bar. Type: BAPICOMREX.

COMMUNICATIONUSAGENON-ADDRESSX

Address usages (address-independent) update bar. Type: BAPIADUSEX.

RETURN

Messages. Type: BAPIRET2.

ADDRESSDUPLICATES

SAP BP: return structure for address duplicates in the BAPI. Type: BAPIBUS1006_ADDRESS_DUPLICATES.

543 BAPI_BUPA_CENTRAL_GETDETAIL

This FM reads the central attributes of a business partner. If no business partner is found, the message is returned in the RETURN table.

Example: You want to display the details of your business partner in your program. You can do so by passing the parameter in this FM and calling this FM.

[↓] Import

BUSINESSPARTNER

Business partner number. Type: BAPIBUS1006_HEAD-BPARTNER.

VALID_DATE

If the date is blank (00000000), then all of the communication data is issued. If no date is entered, then the current date is used, and the data that is valid in each case is selected. If a date is entered, the communication data valid on this date appears in the table. Type: BAPIBUS1006_VALIDITY-VALID_DATE.

[↑] Export

CENTRALDATA

The structure contains general data on a business partner, independent of the partner category. Type: BAPIBUS1006_CENTRAL.

CENTRALDATAPERSON

The structure contains general data on the business partner category. Type: BAPIBUS1006_CENTRAL_PERSON.

CENTRALDATAORGANIZATION

The structure contains general data on the business partner category ORGANIZATION. Type: BAPIBUS1006_CENTRAL_ORGAN.

CENTRALDATAGROUP

The structure contains general data on the business partner category GROUP. Type: BAPIBUS1006_CENTRAL_GROUP.

CENTRALDATAVALIDITY

Validity data of the business partner. Type: BAPIBUS1006_CENTRAL_VALIDITY.

Tables

TELEFONDATANONADDRESS

Telephone numbers (address-independent). Type: BAPIADTEL.

FAXDATANONADDRESS

Fax numbers (address-independent). Type: BAPIADFAX.

TELETEXDATANONADDRESS

Teletex numbers (address-independent). Type: BAPIADTTX.

TELEXDATANONADDRESS

Telex numbers (address-independent). Type: BAPIADTLX.

E_MAILDATANONADDRESS

Email addresses (address-independent). Type: BAPIADSMTP.

RMLADDRESSDATANONADDRESS

RML addresses (address-independent). Type: BAPIADRML.

X400ADDRESSDATANONADDRESS

X.400 addresses (address-independent). Type: BAPIADX400.

RFCADDRESSDATANONADDRESS

RFC addresses (address-independent). Type: BAPIADRFC.

PRTADDRESSDATANONADDRESS

PRT addresses (address-independent). Type: BAPIADPRT.

SSFADDRESSDATANONADDRESS

SSF addresses (address-independent). Type: BAPIADSSF.

URIADDRESSDATANONADDRESS

URI addresses (address-independent). Type: BAPIADURI.

PAGADDRESSDATANONADDRESS

Pager numbers (address-independent). Type: BAPIADPAG.

COMMUNICATIONNOTESNON-ADDRESS

Notes on communication types (address-independent). Type: BAPICOMREM.

COMMUNICATIONUSAGENON-ADDRESS

Communication usages (address-independent, BAS). Type: BAPIADUSE.

RETURN

Messages. Type: BAPIRET2.

544 BAPI_BUPA_CREATE_FROM_DATA

This FM creates a business partner and an address that is related to communication types. The data entered in the interface is checked for consistency and for customizing settings. If this check reveals one or more errors, the errors are logged in the RETURN table, and the business partner isn't created. It's possible to maintain both address-dependent communication data and address-independent communication data for a business partner.

Example: You've started a business with a new business partner, and you want to enter his details into the SAP system. You can pass the parameters in this FM to create the business partner.

[↓] Import

BUSINESSPARTNEREXTERN

Business partner number. Type: BAPIBUS1006_HEAD-BPARTNER.

PARTNERCATEGORY

Business partner category. Type: BAPIBUS1006_HEAD-PARTN_CAT.

PARTNERGROUP

Business partner grouping. Type: BAPIBUS1006_HEAD-PARTN_GRP.

CENTRALDATA

Central data. Type: BAPIBUS1006_CENTRAL.

CENTRALDATAPERSON

Central data, person. Type: BAPIBUS1006_CENTRAL_PERSON.

CENTRALDATAORGANIZATION

Central data, organization. Type: BAPIBUS1006_CENTRAL_ORGAN.

CENTRALDATAGROUP

Central data, group. Type: BAPIBUS1006_CENTRAL_GROUP.

ADDRESSDATA

Address. Type: BAPIBUS1006_ADDRESS.

DUPLICATE_MESSAGE_TYPE

Message type control. Type: BAPIBUS1006_HEAD-CONTROLDUPLICATEMESSAGE.

ACCEPT_ERROR

Accept regional structure check errors with warnings. Type: BAPI4001_1-ACCEPT_ERR.

[↑] Export

BUSINESSPARTNEREXTERN

Business partner number. Type: BAPIBUS1006_HEAD-BPARTNER.

Tables

TELEFONDATA

Telephone numbers. Type: BAPIADTEL.

FAXDATA

Fax numbers. Type: BAPIADFAX.

TELETEXDATA

Teletex numbers. Type: BAPIADTTX.

TELEXDATA

Telex numbers. Type: BAPIADTLX.

E_MAILDATA

Internet addresses. Type: BAPIADSMTP.

RMLADDRESSDATA

Remote mail addresses. Type: BAPIADRML.

X400ADDRESSDATA

X400 addresses. Type: BAPIADX400.

RFCADDRESSDATA

RFC addresses. Type: BAPIADRFC.

PRTADDRESSDATA

Printer addresses. Type: BAPIADPRT.

SSFADDRESSDATA

SSF addresses. Type: BAPIADSSF.

URIADDRESSDATA

URI addresses. Type: BAPIADURI.

PAGADDRESSDATA

Pager numbers. Type: BAPIADPAG.

ADDRESSNOTES

Address notes. Type: BAPIAD_REM.

COMMUNICATIONNOTES

Notes on communication. Type: BAPICOMREM.

COMMUNICATIONUSAGE

Communication usages (BAS). Type: BAPIADUSE.

TELEFONDATANONADDRESS

Telephone numbers (address-independent). Type: BAPIADTEL.

FAXDATANONADDRESS

Fax numbers (address-independent). Type: BAPIADFAX.

TELETEXDATANONADDRESS

Teletex numbers (address-independent). Type: BAPIADTTX.

TELEXDATANONADDRESS

Telex numbers (address-independent). Type: BAPIADTLX.

E_MAILDATANONADDRESS

Email addresses (address-independent). Type: BAPIADSMTP.

RMLADDRESSDATANONADDRESS

RML addresses (address-independent). Type: BAPIADRML.

X400ADDRESSDATANONADDRESS

X.400 addresses (address-independent). Type: BAPIADX400.

RFCADDRESSDATANONADDRESS

RFC addresses (address-independent). Type: BAPIADRFC.

PRTADDRESSDATANONADDRESS

PRT addresses (address-independent). Type: BAPIADPRT.

SSFADDRESSDATANONADDRESS

SSF addresses (address-independent). Type: BAPIADSSF.

URIADDRESSDATANONADDRESS

URI addresses (address-independent). Type: BAPIADURI.

PAGADDRESSDATANONADDRESS

Pager numbers (address-independent). Type: BAPIADPAG.

COMMUNICATIONNOTESNONADDRESS

Notes on communication types (address-independent). Type: BAPICOMREM.

COMMUNICATIONUSAGENONADDRESS

Communication usages (address-independent (BAS). Type: BAPIADUSE.

RETURN

Messages. Type: BAPIRET2.

ADDRESSDUPLICATES

Address duplicates. Type: BAPIBUS1006_ADDRESS_DUPLICATES.

---

545 BAPI_CHARACT_CREATE

This FM creates new object characteristics. Characteristics are the properties for describing and distinguishing an object, such as its length, color, weight, and so on.

Example: You can create the characteristics programmatically by calling this FM to create different options for a material. Next, you should create the class so that you can assign and group these characteristics together in the class. See FM BAPI_CLASS_CREATE (FM 548).

[↓] Import

CHARACTDETAIL

Characteristic attributes. Type: BAPICHARACTDETAIL.

CHANGENUMBER

Change number. Type: BAPICHARACTKEY-CHANGENUM.

KEYDATE

Date. Type: BAPICHARACTKEY-KEYDATE.

Tables

CHARACTDESCR

Characteristic descriptions. Type: BAPICHARACTDESCR.

CHARACTVALUESNUM

Allowed values for NUM characteristics. Type: BAPICHARACTVALUESNUM.

CHARACTVALUESCHAR

Allowed values for CHAR characteristics. Type: BAPICHARACTVALUESCHAR.

CHARACTVALUESCURR

Allowed values for CURR characteristics. Type: BAPICHARACTVALUESCURR.

CHARACTVALUESDESCR

Value descriptions. Type: BAPICHARACTVALUESDESCR.

CHARACTREFERENCES

Entries for reference characteristics. Type: BAPICHARACTREFERENCES.

CHARACTRESTRICTIONS

Restrictions to class types. Type: BAPICHARACTRESTRICTIONS.

RETURN

Error messages. Type: BAPIRET2.

---

546 BAPI_CHARACT_GETDETAIL

This FM returns the details of characteristics, which are identified by name and by date.

Example: You want to get the detail information of a characteristic (such as description, values allowed, authorization object of the characteristics etc.) created for a class. You need to enter the parameter CHARACTNAME and other relevant parameter and call this FM.

[↓] Import

CHARACTNAME

Characteristic name. Type: BAPICHARACTKEY-CHARACTNAME.

KEYDATE

Date. Type: BAPICHARACTKEY-KEYDATE.

LANGUAGE

Language: empty = all languages. Type: BAPIFIELDSCACL-BAPILANGUA.

[↑] Export

CHARACTDETAIL

Characteristic attributes that contain the basic data of the characteristic. Type: BAPICHARACTDETAIL.

Tables

CHARACTDESCR

Characteristic descriptions. Type: BAPICHARACTDESCR.

CHARACTVALUESNUM

Allowed values for NUM characteristics. One of the parameters CHARACTVALUESNUM, CHARACTVALUESCHAR, or CHARACTVALUESCURR must be filled. Type: BAPICHARACTVALUESNUM.

CHARACTVALUESCHAR

Allowed values for CHAR characteristics. Type: BAPICHARACTVALUESCHAR.

CHARACTVALUESCURR

Allowed values for CURR characteristics. Type: BAPICHARACTVALUESCURR.

CHARACTVALUESDESCR

Value descriptions. Type: BAPICHARACTVALUESDESCR.

CHARACTREFERENCES

Entries for reference characteristics. Type: BAPICHARACTREFERENCES.

CHARACTRESTRICTIONS

Restrictions to class types. Type: BAPICHARACTRESTRICTIONS.

RETURN

Error messages table. Type: BAPIRET2.

---

547 BAPI_CLASS_CHANGE

This FM changes the characteristics class and its data. Structures and tables for new data must be maintained correctly. If these structures or tables are empty, the data in them is deleted. If no data is to be changed, the structures and tables ending with NEW must match the current data. To fill structures and tables with current data, use the relevant GETDETAIL module to copy the old data from the class first. The class-specific overwrite data for characteristics and values must be made available using the data copied by calling the GETDETAIL module. If nothing is transferred in these tables, there is no overwriting. You'll use the following FMs in this process:

- BAPI_CLASS_GET_DETAILS (FM 550)
- BAPI_CLASS_GET_CHARACTERISTICS (FM 549)
- BAPI_CLASS_CHANGE
- BAPI_TRANSACTION_COMMIT (FM 644)

Example: You want to change the details of the class. After calling the first FMs listed, fill in the change parameters and pass it to this FM. Your class is changed with the changes you made.

[↓] Import

CLASSNUM

Class. Type: BAPI_CLASS_KEY-CLASSNUM.

CLASSTYPE

Class type. Type: BAPI_CLASS_KEY-CLASSTYPE.

CHANGENUMBER

Change number. Type: BAPI1003_KEY-CHANGENUMBER.

KEYDATE

Valid-from date. Type: BAPI_KEY_DATE-KEY_DATE.

CLASSBASICDATA

Basic data. Type: BAPI1003_BASIC.

CLASSBASICDATANEW

Basic data: new. Type: BAPI1003_BASIC_NEW.

CLASSDOCUMENT

Document information. Type: BAPI1003_DOCU.

CLASSDOCUMENTNEW

Document information: new. Type: BAPI1003_DOCU_NEW.

CLASSADDITIONAL

Additional data. Type: BAPI1003_ADD.

CLASSADDITIONALNEW

Additional data: new. Type: BAPI1003_ADD_NEW.

CLASSSTANDARD

Standards data. Type: BAPI1003_STAND.

CLASSSTANDARDNEW

Standards data: new. Type: BAPI1003_STAND_NEW.

Tables

RETURN

Return parameter. Type: BAPIRET2.

CLASSDESCRIPTIONS

Class description/key words. Type: BAPI1003_CATCH.

CLASSLONGTEXTS

Long texts. Type: BAPI1003_LONGTEXT.

CLASSCHARACTERISTICS

Characteristics. Type: BAPI1003_CHARACT.

CHARACTERISTICOVERWRITE

Overwritten characteristic attributes. Type: BAPI1003_CHARACT_OVERWR.

CHARACTVALUEOVERWRIT

Overwritten characteristic values. Type: BAPI1003_CHARACT_VALUE_OVR.

CHARACTERISTICVALUETEXTOVR

Overwritten characteristic descriptions. Type: BAPI1003_CHARVALTEXT.

CLASSDESCRIPTIONSNEW

Class description/key words: New. Type: BAPI1003_CATCH_NEW.

CLASSLONGTEXTSNEW

Long texts: new. Type: BAPI1003_LONGTEXT.

CLASSCHARACTERISTICSNEW

Characteristics: new. Type: BAPI1003_CHARACT_NEW.

CHARACTERISTICOVERWRITENEW

Overwritten characteristic attributes: new. Type: BAPI1003_CHARACT_OVERWR_NEW.

CHARACTVALUEOVERWRITNEW

Overwritten characteristic values: new. Type: BAPI1003_CHARACT_VALUE_OVR_NEW.

CHARACTERISTICVALUETEXTOVRNEW

Overwritten characteristic descriptions: new. Type: BAPI1003_CHARVALTEXT_NEW.

548 BAPI_CLASS_CREATE

This FM groups similar characteristics together in a class to create a hierarchy. Class hierarchy can look like Pumps -> Rotary Pumps -> Magnetically Coupled Pumps. This FM creates the class if the class doesn't exist.

Example: You're working in a plant and want to create a class for your technical objects, such as pumps. After creating the characteristics (see FM BAPI_CHARACT_CREATE, FM 545), call this FM and pass the relevant parameters. You'll get the new class created. Use FM BAPI_TRANSACTION_COMMIT (FM 644) to commit your new class data to database tables.

[↓] Import

CLASSNUMNEW

Class. Type: BAPI_CLASS_KEY-CLASSNUM.

CLASSTYPENEW

Class type. Type: BAPI_CLASS_KEY-CLASSTYPE.

CHANGENUMBER

Change number. Type: BAPI1003_KEY-CHANGENUMBER.

CLASSBASICDATA

Basic data of class. Type: BAPI1003_BASIC.

CLASSDOCUMENT

Document information of class. Type: BAPI1003_DOCU.

CLASSADDITIONAL

Additional data of class. Type: BAPI1003_ADD.

CLASSSTANDARD

Standards data of class. Type: BAPI1003_STAND.

Tables

RETURN

Return parameter. Type: BAPIRET2.

CLASSDESCRIPTIONS

Class description/key words. Type: BAPI1003_CATCH.

CLASSLONGTEXTS

Long texts. Type: BAPI1003_LONGTEXT.

CLASSCHARACTERISTICS

Characteristics. Type: BAPI1003_CHARACT.

CHARACTOVERWRITE

Overwritten characteristic attributes. Type: BAPI1003_CHARACT_OVERWR.

CHARACTVALUEOVERWRITE

Overwritten characteristic values. Type: BAPI1003_CHARACT_VALUE_OVR.

CHARACTVALUETEXTOVR

Overwritten characteristic descriptions. Type: BAPI1003_CHARVALTEXT

549 BAPI_CLASS_GET_CHARACTERISTICS

This FM reads characteristics and gets the allowed values of characteristics from the database.

Example: You want to get the details of the pump class that you've created. After you pass the relevant parameter in this FM and call this FM, you'll get the details of the class, the details of the characteristics that are under this pump class, and allowed values for the characteristics. See FM BAPI_CLASS_CHANGE (FM 547).

[↓] Import

CLASSNUM

Class number. This parameter along with the class type identifies the class. Type: BAPI_CLASS_KEY-CLASSNUM.

CLASSTYPE

Class type. Type: BAPI_CLASS_KEY-CLASSTYPE.

LANGU_ISO

ISO language key. Type: BAPI_LANGUAGE-LANGUAGE_ISO.

LANGU_INT

Internal language key. Type: BAPI_LANGUAGE-LANGUAGE_INT.

KEY_DATE

Validity date. Type: BAPI_KEY_DATE-KEY_DATE.

WITH_VALUES

If this parameter is set to X, allowed values for the characteristics are read. Type: BAPI_WITH_VALUES-WITH_VAL.

[↑] Export

RETURN

Return parameters for messages. Type: BAPIRETURN1.

Tables

CHARACTERISTICS

This parameter returns the characteristics with its attributes. Type: BAPI_CHAR.

CHAR_VALUES

Allowed values for characteristics. Type: BAPI_CHAR_VALUES.

---

550 BAPI_CLASS_GETDETAIL

This FM returns the details of the class. The class is specified by name, class type, and date.

Example: You've created the class for pumps in FM BAPI_CLASS_CREATE (FM 548). You want to see what all of the class details (such as descriptions, valid from, characteristics under this class, etc.) for this class and display them in the output list of your report program.

[↓] Import

CLASSTYPE

Class type. Type: BAPI_CLASS_KEY-CLASSTYPE.

CLASSNUM

Class. Type: BAPI_CLASS_KEY-CLASSNUM.

LANGUISO

Language key in ISO format. Type: BAPI_LANGUAGE-LANGUAGE_ISO.

LANGUINT

Language key in SAP format. Type: BAPI_LANGUAGE-LANGUAGE_INT.

KEYDATE

Valid-from date. Type: BAPI_KEY_DATE-KEY_DATE.

[↑] Export

CLASSBASICDATA

Basic data. Type: BAPI1003_BASIC.

CLASSDOCUMENT

Document information. Type: BAPI1003_DOCU.

CLASSADDITIONAL

Additional data. Type: BAPI1003_ADD.

CLASSSTANDARD

Standards data. Type: BAPI1003_STAND.

RETURN

Return parameter. Type: BAPIRET2.

Tables

CLASSDESCRIPTIONS

Class description.
Type: BAPI1003_CATCH_R.

CLASSLONGTEXTS

Long texts. Type: BAPI1003_LONGTEXT_R.

CLASSCHARACTERISTICS

Characteristics.
Type: BAPI1003_CHARACT_R.

CLASSCHARVALUES

Characteristic values. Type: BAPI1003_CHAR_VAL_R.

---

551 BAPI_CLASS_GETSTATUS

This FM returns the status information for a class; that is, whether the class can be maintained, objects can be allocated to the class, and the class can be used for selecting objects.

Example: You've created a report where you want to get and display all of the class details with status 1. You can pass the parameters in this FM and call this FM to get the class status details. After you get the class status information, you can pass the parameters in the FM BAPI_CLASS_GETDETAIL (FM 550) to get the class details.

[↓] Import

CLASSTYPE

Class type. Type: BAPI_CLASS_KEY-CLASSTYPE.

CLASSNUM

Class. Type: BAPI_CLASS_KEY-CLASSNUM.

[↑] Export

STATUSFLAGS

Status information. Type: BAPI1003_STATUSINFO.

Tables

RETURN

Return parameter. Type: BAPIRET2.

552 BAPI_COMPANYCODE_GET_PERIOD

This FM allows you to enter a company code and posting date and returns the corresponding fiscal year and posting period.

Example: You want to get the fiscal year and period for a specific company so that you can generate the balance sheet for the period. Pass the parameter in this FM and call this FM.

[↓] Import

COMPANYCODEID

Company code is the smallest organizational unit of financial accounting for which a self-contained account can be received for reporting purposes. Type: BAPI0002_2-COMP_CODE.

POSTING_DATE

Posting date. It's the date when the document is entered in financial accounting. Type: BAPI0002_4-POSTING_DATE.

[↑] Export

FISCAL_YEAR

Fiscal year. It's the period for which the company creates its balance sheet and inventory. Type: BAPI0002_4-FISCAL_YEAR.

FISCAL_PERIOD

Fiscal period. Fiscal year is divided into fiscal periods. Account transactions are updated per period within the fiscal year. Type: BAPI0002_4-FISCAL_PERIOD.

RETURN

Return message. Type: BAPIRETURN1.

553 BAPI_COMPANYCODE_GETLIST

This FM returns the key and name of the company code.

Example: You have a company code on a selection screen and want to use the F4 help to display the list of company codes. You can pass the parameter in this FM and call this FM to get all of the company codes in the table COMPANYCODE_LIST.

[↑] Export

RETURN

Return code with error details. Type: BAPIRETURN.

Tables

COMPANYCODE_LIST

List of company codes returned in this parameter. Type: BAPI0002_1.

554 BAPI_CONTRACT_CHANGE

This FM allows you to make changes to the purchasing contract. A purchasing contract is a purchasing document that defines the purchasing and supply of goods at an agreed price, along with other conditions between the purchaser and vendor.

Example: You've created a purchase contract with the vendor that you'll purchase 10 pieces of an item, ITEM1. You want to change the quantity to 15 based on the business requirement, so you need to change the purchasing contract. After you get the details of your contract with FM BAPI_CONTRACT_GETDETAIL (FM 556), use this FM to change the details and the contract.

[↓] Import

PURCHASINGDOCUMENT

Purchasing document number. Type: BAPIMEOUTHEADER-NUMBER.

HEADER

Header data of outline agreement or purchasing contract. Type: BAPIMEOUTHEADER.

HEADERX

Header data change parameter of outline agreement. Type: BAPIMEOUTHEADERX.

VENDOR_ADDRESS

Vendor address if it's different from the master record. Type: BAPIMEOUTADDRVENDOR.

TESTRUN

The parameter is used to check the information prior to actual creation of the object instance in the database. In the process, the creation of the object instance is only simulated, and the data isn't updated. Possible values:

- X: Describes the simulation mode.
- [Space]: Describes the update run.

Type: BAPIFLAG-BAPIFLAG.

TECHNICAL_DATA

Technical parameters for outline agreement BAPIs. Type: BAPIMEOUTTECH.

[↑] Export

EXP_HEADER

Header data purchasing contract. Type: BAPIMEOUTHEADER.

Tables

ITEM

Item data of the purchasing contract. Type: BAPIMEOUTITEM.

ITEMX

Change parameter for item data. Type: BAPIMEOUTITEMX.

ACCOUNT

Account assignment data for the contract item. It's necessary to populate the item number ITEM_NO and the number of the account assignment SERIAL_NO; otherwise, an assignment between the database information (EKKN) and the BAPI interface isn't possible. Type: BAPIMEOUTACCOUNT.

ACCOUNTPROFITSEGMENT

The parameter contains the data on the basis of which a new profitability segment is created for the account assignment.

Upon creation, both the characteristics and the characteristic values are validated against the information specified in the system. Type: BAPIMEOUTPROFITSEGMENT.

ACCOUNTX

Change parameter for account assignment data. Type: BAPIMEOUTACCOUNTX.

DELIVERY_ADDRESS

Delivery address. The system creates a separate address in Central Address Management for each address that is passed on. Type: BAPIMEOUTADDRDELIVERY.

ITEM_COND_VALIDITY

The parameter ITEM_COND_VALIDITY contains the validity periods for conditions for outline agreement items. This corresponds to the tables A016 and A068. The field ITEM_NO contains the item number of the associated item of the outline agreement. The field SERIAL_ID serves as a unique key enabling the representation of several condition periods. If plant-specific conditions are involved, the PLANT field is populated (corresponds to table A068). Otherwise, it's empty (corresponds to table A016).

Several entries in the table ITEM_CONDITION can belong to an entry in ITEM_COND_VALIDITY. These are linked to the entry in the table ITEM_COND_VALIDITY via the fields ITEM_NO and SERIAL_ID. The field COND_COUNT serves to uniquely identify condition items. Type: BAPIMEOUTVALIDITY.

ITEM_COND_VALIDITYX

Change parameter: validity area of item condition. Type: BAPIMEOUTVALIDITYX.

ITEM_CONDITION

The parameter ITEM_CONDITION contains data such as the condition amount and condition currency. It corresponds to KONP. Note: An analogous table to KONH doesn't exist. Type: BAPIMEOUTCONDITION.

ITEM_CONDITIONX

Change parameter: item condition. Type: BAPIMEOUTCONDITIONX.

ITEM_COND_SCALE_VALUE

The parameter ITEM_COND_SCALE_VALUE contains the value scales for a condition of an outline agreement item (i.e., a condition from the table ITEM_CONDITION). Type: BAPIMEOUTITEMSCALEVAL.

ITEM_COND_SCALE_QUAN

The parameter ITEM_COND_SCALE_QUAN contains the quantity scales for a condition of an outline agreement item (i.e., a condition from the table ITEM_CONDITION). The scales for item conditions are assigned to a condition via the fields ITEM_NO, SERIAL_ID, and COND_COUNT. Type: BAPIMEOUTITEMSCALEQUAN.

ITEM_TEXT

Item text. Type: BAPIMEOUTITEMTEXT.

HEADER_TEXT

Header text. Type: BAPIMEOUTTEXT.

HEAD_COND_VALIDITY

The parameter HEAD_COND_VALIDITY contains the validity periods for conditions for the document header. It corresponds to table A019. The SERIAL_ID field serves as a unique key to represent several condition validity periods. Several entries in the table HEAD_CONDITION can belong to an entry in HEAD_COND_VALIDITY. These are linked to the entry in the table HEAD_COND_VALIDITY via the field SERIAL_ID. The COND_COUNT field serves to uniquely identify condition items. Type: BAPIMEOUTHEADVALIDITY.

HEAD_COND_VALIDITYX

Change parameter: validity area of header condition. Type: BAPIMEOUTHEADVALIDITYX.

HEAD_CONDITION

The parameter HEAD_CONDITION contains data such as condition amount and condition currency. It corresponds to KONP. Note: There is no analogous table to KONH. Type: BAPIMEOUTHEADERCOND.

HEAD_CONDITIONX

Change parameter: header condition. Type: BAPIMEOUTHEADERCONDX.

HEAD_COND_SCALE_VAL

The parameter HEAD_COND_SCALE_VAL contains the value scales for a header condition (i.e., a condition from the table HEAD_CONDITION). Type: BAPIMEOUTSCALEVALUE.

HEAD_COND_SCALE_QUAN

The parameter HEAD_COND_SCALE_QUAN contains the quantity scales for a header condition (i.e., a condition from the table HEAD_CONDITION). The scales for header conditions are assigned to a condition via the fields SERIAL_ID and COND_COUNT. Type: BAPIMEOUTSCALE.

PARTNER

The parameter contains the partner data for a purchasing contract. You can specify the partners at the following levels:

- Purchasing organization
- Purchasing organization and plant
- Purchasing organization, plant, and vendor subrange

Type: BAPIMEOUTPARTNER.

PARTNERX

Change parameter for partner outline agreement. Type: BAPIMEOUTPARTNERX.

RELEASE_DOCU

The parameter returns the release order documentation for the current purchasing contract on an item-wise basis. Type: BAPIMEOUTRELEASEINFO.

EXTENSIONIN

This parameter is used to enhance the interface of a BAPI without modification to achieve the automatic processing of customer-specific data. Type: BAPIPAREX.

EXTENSIONOUT

This parameter is used to enhance the interface of a BAPI without modification to achieve the automatic processing of customer-specific data. Type: BAPIPAREX.

RETURN

Return table. Type: BAPIRET2.

555 BAPI_CONTRACT_CREATE

This FM creates a purchasing contract. A purchasing contract is a document that defines the purchasing and supply of goods at an agreed price, along with other conditions between the purchaser and vendor.

Example: You want to purchase an item from a vendor. You've finalized the price and other details. You need to sign a contract for the legality. You can pass the relevant parameter in this FM to create a purchasing contract.

[↓] Import

HEADER

Header data of outline agreement or purchasing contract. Type: BAPIMEOUTHEADER.

HEADERX

Header data change parameter of the outline agreement. Type: BAPIMEOUTHEADERX.

VENDOR_ADDRESS

Vendor address if it's different from the master record. Type: BAPIMEOUTADDRVENDOR

TESTRUN

The parameter is used to check the information prior to actual creation of the object instance in the database. In the process, the creation of the object instance is only simulated, and the data isn't updated. Possible values:

- X: Describes the simulation mode.
- [Space]: Describes the update run.

Type: BAPIFLAG-BAPIFLAG.

TECHNICAL_DATA

Technical parameters for outline agreement BAPIs. Type: BAPIMEOUTTECH.

[↑] Export

PURCHASINGDOCUMENT

Purchasing document number. Type: BAPIMEOUTHEADER-NUMBER.

EXP_HEADER

Header data purchasing contract. Type: BAPIMEOUTHEADER.

Tables

RETURN

Return table. Type: BAPIRET2.

ITEM

Item data of purchasing contract. Type: BAPIMEOUTITEM.

ITEMX

Change parameter for item data. Type: BAPIMEOUTITEMX.

ACCOUNT

Account assignment data for the contract item. It's necessary to populate the item number ITEM_NO and the number of the account assignment SERIAL_NO; otherwise, an assignment between database information (EKKN) and BAPI interface isn't possible. Type: BAPIMEOUTACCOUNT.

ACCOUNTPROFITSEGMENT

The parameter contains the data on the basis of which a new profitability segment is created for the account assignment.

Upon creation, both the characteristics and the characteristic values are validated against the information specified in the system. Type: BAPIMEOUTPROFITSEGMENT.

ACCOUNTX

Change parameter for account assignment data. Type: BAPIMEOUTACCOUNTX.

DELIVERY_ADDRESS

Delivery address. The system creates a separate address in Central Address Management for each address that is passed on. Type: BAPIMEOUTADDRDELIVERY.

ITEM_COND_VALIDITY

The parameter ITEM_COND_VALIDITY contains the validity periods for conditions for outline agreement items. This corresponds to the tables A016 and A068. The field ITEM_NO contains the item number of the associated item of the outline agreement. The field SERIAL_ID serves as a unique key enabling the representation of several condition periods.

If plant-specific conditions are involved, the PLANT field is populated (corresponds to table A068). Otherwise, it's empty (corresponds to table A016).

Several entries in the table ITEM_CONDITION can belong to an entry in ITEM_COND_VALIDITY. These are linked to the entry in the table ITEM_COND_VALIDITY via the fields ITEM_NO and SERIAL_ID. The field COND_COUNT serves to uniquely identify condition items. Type: BAPIMEOUTVALIDITY.

ITEM_COND_VALIDITYX

Change parameter: validity area of item condition. Type: BAPIMEOUTVALIDITYX.

ITEM_CONDITION

Contains data such as the condition amount and condition currency. It corresponds to KONP. Note: An analogous table to KONH doesn't exist. Type: BAPIMEOUTCONDITION.

ITEM_CONDITIONX

Change parameter: item condition. Type: BAPIMEOUTCONDITIONX.

ITEM_COND_SCALE_VALUE

Contains the value scales for a condition of an outline agreement item (i.e., a condition from the table ITEM_CONDITION). Type: BAPIMEOUTITEMSCALEVAL.

ITEM_COND_SCALE_QUAN

This parameter contains the quantity scales for a condition of an outline agreement item (i.e., a condition from the table ITEM_CONDITION).

The scales for item conditions are assigned to a condition via the fields ITEM_NO, SERIAL_ID, and COND_COUNT. Type: BAPIMEOUTITEMSCALEQUAN.

ITEM_TEXT

Item text. Type: BAPIMEOUTITEMTEXT.

HEADER_TEXT

Header text. Type: BAPIMEOUTTEXT.

HEAD_COND_VALIDITY

Contains the validity periods for conditions for the document header. It corresponds to table A019. The SERIAL_ID field serves as a unique key to represent several condition validity periods.

Several entries in the table HEAD_CONDITION can belong to an entry in HEAD_COND_VALIDITY. These are linked to the entry in the table HEAD_COND_VALIDITY via the field SERIAL_ID. The COND_COUNT field serves to uniquely identify condition items. Type: BAPIMEOUTHEADVALIDITY.

HEAD_COND_VALIDITYX

Change parameter: validity area of header condition. Type: BAPIMEOUTHEADVALIDITYX.

HEAD_CONDITION

The parameter HEAD_CONDITION contains data such as condition amount and condition currency. It corresponds to KONP. Note: There is no analogous table to KONH. Type: BAPIMEOUTHEADERCOND.

HEAD_CONDITIONX

Change parameter: header condition. Type: BAPIMEOUTHEADERCONDX.

HEAD_COND_SCALE_VAL

The parameter HEAD_COND_SCALE_VAL contains the value scales for a header condition (i.e., a condition from the table HEAD_CONDITION). Type: BAPIMEOUTSCALEVALUE.

HEAD_COND_SCALE_QUAN

The parameter HEAD_COND_SCALE_QUAN contains the quantity scales for a header condition (i.e., a condition from the table HEAD_CONDITION).

The scales for header conditions are assigned to a condition via the fields SERIAL_ID and COND_COUNT. Type: BAPIMEOUTSCALE.

PARTNER

The parameter contains the partner data for a purchasing contract. You can specify the partners at the following levels:

- Purchasing organization
- Purchasing organization and plant
- Purchasing organization, plant, and vendor subrange

Type: BAPIMEOUTPARTNER.

PARTNERX

Change parameter for partner outline agreement. Type: BAPIMEOUTPARTNERX.

EXTENSIONIN

This parameter is used to enhance the interface of a BAPI without modification to achieve the automatic processing of customer-specific data. Type: BAPIPAREX.

EXTENSIONOUT

This parameter is used to enhance the interface of a BAPI without modification to achieve the automatic processing of customer-specific data. Type: BAPIPAREX.

556 BAPI_CONTRACT_GETDETAIL

This FM returns the details of a purchasing contract.

Example: You've signed a contract to purchase a material from a vendor. To view the contract details and its conditions, pass the parameter and call this FM. See FM BAPI_CONTRACT_CHANGE (FM 554).

[↓] Import

PURCHASINGDOCUMENT

Purchasing document number. Type: BAPIMEOUTHEADER-NUMBER.

ITEM_DATA

Item data. Type: BAPIFLAG-BAPIFLAG.

ACCOUNT_DATA

Account assignment data. Type: BAPIFLAG-BAPIFLAG.

CONDITION_DATA

Conditions. Type: BAPIFLAG-BAPIFLAG.

TEXT_DATA

Texts (header and item). Type: BAPIFLAG-BAPIFLAG.

PARTNER_DATA

Partner. Type: BAPIFLAG-BAPIFLAG.

RELEASE_DATA

Release order documentation. Type: BAPIFLAG-BAPIFLAG.

[↑] Export

HEADER

Header data purchasing contract. Type: BAPIMEOUTHEADER.

Tables

ITEM

Item data of purchasing contract. Type: BAPIMEOUTITEM.

ACCOUNT

Account assignment data for the contract item. It's necessary to populate the item number ITEM_NO and the number of the account assignment SERIAL_NO; otherwise, an assignment between the database information (EKKN) and the BAPI interface isn't possible. Type: BAPIMEOUTACCOUNT.

DELIVERY_ADDRESS

Delivery address. The system creates a separate address in Central Address Management for each address that is passed on. Type: BAPIMEOUTADDRDELIVERY.

ITEM_COND_VALIDITY

The parameter ITEM_COND_VALIDITY contains the validity periods for conditions for outline agreement items. This corresponds to the tables A016 and A068. The field ITEM_NO contains the item number of the associated item of the outline agreement. The field SERIAL_ID serves as a unique key enabling the representation of several condition periods.

If plant-specific conditions are involved, the PLANT field is populated (corresponds to table A068). Otherwise, it's empty (corresponds to table A016).

Several entries in the table ITEM_CONDITION can belong to an entry in ITEM_COND_VALIDITY. These are linked to the entry in the table ITEM_COND_VALIDITY via the fields ITEM_NO and SERIAL_ID. The field COND_COUNT serves to uniquely identify condition items. Type: BAPIMEOUTVALIDITY.

ITEM_CONDITION

The parameter ITEM_CONDITION contains data such as the condition amount and condition currency. It corresponds to KONP. Note: An analogous table to KONH doesn't exist. Type: BAPIMEOUTCONDITION.

ITEM_COND_SCALE_VALUE

The parameter ITEM_COND_SCALE_VALUE contains the value scales for a condition of an outline agreement item (i.e., a condition from the table ITEM_CONDITION). Type: BAPIMEOUTITEMSCALEVAL.

ITEM_COND_SCALE_QUAN

The parameter ITEM_COND_SCALE_QUAN contains the quantity scales for a condition of an outline agreement item (i.e., a condition from the table ITEM_CONDITION).

The scales for item conditions are assigned to a condition via the fields ITEM_NO, SERIAL_ID, and COND_COUNT. Type: BAPIMEOUTITEMSCALEQUAN.

ITEM_TEXT

Item text. Type: BAPIMEOUTITEMTEXT.

HEADER_TEXT

Header text. Type: BAPIMEOUTTEXT.

HEAD_COND_VALIDITY

The parameter HEAD_COND_VALIDITY contains the validity periods for conditions for the document header. It corresponds to table A019. The SERIAL_ID field serves as a unique key to be able to represent several condition validity periods.

Several entries in the table HEAD_CONDITION can belong to an entry in HEAD_COND_VALIDITY. These are linked to the entry in the table HEAD_COND_VALIDITY via the field SERIAL_ID. The COND_COUNT field serves to uniquely identify condition items. Type: BAPIMEOUTHEADVALIDITY.

HEAD_COND_VALIDITYX

Change parameter: validity area of header condition. Type: BAPIMEOUTHEADVALIDITYX.

HEAD_CONDITION

The parameter HEAD_CONDITION contains data such as the condition amount and condition currency. It corresponds to KONP. Note: There is no analogous table to KONH. Type: BAPIMEOUTHEADERCOND.

HEAD_CONDITIONX

Change parameter: header condition. Type: BAPIMEOUTHEADERCONDX.

HEAD_COND_SCALE_VAL

The parameter HEAD_COND_SCALE_VAL contains the value scales for a header condition (i.e., a condition from the table HEAD_CONDITION). Type: BAPIMEOUTSCALEVALUE.

HEAD_COND_SCALE_QUAN

The parameter HEAD_COND_SCALE_QUAN contains the quantity scales for a header condition (i.e., a condition from the table HEAD_CONDITION).

The scales for header conditions are assigned to a condition via the fields SERIAL_ID and COND_COUNT. Type: BAPIMEOUTSCALE.

PARTNER

The parameter contains the partner data for a purchasing contract. You can specify the partners at the following levels:

- Purchasing organization
- Purchasing organization and plant
- Purchasing organization, plant, and vendor subrange

Type: BAPIMEOUTPARTNER.

PARTNERX

Change parameter for partner outline agreement. Type: BAPIMEOUTPARTNERX.

RELEASE_DOCU

The parameter returns the release order documentation for the current purchasing contract on an item-wise basis. Type: BAPIMEOUTRELEASEINFO.

EXTENSIONOUT

This parameter is used to enhance the interface of a BAPI without modification to achieve the automatic processing of customer-specific data. Type: BAPIPAREX.

RETURN

Return table. Type: BAPIRET2.

557 BAPI_COPAACTUALS_POSTCOSTDATA

This FM is used to post the actual data to the SAP system costing based on Controlling–Profitability Analysis (CO-PA). This FM isn't used to transfer mass data. The data you want to post is transferred in parameter INPUTDATA. To assure the consistency of the transferred data, the parameter FIELDLIST must also be filled with the number of field names that are transferred in INPUTDATA. The FM BAPI_TRANSACTION_COMMIT must be called to perform the COMMIT.

Example: You want to post the billing document from SD to determine the revenues to CO-PA programmatically. You can pass the desired parameters in this FM and can call this FM to post the actual data.

[↓] Import

OPERATINGCONCERN

This parameter represents the operating concern. It's a representation of an organization for which the sales market is structured in a uniform manner. Type: BAPI0017-OP_CONCERN.

TESTRUN

Switch to simulation session for writing the BAPI. Value X doesn't update the data to database. Value [Space] updates (i.e., posts) the database. Type: BAPI0017-TESTRUN.

Tables

INPUTDATA

Input: data to be posted. Each record in this parameter contains a field for a CO-PA record. All fields belonging to a CO-PA record must have the same record number. All value fields belonging to a CO-PA record must have the same currency. Type: BAPI_COPA_DATA.

FIELDLIST

List of field names filled in the parameter INPUTDATA. For each CO-PA record to be written, all fields from the parameter FIELDLIST must also appear in INPUTDATA. Type: BAPI_COPA_FIELD.

RETURN

List of generated error messages. Type: BAPIRET2.

---

558 BAPI_COPAPLANNING_GETDATA

This FM reads planning data for CO-PA.

Example: You want to get the sales volume and the revenue for all products in division 01 for the period 001.2013 through 012.2013 from CO-PA to SAP. You can pass these parameters along with other relevant parameters in this FM to get the planning data details. The next step involves FM BAPI_COPAPLANNING_POSTDATA (FM 559).

[↓] Import

OPERATINGCONCERN

This parameter represents the operating concern. It's a representation of an organization for which the sales market is structured in a uniform manner. Type: BAPI0017-OP_CONCERN.

TYPEOFPROFITANALYSIS

Type of Profitability Analysis. Value 1 represents costing-based Profitability Analysis, and value 2 represents account-based Profitability Analysis. Type: BAPI1167-TYPEOFPA.

Tables

SELECTEDFIELDS

The characteristics and value fields to be read are placed in the parameter. In CO-PA, the characteristics VERSI, VRGAR, PALEDGER, and PERIO (or ALTPERIO) are mandatory fields. In account-based Profitability Analysis, the characteristics VERSI, PERIO, KOKRS (controlling area), and KSTAR are mandatory fields. Type: BAPI_COPA_FIELD.

SELECTION

Selection criteria for characteristics. Type: BAPI_COPA_SELECTION.

DATA

Export table containing the data that has been read. Type: BAPI_COPA_DATA.

RETURN

Messages. Type: BAPIRET2.

559 BAPI_COPAPLANNING_POSTDATA

This FM is used to return planning data from Profitability Analysis back into the system. This overwrites existing planning data in the system (or, more precisely, the delta value between the old and new values is determined and updated as a line item).

Example: From the previous step, after you've received the planning data for the sales volume and revenues, you need to post that data from COPA. You can do so by calling this FM.

[↓] Import

OPERATINGCONCERN

This parameter represents the operating concern. It's a representation of an organization for which the sales market is structured in a uniform manner. Type: BAPI0017-OP_CONCERN.

TYPEOFPROFITANALYSIS

Type of Profitability Analysis. Value 1 represents costing-based Profitability Analysis, and value 2 represents account-based Profitability Analysis. Type: BAPI1167-TYPEOFPA.

TESTRUN

Switch to the simulation session for writing BAPI. Value X doesn't update the data to database. Value [Space] updates (i.e., posts) the database. Type: BAPI0017-TESTRUN.

Tables

SELECTEDFIELDS

The characteristics and value fields to be read are placed in the parameter. In costing-based Profitability Analysis, the characteristics VERSI, VRGAR, PALEDGER, and PERIO (or ALTPERIO) are mandatory fields. In account-based Profitability Analysis, the characteristics VERSI, PERIO, KOKRS (controlling area), and KSTAR are mandatory fields. Type: BAPI_COPA_FIELD.

SELECTION

Selection criteria for characteristics. Type: BAPI_COPA_SELECTION.

DATA

Export table containing the data that has been read. Type: BAPI_COPA_DATA.

RETURN

Messages. Type: BAPIRET2.

560 BAPI_CURRENCY_CONV_TO_EXTERNAL

This FM converts the currency amount into an external format.

Example: You have a currency field in the display. You get the currency amount from the table and want to display it in a user format. You can pass the internal format currency in this FM to convert it to a user external format and display it.

[↓] Import

CURRENCY

Currency. Type: TCURC-WAERS.

AMOUNT_INTERNAL

Currency amount (SAP): internal data format.

[↑] Export

AMOUNT_EXTERNAL

Currency amount: external data format. Type: BAPICURR-BAPICURR.

---

561 BAPI_CURRENCY_CONV_TO_INTERNAL

This FM converts the currency amount into an internal format.

Example: You have a currency field in the display, and a user entered the amount in this field. While saving, you need to convert the field from an external to an internal format.

[↓] Import

CURRENCY

Currency. Type: TCURC-WAERS.

AMOUNT_EXTERNAL

External currency amount. Type: BAPICURR-BAPICURR.

MAX_NUMBER_OF_DIGITS

This parameter is used to delimit value ranges. The maximum number of places before and after the decimal point is 23. If a domain in an internal field is 11 digits long, then the maximum number of places before and after the decimal point is 11. For all external amounts that are more than 11 digits, the error code HV041 "External currency amount too large to be converted" is returned in the parameter RETURN.

[↑] Export

AMOUNT_INTERNAL

Converted internal currency amount. Type: I.

RETURN

Return messages. Type: BAPIRETURN.

---

562 BAPI_CUSTOMER_CHANGEFROMDATA1

This FM is used to change the personal or company data of the customer. Only one structure (either personal or company) can be passed on to the BAPI.

Example: You want to change the name of the customer. You can pass the relevant parameters in this FM and call this FM for changing the name.

[↓] Import

PI_PERSONALDATA

Customer personal data to be changed for the customer. Type: BAPIKNA101_1.

PI_PERSONALDATAX

Corresponding checkbox (X fields) for the optional personal data to be changed for an existing customer. Only the fields marked with X are updated. Type: BAPIKNA101_1X.

PI_OPT_PERSONALDATA

Customer personal data of an existing customer. It this parameter is filled with data, don't fill the import parameter for the optional company address data. Type: BAPIKNA105.

PI_OPT_PERSONALDATAX

Corresponding checkboxes (X fields) for an existing customer's company address data that is to be changed. Only those fields marked "X" are transferred. Type: BAPIKNA105X.

PI_COMPANYDATA

Customer company data. If this parameter is filled with data, the import structure for the personal data can't be filled. Type: BAPIKNA106.

PI_COMPANYDATAX

Customer company data structure that contains the data to be changed. Only fields marked as X are transferred. Type: BAPIKNA106X.

PI_OPT_COMPANYDATA

Optional customer company data. Type: BAPIKNA105.

PI_OPT_COMPANYDATAX

Optional customer company data structure which is to be changed. Only fields marked as X are transferred. Type: BAPIKNA105X.

PI_SALESORG

Sales organization. Type: BAPIKNA102-SALESORG.

PI_DISTR_CHAN

Distribution channel. Type: BAPIKNA102-DISTR_CHAN.

PI_DIVISION

Division. Type: BAPIKNA102-DIVISION.

CUSTOMERNO

Number of the customer to be changed. Type: BAPIKNA103-CUSTOMER.

[↑] Export

RETURN

Return value. Type: BAPIRETURN1.

---

563 BAPI_CUSTOMER_CREATE

This FM is used to create a new customer.

Example: You have orders from a new customer, so you need to create and maintain the details in your SAP system. You can pass the parameters in this FM and call this FM to create the new customer.

[↓] Import

INIT_KUNNR

Single-character indicator. Type: BAPIFLAG-BAPIFLAG.

[↑] Export

CUSTOMERNO

Customer number field of newly created customer. Type: BAPICUSTOMER_01-CUSTOMER.

CUSTOMER

Customer number field of newly created customer. Type: BAPICUSTOMER_01-CUSTOMER.

RETURN

Return parameter. Type: BAPIRET1.

---

564 BAPI_CUSTOMER_GETLIST

This FM reads the customer and its address based on the range select options provided and returns those details in the tables parameter.

Example: You have a field of customers in the selection screen and want the list of customer in the F4 help. You can pass the relevant parameter in this FM and call this FM to display the list.

[↓] Import

MAXROWS

Maximum number of customers for which address data is returned. Type: BAPIF4A-MAX_ROWS.

CPDONLY

Read one-time customers only. Value [Space] selects all of the customers, but the value X selects only one-time customers. Type: BAPICUSTOMER_CPD-CPD_ONLY.

[↑] Export

RETURN

Return messages from the method call. Type: BAPIRETURN1.

Tables

IDRANGE

Range table to find customer numbers. Possible values for field SIGN include I (inclusive) or E (exclusive), and values for field OPTIONS include EQ, NE, GT, GE, LE, LT, BT, NB, CP, and NP. Type: BAPICUSTOMER_IDRANGE.

ADDRESSDATA

Customer address data returned. Type: BAPICUSTOMER_ADDRESSDATA.

SPECIALDATA

Special customer data returned. Type: BAPICUSTOMER_SPECIALDATA.

565 BAPI_CUSTOMERINQUIRY_CHANGE

This FM is used to change or delete customer inquiries. The header, item, schedule lines, and configuration data can be changed for customer inquiries. The following rules should be followed:

- Only enter the fields you want to change.
- Select the fields you want to change with an X.
- Enter "U" in the UPDATEFLAG field. (Other possible values include D for delete and I for insert.)
- Always enter the key fields when you make a change.

- An exception is a configuration type. If you want to change this, you must fill it in completely.
- Maintain the quantities and dates via the schedule line data.

Example: You've inquired for product PROD1 and PROD2, and now you want to change the inquiry to PROD3. Pass the relevant parameters in this FM and call it.

[↓] Import

SALESDOCUMENT

Sales document number of customer inquiry. Type: BAPIVBELN-VBELN.

INQUIRY_HEADER_IN

Sales document header. Type: BAPISDH1.

INQUIRY_HEADER_INX

If the UPDATEFLAG field of this structure is activated, the system only copies those fields from the SALES_HEADER_IN parameter that have been activated with X. Type: BAPISDH1X.

SIMULATION

Simulation mode. Type: BAPIFLAG-BAPIFLAG.

BEHAVE_WHEN_ERROR

This parameter controls how the program behaves when an error occurs. Possible values:

- [Space]: If an error occurs, processing stops and the sales document isn't saved.
- P: If an error occurs, the sales document can be saved. Problematic items aren't processed, but they are logged.
- R: Same as for, P but the document isn't saved.

Type: BAPIFLAG-BAPIFLAG.

INT_NUMBER_ASSIGNMENT

Internal item number assignment. Type: BAPIFLAG-BAPIFLAG.

LOGIC_SWITCH

This parameter controls internal program behavior. The parameter is intended to be used to switch off specific processing because data is to be transferred externally. Type: BAPISDLS.

NO_STATUS_BUF_INIT

If this parameter is set to X, status buffer isn't refreshed. Type: BAPIFLAG-BAPIFLAG.

Tables

RETURN

Return code. Type: BAPIRET2.

INQUIRY_ITEM_IN

Inquiry items. Type: BAPISDITM.

INQUIRY_ITEM_INX

Check table for inquiry items. Type: BAPISDITMX.

PARTNERS

Communications fields: SD document partner: WWW. Type: BAPIPARNR.

PARTNERCHANGES

Partner changes. Type: BAPIPARNRC.

PARTNERADDRESSES

BAPI reference structure for addresses (org./company). Type: BAPIADDR1.

INQUIRY_CFGS_REF

Configuration: reference data. Type: BAPICUCFG.

INQUIRY_CFGS_INST

Configuration: instances. Type: BAPICUINS.

INQUIRY_CFGS_PART_OF

Configuration: part-of specifications. Type: BAPICUPRT.

INQUIRY_CFGS_VALUE

Configuration: characteristic values. Type: BAPICUVAL.

INQUIRY_CFGS_BLOB

Internal configuration data (SCE). Type: BAPICUBLB.

INQUIRY_CFGS_VK

Configuration: variant condition key. Type: BAPICUVK.

INQUIRY_CFGS_REFINST

Configuration: reference item/instance. Type: BAPICUREF.

SCHEDULE_LINES

Schedule lines. Type: BAPISCHDL.

SCHEDULE_LINESX

Check table for schedule lines. Type: BAPISCHDLX.

INQUIRY_TEXT

Texts. Type: BAPISDTEXT.

INQUIRY_KEYS

Output table of reference keys. Type: BAPISDKEY.

CONDITIONS_IN

Conditions. Type: BAPICOND.

CONDITIONS_INX

Conditions checkbox. Type: BAPICONDX.

EXTENSIONIN

Customer enhancement for VBAK, VBAP, and VBEP. Type: BAPIPAREX.

NFMETALLITMS

BAPI communication structure. Type: /NFM/BAPIDOCITM.

566 BAPI_CUSTOMERQUOTATION_CHANGE

This FM is used to change or delete customer quotations. Customer header, item, schedule lines, and configuration data can be changed. The following rules should be followed:

- Only enter the fields you want to change.
- Select the fields you want to change with an X.
- Enter "U" in the UPDATEFLAG field. (Other values include D for delete and I for insert.)
- Always enter the key fields when you make a change.
- An exception is a configuration type. If you want to change this, you must fill it in completely.
- Maintain the quantities and dates via the schedule line data.

Example: To delete an item of a quotation, you can pass the relevant parameters and call this FM.

[↓] Import

SALESDOCUMENT

Sales document number of customer inquiry. Type: BAPIVBELN-VBELN.

QUOTATION_HEADER_IN

Inquiry header. Type: BAPISDH1.

QUOTATION_HEADER_INX

If the UPDATEFLAG field has been activated, the system only copies those fields from the SALES_HEADER_IN parameter that have been activated with X. Type: BAPISDH1X.

SIMULATION

Simulation mode. Type: BAPIFLAG-BAPIFLAG.

BEHAVE_WHEN_ERROR

This parameter controls how the program behaves when an error occurs. Possible values:

- [Space]: If an error occurs, processing stops, and the sales document isn't saved.
- P: If an error occurs, the sales document can be saved. Problematic items aren't processed, but they are logged.
- R: Same as for P, but the document isn't saved.

Type: BAPIFLAG-BAPIFLAG.

INT_NUMBER_ASSIGNMENT

Internal item number assignment. Type: BAPIFLAG-BAPIFLAG.

LOGIC_SWITCH

This parameter controls internal program behavior. The parameter is intended to be used to switch off specific processing because data is to be transferred externally. Type: BAPISDLS.

NO_STATUS_BUF_INIT

If this parameter is set to X, status buffer isn't refreshed. Type: BAPIFLAG-BAPIFLAG.

Tables

RETURN

Return code. Type: BAPIRET2.

QUOTATION_ITEM_IN

Quotation items. Type: BAPISDITM.

QUOTATION_ITEM_INX

Quotation items checkbox. Type: BAPISDITMX.

PARTNERS

Communications fields: SD document partner: WWW. Type: BAPIPARNR.

PARTNERCHANGES

Partner changes. Type: BAPIPARNRC.

PARTNERADDRESSES

BAPI reference structure for addresses (org./company). Type: BAPIADDR1.

CONDITIONS_IN

Conditions. Type: BAPICOND.

CONDITIONS_INX

Conditions checkbox. Type: BAPICONDX.

QUOTATION_CFGS_REF

Configuration: reference data. Type: BAPICUCFG.

QUOTATION_CFGS_INST

Configuration: instances. Type: BAPICUINS.

QUOTATION_CFGS_PART_OF

Configuration: part-of specifications. Type: BAPICUPRT.

QUOTATION_CFGS_VALUE

Configuration: characteristic values. Type: BAPICUVAL.

QUOTATION_CFGS_BLOB

Configuration: BLOB internal data (SCE). Type: BAPICUBLB.

QUOTATION_CFGS_VK

Configuration: variant condition key. Type: BAPICUVK.

QUOTATION_CFGS_REFINST

Configuration: reference item/instance. Type: BAPICUREF.

SCHEDULE_LINES

Schedule lines. Type: BAPISCHDL.

SCHEDULE_LINESX

Check table for schedule lines. Type: BAPISCHDLX.

QUOTATION_TEXT

Texts. Type: BAPISDTEXT.

QUOTATION_KEYS

Output table of reference keys. Type: BAPISDKEY.

EXTENSIONIN

Customer enhancement for VBAK, VBAP, and VBEP. Type: BAPIPAREX.

NFMETALLITMS

BAPI communication structure. Type: /NFM/BAPIDOCITM.

567 BAPI_ENTRYSHEET_CREATE

This FM is used to create service entry sheets. A service entry sheet is created for the services that are provided by the vendor for maintaining or servicing your material, machines, or services. Service entry sheets are created in Transaction ML81N. In this FM, authorization objects M_BEST_BSA (document type of the purchase order on the entry sheet), M_BEST_EKG (purchasing group of the purchase order), M_BEST_EKO (purchasing organization of the purchase order), and M_BEST_WRK (plant of the purchase order) with the activity 76 are checked.

Example: You've called a vendor to maintain your casting machine. To receive maintenance services, you need to create the service entry sheet. You can programmatically create the service list by passing the parameters in this FM and calling this FM.

[↓] Import

ENTRYSHEETHEADER

Header data of entry sheet header such as short text, PO number, item number, location service performed, posting date, and so on. Type: BAPIESSRC.

TESTRUN

Whether this method should be used in the simulation run (when TESTRUN is X) or in update run (when TESTRUN is [Space]). Type: BAPI2091-TESTRUN.

NO_COMMIT

If this parameter is set to X, changes aren't committed to the database. Type: BAPIFLAG-BAPIFLAG.

[↑] Export

ENTRYSHEET

Entry sheet number. Type: BAPIESSR-SHEET_NO.

Tables

ENTRYSHEETACCOUNTASSIGNMENT

Account assignments of the entry sheet. Type: BAPIESKNC.

ENTRYSHEETSERVICES

This table contains the services and service specification outlines of the entry sheet. Type: BAPIESLLC.

ENTRYSHEETSRVACCASSVALUES

This table contains the values of the account assignments at the service level and a reference to the overall account assignment of the entry sheet. Type: BAPIESKLC.

RETURN

Processing errors that occurred. Type: BAPIRET2.

ENTRYSHEETSERVICESTEXTS

Long texts for the service line are passed in this table. Type: BAPIESLLTX.

ENTRYSHEETHEADERTEXT

Long text for the entry sheet header. Type: BAPIESSRTX.

---

568 BAPI_EQUI_CHANGE

This FM is used to change details for equipment. Equipment refers to the individual physical objects that can be maintained independently. Equipment can be installed in the technical system or as a part of a technical system.

Example: You've created an equipment item for a production resource tool. You want to change the equipment details such as its description, size, weight, and so on. You can pass the parameter in this FM and call this FM to change the equipment details.

[↓] Import

EQUIPMENT

Equipment number. Type: BAPI_ITOB_PARMS-EQUIPMENT.

DATA_GENERAL

This parameter contains the master data of equipment such as object number, size, weight, acquisition date, manufacturing serial number, model number, and so on. Type: BAPI_ITOB.

DATA_GENERALX

If you want to change the master data, you should mark it as X on the field you want to change in this parameter. Type: BAPI_ITOBX.

DATA_SPECIFIC

Equipment-specific data that can be created by this FM. Type: BAPI_ITOB_EQ_ONLY.

DATA_SPECIFICX

If you want to change the equipment-specific data, you should mark it as X on the field you want to change in this parameter. Type: BAPI_ITOB_EQ_ONLYX.

DATA_FLEET

Vehicle-specific data that can be created by this BAPI. Type: BAPI_FLEET.

DATA_FLEETX

If you want to change the vehicle specific data, you should mark it as X on the field you want to change in this parameter. Type: BAPI_FLEETX.

VALID_DATE

If changing the equipment leads to the creation of a new equipment usage period, then the date for its validity can be determined using the VALID_DATE parameter. The default value is SY-DATUM. Type: BAPI_ITOB_PARMS-INST_DATE.

VALID_TIME

If changing a piece of equipment leads to the creation of a new equipment usage period, then the time for its validity can be determined using the VALID_TIME parameter. The default value is SY-DATUM. Type: BAPI_ITOB_PARMS-INST_TIME.

[↑] Export

DATA_GENERAL_EXP

General data for technical objects are returned in this parameter. Type: BAPI_ITOB.

DATA_SPECIFIC_EXP

Equipment-specific data are returned in this parameter. Type: BAPI_ITOB_EQ_ONLY.

DATA_FLEET_EXP

Vehicle-specific data are returned in this parameter. Type: BAPI_FLEET.

RETURN

Return parameter with messages. Type: BAPIRET2.

---

569 BAPI_EQUI_CREATE

This FM is used to create equipment. Equipment refers to the individual physical objects that can be maintained independently. Equipment can be installed in the technical system or as a part of a technical system.

Example: You want to create an entry for an equipment item (such as a hammer) that you want to use in your manufacturing process. Pass the parameters and call this FM.

[↓] Import

EXTERNAL_NUMBER

Equipment number to be created. Type: BAPI_ITOB_PARMS-EQUIPMENT.

DATA_GENERAL

This parameter contains the master data of equipment such as object number, size, weight, acquisition date, manufacturing serial number, model number, and so on. Type: BAPI_ITOB.

DATA_SPECIFIC

Equipment-specific data that can be created by this FM. Type: BAPI_ITOB_EQ_ONLY.

DATA_FLEET

Vehicle-specific data that can be created by this FM. Type: BAPI_FLEET.

VALID_DATE

If changing the equipment leads to the creation of a new equipment usage period, then the date for its validity can be determined using the VALID_DATE parameter. The default value is SY-DATUM. Type: BAPI_ITOB_PARMS-INST_DATE.

DATA_INSTALL

Installation date for equipment. Type: BAPI_ITOB_EQ_INSTALL.

[↑] Export

EQUIPMENT

Equipment number returned. Type: BAPI_ITOB_PARMS-EQUIPMENT.

DATA_GENERAL_EXP

General data for technical objects are returned in this parameter. Type: BAPI_ITOB.

DATA_SPECIFIC_EXP

Equipment-specific data are returned in this parameter. Type: BAPI_ITOB_EQ_ONLY.

DATA_FLEET_EXP

Vehicle-specific data are returned in this parameter. Type: BAPI_FLEET.

RETURN

Return parameter with messages. Type: BAPIRET2.

---

570 BAPI_FIXEDASSET_CHANGE

This FM allows you to change the asset master record. To change the asset master from this FM, you need to have the same authorization that's needed to change an asset in a dialog mode.

Example: You've purchased an asset such as a laptop for your organization and want to change details that were incorrectly entered, such as the serial number or memory capacity. You can pass the parameter in this FM and call it for the change in asset.

[↓] Import

COMPANYCODE

Company code. This is a mandatory parameter so that the system can know from which company code the asset is required to be changed. Type: BAPI1022_1-COMP_CODE.

ASSET

Main asset number. This is a mandatory parameter. Type: BAPI1022_1-ASSETMAINO.

SUBNUMBER

Asset subnumber. This is a mandatory parameter. Type: BAPI1022_1-ASSETSUBNO.

GROUPASSET

This parameter is set to X if the asset master record to be changed is a group asset. Type: BAPI1022_MISC-XANLGR.

GENERALDATA

General data. To change the general data, the corresponding field must be set in the change parameter it belongs to. Type: BAPI1022_FEGLG001.

GENERALDATAX

This parameter is used to specify whether the contents of the field should be transferred to the corresponding field of the data parameter in the asset. Possible values:

- Space: Field contents of the corresponding field of the data parameter aren't transferred.
- X: Field contents of the corresponding field of the data parameter are transferred.

Type: BAPI1022_FEGLG001X.

INVENTORY

Inventory. Type: BAPI1022_FEGLG011.

INVENTORYX

Change parameters for inventory. This field must be set to X if you want to change the inventory data. Type: BAPI1022_FEGLG011X.

POSTINGINFORMATION

Posting information. Type: BAPI1022_FEGLG002.

POSTINGINFORMATIONX

Change parameters for posting information. This field must be set to X if you want to change the data. Type: BAPI1022_FEGLG002X.

TIMEDEPENDENTDATA

Time-dependent data. Type: BAPI1022_FEGLG003.

TIMEDEPENDENTDATAX

Change parameters for time-dependent data. This field must be set to X if you want to change the data. Type: BAPI1022_FEGLG003X.

ALLOCATIONS

Allocations. Type: BAPI1022_FEGLG004.

ALLOCATIONSX

Change parameters for allocations. This field must be set to X if you want to change the data. Type: BAPI1022_FEGLG004X.

ORIGIN

Origin. Type: BAPI1022_FEGLG009.

ORIGINX

Change parameters for origin. This field must be set to X if you want to change the data. Type: BAPI1022_FEGLG009X.

INVESTACCTASSIGNMNT

Account assignment for investment. Type: BAPI1022_FEGLG010.

INVESTACCTASSIGNMNTX

Change parameters for investment account assignments. This field must be set to X if you want to change the data. Type: BAPI1022_FEGLG010X.

NETWORTHVALUATION

Net worth valuation. Type: BAPI1022_FEGLG006.

NETWORTHVALUATIONX

Change parameters for net worth valuation. This field must be set to X if you want to change the data. Type: BAPI1022_FEGLG006X.

REALESTATE

Real estate and similar rights. Type: BAPI1022_FEGLG007.

REALESTATEX

Change parameters for real estate. This field must be set to X if you want to change the data. Type: BAPI1022_FEGLG007X.

INSURANCE

Insurance. Type: BAPI1022_FEGLG008.

INSURANCEX

Change parameters for insurance. This field must be set to X if you want to change the data. Type: BAPI1022_FEGLG008X.

LEASING

Leasing. Type: BAPI1022_FEGLG005.

LEASINGX

Change parameters for leasing. This field must be set to X if you want to change the data. Type: BAPI1022_FEGLG005X.

[↑] Export

RETURN

Information about errors that occurred. Type: BAPIRET2.

Tables

DEPRECIATIONAREAS

Depreciation areas. Type: BAPI1022_DEP_AREAS.

DEPRECIATIONAREASX

Change parameters for depreciation areas. This field must be set to X if you want to change the data. Type: BAPI1022_DEP_AREASX.

INVESTMENT_SUPPORT

Used to add, change, or delete existing investment support depreciation areas on a fixed asset. Type: BAPI1022_INV_SUPPORT.

EXTENSIONIN

Used for customer enhancement. You can transfer user fields from the asset master record. Only user fields that were defined using SAP enhancement AIST0002 and automatically updated in the table ANLU are updated. Type: BAPIPAREX.

571 BAPI_FIXEDASSET_CREATE1

This FM is used to create asset master records (or fixed assets). To create the asset master from this BAPI, you need to have the same authorization as for changing an asset in a dialog mode.

Example: You've purchased an asset such as a laptop for your organization and want to maintain its details as an asset. Pass the parameter and call this FM to do so.

[↓] Import

KEY

Key of asset being created. Type: BAPI1022_KEY.

REFERENCE

You can specify the company code, asset number, and asset subnumber, and you can specify the asset that can be used as a reference for the asset being created. The system copies the details of the reference asset. Type: BAPI1022_REFERENCE.

CREATESUBNUMBER

This parameter specifies if you want to create and return the asset subnumber. Type: BAPI1022_MISC-XSUBNO.

POSTCAP

If you set this parameter to X, you can set the asset for post-capitalization that is you can make corrections in the acquisition and production cost of the asset. Type: BAPI1022_MISC-POSTCAP.

CREATEGROUPASSET

This parameter is set to X if the asset master record to be created is a group asset. Type: BAPI1022_MISC-XANLGR.

TESTRUN

This parameter checks the specifications of the instance in write BAPIs before the object instance is actually created in the database. The creation of the object instance is simulated, and the data isn't updated. Possible values:

- [Space]: No value means that updating takes place. This is the standard setting and is what the caller usually expects.
- X: This value means that this BAPI is called only in simulation mode and that data isn't updated when the BAPI is called.

Type: BAPI1022_MISC-TESTRUN.

GENERALDATA

General data. To change the general data, the corresponding field must be set in the change parameter it belongs to. Type: BAPI1022_FEGLG001.

GENERALDATAX

This parameter is used to specify whether the contents of the field should be transferred to the corresponding field of the data parameter in the asset. Possible values:

- [Space]: Field contents of the corresponding field of the data parameter aren't transferred.

- X: Field contents of the corresponding field of the data parameter are transferred.

Type: BAPI1022_FEGLG001X.

INVENTORY

Inventory. Type: BAPI1022_FEGLG011.

INVENTORYX

Change parameters for inventory. This field must be set to X if you want to change the inventory data. Type: BAPI1022_FEGLG011X.

POSTINGINFORMATION

Posting information. Type: BAPI1022_FEGLG002.

POSTINGINFORMATIONX

Change parameters for posting information. This field must be set to X if you want to change the data. Type: BAPI1022_FEGLG002X.

TIMEDEPENDENTDATA

Time-dependent data. Type: BAPI1022_FEGLG003.

TIMEDEPENDENTDATAX

Change parameters for time-dependent data. This field must be set to X if you want to change the data. Type: BAPI1022_FEGLG003X.

ALLOCATIONS

Allocations. Type: BAPI1022_FEGLG004.

ALLOCATIONSX

Change parameters for allocations. This field must be set to X if you want to change the data. Type: BAPI1022_FEGLG004X.

ORIGIN

Origin. Type: BAPI1022_FEGLG009.

ORIGINX

Change parameters for origin. This field must be set to X if you want to change the data. Type: BAPI1022_FEGLG009X.

INVESTACCTASSIGNMNT

Account assignment for investment. Type: BAPI1022_FEGLG010.

INVESTACCTASSIGNMNTX

Change parameters for investment account assignments. This field must be set to X if you want to change the data. Type: BAPI1022_FEGLG010X.

NETWORTHVALUATION

Net worth valuation. Type: BAPI1022_FEGLG006.

NETWORTHVALUATIONX

Change parameters for net worth valuation. This field must be set to X if you want to change the data. Type: BAPI1022_FEGLG006X.

REALESTATE

Real estate and similar rights. Type: BAPI1022_FEGLG007.

REAESTATEX

Change parameters for real estate. This field must be set to X if you want to change the data. Type: BAPI1022_FEGLG007X.

INSURANCE

Insurance. Type: BAPI1022_FEGLG008.

INSURANCEX

Change parameters for insurance. This field must be set to X if you want to change the data. Type: BAPI1022_FEGLG008X.

LEASING

Leasing. Type: BAPI1022_FEGLG005.

LEASINGX

Change parameters for leasing. This field must be set to X if you want to change the data. Type: BAPI1022_FEGLG005X.

[↑] Export

COMPANYCODE

Company code of the asset created is returned in this parameter. Type: BAPI1022_1-COMP_CODE.

ASSET

Main asset number is returned in this parameter. Type: BAPI1022_1-ASSETMAINO.

SUBNUMBER

Asset subnumber is returned in this parameter. Type: BAPI1022_1-ASSETSUBNO.

ASSETCREATED

Asset created. This structure contains the company code, asset, and subnumber of the asset created. Type: BAPI1022_REFERENCE.

RETURN

Information about errors that occurred. Type: BAPIRET2.

Tables

DEPRECIATIONAREAS

You can transfer data of depreciation areas using this parameter. Transfer a row in this table parameter for each depreciation area in which data is to be transferred. Type: BAPI1022_DEP_AREAS.

DEPRECIATIONAREASX

Change parameters for depreciation areas. This field must be set to X if you want to change the data. Type: BAPI1022_DEP_AREASX.

INVESTMENT_SUPPORT

Using this table, you can create investment support depreciation areas on the new asset. For each investment support depreciation area, you need a row. For each row, you fill the INV_SUPPORT_KEY field with an investment support key that is valid in the chart of depreciation of the asset being created. The CHANGE_IND field isn't used when creating assets, and therefore can be left blank. Type: BAPI1022_INV_SUPPORT.

EXTENSIONIN

This parameter is used for customer enhancement. You can transfer user fields from the asset master record. Only user fields that were defined using SAP enhancement AIST0002 and are automatically updated in the table ANLU are updated. Type: BAPIPAREX.

572 BAPI_GOODSMVT_CANCEL

This FM is used to reverse the goods movement (i.e., material documents can be cancelled by calling this FM). Only one material document is cancelled each time this transaction is called. For reversal, the system doesn't delete the original document but creates a new reversal document.

Example: You've created a goods movement to move the material from one location to another location. You've analyzed that the goods movement isn't required now because the order is cancelled. You can cancel or reverse the goods movement. After you use FM BAPI_GOODSMVMT_GETITEMS (FM 574) to get the material document you want to change, use this FM to cancel the goods movement.

[↓] Import

MATERIALDOCUMENT

Material document number. Type: BAPI2017_GM_HEAD_02-MAT_DOC.

MATDOCUMENTYEAR

Material document year. Type: BAPI2017_GM_HEAD_02-DOC_YEAR.

GOODSMVT_PSTNG_DATE

Posting date is the date on which the document is entered in FI. Type: BAPI2017_GM_HEAD_02-PSTNG_DATE.

GOODSMVT_PR_UNAME

User name is required if you want to print a goods receipt when the material is posted. Type: BAPI2017_GM_HEAD_01-PR_UNAME.

[↑] Export

GOODSMVT_HEADRET

Material document number and material document year are returned in this parameter. Type: BAPI2017_GM_HEAD_RET.

Tables

RETURN

Return messages. Type: BAPIRET2.

GOODSMVT_MATDOCITEM

Items details that you want to reverse. If this table isn't filled, the system cancels all of the items in the material document. Type: BAPI2017_GM_ITEM_04.

573 BAPI_GOODSMVT_CREATE

This FM is used to post the goods movement (i.e., material documents can be created for the goods movement by calling this FM). You should be sure to pass the data in the same way it should be saved in the database; for example, a material number should be 18 characters with leading zeros, batches should be in uppercase letters, and so on. You can post goods movement codes 01 through 07.

For details of which field should be populated for which goods movement code, refer to the FM documentation in the SAP system.

Example: To move a material from one location to another, pass the parameter in the FM and call the FM for creation of the goods movement.

[↓] Import

GOODSMVT_HEADER

Material document header data. Type: BAPI2017_GM_HEAD_01.

GOODSMVT_CODE

Assign code to transaction for goods movement. Type: BAPI2017_GM_CODE.

TESTRUN

Simulate goods movement. Type: BAPI2017_GM_GEN-TESTRUN.

GOODSMVT_REF_EWM

BAPI communication structure: material document EWM reference. Type: /SPE/BAPI2017_GM_REF_EWM.

[↑] Export

GOODSMVT_HEADRET

Material document number/material document year. Type: BAPI2017_GM_HEAD_RET.

MATERIALDOCUMENT

Number of material document. Type: BAPI2017_GM_HEAD_RET-MAT_DOC.

MATDOCUMENTYEAR

Material document year. Type: BAPI2017_GM_HEAD_RET-DOC_YEAR.

Tables

GOODSMVT_ITEM

Material document items. Type: BAPI2017_GM_ITEM_CREATE.

GOODSMVT_SERIALNUMBER

Serial number. Type: BAPI2017_GM_SERIALNUMBER.

RETURN

Return messages. Type: BAPIRET2.

GOODSMVT_SERV_PART_DATA

This parameter contains the fields of the "Service Parts for SPM" scenario. Type: BAPIPAREX /SPE/BAPI2017_SERVICEPART_DATA.

EXTENSIONIN

Reference structure for BAPI parameters EXTENSIONIN/EXTENSIONOUT.

GOODSMVT_ITEM_CWM

BAPI communication structure: create item data for CW. Type: /CWM/BAPI2017_GM_ITEM_CREATE.

---

574 BAPI_GOODSMVT_GETITEMS

This FM reads and displays the header and item data of all material documents that meet the selection criteria entered in a ranges table. Authorization objects M_MSEG_WMB and M_MSEG_BMB are checked for reading material document from this FM.

Example: You want to cancel the goods movement. You need to read the goods movement and get its details before you cancel it. Pass the parameter in this FM and call this FM to do so.

Tables

MATERIAL_RA

Range for material. Type: BAPI2017_GM_MATERIAL_RA.

PLANT_RA

Range for plant. Type: BAPI2017_GM_PLANT_RA.

STGE_LOC_RA

Range for storage location. Type: BAPI2017_GM_STGE_LOC_RA.

BATCH_RA

Range for batch. Type: BAPI2017_GM_BATCH_RA.

MOVE_TYPE_RA

Range for movement type. Type: BAPI2017_GM_MOVE_TYPE_RA.

SPEC_STOCK_RA

Range for special stock indicator. Type: BAPI2017_GM_SPEC_STOCK_RA.

TR_EV_TYPE_RA

Range for transaction or event type. Type: BAPI2017_GM_TR_EV_TYPE_RA.

PSTNG_DATE_RA

Range for posting date. Type: BAPI2017_GM_PSTNG_DATE_RA.

VENDOR_RA

Range for vendor. Type: BAPI2017_GM_VENDOR_RA.

USERNAME_RA

Range for user name. Type: BAPI2017_GM_USERNAME_RA.

GOODSMVT_HEADER

This table contains the header data for the material document. Type: BAPI2017_GM_HEAD_02.

GOODSMVT_ITEMS

This table contains the item data for the material document. Type: BAPI2017_GM_ITEM_SHOW.

RETURN

Return messages. Type: BAPIRET2.

GOODSMVT_ITEMS_CWM

BAPI communication structure: display CW item data. Type: /CWM/BAPI2017_GM_ITEM_SHOW.

---

575 BAPI_HELPVALUES_GET

This FM tells you what the valid F4 help is for a specific parameter or field of BAPI structure parameters. The help values are based on check tables, fixed domain values, or value helps defined in a data element. If no search help of this type is defined, no help values can be specified.

This method can only be used for BAPI parameters. A check is made to see whether the field exists in the named parameter, in the method, and in the business object.

Example: You want to get the F4 help for the country code (field NATIONAL). This field is present in the BAPI EMPLOYEE_GETLIST in the personal DATA parameter. You can pass the parameters as OBJ TYPE as EMPLOYEE, OBJ NAME as GET LIST, PARAMETER as PERSONALDATA, and FIELD as NATIONAL; then call this FM for the F4 help.

[↓] Import

OBJTYPE

Object type. This parameter is the key field for identifying the business object. Either OBJTYPE or OBJNAME is to be specified to identify the business object; for example, object type BUS1065 is for object EMPLOYEE. Type: BAPIF4A-OBJTYPE.

OBJNAME

Object name in the BOR. The business object name is the description of the business object. The object name is case sensitive so you should enter the name exactly as it appears. Either OBJTYPE or OBJNAME is to be specified to identify the business object; for example, object type BUS1065 is for object EMPLOYEE. Type: BAPIF4A-OBJNAME.

METHOD

Name of the business object method, for example, GETLIST, containing the parameter and the field for which input help (F4 help) has been requested. Type: BAPIF4A-METHOD.

PARAMETER

Name of the parameter of a method, for example PERSONALDATA, containing the field for which input help (F4 help) is requested. Type: BAPIF4A-PARAM.

FIELD

Field name in the parameter for F4 values. To find a field name, find the reference structure of the parameter in the BOR. You can use the reference structure in the ABAP Data Dictionary to display information about the individual fields of the parameter.

If the method parameter that contains the field doesn't consist of a structure or table (which means it's a field), you must make sure of the following:

- The name of the field is specified in the PARAMETERfield of the `Helpvalues.GetList` method.
- The FIELD parameter of the `Helpvalues.GetList` method is set to BLANK.

Type: BAPIF4A-FIELD.

EXPLICIT_SHLP

You can transfer unique search help for a field. Only basic search help can be passed. Collective search help can't be processed by this method. Type: BAPISHLP.

MAX_OF_ROWS

Maximum number of input values for the selected fields. Type: BAPIF4A-MAX_ROWS.

DESCRIPTIONONLY

If this parameter is set to X, only the description of the help values is retrieved and returned. Type: BAPIF4A-DESCRIP.

[↑] Export

RETURN

Return messages for error. Type: BAPIRETURN.

Tables

SELECTION_FOR_HELPVALUES

You can use this parameter to restrict the hit list by specifying selection conditions. You can also make foreign key checks. Type: BAPIF4B.

HELPVALUES

Hit list matching selection criteria are returned in this parameter. Type: BAPIF4C.

VALUES_FOR_FIELD

Hit list for the field selected. Type: BAPIF4D.

DESCRIPTION_FOR_HELPVALUES

This parameter contains the metadata (description) of the values in the parameter HELPVALUES. This data is used to define the selection conditions in the parameter SELECTION_FOR_HELPVALUES. Type: BAPIF4E.

576 BAPI_HU_CREATE

This FM creates the nonassigned handling unit without stock information. The handling units may also be created with items and a serial number. A handling unit (HU) is a physical unit that consists of the packaging material and the goods in it. All of the information, such as serial numbers and batches, is available in a handling unit.

Example: You want to create a nonassigned handling unit with a packaging material PALLAT with the planned HU physical status. You can pass the relevant parameters in this FM and call this FM for creation of the HU.

[↓] Import

HEADERPROPOSAL

Default data for creating a handling unit. Type: BAPIHUHDRPROPOSAL.

[↑] Export

HUHEADER

Handling unit header data is returned in this parameter. Type: BAPIHUHEADER.

HUKEY

External handling unit identification key. You can use this key for your other handling unit functions. Type: BAPIHUKEY-HU_EXID.

Tables

ITEMSPROPOSAL

Default tables for the item to be created. Type: BAPIHUITMPROPOSAL.

ITEMSSERIALNO

Serial numbers for HU items. Type: BAPIHUITMSERIALNO.

RETURN

Return parameter. Type: BAPIRET2.

HUITEM

Table of handling unit items that were created. Type: BAPIHUITEM.

CWM_ITEMSPROPOSAL

CWM enhancement: HU items to be created. Type: /CWM/BAPIHUITMPROPOSAL.

CWM_HUITEM

CWM enhancement: structure for handling unit item. Type: /CWM/BAPIHUITEM.

577 BAPI_HU_PACK

This FM is used to pack an item into a handling unit (HU). If the handling unit already contains the material that is to be packed with the same stock category, the existing item is increased by the corresponding quantity. If the material that is to be packed doesn't exist yet in the handling, the system adds a new item.

Example: You want to pack the handling unit 1631 to the handling unit with external identification 1630. You can pass the relevant parameters in this FM and call this FM for packing the HU.

[↓] Import

HUKEY

External handling unit identification key. Type: BAPIHUKEY-HU_EXID.

ITEMPROPOSAL

Default for HU item creation. Type: BAPIHUITMPROPOSAL.

CWM_ITEMPROPOSAL

CWM: HU items to be created. Type: /CWM/BAPIHUITMPROPOSAL.

[↑] Export

HUITEM

Handling unit item that was created. Type: BAPIHUITEM.

HUHEADER

Handling unit header information is returned in this parameter. Type: BAPIHUHEADER.

CWM_HUITEM

CWM: structure for handling unit item. Type: /CWM/BAPIHUITEM.

Tables

SERIALNUMBERS

Serial numbers for a handling unit item. Type: BAPIHUITMSERIAL.

RETURN

Return parameters. Type: BAPIRET2.

---

578 BAPI_HU_UNPACK

This FM is used to unpack material from a handling unit. Lower-level handling units, material items, or auxiliary packaging materials can be unpacked. Even partial quantities can be unpacked from material items and auxiliary packaging materials.

Example: You want to unpack the handling unit 1630, which contains the handling unit 1631 and 1632. You can pass the relevant parameters in this FM and call this FM for unpacking the HU.

[↓] Import

HUKEY

External handling unit identification key. Type: BAPIHUKEY-HU_EXID.

ITEMUNPACK

Handling unit item to be unpacked. Type: BAPIHUITMUNPACK.

CWM_ITEMUNPACK

CWM: handling unit item to be unpacked. Type: /CWM/BAPIHUITMUNPACK.

[↑] Export

HUHEADER

Handling unit header information is returned in this parameter. Type: BAPIHUHEADER.

Tables

SERIALNUMBERS

Serial numbers for a handling unit item. Type: BAPIHUITMSERIAL.

RETURN

Return parameters. Type: BAPIRET2.

579 BAPI_IBAN_CHANGE

This FM is used to change an IBAN (International Bank Account Number). IBAN is used for standardized and uniform representation of complete bank details according to the European Committee for Banking Standards (ECBS). IBAN consists of a bank country key (ISO code), two check digits, and a country-specific account number. The maximum length of IBAN is 34 characters.

Example: You want to change the details of the IBAN bank account because the customer has changed his bank. Pass the relevant parameters in this FM and call it.

[↓] Import

BANKCOUNTRY

Bank country key. This parameter identifies the country in which the bank is located. Type: BAPI1013_KEY-BANK_CTRY.

BANKKEY

Bank key that uniquely identifies the bank. Type: BAPI1013_KEY-BANK_KEY.

BANKACCOUNTNUMBER

Bank account number. Type: BAPI1013_KEY-BANK_ACCT.

BANKCONTROLKEY

Bank control key. This field contains a key for checking the combination of bank number and bank account number. For example, the field can determine whether you're dealing with a savings account or checking account. Type: BAPI1013_KEY-CTRL_KEY.

DETAIL

IBAN detail structure. Type: BAPI1013_DETAIL.

DETAILX

Change information on IBAN: detail. If you want to change any field in IBAN details, you need to mark the corresponding field as X. Type: BAPI1013_DETAILX.

I_CHECK_BEFORE_SAVE

Check only; no data is saved. Type: XFELD.

[↑] Export

RETURN

Return parameter. Type: BAPIRET2.

---

580 BAPI_INB_DELIVERY_CONFIRM_DEC

This FM is used to verify inbound delivery from a decentralized system. This method reports the inbound deliveries from a WM system back to an SAP ERP system. You must pass the header data of the delivery that is to be reported back, delivery items (i.e., the items that are changed), and the packaging data if they exist. Packaging data that isn't reported back is deleted.

Example: You've received the raw material for your manufacturing product. You need to verify the raw material and report back the inbound delivery of your raw material from WM to the SAP ERP system. You can pass the relevant parameters in this FM and call this FM for reporting.

[↓] Import

HEADER_DATA

Header data of the inbound delivery. Type: BAPIIBDLVHDRCON.

HEADER_CONTROL

This parameter contains the control data for confirmation header data. Type: BAPIIBDLVHDRCTRLCON.

DELIVERY

Inbound delivery number. Type: BAPIIBDLVHDRCON-DELIV_NUMB.

TECHN_CONTROL

Technical control data, for example, you can set the DEBUG_FLG as X for the abnormal behavior in the receiving system so that you can reimport the IDoc into the receiving system with the respective BREAKPOINTS. Type: BAPIDLVCONTROL.

HEADER_CONTROL_SPL

Additional control data for the completion confirmation of the header data. Type: /SPE/BAPIIBDLVHDRCTRLCONF.

Tables

HEADER_PARTNER

This parameter is used to enter changes to partners. If business partners aren't entered as mandatory partners in the delivery, you can delete business partners by making an entry in the HEADER_PARTNER table with indicator UPD_MODE_PARTN = D.

You can change the data for existing business partners by making an entry in the HEADER_PARTNER table with indicator UPD_MODE_PARTN = U. Type: BAPIDLVPARTNERCHG.

HEADER_PARTNER_ADDR

This parameter is used for entering an address change for a partner. You must set the UPD_MODE_ADR indicator to U if you want to change the address. Type: BAPIDLVPARTNADDRCHG.

HEADER_DEADLINES

Header deadlines are passed in this parameter. Type: BAPIDLVDEADLN.

ITEM_DATA

Item data. Type: BAPIIBDLVITEMCON.

ITEM_CONTROL

Control data confirmation item data. Type: BAPIIBDLVITEMCTRLCON.

ITEM_SERIAL_NO

Entry of serial numbers for delivery items.

For materials with a serial number requirement, you must enter the serial numbers here. The system replaces the serial numbers assigned to the delivery item with the new serial numbers.

If you don't enter any serial numbers for the delivery item, the system deletes any serial numbers that had already been assigned. Type: BAPIDLVITMSERNO.

HANDLING_UNIT_HEADER

Handling unit header data. Type: BAPIDLVHDUNHDR.

HANDLING_UNIT_ITEM

Handling unit item data. Type: BAPIDLVHDUNITM.

HANDLING_UNIT_SERNO

Handling unit serial numbers. Type: BAPIDLVHDUNSERNO.

EXTENSION1

Customer enhancement segment with unstructured data. Type: BAPIEXTC.

EXTENSION2

Customer enhancement segment with structured data. Type: BAPIEXT.

RETURN

Return parameters. Type: BAPIRET2.

TOKENREFERENCE

The partner system uses this parameter to deliver the token reference for the CSL. The token is used to issue authorization to change deliveries across systems. The delivery is blocked in one system if the partner system opens the delivery in the change transaction, or if changes still need to be copied in the partner system to achieve a consistent status between the two systems. Type: BAPITOKENREFERENCE.

ITEM_DATA_SPL

Inbound delivery additional item data. Type: /SPE/BAPIIBDLVITEMCONF.

CWM_ITEM_DATA

BAPI communication structure: addition for CWM. Type: /CWM/BAPIOBDLVITEM.

CWM_HU_ITEM

Handling unit item: addition for CWM. Type: /CWM/BAPIDLVHDUNITM.

---

581 BAPI_INSPECTIONPLAN_CREATE

This FM creates an inspection plan and its subobjects, such as inspection characteristics. You can also create a key date view and the complete history of an inspection plan with this FM.

Example: You want to create an inspection plan with operations and inspection characteristics for a key date of 16.11.2013. You can pass the relevant parameter in this FM and can call this FM for the same purpose.

[↓] Import

TESTRUN

You can use this parameter to define whether the transferred data is only checked for consistency or also saved to the database if the data is consistent. If the value of this parameter is Space, data is saved to the database as well; otherwise, if this parameter has the value X, it's only checked. Type: BAPIFLAG.

PROFILE

Profile to be used in the inspection plan. Type: BAPI1012_CONTROL_DATA-PROFILE.

[↑] Export

GROUP

When data is transferred successfully, the ID of the created inspection plan is displayed in this parameter. You can use this ID later to access the inspection plan. Type: BAPI1191_TSK_C-TASK_LIST_GROUP.

GROUPCOUNTER

This parameter returns the group counter for the created inspection plan. Type: BAPI1191_TSK_C-GROUP_COUNTER.

Tables

TASK

Task list header data to be created are transferred in this parameter. Fields TASK_LIST_USAGE, PLANT, TASK_LIST_STATUS, and TASK_MEASURE_UNIT must be passed in this table. Type: BAPI1191_TSK_C.

MATERIALTASKALLOCATION

Material task list assignment to be created in the inspection plan is passed in this parameter. Field material and plant must be passed in this table. Task list group can only be specified with the external assignment. For internal assignments, the task list group is programmatically determined. Type: BAPI1191_MTK_C.

OPERATION

The data are transferred for the operations to be created. Create one entry in this table for each operation. Operation ID must be unique to identify the operation. Fields ACTIVITY, CONTROL_KEY, WORK_CNTR, and PLANT must be passed. The task list group can only be specified with the external assignment. For internal assignment, the task list group is programmatically determined. Type: BAPI1191_OPR_C.

REFERENCEOPERATION

In this table, the data are transferred for the reference operation set references that are to be created. You specify the reference operation set to be referenced in the fields REFERENCED_TASK_LIST_GROUP (task list group) and REFERENCED_GROUP_COUNTER (group counter).

You specify the operation number for the reference operation within the inspection plan in the field ACTIVITY. Type: BAPI1191_REF_OPR_C.

PRODUCTIONRESOURCE

This table passed the data for the production resource/tool assignments. Type: BAPI1191_PRT_C.

INSPCHARACTERISTIC

Inspection characteristics data to be created are passed in this table. Inspection characteristic numbers can be passed in field INSPCHAR. Type: BAPI1191_CHA_C.

TEXTALLOCATION

Data from this table can be used to assign long texts when creating or changing inspection plans. You need to define the respective object ID for which you need to assign the long text in attribute OBJECT_TYPE. Possible values:

- 10: Inspection plan header.
- 13: Operation.
- 15: Reference for reference operation set.
- 17: Production resource/tool.
- 19: Inspection characteristics.

Type: BAPI1191_TXT_HDR_C.

TEXT

Long text lines for inspection plans.
Long text are determined by field LINE_FROM and LINE_TO of parameter TEXTALLOCATION.
Type: BAPI1012_TXT_C.

RETURN

Return parameter. Type: BAPIRET2.

INSP_CHAR_VALUES

The dependent characteristic specifications for the task list to be created are transferred in this table. Fields OPERTION_ID, INSPCHAR, and NO_OF_SPEC_VALUE must be passed. Type: BAPI1191_CHV_C.

582 BAPI_INTERNALORDER_CREATE

This FM creates an internal order from transferred data. An internal order is a virtual place for collecting the cost of a particular task or service. You can transfer master data and settlement rules in this FM to create an internal order.

Example: One department is short of some engines for its product and asks for another department to provide the engines. For this, create an internal order by passing the parameter and calling this FM.

[↓] Import

I_MASTER_DATA

Master data of internal order to be created. Type: BAPI2075_7.

TESTRUN

Switch to simulation session for write BAPIs. Type: BAPI0012_GEN-TESTRUN.

[↑] Export

E_MASTER_DATA

Return structure that contains the master data of the order created. Type: BAPI2075_2.

ORDERID

Order number of the internal order created. Type: BAPI2075_2-ORDER.

Tables

SRULES

Settlement rules from which the order is created. Type: BAPI2075_6.

RETURN

Table with return information. Type: BAPIRET2.

EXTENSIONIN

You can use this parameter to transfer user-defined fields from the order master records. However, no fields are supported that are updated in user-defined tables. Only the user-defined fields created within the framework of the SAP enhancement COOPA003 that are updated automatically in table AUFK are supported. This means you must use the table extension BAPI_TE_AUFK. Type: BAPIPAREX.

583 BAPI_MATERIAL_AVAILABILITY

This FM determines the receipt quantity that is still available for a material in a plant according to available-to-promise logic. The availability check is carried out on transferring the material number, plant, and input table WMDVSX.

Example: You want to know if the product is available or not. You can pass the relevant parameter in this FM and call this FM to get the ATP quantity.

[↓] Import

PLANT

Plant. Type: BAPIMATVP-WERKS.

MATERIAL

Material number. Type: BAPIMATVP-MATNR.

UNIT

Unit of measure for display. Type: BAPIADMM-UNIT.

CHECK_RULE

The check is made against the checking rule about which stock or receipts are to be included. Type: BAPIT441V-PRREG.

STGE_LOC

Storage location. Type: BAPICM61V-LGORT.

BATCH

Batch. Type: BAPICM61V-CHARG.

CUSTOMER

Customer number. Type: BAPIKNVVKY-CUSTOMER.

DOC_NUMBER

Document number. Type: BAPISDHEAD-DOC_NUMBER.

ITM_NUMBER

Item number. Type: BAPISDITM-ITM_NUMBER.

WBS_ELEM

WBS element. Type: BAPISDITM-WBS_ELEM.

STOCK_IND

Special stock indicator. Type: BAPICM61V-SOBKZ.

DEC_FOR_ROUNDING

Number of decimal places to which rounding should be performed. Type: BAPICM61M-ANDEC.

DEC_FOR_ROUNDING_X

Updated information in related user data field. Type: BAPIUPDATE.

READ_ATP_LOCK

Control indicator for availability check. Type: BAPICM61V-VFPST.

READ_ATP_LOCK_X

Updated information in related user data field. Type: BAPIUPDATE.

MATERIAL_EVG

This parameter is used to transfer long material numbers that are allowed to have a maximum length of 40 characters. This parameter can only be used when the business function LAMA (i.e., long material number) is called. Type: BAPIMGVMATNR.

[↑] Export

ENDLEADTME

End of replenishment lead time. Type: BAPICM61M-WZTER.

AV_QTY_PLT

Quantity available at plant level. Type: BAPICM61V-WKBST.

DIALOGFLAG

This parameter can have the following values:

- [Space]: Quantity completely available.
- X: Partial or no quantity is available.
- N: Material isn't included in the availability check.

Type: BAPICM61V-DIAFL.

RETURN

Return table with error messages. Type: BAPIRETURN.

Tables

WMDVSX

Input table with date and quantity to be checked. Type: BAPIWMDVS.

WMDVEX

Output table with date and available receipt quantity.

If no date and no quantity is transferred, the system displays the ATP situation from today's date into the future as the result.

If only a date and no quantity is transferred, the system displays the ATP situation from the corresponding date as the result.

If both a date and a quantity are transferred, the system calculates the availability situation for the quantity specified. Type: BAPIWMDVE.

584 BAPI_MATERIAL_BOM_GROUP_CREATE

This FM creates a material BOM along with its subobjects, for instance, all variants of a variant BOM or an alternative of a material BOM, material history, material assignment, all items, and subitems.

Example: You want to transfer the BOM group and its complete history from the legacy system to a SAP system (i.e., you want all of the subitems and objects, variants or alternatives, and their history to be transferred and created). You can pass the relevant parameters in this FM and can call this FM to create the material BOM group.

[↓] Import

TESTRUN

Switch to the simulation session for writing BAPIs. Type: BAPIFLAG.

ALL_ERROR

If this parameter is set to X, all errors are returned in the RETURN parameter. Type: XFLAG.

[↑] Export

BOMGROUP

Basic data of the BOM group is transferred in this table. Type: BAPI1080_BGR_C.

VARIANTS

Alternative and variant specific data is transferred in this table. Field BOM_GROUP_IDENTIFICAION of BOMGROUP must have the value to assign BOM group items. Type: BAPI1080_BOM_C.

ITEMS

This table contains the BOM item data. Field BOM_GROUP_IDENTIFICAION of BOMGROUP must have the value to assign BOM group items. Type: BAPI1080_ITM_C.

SUBITEMS

Subitem data of BOM is passed in this table. Type: BAPI1080_SUI_C.

MATERIALRELATIONS

This structure contains the assignment of the assembly material to the BOM group. Field MATERIAL and the alternative BOM must be passed in this table. Type: BAPI1080_MBM_C.

ITEMASSIGNMENTS

This table contains the assignment of items to the variant and alternative BOM. Type: BAPI1080_REL_ITM_BOM_C.

SUBITEMASSIGNMENTS

This table contains the assignment of the subitems to the items. Type: BAPI1080_REL_SUI_ITM_C.

TEXTS

This table contains the long text for the objects. Type: BAPI1080_TXT_C.

RETURN

Return parameter with error messages received from this FM. Type: BAPIRET2.

585 BAPI_MATERIAL_DELETE

This FM is used to flag a material for deletion or to cancel the deletion flag of a material.

Example: To mark the material for deletion, pass the relevant parameters in this FM and call this FM.

[↓] Import

MATERIAL

Material to be flagged for deletion. Type: BAPIMATALL-MATERIAL.

MATERIAL_EVG

This parameter can be used to transfer long material numbers that can contain a maximum of 40 characters. To use this parameter, you must activate the switch for the LAMA (long material number) business function in the system. The LAMA business function is available as part of the business function set for DIMP. Type: BAPIMGVMATNR.

[↑] Export

RETURN

Return parameter. Type: BAPIRET1.

---

586 BAPI_MATERIAL_EDIT

This FM is used to edit a material. This FM calls Transaction MM02 or Transaction MM42 to change the material.

Example: You have a report that includes all of the materials. On selecting one material, you want to edit it. You can pass the parameter in this FM and call this FM to open the material for editing via one of the transaction codes.

[↓] Import

MATERIAL

Material to be changed. Type: BAPIMATALL-MATERIAL.

SKIP_1ST_SCREEN

Skip initial screen of Transaction MM02 or Transaction MM42. The default value is X. Type: BAPIMATALL-SKIP_1ST_SCREEN.

MATERIAL_EVG

This parameter can be used to transfer long material numbers that can contain a maximum of 40 characters. To use this parameter, you must activate the switch for the LAMA (long material number) business function in the system. The LAMA business function is available as part of the business function set for DIMP. Type: BAPIMGVMATNR.

[↑] Export

RETURN

Return parameter. Type: BAPIRET1.

587 BAPI_MATERIAL_GET_ALL

This FM is the follow-on method from BAPI_MATERIAL_GET_DETAIL (FM 588). In contrast to the original method, this method can read all of the material master data for a specified combination of organizational units. Depending on the organizational unit entered, the system returns the corresponding material master data. If you only enter the material number (the only required entry parameter), the method only returns client-specific material data for the material entered.

Example: To get the details of many materials that have different company codes, pass the relevant parameter in this FM and call this FM.

[↓] Import

MATERIAL

Material number. Type: BAPI_MARA_GA-MATERIAL.

COMP_CODE

Company code. Type: BAPI0002_1-COMP_CODE.

VAL_AREA

Valuation area is the organization level at which the material is valuated. Valuation area can be a plant or a company code. Type: BAPI_MBEW_GA-VAL_AREA.

VAL_TYPE

Valuation type identifies separately valuated stocks of material. Type: BAPI_MBEW_GA-VAL_TYPE.

PLANT

Plant. Type: BAPI_MARC_GA-PLANT.

STGE_LOC

Storage location. Type: BAPI_MARD_GA-STGE_LOC.

SALESORG

Sales organization is the organization unit responsible for the sale of a certain product or services. Type: BAPI_MVKE_GA-SALES_ORG.

DISTR_CHAN

Distribution channel is the way the product reaches the customer such as wholesale, retail, or direct sale. Type: BAPI_MVKE_GA-DISTR_CHAN.

WHSENUMBER

Warehouse number. Type: BAPI_MLGN_GA-WHSE_NO.

STGE_TYPE

Storage type is the subdivision of the physical warehouse. Type: BAPI_MLGT_GA-STGE_TYPE.

LIFO_VALUATION_LEVEL

LIFO valuation level. Type: BAPI_MYMS_GA-LIFO_VALUATION_LEVEL.

KZRFB_ALL

This parameter resets the buffer for material prefetch and will be fetched again. Type: MTCOM-KZRFB.

[↑] Export

CLIENTDATA

Material data at the client level is returned. Type: BAPI_MARA_GA.

PLANTDATA

Material data at the plant level is returned. Type: BAPI_MARC_GA.

FORECASTPARAMETERS

This structure is used to return the plant-specific forecast data for a material to the calling program. Type: BAPI_MPOP_GA.

PLANNINGDATA

This structure is used to return the plant-specific planning data for a material to the calling program. Type: BAPI_MPGD_GA.

STORAGELOCATIONDATA

Storage location specific data is returned in this parameter. Type: BAPI_MARD_GA.

VALUATIONDATA

This structure returns the valuation area specific and valuation type specific data for a material. Type: BAPI_MBEW_GA.

WAREHOUSENUMBERDATA

Warehouse number data for each warehouse for a material is returned in this parameter. Type: BAPI_MLGN_GA.

SALESDATA

Sales data of a material is returned in this parameter. Type: BAPI_MVKE_GA.

STORAGETYPEDATA

Storage type data of a material is returned in this parameter. Type: BAPI_MLGT_GA.

PRTDATA

Production resource tool (PRT) fields for a material. Type: BAPI_MFHM_GA.

LIFOVALUATIONDATA

LIFO-relevant material data is returned in this parameter. Type: BAPI_MYMS_GA.

IS_MP_CLIENTDATA

Material data on the client level. Type: /SAPMP/BAPI_MARA.

Tables

MATERIALDESCRIPTION

Material descriptions are returned in this table. Type: BAPI_MAKT_GA.

UNITSOFMEASURE

Units of measure are returned in this table. Type: BAPI_MARM_GA.

INTERNATIONALARTNOS

International article numbers (EANs) of a material are returned in this table. Type: BAPI_MEAN_GA.

MATERIALLONGTEXT

Long texts. Type: BAPI_MLTX_GA.

TAXCLASSIFICATIONS

Tax data of the material is returned in this parameter. Type: BAPI_MLAN_GA.

EXTENSIONOUT

Reference structure for BAPI parameters EXTENSIONIN/EXTENSIONOUT. Type: BAPIPAREX.

RETURN

Return parameters. Type: BAPIRETURN.

NFMCHARGEWEIGHTS

Structural weight for a material is returned in this parameter. Type: /NFM/BAPITVGW.

NFMSTRUCTURALWEIGHTS

Charge weights for materials on the organization level are returned in this parameter. Type: /NFM/BAPITKGW.

---

588 BAPI_MATERIAL_GET_DETAIL

This FM reads and returns the details of a material. If only the material number is provided, only client-specific details are returned. If you want other details, such as plant-specific details and valuation-specific details, you need to pass the parameters for plant, valuation area, and valuation type, respectively.

Example: To get the details of the material such as its general data, valuation, or plant data, pass the relevant parameter in this FM and call this FM.

[↓] Import

MATERIAL

Material number. Type: BAPIMATDET-MATERIAL.

PLANT

Plant. Type: BAPIMATALL-PLANT.

VALUATIONAREA

Valuation area is the organization level at which the material is valuated. Valuation area can be a plant or a company code. Type: BAPIMATALL-VAL_AREA.

VALUATIONTYPE

Valuation type identifies separately valuated stocks of material. Type: BAPIMATALL-VAL_TYPE.

MATERIAL_EVG

This parameter can be used to transfer long material numbers that can contain a maximum of 40 characters. To use this parameter, you must activate the switch for the LAMA (long material number) business function in the system. The LAMA business function is available as part of the business function set for DIMP. Type: BAPIMGVMATNR.

[↑] Export

MATERIAL_GENERAL_DATA

General material data is returned in this parameter. Type: BAPIMATDOA.

RETURN

Return parameter with error messages. Type: BAPIRETURN.

MATERIALPLANTDATA

Plant-specific data of a material is returned in this parameter. Type: BAPIMATDOC.

MATERIALVALUATIONDATA

Valuation area-specific data of a material is returned in this parameter. Type: BAPIMATDOBEW.

589 BAPI_MATERIAL_GETLIST

This FM filters and returns the list of materials with material descriptions for a logon language based on the specified selection criteria.

Example: You have plant and material fields in a selection screen, and you want to display the F4 help for the material present in the plant. You can pass the plant and other parameter in this FM as a ranges table for selection criteria, and you'll get the list of materials. A description of the material is returned in the logon language.

[↓] Import

MAXROWS

Maximum number of materials to be selected. Type: BAPIF4A-MAX_ROWS.

Tables

MATNRSELECTION

Selection options for material number. Type: BAPIMATRAM.

MATERIALSHORTDESCSEL

Selection options for material description. Type: BAPIMATRAS.

MANUFACTURERPARTNUMB

Material is determined from the manufacturer number and manufacturing part number. Field MANU_MAT contains the manufacturer part number, and field MFR_NO contains the manufacturer number. Type: BAPIMATMFRPN.

PLANTSELECTION

Selection options for plants. Type: BAPIMATRAW.

STORAGELOCATIONSELECT

Selection options for storage locations. Type: BAPIMATRAL.

SALESORGANISATIONSELECTION

Selection options for sales organization. Type: BAPIMATRASO.

DISTRIBUTIONCHANNELSELECTION

Selection options for distribution channel. Type: BAPIMATRADC.

MATNRLIST

List of material numbers with material description are returned in this parameter. Type: BAPIMATLST.

RETURN

Return parameter with messages. Type: BAPIRET2.

590 BAPI_MATERIAL_SAVEDATA

This FM is used to create new material master data or change existing material master data and also save the changes. When creating material master data, you must transfer the material number, the material type, and the industry sector to the method. You must also enter a material description and its language.

When changing material master data, you need enter only the material number. In the header data, you must select at least one view for which data is to be created. Depending on the view selected, you must maintain other required parameters. If you don't enter values for all of the required parameters, the method is ended with an error message.

Example: You want to change the existing material programmatically. You can pass the material number in this FM along with the other details and call this FM to change the material details.

[↓] Import

HEADDATA

The fundamental data necessary for maintaining a material is defined in this structure. The fields essentially correspond to the fields available on the initial screen in dialog maintenance when creating a material master record.

When creating material master data, you can create or change only single materials in this structure. Type: BAPIMATHEAD.

CLIENTDATA

This structure is used to transfer the general basic data on a material. Type: BAPI_MARA.

CLIENTDATAX

This (checkbox) structure determines which fields in the assigned user data structure (CLIENTDATA) contain change-relevant information.

This makes it possible to supply valid values only to the relevant fields in the assigned user data structure. These fields are then selected in the checkbox structure, ensuring that all nonselected fields retain their current value in the database. Type: BAPI_MARAX.

PLANTDATA

Plant-specific material data. Type: BAPI_MARC.

PLANTDATAX

Information on update for PLANTDATA. Type: BAPI_MARCX.

FORECASTPARAMETERS

Forecast parameters. Type: BAPI_MPOP.

FORECASTPARAMETERSX

Information on update for FORECASTDATA. Type: BAPI_MPOPX.

PLANNINGDATA

Planning data. Type: BAPI_MPGD.

PLANNINGDATAX

Information on update for PLANNINGDATA. Type: BAPI_MPGDX.

STORAGELOCATIONDATA

Storage-location-specific material data. Type: BAPI_MARD.

STORAGELOCATIONDATAX

Information on update for STORAGELOCATIONDATA. Type: BAPI_MARDX.

VALUATIONDATA

Valuation data. Type: BAPI_MBEW.

VALUATIONDATAX

Information on update for VALUATIONDATA. Type: BAPI_MBEWX.

WAREHOUSENUMBERDATA

Warehouse-number-specific material data. Type: BAPI_MLGN.

WAREHOUSENUMBERDATAX

Information on update for WAREHOUSEDATA. Type: BAPI_MLGNX.

SALESDATA

Sales data. Type: BAPI_MVKE.

SALESDATAX

Information on update for SALESDATA. Type: BAPI_MVKEX.

STORAGETYPEDATA

Storage-type-specific material data. Type: BAPI_MLGT.

STORAGETYPEDATAX

Information on update for STORAGETYPEDATA. Type: BAPI_MLGTX.

FLAG_ONLINE

No ALE field selection. Type: BAPIE1GLOBAL_DATA-TESTRUN.

FLAG_CAD_CALL

Call from CAD system. Type: BAPIE1GLOBAL_DATA-TESTRUN.

NO_DEQUEUE

Screens for displaying user entry. Type: BAPIE1GLOBAL_DATA-TESTRUN.

NO_ROLLBACK_WORK

Override rollback if error occurs. Type: BAPIE1GLOBAL_DATA-TESTRUN.

[↑] Export

RETURN

Return parameters with error messages. Type: BAPIRET2.

Tables

MATERIALDESCRIPTION

Material descriptions. Type: BAPI_MAKT.

UNITSOFMEASURE

Units of measure for the material is transferred in this table. Type: BAPI_MARM.

UNITSOFMEASUREX

Information on update for UNITSOFMEASURE. Type: BAPI_MARMX.

INTERNATIONALARTNOS

International article numbers (EANs) of a material. Type: BAPI_MEAN.

MATERIALLONGTEXT

Long texts of a material. Type: BAPI_MLTX.

TAXCLASSIFICATIONS

Tax classifications of a material. Type: BAPI_MLAN.

RETURNMESSAGES

All messages are displayed in this table. Type: BAPI_MATRETURN2.

PRTDATA

Production resource/tool (PRT) fields in the material master. Type: BAPI_MFHM.

PRTDATAX

Information on update for PRTs. Type: BAPI_MFHMX.

EXTENSIONIN

This parameter is used to transfer the material's customer-defined fields. Type: BAPIPAREX.

EXTENSIONINX

Changes in data to be saved in the database for the customer-defined fields. Type: BAPIPAREXX.

NFMCHARGEWEIGHTS

Structural weight for a material is returned in this parameter. Type: /NFM/BAPITVGW.

NFMCHARGEWEIGHTSX

Information on update for NFMCHARGEWEIGHTS. Type: /NFM/BAPITVGWX.

NFMSTRUCTURALWEIGHTS

Charge weights for materials on the organization level are returned in this parameter. Type: /NFM/BAPITKGW.

NFMSTRUCTURALWEIGHTSX

Information on update for NFMSTRUCTURALWEIGHTS. Type: /NFM/BAPITKGWX.

591 BAPI_MATERIAL_SAVEREPLICA

This FM is used for mass maintenance or creation of material data. When new material master data is created, the material number, the material type, and the sector must be sent to the method. Furthermore, a short text and the language in which the short text has been created must be entered.

When data is being changed, only the material number needs to be entered. In the header data, at least one view has to be selected for which the data is to be created. Depending on the selected view, additional mandatory parameters that have been defined as such in customizing have to be created. If not all mandatory fields are filled with data, the method ends with an error message being displayed.

Example: You've made changes on many materials that belong to different plants. To save those changes in the ABAP Data Dictionary, pass the parameter in this FM.

[↓] Import

NOAPPLLOG

This checkbox structure can be used to specify whether an application log is generated or not. Type: BAPIE1GLOBAL_DATA-NO_APPL_LOG.

NOCHANGEDOC

This checkbox structure can be used to specify whether change documents should be created during processing of material data. Type: BAPIE1GLOBAL_DATA-NO_CHANGE_DOC.

TESTRUN

Switch to the simulation session for write BAPIs. Type: BAPIE1GLOBAL_DATA-TESTRUN.

INPFLDCHECK

This parameter should be set to X if you want to receive a message if there are fields that aren't ready for input. Type: BAPIE1GLOBAL_DATA-INP_FLD_CHECK.

FLAG_CAD_CALL

If this BAPI is called from a CAD system, this parameter must be set. Type: BAPIE1GLOBAL_DATA-TESTRUN.

NO_ROLLBACK_WORK

This parameter must be set to X if you want to override rollback work. Type: BAPIE1GLOBAL_DATA-TESTRUN.

FLAG_ONLINE

If this parameter is set to X, you can't select fields from ALE. Type: BAPIE1GLOBAL_DATA-TESTRUN.

[↑] Export

RETURN

Return parameter. Type: BAPIRET2.

Tables

HEADDATA

The fundamental data necessary for maintaining a material is defined in this structure.

The fields essentially correspond to the fields available on the initial screen in dialog maintenance when creating a material master record.

When creating material master data, you can create or change only single materials in this structure. Type: BAPIMATHEAD.

CLIENTDATA

This structure is used to transfer the general basic data on a material. Type: BAPI_MARA.

CLIENTDATAX

This (checkbox) structure determines which fields in the assigned user data structure (CLIENTDATA) contain change-relevant information.

This makes it possible to supply valid values only to the relevant fields in the assigned user data structure. These fields are then selected in the checkbox structure, ensuring that all nonselected fields retain their current value in the database. Type: BAPI_MARAX.

PLANTDATA

Plant-specific material data. Type: BAPI_MARC.

PLANTDATAX

Information on update for PLANTDATA. Type: BAPI_MARCX.

FORECASTPARAMETERS

Forecast parameters. Type: BAPI_MPOP.

FORECASTPARAMETERSX

Information on update for FORECASTDATA. Type: BAPI_MPOPX.

PLANNINGDATA

Planning data. Type: BAPI_MPGD.

PLANNINGDATAX

Information on update for PLANNINGDATA. Type: BAPI_MPGDX.

STORAGELOCATIONDATA

Storage-location-specific material data. Type: BAPI_MARD.

STORAGELOCATIONDATAX

Information on update for STORAGELOCATIONDATA. Type: BAPI_MARDX.

VALUATIONDATA

Valuation data. Type: BAPI_MBEW.

VALUATIONDATAX

Information on update for VALUATIONDATA. Type: BAPI_MBEWX.

WAREHOUSENUMBERDATA

Warehouse-number-specific material data. Type: BAPI_MLGN.

WAREHOUSENUMBERDATAX

Information on update for WAREHOUSEDATA. Type: BAPI_MLGNX.

SALESDATA

Sales data. Type: BAPI_MVKE.

SALESDATAX

Information on update for SALESDATA. Type: BAPI_MVKEX.

STORAGETYPEDATA

Storage-type-specific material data. Type: BAPI_MLGT.

STORAGETYPEDATAX

Information on update for STORAGETYPEDATA. Type: BAPI_MLGTX.

MATERIALDESCRIPTION

Material descriptions. Type: BAPI_MAKT.

UNITSOFMEASURE

Units of measure for the material is transferred in this table. Type: BAPI_MARM.

UNITSOFMEASUREX

Information on update for UNITSOFMEASURE. Type: BAPI_MARMX.

INTERNATIONALARTNOS

International article numbers (EANs) of a material. Type: BAPI_MEAN.

MATERIALLONGTEXT

Long texts of a material. Type: BAPI_MLTX.

TAXCLASSIFICATIONS

Tax classifications of a material. Type: BAPI_MLAN.

PRTDATA

Production resource/tool (PRT) fields in the material master. Type: BAPI_MFHM.

PRTDATAX

Information on update for PRTs. Type: BAPI_MFHMX.

EXTENSIONIN

This parameter is used to transfer the material's customer-defined fields. Type: BAPIPAREX.

EXTENSIONINX

Changes in data to be saved in database for the customer-defined fields. Type: BAPIPAREXX.

FORECASTVALUES

Data transfer for forecast values. Type: BAPIE1MPRW.

UNPLNDCONSUMPTION

Data transfer for unplanned consumption. Type: BAPIE1MVEU.

TOTALCONSUMPTION

Data transfer for total consumption of material. Type: BAPIE1MVEG.

RETURNMESSAGES

Substitute structure for return parameter BAPIRET2. Type: BAPIE1RET2.

592 BAPI_MATERIAL_STOCK_REQ_LIST

This FM is used to select the current stock or requirement list of a material. The stock, requirement, and receipt information for the material are based on MRP and master production scheduling (MPS). This FM can be used for detailed display of the MRP elements, single line display of the requirements and stock list, and period total display of the requirement and stock list.

Example: You've forecasted the requirement of the S series model of your car from MRP and MPS. You've displayed the list in a program. The end user wants to select one entry from the list and see the details of that requirement entry. To add that functionality to the click of a display button, you need to pass the relevant parameters in this FM.

[↓] Import

MATERIAL

Material number. Type: BAPI_MRP_MAT_PARAM-MATERIAL.

PLANT

Plant. Type: BAPI_MRP_MAT_PARAM-PLANT.

MRP_AREA

MRP area is the organizational unit for which you can perform MRP. Type: BAPI_MRP_MAT_PARAM-MRP_AREA.

PLAN_SCENARIO

Planning scenario of long-term planning. This includes the plant where long-term planning is required, the planning period, and the independent requirement versions. Type: BAPI_MRP_MAT_PARAM-PLAN_SCENARIO.

SELECTION_RULE

A selection rule consists of particular time information, MRP elements, and stocks that are grouped together in a rule. The selection rule controls the selection of the data in the stock/requirements list and the planning table in repetitive manufacturing. Type: BAPI_MRP_LIST_PARAM-SELECTION_RULE.

DISPLAY_FILTER

A display filter consists of a group of certain time information, MRP elements, and MRP segments. The filter controls the display of the lists for the planning result. Type: BAPI_MRP_LIST_PARAM-DISPLAY_FILTER.

PERIOD_INDICATOR

The period indicator specifies the type of period used in the period totals display of the current requirements/stock list and MRP list of the material. Possible values:

- T: Days.
- W: Weeks.
- M: Months.
- P: Posting periods according to the fiscal year variant.
- K: Periods according to the planning calendar of the material.
- R: Individual period split of the plant.

Type: BAPI_MRP_LIST_PARAM-PERIOD_INDICATOR.

GET_ITEM_DETAILS

This indicator shows whether the current requirement/stock list or MRP list for a material should be output in the detailed display of the MRP elements. The detailed display contains the planning-relevant parameters for all MRP elements that are to be taken into account in the planning environment in question.

Possible values:

- X: The detailed display of the MRP elements should be output.
- [Space]: The detailed display of the MRP elements should not be output.

Type: BAPI_MRP_LIST_PARAM-GET_ITEM_DETAILS.

GET_IND_LINES

This indicator specifies whether the current requirements/stock list or MRP list of a material should be output in the individual line display.

The individual line display contains all of the information that is displayed in the individual line display of the corresponding dialog transactions for the display of the requirements/stock list or MRP list.

The following characteristic values are possible for this indicator:

- X: The result should be output in the individual line display.
- [Space]: The result should not be output in the individual line display.

Type: BAPI_MRP_LIST_PARAM-GET_IND_LINES.

GET_TOTAL_LINES

This indicator specifies whether the current requirements/stock list or MRP list for material should be output in the period totals display.

The period totals display contains all information that is displayed in the period totals display of the corresponding dialog transactions for the display of the requirements/stock list or MRP list. Possible values:

- X: The result should be output in the period totals display.
- [Space]: The result should not be output in the period totals display.

Type: BAPI_MRP_LIST_PARAM-GET_TOTAL_LINES.

IGNORE_BUFFER

This parameter specifies in which of the following ways the current stock/requirements list is to be determined in the method:

- The relevant MRP elements are read directly from the database.
- The buffered requirements list/stock list from the main memory is used.

Possible values:

- X: All relevant MRP elements are read from the database.
- [Space]: The stock/requirements list from the main memory buffer is used if this buffer had been filled. Otherwise, all MRP elements are read from the database.

Type: BAPI_MRP_LIST_PARAM-IGNORE_BUFFER.

MATERIAL_EVG

Parameter for the long material number. Type: BAPIMGVMATNR.

[↑] Export

MRP_LIST

MRP planning parameters are returned in this parameter. Type: BAPI_MRP_LIST.

MRP_CONTROL_PARAM

Control parameters for planning a material received from the material master and customized from MRP are returned in this parameter. Type: BAPI_MRP_CONTROL_PARAM.

MRP_STOCK_DETAIL

This parameter contains the stock statistics for the material. Type: BAPI_MRP_STOCK_DETAIL

RETURN

Return parameters. Type: BAPIRET2.

Tables

MRP_ITEMS

This table contains the detailed display of all of the MRP elements that are relevant for the current requirements/stock list or that were relevant at the time of the last planning. Type: BAPI_MRP_ITEMS.

MRP_IND_LINES

This table contains the single line display of the current stock/requirements list or MRP list.

In the single line display, all MRP elements are listed individually. This type of display allows you to see MRP elements for which there is currently a material shortage. Type: BAPI_MRP_IND_LINES.

MRP_TOTAL_LINES

This table contains the period totals display of the current requirements/stock list or MRP list.

In the period totals display, MRP elements are grouped by period. This type of display gives you a quick overview of how the availability of the material has developed over a period of time. Type: BAPI_MRP_TOTAL_LINES.

EXTENSIONOUT

This parameter is used to transfer the customized defined fields. Type: BAPIPAREX.

593 BAPI_NETWORK_COMP_ADD

This FM adds a list of material components to the network. It's also possible to create multiple network components for the network object with one single call. It isn't possible to create a long text for a material component using this FM. You can't add configurable material, discontinued material, or alternative items to the network.

Example: You want to add the material component to the network structure. You can pass the relevant parameter in this FM and call this FM to add the materials to the network you want.

[↓] Import

NUMBER

Unique network number for network identification. Type: BAPI_NETWORK_LIST-NETWORK.

[↑] Export

RETURN

Return parameter with error messages. Type: BAPIRET2.

Tables

I_COMPONENTS_ADD

Detailed information of the material to be added to the network is passed in this table. Type: BAPI_NETWORK_COMP_ADD.

E_MESSAGE_TABLE

Message table that contains all of the errors. Type: BAPI_METH_MESSAGE.

594 BAPI_OBJCL_CHANGE

This FM changes the existing classification for the materials or creates the classification if they don't exist. The classification is defined by the required entry fields for the object key (OBJECTKEY), object table (OBJECTTABLE), class number (ALLOCDETAIL-CLASS_NUM), and class type (ALLOCDETAIL-CLASS_TYPE).

Example: You've assigned characteristics to your customer. To change the characteristics of the classification for that customer, pass the parameter in this FM and execute it.

[↓] Import

OBJECTKEY

Object key, concatenated. Type: BAPI1003_KEY-OBJECT.

OBJECTTABLE

Object table. Type: BAPI1003_KEY-OBJECTTABLE.

CLASSNUM

Class number. Type: BAPI1003_KEY-CLASSNUM.

CLASSTYPE

Class type. Type: BAPI1003_KEY-CLASSTYPE.

STATUS

Classification status. Type: BAPI1003_KEY-STATUS.

STANDARDCLASS

Indicator: standard class. Type: BAPI1003_KEY-STDCLASS.

CHANGENUMBER

Change number. Type: BAPI1003_KEY-CHANGENUMBER.

KEYDATE

Date on which all classification are valid. Type: BAPI1003_KEY-KEYDATE.

NO_DEFAULT_VALUES

Default values. Type: BAPI1003_KEY-FLAG.

[↑] Export

CLASSIF_STATUS

Classification status. Type: BAPI1003_KEY-STATUS.

Tables

ALLOCVALUESNUMNEW

Table containing new NUMERIC, DATE, and TIME values. Type: BAPI1003_ALLOC_VALUES_NUM.

ALLOCVALUESCHARNEW

Table containing new assigned CHARACTER and BOOLEAN values. Type: BAPI1003_ALLOC_VALUES_CHAR.

ALLOCVALUESCURRNEW

Table with new assigned CURRENCY values. Type: BAPI1003_ALLOC_VALUES_CURR.

RETURN

Error messages. Type: BAPIRET2.

---

595 BAPI_OBJCL_CREATE

This FM creates the classification with the assigned values if the classification doesn't exist. The classification is defined by the required entry fields for the object key (OBJECTKEYNEW), object table (OBJECTTABLENEW), class number (ALLOCDETAIL-CLASS_NUM), and class type (ALLOCDETAIL-CLASS_TYPE).

Example: You want to create a material classification so that later you can create characteristics and group them in a class. You can pass the parameter in this FM and execute this FM to create the classification.

[↓] Import

OBJECTKEYNEW

Object key, concatenated. Type: BAPI1003_KEY-OBJECT.

OBJECTTABLENEW

Object table. Type: BAPI1003_KEY-OBJECTTABLE.

CLASSNUMNEW

Class number. Type: BAPI1003_KEY-CLASSNUM.

CLASSTYPENEW

Class type. Type: BAPI1003_KEY-CLASSTYPE.

STATUS

Classification status. Type: BAPI1003_KEY-STATUS.

STANDARDCLASS

Standard class indicator. Type: BAPI1003_KEY-STDCLASS.

CHANGENUMBER

Change number. Type: BAPI1003_KEY-CHANGENUMBER.

KEYDATE

Date. Type: BAPI1003_KEY-KEYDATE.

NO_DEFAULT_VALUES

Default values. Type: BAPI1003_KEY-FLAG.

[↑] Export

CLASSIF_STATUS

Classification status. Type: BAPI1003_KEY-STATUS.

Tables

ALLOCVALUESNUM

Table containing new NUMERIC, DATE, and TIME values. Type: BAPI1003_ALLOC_VALUES_NUM.

ALLOCVALUESCHAR

Table containing new assigned CHARACTER and BOOLEAN values. Type: BAPI1003_ALLOC_VALUES_CHAR.

ALLOCVALUESCURR

Table with new assigned CURRENCY values. Type: BAPI1003_ALLOC_VALUES_CURR.

RETURN

Error messages. Type: BAPIRET2.

596 BAPI_OBJCL_GETCLASSES

This FM returns all of the classes for a specific object in which the object is classified. The object is determined by the object key (OBJECTKEY) and the object table (OBJECTTABLE). You must also enter the class type (CLASSTYPE).

Example: You've created a material classification, based on the material object, MAT1, and you want to get all of the classes for which it's assigned. You can pass the relevant parameters in this FM and call this FM to get the class details.

[↓] Import

OBJECTKEY_IMP

Key of object. Type: BAPI1003_KEY-OBJECT.

OBJECTTABLE_IMP

Name of database table for object. Type: BAPI1003_KEY-OBJECTTABLE.

CLASSTYPE_IMP

Class type. Type: BAPI1003_KEY-CLASSTYPE.

READ_VALUATIONS

Read assigned values. Type: BAPI1003_KEY-FLAG.

KEYDATE

Key date. Type: BAPI1003_KEY-KEYDATE.

LANGUAGE

Language key. Type: BAPIFIELDSCACL-BAPILANGUA.

Tables

ALLOCLIST

This table contains the details of classes, class type, classification status, and object information. Type: BAPI1003_ALLOC_LIST.

ALLOCVALUESCHAR

Table containing new assigned CHARACTER and BOOLEAN values. Type: BAPI1003_ALLOC_VALUES_CHAR.

ALLOCVALUESCURR

Table with new assigned CURRENCY values. Type: BAPI1003_ALLOC_VALUES_CURR.

ALLOCVALUESNUM

Table containing new NUMERIC, DATE, and TIME values. Type: BAPI1003_ALLOC_VALUES_NUM.

RETURN

Return parameters. Type: BAPIRET2.

---

597 BAPI_OBJCL_GETDETAIL

This FM returns the detailed information of a classification. The object key (OBJECTKEY), object table (OBJECTTABLE), class number (CLASSNUM), class type (CLASSTYPE), and (if required) a date are imported. The default date is the system date.

Example: You want to get the details such as class number, characteristics value, and so on of the object classification so that you can use it for changing the classification from FM BAPI_OBJCL_CHANGE (FM 594). You can pass the relevant parameter in this FM and execute this FM.

[↓] Import

OBJECTKEY

Object key, concatenated. Type: BAPI1003_KEY-OBJECT.

OBJECTTABLE

Classification BAPI: key fields. Type: BAPI1003_KEY-OBJECTTABLE.

CLASSNUM

Class number. Type: BAPI1003_KEY-CLASSNUM.

CLASSTYPE

Class type. Type: BAPI1003_KEY-CLASSTYPE.

KEYDATE

Date. Type: BAPI1003_KEY-KEYDATE.

UNVALUATED_CHARS

Output characteristics without assigned values. Type: FLAG.

LANGUAGE

Language key. Type: BAPIFIELDSCACL-BAPILANGUA.

[↑] Export

STATUS

Classification status. Type: BAPI1003_KEY-STATUS.

STANDARDCLASS

Standard class for indicators. Type: BAPI1003_KEY-STDCLASS.

Tables

ALLOCVALUESNUM

Table containing new NUMERIC, DATE, and TIME values. Type: BAPI1003_ALLOC_VALUES_NUM.

ALLOCVALUESCHAR

Table containing new assigned CHARACTER and BOOLEAN values. Type: BAPI1003_ALLOC_VALUES_CHAR.

ALLOCVALUESCURR

Table with new assigned CURRENCY values. Type: BAPI1003_ALLOC_VALUES_CURR.

RETURN

Error messages. Type: BAPIRET2.

---

598 BAPI_OBJCL_GETDETAIL_KEY

This FM returns detailed information on a classification. The object key (OBJECT), class number (CLASSNUM), and a date, if required, are imported. The default value for the date is the system date.

Example: You want to get all of the details of the classification object so that you can pass it to the FM BAPI_OBJCL_CHANGE (FM 594). You can pass the relevant parameters in this FM and call this FM to get the details of the object classification.

[↓] Import

CLOBJECTKEY

Internal object number. Type: BAPI1003_KEY-OBJECT_GUID.

CLASSNUM

Class number. Type: BAPI1003_KEY-CLASSNUM.

KEYDATE

Date. Type: BAPI1003_KEY-KEYDATE.

LANGUAGE

Language key. Type: BAPIFIELDSCACL-BAPILANGUA.

[↑] Export

STATUS

Classification status. Type: BAPI1003_KEY-STATUS.

STANDARDCLASS

Indicator: standard class. Type: BAPI1003_KEY-STDCLASS.

Tables

ALLOCVALUESNUM

Table containing new NUMERIC, DATE, and TIME values. Type: BAPI1003_ALLOC_VALUES_NUM.

ALLOCVALUESCHAR

Table containing new assigned CHARACTER and BOOLEAN values. Type: BAPI1003_ALLOC_VALUES_CHAR.

ALLOCVALUESCURR

Table with new assigned CURRENCY values. Type: BAPI1003_ALLOC_VALUES_CURR.

RETURN

Error messages. Type: BAPIRET2.

---

599 BAPI_OBJCL_GETSTATUS

This FM returns the status of a classification, which is described by the object key (OBJECTKEY), object table (OBJECTTABLE), class number (CLASSNUM), class type (CLASSTYPE), and a date if required (KEYDATE). The default date is the system date.

Example: You want to get the status of the classification if the classification is released, interpreted, or locked. You can pass the relevant parameter in this FM and call this FM to get the object classification status.

[↓] Import

OBJECTKEY

Object key, concatenated. Type: BAPI1003_KEY-OBJECT.

OBJECTTABLE

Name of database table for object. Type: BAPI1003_KEY-OBJECTTABLE.

CLASSNUM

Class number. Type: BAPI1003_KEY-CLASSNUM.

CLASSTYPE

Class type. Type: BAPI1003_KEY-CLASSTYPE.

KEYDATE

Date. Type: BAPI1003_KEY-KEYDATE.

[↑] Export

STATUS

Classification status. Type: BAPI1003_KEY-STATUS.

Tables

RETURN

Error messages. Type: BAPIRET2.

---

600 BAPI_OUTB_DELIVERY_CONFIRM_DEC

This FM is used to verify outbound delivery from a decentralized system. This method is used to report back the outbound deliveries from a WM system to an SAP ERP system. You must pass the header data of the delivery that is to be reported back, delivery items (i.e., the items that are changed), and packaging data if they exist. Packaging data that exists and isn't reported back means there is no change in the items.

Example: You want to send the finished good to your customer. You need to verify the product before sending it and then report the outbound delivery from WM to the SAP ERP system. Pass the relevant parameters in this FM and call this FM.

[↓] Import

HEADER_DATA

Header data of the inbound delivery. Type: BAPIIBDLVHDRCON.

HEADER_CONTROL

This parameter contains the control data for confirmation header data. Type: BAPIIBDLVHDRCTRLCON.

DELIVERY

Outbound delivery number. Type: BAPIIBDLVHDRCON-DELIV_NUMB.

TECHN_CONTROL

Technical control data; for example, you can set the DEBUG_FLG as X for the abnormal behavior in the receiving system so that you can reimport the IDoc into the receiving system with the BREAKPOINTS. Type: BAPIDLVCONTROL.

HEADER_DATA_SPL

Confirmation of outbound delivery picking data. Outbound delivery header data is passed in this structure. Type: /SPE/BAPIOBDLVHDRCONF.

HEADER_CONTROL_SPL

Additional control data for the completion confirmation of the header data. Type: /SPE/BAPIIBDLVHDRCTRLCONF.

Tables

HEADER_PARTNER

This parameter is used to enter changes to partners. If business partners aren't entered as mandatory partners in the delivery, you can delete business partners by making an entry in the HEADER_PARTNER table with indicator UPD_MODE_PARTN = D.

You can change data for existing business partners by making an entry in the HEADER_PARTNER table with indicator UPD_MODE_PARTN = U.Type: BAPIDLVPARTNERCHG.

HEADER_PARTNER_ADDR

This parameter is used for entering address changes for partners. You must set the UPD_MODE_ADR indicator to U if you want to change addresses. Type: BAPIDLVPARTNADDRCHG.

HEADER_DEADLINES

Header deadlines are passed in this parameter. Type: BAPIDLVDEADLN.

ITEM_DATA

Item data. Type: BAPIIBDLVITEMCON.

ITEM_CONTROL

Control data confirmation item data. Type: BAPIIBDLVITEMCTRLCON.

ITEM_CODING_BLOCK

Delivery account assignment data item. Type: BAPIDLVCOBLITEM.

ITEM_SERIAL_NO

Entry of serial numbers for delivery items.

For materials with a serial number requirement, you must enter the serial numbers here. The system replaces the serial numbers assigned to the delivery item with the new serial numbers.

If you don't enter any serial numbers for the delivery item, the system deletes any serial numbers that had already been assigned. Type: BAPIDLVITMSERNO.

SUPPLIER_CONS_DATA

Vendor consignment goods data. Type: BAPIOBDLVITEMSUPCON.

HANDLING_UNIT_HEADER

Handling unit header data. Type: BAPIDLVHDUNHDR.

HANDLING_UNIT_ITEM

Handling unit item data. Type: BAPIDLVHDUNITM.

HANDLING_UNIT_HEADER_EPC

Structure for handling unit EPC data. Type: /SPE/BAPI_HU_HEADER_EPC.

HANDLING_UNIT_ITEMS_EPC

Structure for handling unit item EPCs. Type: /SPE/BAPI_HU_ITEMS_EPC.

HANDLING_UNIT_SERNO

Handling unit serial numbers. Type: BAPIDLVHDUNSERNO.

EXTENSION1

Customer enhancement segment with unstructured data. Type: BAPIEXTC.

EXTENSION2

Customer enhancement segment with structured data. Type: BAPIEXT.

RETURN

Return parameters. Type: BAPIRET2.

TOKENREFERENCE

The partner system uses this parameter to deliver the token reference for the CSL. The token is used to issue authorization to change deliveries across systems. The delivery is blocked in one system if the partner system opens the delivery in the change transaction or if changes still need to be copied in the partner system to achieve a consistent status between the two systems. Type: BAPITOKENREFERENCE.

HANDLING_UNIT_HEADER_CROSS

Handling unit header data for which multiple deliveries are assigned for cross-delivery HUs. Type: BAPIDLVHDUNHDR.

ITEM_DATA_SPL

Confirmation outbound delivery picking data item level (SPE). Type: /SPE/BAPIOBDLVITEMCONF.

HANDLING_UNIT_IDENTIFIERS

Handling unit IDs are transferred in this table. Type: /SPE/BAPIDLVHDUNID.

HANDLING_UNIT_ITEM_SPL

Handling unit item. Type: /SPE/BAPIDLVHDUNITM.

ITEM_DATA_DOCU_BATCH

Assigned documentary batches per delivery item. Type: /SPE/DOCUBATCH_DLV_ITEM_STY.

NEW_ITEM_DATA

Outbound delivery item. Type: BAPIOBDLVITEM.

NEW_ITEM_DATA_SPL

Outbound deliver item fields/service parts management. Type: /SPE/BAPIOBDLVITM.

NEW_ITEM_ORG

Outbound delivery item organizational units. Type: BAPIOBDLVITEMORG.

TEXT_HEADER

Delivery text header. Type: BAPIDLVTXTHDR.

TEXT_LINES

Delivery text lines. Type: BAPIDLVTXTITEM.

CWM_ITEM_DATA

BAPI communication structure: addition for CWM. Type: /CWM/BAPIOBDLVITEM.

CWM_HU_ITEM

Handling unit item: addition for CWM. Type: /CWM/BAPIDLVHDUNITM.

---

601 BAPI_OUTB_DELIVERY_CREATE_SLS

This FM creates delivery documents for the sales order and sales order item.

Example: To create an outbound delivery for the sales order, pass the relevant parameters in this FM and call it.

[↓] Import

SHIP_POINT

Shipping point or receiving point. If the delivery location isn't specified, the first shipping point is selected, which can be determined under the referenced contract positions. In case of returns, sales order is meant by the shipping point as the goods receiving point of the returned goods. Type: BAPIDLVCREATEHEADER-SHIP_POINT.

DUE_DATE

Delivery creation date. Type: BAPIDLVCREATEHEADER-DUE_DATE.

DEBUG_FLG

For internal use only. The DEBUG_FLG can be used in the ALE inbound processing to force a termination, thereby making the analysis of processing. Type: BAPIDLVCONTROL-DEBUG_FLG.

NO_DEQUEUE

This parameter won't unlock the lock. Type: BAPIFLAG-BAPIFLAG.

[↑] Export

DELIVERY

Delivery number. If more than one delivery is created due to a delivery split, the first delivery number is returned. Type: BAPISHPDELIVNUMB-DELIV_NUMB.

NUM_DELIVERIES

Number of created documents. Possible values:

- 0,1: No delivery or one delivery generated.
- >1: More than one delivery is generated due to a delivery split.

Type: BAPIDLVCREATEHEADER-NUM_DELIVERIES.

Tables

SALES_ORDER_ITEMS

Delivery item with reference to a sales order. This table has the following fields: REF_DOC (sales order), REF_ITEM (sales order item), DLV_QTY (delivery quantity), SALES_UNIT (sales unit), SALES_UNIT_ISO (sales unit of ISO format), and DELIV_NUMB (external delivery number).

If you want to deliver the order number, you need to specify the order number in an entry of the table parameter. Delivery quantities are calculated from the schedule lines and the delivery creation date. If the delivery quantity is specified, the sales unit must be specified in SALES_UNIT or SALES_UNIT_ISO. Type: BAPIDLVREFTOSALESORDER.

SERIAL_NUMBERS

A serial number is assigned to the delivery item that refers to the specified sales order item. Type: BAPIDLVSERIALNUMBER.

EXTENSION_IN

Additional data for enhancements are provided in this table. Data can be processed from the BAdI.

- Enhancement spot: ES_SAPVL50I_BADI.
- BADI definition: BADI_DLV_CREATE_SLS_EXTIN.
- Method: ADDITIONAL_INPUT.

Type: BAPIPAREX.

DELIVERIES

Generated delivery number. Type: BAPISHPDELIVNUMB.

CREATED_ITEMS

Generated Delivery Items. Type: BAPIDLVITEMCREATED.

EXTENSION_OUT

Additional data for synchronous caller enhancements are provided in this table. Data can be processed from the BAdI.

- Enhancement spot: ES_SAPVL50I_BADI.
- BADI definition: BADI_DLV_CREATE_SLS_EXTOUT.
- Method: ADDITIONAL_OUTPUT.

Type: BAPIPAREX.

RETURN

Return parameters with error messages. Type: BAPIRET2.

602 BAPI_OUTB_DELIVERY_CREATE_STO

This FM creates delivery documents for a stock transport order or for individual stock transport order items. A delivery split can mean that several deliveries are created.

Example: You're sending your stock items to the customer for the sales order. To create a delivery document for the same, pass the relevant parameters in this FM and call this FM.

[↓] Import

SHIP_POINT

Shipping point or receiving point. If the delivery location isn't specified, the first shipping point is selected, which can be determined under the referenced contract positions. In case of returns, sales order is meant by the shipping point as the goods receiving point of returned goods. Type: BAPIDLVCREATEHEADER-SHIP_POINT.

DUE_DATE

Delivery creation date. Type: BAPIDLVCREATEHEADER-DUE_DATE.

DEBUG_FLG

For internal use only. The DEBUG_FLG can be used in the ALE inbound processing to force a termination, thereby making the analysis of processing. Type: BAPIDLVCONTROL-DEBUG_FLG.

NO_DEQUEUE

This parameter won't unlock the lock. Type: BAPIFLAG-BAPIFLAG.

[↑] Export

DELIVERY

Delivery number. If more than one delivery is created due to a delivery split, the first delivery number is returned. Type: BAPISHPDELIVNUMB-DELIV_NUMB.

NUM_DELIVERIES

Number of created documents. Possible values:

- 0,1: No delivery or one delivery generated.
- >1: More than one delivery is generated due to the delivery split.

Type: BAPIDLVCREATEHEADER-NUM_DELIVERIES.

Tables

STOCK_TRANS_ITEMS

Delivery item with reference to a stock transport order. This table has the following fields: REF_DOC (stock transport order), REF_ITEM (stock transport item), DLV_QTY (delivery quantity), SALES_UNIT (sales unit), SALES_UNIT_ISO (sales unit of ISO format), and DELIV_NUMB (external delivery number).

If you want to deliver the order number, you need to specify the order number in this table parameter. Delivery quantities are calculated from the schedule lines and the delivery creation date. If the delivery quantity is specified, the sales unit must be specified in SALES_UNIT or SALES_UNIT_ISO. Type: BAPIDLVREFTOSTO.

SERIAL_NUMBERS

A serial number is assigned to the delivery item that refers to the specified sales order item. Type: BAPIDLVSERIALNUMBER.

EXTENSION_IN

Additional data for enhancements are provided in this table. Data can be processed from the BAdI.

- Enhancement spot: ES_SAPVL50I_BADI.
- BADI definition: BADI_DLV_CREATE_SLS_EXTIN.
- Method: ADDITIONAL_INPUT.

Type: BAPIPAREX.

DELIVERIES

Generated delivery number. Type: BAPISHPDELIVNUMB.

CREATED_ITEMS

Generated delivery items. Type: BAPIDLVITEMCREATED.

EXTENSION_OUT

Additional data for synchronous caller enhancements are provided in this table. Data can be processed from the BAdI.

- Enhancement spot: ES_SAPVL50I_BADI.
- BADI definition: BADI_DLV_CREATE_SLS_EXTOUT.
- Method: ADDITIONAL_OUTPUT.

Type: BAPIPAREX.

RETURN

Return parameters with error messages. Type: BAPIRET2.

603 BAPI_PLANNEDORDER_CHANGE

This FM is used to change the header data of the planned order. In terms of MRP, a planned order can be defined as a request created in a planning run for a plant to trigger some procurement of the plant material to a certain quantity for a given date. In terms of Production Planning, a plan order is the request for a plant to trigger the production of a product for a specific quantity at a given date.

Example: You want to change the total plan order quantity of the planned order. You can fill up the parameters of this FM and call this FM to change the quantity of the planned order. Also refer to Transaction MDBA (program MDBAPI01) for detailed use of planned order FMs.

[↓] Import

PLANNEDORDER

Planned order number. Type: BAPI_PLDORD-PLDORD_NUM.

HEADERDATA

Header data of the panned order. Type: BAPIPLAF_I2.

HEADERDATAX

Change header data indicator fields. If you want to change any field of the header data of the planned order, you need to pass X on the respective field of this parameter. HEADER_DATAX-TOTAL_QTY should be X if you want to change the quantity of the planned order. Type: BAPIPLAF_I2X.

[↑] Export

RETURN

Return messages from this FM call. Type: BAPIRETURN1.

CHANGEDHEADERDATA

This parameter has the changed planned order header data. Type: BAPIPLAF_E1.

CAPACITYHEADERDATA1

This parameter has the changed planned order capacity data if detailed scheduling data exists. Type: BAPI_KBKO.

CAPACITYHEADERDATA2

This parameter has the changed planned order capacity data if rate-based scheduling data exists. Type: BAPI_KBKO.

CAPACITYHEADERDATA3

This parameter has the changed planned order capacity data if rough-cut scheduling data exists. Type: BAPI_KBKO.

Tables

COMPONENTSDATA

If a planned order has been successfully changed and has material components, all material components data is exported in this table parameter. Type: BAPI_PLDORDCOMP_E1.

CAPACITYDATA1

This table has the changed planned order capacity data if detailed scheduling data exists. Type: BAPI_PLDORDCAPA_E1.

CAPACITYDATA2

This table has the changed planned order capacity data if rate-based scheduling data exists. Type: BAPI_PLDORDCAPA_E1.

CAPACITYDATA3

This table has the changed planned order capacity data if rough-cut scheduling data exists. Type: BAPI_PLDORDCAPA_E1.

604 BAPI_PLANNEDORDER_CREATE

This FM creates planned orders with BOM explosions or with manual entry of components.

Example: You want to create a planned order for the material of car model S123. You also want to create a BOM for the car model that includes all of the components you want to make this model. You can fill in the parameters of this FM and call this FM to create the planned order. Also refer to Transaction MDBA (program MDBAPI01) for detailed use of planned order FMs.

[↓] Import

HEADERDATA

Header data of the panned order. Type: BAPIPLAF_I1.

[↑] Export

RETURN

Return messages from this FM call. Type: BAPIRETURN1.

PLANNEDORDER

Planned order number. Type: BAPI_PLDORD-PLDORD_NUM.

CHANGEDHEADERDATA

This parameter has the changed planned order header data. Type: BAPIPLAF_E1.

CAPACITYHEADERDATA1

This parameter has the changed planned order capacity data if detailed scheduling data exists. Type: BAPI_KBKO.

CAPACITYHEADERDATA2

This parameter has the changed planned order capacity data if rate-based scheduling data exists. Type: BAPI_KBKO.

CAPACITYHEADERDATA3

This parameter has the changed planned order capacity data if rough-cut scheduling data exists. Type: BAPI_KBKO.

Tables

COMPONENTSDATA

If a planned order has been successfully changed and has material components, all material components data is exported in this table parameter. Type: BAPI_PLDORDCOMP_E1.

CREATEDCOMPONENTSDATA

Created component data. Type: BAPI_PLDORDCOMP_E1.

CAPACITYDATA1

This table has the changed planned order capacity data if detailed scheduling data exists. Type: BAPI_PLDORDCAPA_E1.

CAPACITYDATA2

This table has the changed planned order capacity data if rate-based scheduling data exists. Type: BAPI_PLDORDCAPA_E1.

CAPACITYDATA3

This table has the changed planned order capacity data if rough-cut scheduling data exists. Type: BAPI_PLDORDCAPA_E1.

---

605 BAPI_PLANNEDORDER_DELETE

This FM deletes the planned order from the database. This FM also deletes all of the corresponding data such as capacity and components. Authorization objects M_PLAF_ORG and M_LFPL_ORG are checked when this FM is executed.

Example: You want to delete the planned order. You can fill in the parameters of this FM and call this FM to delete the planned order. Also refer to Transaction MDBA (program MDBAPI01) for the detailed use of planned order FMs.

[↓] Import

PLANNEDORDER

Planned order number. Type: BAPI_PLDORD-PLDORD_NUM.

USE_COLL_UPDATE

During mass processing of planned orders, you can execute a collective update of planning file entries. You can activate the collective update by setting this parameter with value X. Type: BAPIPLAF_I1-USE_COLL_UPD.

LAST_ORDER

If the planned order is the last order for the collective update, you should mark the parameter is X. The system transfers all of the planned file entries collected to the database table when this parameter is found as X. Type: BAPIPLAF_I1-LAST_ORDER.

[↑] Export

RETURN

Return messages from this FM call. Type: BAPIRETURN1.

---

606 BAPI_PLANNEDORDER_EXIST_CHECK

This FM checks whether the specified planned order exists in the database or not. Authorization objects M_PLAF_ORG and M_LFPL_ORG are checked when this FM is executed.

Example: You have a planned order field in the selection screen. To validate the user input to check if the planned order exists in the database or not, you can pass the relevant parameters and call this FM. Also refer to Transaction MDBA (program MDBAPI01) for the detailed use of planned order FMs.

[↓] Import

PLANNEDORDER

Planned order number. Type: BAPI_PLDORD-PLDORD_NUM.

[↑] Export

RETURN

Return messages from this FM call. Type: BAPIRETURN1.

---

607 BAPI_PLANNEDORDER_GET_DETAIL

This FM reads the planned order and gets its details. The details include the header data, material components, capacity header data, and capacity data. Authorization objects M_PLAF_ORG and M_LFPL_ORG are checked when this FM is executed.

Example: You want to get the details of the planned order so that you can use those details for other purposes, such as for changing the planned order. Pass the relevant parameters in this FM and call this FM. Also refer to Transaction MDBA (program MDBAPI01) for the detailed use of planned order FMs.

[↓] Import

PLANNEDORDER

Planned order number. Type: BAPI_PLDORD-PLDORD_NUM.

[↑] Export

RETURN

Return messages from this FM call. Type: BAPIRETURN1.

HEADERDATA

This parameter has the changed planned order header data. Type: BAPIPLAF_E1.

CAPACITYHEADERDATA1

This parameter has the changed planned order capacity data if detailed scheduling data exists. Type: BAPI_KBKO.

CAPACITYHEADERDATA2

This parameter has the changed planned order capacity data if rate-based scheduling data exists. Type: BAPI_KBKO.

CAPACITYHEADERDATA3

This parameter has the changed planned order capacity data if rough-cut scheduling data exists. Type: BAPI_KBKO.

Tables

COMPONENTSDATA

If a planned order has been successfully changed and has material components, all material components data is exported in this table parameter. Type: BAPI_PLDORDCOMP_E1.

CAPACITYDATA1

This table has the changed planned order capacity data if detailed scheduling data exists. Type: BAPI_PLDORDCAPA_E1.

CAPACITYDATA2

This table has the changed planned order capacity data if rate-based scheduling data exists. Type: BAPI_PLDORDCAPA_E1.

CAPACITYDATA3

This table has the changed planned order capacity data if rough-cut scheduling data exists. Type: BAPI_PLDORDCAPA_E1.

608 BAPI_PLANNEDORDER_GET_DET_LIST

This FM is used to get the detailed list of planned orders that is called using the selection criteria. Authorization objects M_PLAF_ORG and M_LFPL_ORG are checked when this FM is executed.

Example: You have a planned order field in the selection screen of your report. Based on the other input parameters, you want to display the F4 help of the planned order. You can pass the relevant parameters in this FM and call this FM to get the list of the planned order, which you can use for F4 help. Also refer to Transaction MDBA (program MDBAPI01) for the detailed use of planned order FMs.

[↓] Import

SELECTIONCRITERIA

You can use this parameter to specify selection criteria to select planned orders. There are certain restrictions for specifying selection criteria:

- At least one selection criteria has to be specified.
- A plant has to be specified if you use a material as the selection criteria.
- A plant has to be specified if you use an MRP controller as the selection criteria.
- Both a plant and a material have to be specified if you use a production version as the selection criteria.
- You must specify a planning scenario number if you enter X in the SIMULATIVE_ORD field of this parameter.

Type: BAPIPLAF_I3.

[↑] Export

RETURN

Return messages from this FM call. Type: BAPIRETURN1.

Table

DETAILEDLIST

If one or more planned orders are selected with the selection criteria that you had given by the import parameter, you'll have a list of planned orders in this parameter. Not only the planned order number but also the header data of each planned order are shown in each line of the list. Type: BAPIPLAF_E1.

---

609 BAPI_PO_CHANGE

This FM is used to change the purchase order (PO). This change method uses the same logic that is behind the online Transaction ME22N. When you change the PO, authorization objects M_BEST_BSA (document type authorization in PO), M_BEST_EKG (purchasing group authorization in PO), M_BEST_EKO (purchasing organization authorization in PO), and M_

BEST_WRK (plant authorization in PO) are checked. Note that you can't use this FM to create subcontracting components, create configurations, change message records from table NAST, attach documents to POs, change foreign trade data, change service data, change or reexplode BOMs, or change PO items with invoicing plans.

Example: To change the header or item data of the purchase order programmatically, you can pass the relevant parameter in this FM and call this FM.

[↓] Import

PURCHASEORDER

Purchasing document number. Type: BAPIMEPOHEADER-PO_NUMBER.

POHEADER

Header data of the purchase order such as purchase organization, vendor, currency, terms of payment, and so on. Type: BAPIMEPOHEADER.

POHEADERX

Change structure of purchase order header. Type: BAPIMEPOHEADERX.

POADDRVENDOR

Address of vendor. The address can only be created and can't be changed. Type: BAPIMEPOADDRVENDOR.

TESTRUN

If this parameter has value X, the purchase order isn't saved to the database; otherwise, it is saved. Type: BAPIFLAG-BAPIFLAG.

MEMORY_UNCOMPLETE

You can use this parameter to control whether the changes made to a purchase order are put on hold if the purchase order is faulty or incomplete.

If the value of this parameter is X, the purchase order is put on hold. If the value of this parameter is [Space], the changes to the purchase order aren't saved. Type: BAPIFLAG-BAPIFLAG.

MEMORY_COMPLETE

You can use this parameter to control whether the changed purchase order is put on hold or saved if it's free of errors and complete.

If the value of this parameter is X, the changed purchase order is put on hold. If the value of this parameter is [Space], the changed purchase order is saved. Type: BAPIFLAG-BAPIFLAG.

POEXPIMPHEADER

This parameter contains foreign trade data for the PO header. Type: BAPIEIKP.

POEXPIMPHEADERX

Change structure of foreign trade data for the PO header. Type: BAPIEIKPX.

VERSIONS

Version management. Type: BAPIMEDCM.

NO_MESSAGING

This parameter activates/deactivates message determination for this call. Possible values:

- X: Don't carry out message determination.

- [Space]: Carry out message determination.

Type: BAPIFLAG-BAPIFLAG.

NO_MESSAGE_REQ

This parameter controls the output of the system message that appears if no messages (print format versions of a document for transmission to a recipient) can be generated for a purchase order. Possible values:

- X: No messages are needed; the system warning message isn't issued.
- [Space]: Messages (print-format documents) are to be generated; the system warning message is issued if no such messages have been generated.

Type: BAPIFLAG-BAPIFLAG.

NO_AUTHORITY

No authorization check. Possible values:

- X: Don't carry out authorization check.
- [Space]: Carry out authorization check.

Type: BAPIFLAG-BAPIFLAG.

NO_PRICE_FROM_PO

This parameter controls price adoption from the last document. Possible values:

- X: Price isn't to be adopted from the last purchase order.
- [Space]: Price is to be adopted from the last purchase order if no price is found in the master records. Type: BAPIFLAG-BAPIFLAG.

PARK_UNCOMPLETE

Park document if faulty. Type: BAPIFLAG-BAPIFLAG.

PARK_COMPLETE

Park document if *not* faulty. Type: BAPIFLAG-BAPIFLAG.

[↑] Export

EXPHEADER

Purchase order header data is returned in this parameter. Type: BAPIMEPOHEADER.

EXPPOEXPIMPHEADER

Export trade header data is returned in this parameter. Type: BAPIEIKP.

Tables

RETURN

Return parameter. Type: BAPIRET2.

POITEM

Item data of the purchase order. Type: BAPIMEPOITEM.

POITEMX

Change structure of the Item data of the purchase order. Type: BAPIMEPOITEMX.

POADDRDELIVERY

This table contains the delivery address for the PO item. The Central Address Management fields are used for this purpose. Type: BAPIMEPOADDRDELIVERY.

POSCHEDULE

This table contains the scheduling data for the purchase order item (scheduled quantity and delivery date per schedule line). Type: BAPIMEPOSCHEDULE

POSCHEDULEX

Change structure for delivery schedule. Type: BAPIMEPOSCHEDULX.

POACCOUNT

Account assignment data for the purchase order item. Type: BAPIMEPOACCOUNT.

POACCOUNTPROFITSEGMENT

This table contain the characteristics and characteristic value for the profitability segment. Type: BAPIMEPOACCOUNTPROFITSEGMENT.

POACCOUNTX

Account assignment fields change structure. Type: BAPIMEPOACCOUNTX.

POCONDHEADER

Conditions header table. Type: BAPIMEPOCONDHEADER.

POCONDHEADERX

Conditions header change structure. Type: BAPIMEPOCONDHEADERX.

POCOND

Conditions items table. Type: BAPIMEPOCOND.

POCONDX

Conditions items, change structure. Type: BAPIMEPOCONDX.

POLIMITS

This table contains the purchasing order limits. Type: BAPIESUHC.

POCONTRACTLIMITS

This table contains the limits for purchase orders with reference to a contract. Type: BAPIESUCC.

POSERVICES

This table contains the services and services specifications outline for the purchase order. Type: BAPIESLLC.

POSRVACCESSVALUES

Account assignment distribution for service lines. Type: BAPIESKLC.

POSERVICESTEXT

Service long text. Type: BAPIESLLTX.

EXTENSIONIN

Customer's own fields. This table allows you to add to the import parameters of the BAPI without modifying it. Type: BAPIPAREX.

EXTENSIONOUT

Customer's own fields. This table allows you to add to the export parameters of the BAPI without modifying it. Type: BAPIPAREX.

POEXPIMPITEM

Foreign trade: item data. Type: BAPIEIPO.

POEXPIMPITEMX

Foreign trade: change parameter: item data. Type: BAPIEIPOX.

POTEXTHEADER

Header texts. Type: BAPIMEPOTEXTHEADER.

POTEXTITEM

Item texts. Type: BAPIMEPOTEXT.

ALLVERSIONS

All versions (export parameter). Type: BAPIMEDCM_ALLVERSIONS.

POPARTNER

Partner table. Type: BAPIEKKOP.

POCOMPONENTS

BAPI structure for components. Type: BAPIMEPOCOMPONENT.

POCOMPONENTSX

Update information for components in the BUS2012 API. Type: BAPIMEPOCOMPONENTX.

POSHIPPING

BAPI shipping data for stock transport orders. Type: BAPIITEMSHIP.

POSHIPPINGX

BAPI shipping data change parameter. Type: BAPIITEMSHIPX.

POSHIPPINGEXP

Export structure for shipping data. Type: BAPIMEPOSHIPPEXP.

POHISTORY

Transfer structure for PO history. Type: BAPIEKBE.

POHISTORY_TOTALS

Transfer structure for PO history: totals. Type: BAPIEKBES.

POCONFIRMATION

Transfer structure display/list vendor confirmation. Type: BAPIEKES.

SERIALNUMBER

Serial numbers in purchase order BAPIs. Type: BAPIMEPOSERIALNO.

SERIALNUMBERX

Change parameter: serial number in purchase order BAPIs. Type: BAPIMEPOSERIALNOX.

INVPLANHEADER

Invoicing plan header data. Type: BAPI_INVOICE_PLAN_HEADER.

INVPLANHEADERX

Invoicing plan header data (change parameter). Type: BAPI_INVOICE_PLAN_HEADERX.

INVPLANITEM

Invoicing plan item data. Type: BAPI_INVOICE_PLAN_ITEM.

INVPLANITEMX

Invoicing plan item data (change parameter). Type: BAPI_INVOICE_PLAN_ITEMX.

POHISTORY_MA

Transfer structure for purchase order history account assignment level. Type: BAPIEKBE_MA.

NFMETALLITMS

BAPI communication structure. Type: /NFM/BAPIDOCITM.

610 BAPI_PO_CREATE1

This FM creates a purchase order (PO). This change method uses the same logic that is behind the online Transaction ME21N. Authorization objects M_BEST_BSA (document type authorization in PO), M_BEST_EKG (purchasing group authorization in PO), M_BEST_EKO (purchasing organization authorization in PO), and M_BEST_WRK (plant authorization in PO) are checked. You can't use this FM to create subcontracting components, create configurations, change message records from table NAST, or attach documents to POs.

Example: To create a PO programmatically, you can pass the relevant parameter and call this FM.

[↓] Import

POHEADER

Header data of the purchase order such as purchase organization, vendor, currency, terms of payment, and so on. Type: BAPIMEPOHEADER.

POHEADERX

Change structure of purchase order header. Type: BAPIMEPOHEADERX.

POADDRVENDOR

Address of vendor. Addresses can only be created and can't be changed. Type: BAPIMEPOADDRVENDOR.

TESTRUN

If this parameter has the value X, the purchase order isn't saved to the database; otherwise, it is saved. Type: BAPIFLAG-BAPIFLAG.

MEMORY_UNCOMPLETE

You can use this parameter to control whether the changes made to a purchase order are put on hold if the purchase order is faulty or incomplete.

If the value of this parameter is X, the purchase order is put on hold.

If the value of this parameter is [Space], the changes to the purchase order aren't saved. Type: BAPIFLAG-BAPIFLAG.

MEMORY_COMPLETE

You can use this parameter to control whether the changed purchase order is put on hold or saved if it's free of errors and complete. Possible values:

- X: The changed purchase order is put on hold.
- Space: The changed purchase order is saved.

Type: BAPIFLAG-BAPIFLAG.

POEXPIMPHEADER

This parameter contains foreign trade data for the PO header. Type: BAPIEIKP.

POEXPIMPHEADERX

Change structure of foreign trade data for the PO header. Type: BAPIEIKPX.

VERSIONS

Version Management. Type: BAPIMEDCM.

NO_MESSAGING

This parameter activates/deactivates message determination for this call. Possible values:

- X: Don't carry out message determination.
- Space: Carry out message determination.

Type: BAPIFLAG-BAPIFLAG.

NO_MESSAGE_REQ

This parameter controls the output of the system message that appears if no messages (print format versions of a document for transmission to a recipient) can generated for a purchase order. Possible values:

- X: No messages are needed; the system warning message isn't issued.
- Space: Messages (print format documents) are to be generated; the system warning message is issued if no such messages have been generated.

Type: BAPIFLAG-BAPIFLAG.

NO_AUTHORITY

No authorization check. Possible values:

- X: Don't carry out authorization check.
- Space: Carry out authorization check.

Type: BAPIFLAG-BAPIFLAG.

NO_PRICE_FROM_PO

This parameter controls price adoption from the last document. Possible values:

- X: Price isn't to be adopted from the last purchase order.
- Space: Price is to be adopted from the last purchase order if no price can be found in the master records.

Type: BAPIFLAG-BAPIFLAG.

PARK_UNCOMPLETE

Park document if faulty. Type: BAPIFLAG-BAPIFLAG.

PARK_COMPLETE

Park document if *not* faulty. Type: BAPIFLAG-BAPIFLAG.

[↑] Export

EXPPURCHASEORDER

Purchasing document number. Type: BAPIMEPOHEADER-PO_NUMBER.

EXPHEADER

Purchase order header data is returned in this parameter. Type: BAPIMEPOHEADER.

EXPPOEXPIMPHEADER

Export trade header data is returned in this parameter. Type: BAPIEIKP.

Tables

RETURN

Return parameter. Type: BAPIRET2.

POITEM

Item data of the purchase order. Type: BAPIMEPOITEM.

POITEMX

Change structure of the item data of the purchase order. Type: BAPIMEPOITEMX.

POADDRDELIVERY

This table contains the delivery address for the PO item. The Central Address Management fields are used for this purpose. Type: BAPIMEPOADDRDELIVERY.

POSCHEDULE

This table contains the scheduling data for the purchase order item (scheduled quantity and delivery date per schedule line). Type: BAPIMEPOSCHEDULE.

POSCHEDULEX

Change structure for delivery schedule. Type: BAPIMEPOSCHEDULX.

POACCOUNT

Account assignment data for the purchase order item. Type: BAPIMEPOACCOUNT.

POACCOUNTPROFITSEGMENT

This table contains the characteristics and characteristic value for the profitability segment. Type: BAPIMEPOACCOUNTPROFITSEGMENT.

POACCOUNTX

Account assignment fields change structure. Type: BAPIMEPOACCOUNTX.

POCONDHEADER

Conditions header table. Type: BAPIMEPOCONDHEADER.

POCONDHEADERX

Conditions header change structure. Type: BAPIMEPOCONDHEADERX.

POCOND

Conditions items table. Type: BAPIMEPOCOND.

POCONDX

Conditions items, change structure. Type: BAPIMEPOCONDX.

POLIMITS

This table contains the purchasing order limits. Type: BAPIESUHC.

POCONTRACTLIMITS

This table contains the limits for purchase orders with reference to a contract. Type: BAPIESUCC.

POSERVICES

This table contains the services and services specifications outline for the purchase order. Type: BAPIESLLC.

POSRVACCESSVALUES

Account assignment distribution for service lines. Type: BAPIESKLC.

POSERVICESTEXT

Service long text. Type: BAPIESLLTX.

EXTENSIONIN

Customer's own fields. This table allows you to add to the import parameters of the BAPI without modifying it. Type: BAPIPAREX.

EXTENSIONOUT

Customer's own fields. This table allows you to add to the export parameters of the BAPI without modifying it. Type: BAPIPAREX.

POEXPIMPITEM

Foreign trade: item data. Type: BAPIEIPO.

POEXPIMPITEMX

Foreign trade: change parameter: item data. Type: BAPIEIPOX.

POTEXTHEADER

Header texts. Type: BAPIMEPOTEXTHEADER.

POTEXTITEM

Item texts. Type: BAPIMEPOTEXT.

ALLVERSIONS

All versions (export parameter). Type: BAPIMEDCM_ALLVERSIONS.

POPARTNER

Partner table. Type: BAPIEKKOP.

POCOMPONENTS

BAPI structure for components. Type: BAPIMEPOCOMPONENT.

POCOMPONENTSX

Update information for components in the BUS2012 API. Type: BAPIMEPOCOMPONENTX.

POSHIPPING

BAPI shipping data for stock transport orders. Type: BAPIITEMSHIP.

POSHIPPINGX

BAPI shipping data change parameter. Type: BAPIITEMSHIPX.

POSHIPPINGEXP

Export structure for shipping data. Type: BAPIMEPOSHIPPEXP.

SERIALNUMBER

Serial numbers in purchase order BAPIs. Type: BAPIMEPOSERIALNO.

SERIALNUMBERX

Change parameter: serial number in purchase order BAPIs. Type: BAPIMEPOSERIALNOX.

INVPLANHEADER

Invoicing plan header data. Type: BAPI_INVOICE_PLAN_HEADER.

INVPLANHEADERX

Invoicing plan header data (change parameter). Type: BAPI_INVOICE_PLAN_HEADERX.

INVPLANITEM

Invoicing plan item data. Type: BAPI_INVOICE_PLAN_ITEM.

INVPLANITEMX

Invoicing plan item data (change parameter). Type: BAPI_INVOICE_PLAN_ITEMX.

NFMETALLITMS

BAPI communication structure. Type: /NFM/BAPIDOCITM.

---

611 BAPI_PO_GETDETAIL1

This FM returns the detailed information for the purchase order. This change method uses the same logic that is behind the online Transaction ME21N. Authorization objects M_BEST_BSA (document type authorization in PO), M_BEST_EKG (purchasing group authorization in PO), M_BEST_EKO (purchasing organization authorization in PO), and M_BEST_WRK (plant authorization in PO) are checked.

Example: You want to get the detail of the purchase order so that you can change it from FM BAPI_PO_CHANGE (FM 609). You can pass the relevant parameter in this FM and call this FM to get the purchase order details.

[↓] Import

PURCHASEORDER

Purchasing document number. Type: BAPIMEPOHEADER-PO_NUMBER.

ACCOUNT_ASSIGNMENT

Account assignment data. Type: BAPIMMPARA-SELECTION.

ITEM_TEXT

Item text. Type: BAPIMMPARA-SELECTION.

HEADER_TEXT

Header text. Type: BAPIMMPARA-SELECTION.

DELIVERY_ADDRESS

Delivery address. Type: BAPIMMPARA-SELECTION.

VERSION

Version management. Type: BAPIMMPARA-SELECTION.

SERVICES

External service data. Type: BAPIMMPARA-SELECTION.

SERIALNUMBERS

Serial numbers. Type: BAPIMMPARA-SELECTION.

INVOICEPLAN

Invoicing plan. Type: BAPIMMPARA-SELECTION.

[↑] Export

POHEADER

Purchase order header data. Type: BAPIMEPOHEADER.

POEXPIMPHEADER

Foreign trade: export/import: header data. Type: BAPIEIKP.

Tables

RETURN

Return parameter. Type: BAPIRET2.

POITEM

Item data of the purchase order. Type: BAPIMEPOITEM.

POITEMX

Change structure of the item data of the purchase order. Type: BAPIMEPOITEMX.

POADDRDELIVERY

This table contains the delivery address for the PO item. The Central Address Management fields are used for this purpose. Type: BAPIMEPOADDRDELIVERY.

POSCHEDULE

This table contains the scheduling data for the purchase order item (scheduled quantity and delivery date per schedule line). Type: BAPIMEPOSCHEDULE.

POACCOUNT

Account assignment data for the purchase order item. Type: BAPIMEPOACCOUNT.

POCONDHEADER

Conditions header table. Type: BAPIMEPOCONDHEADER.

POCOND

Conditions items table. Type: BAPIMEPOCOND.

POLIMITS

This table contains the purchasing order limits. Type: BAPIESUHC.

POCONTRACTLIMITS

This table contains the limits for purchase orders with reference to a contract. Type: BAPIESUCC.

POSERVICES

This table contains the services and services specifications outline for the purchase order. Type: BAPIESLLC.

POSRVACCESSVALUES

Account assignment distribution for service lines. Type: BAPIESKLC.

POTEXTHEADER

Header texts. Type: BAPIMEPOTEXTHEADER.

POTEXTITEM

Item texts. Type: BAPIMEPOTEXT.

POEXPIMPITEM

Foreign trade: item data. Type: BAPIEIPO.

POCOMPONENTS

BAPI structure for components. Type: BAPIMEPOCOMPONENT.

POSHIPPINGEXP

Export structure for shipping data. Type: BAPIMEPOSHIPPEXP.

POHISTORY

Purchase order history. Type: BAPIEKBE.

POHISTORY_TOTALS

Purchase order history: totals. Type: BAPIEKBES.

POCONFIRMATION

Vendor confirmation. Type: BAPIEKES.

ALLVERSIONS

Version data. Type: BAPIMEDCM_ALLVERSIONS.

POPARTNER

Partner. Type: BAPIEKKOP.

EXTENSIONOUT

This parameter is used to enhance the interface of a BAPI without modification to achieve the automatic processing of purchase order. Type: BAPIPAREX.

SERIALNUMBER

Serial numbers in purchase order BAPIs. Type: BAPIMEPOSERIALNO.

INVPLANHEADER

Invoicing plan: header data. Type: BAPI_INVOICE_PLAN_HEADER.

INVPLANITEM

Invoicing plan: item data. Type: BAPI_INVOICE_PLAN_ITEM.

POHISTORY_MA

Transfer structure for purchase order history account assignment level. Type: BAPIEKBE_MA.

612 BAPI_PO_GETITEMS

This FM returns a list of purchase order items based on the selection criteria you enter, such as purchase groups, plant vendors, and so on, in the import parameter. Authorization objects M_BEST_BSA (document type authorization in PO), M_BEST_EKG (purchasing group authorization in PO), M_BEST_EKO (purchasing organization authorization in PO), and M_BEST_WRK (plant authorization in PO) are checked.

Example: You have a PO field in your selection screen. Based on the other selection criteria on the selection screen, you want to get the list of POs in the F4 field of the PO. You can pass the relevant parameter in this FM and call this FM to get the list of purchase orders.

[↓] Import

PURCHASEORDER

Purchase order number. Type: BAPIEKKO-PO_NUMBER.

DOC_TYPE

Order type. Type: BAPIEKKO-DOC_TYPE.

DOC_DATE

As of PO date. Type: BAPIEKKO-DOC_DATE.

PUR_GROUP

Purchasing group. Type: BAPIEKKO-PUR_GROUP.

PURCH_ORG

Purchasing organization. Type: BAPIEKKO-PURCH_ORG.

VENDOR

Vendor. Type: BAPIEKKO-VENDOR.

SUPPL_PLANT

Supplying plant. Type: BAPIEKKO-SUPPL_PLNT.

MATERIAL

Material. Type: BAPIEKPO-MATERIAL.

MAT_GRP

Material group. Type: BAPIEKPO-MAT_GRP.

ITEM_CAT

Item category. Type: BAPIEKPO-ITEM_CAT.

ACCTASSCAT

Account assignment category. Type: BAPIEKPO-ACCTASSCAT.

PLANT

Plant. Type: BAPIEKPO-PLANT.

TRACKINGNO

Requirement tracking number. Type: BAPIEKPO-TRACKINGNO

SHORT_TEXT

Short text. Type: BAPIEKPO-SHORT_TEXT.

CREATED_BY

Creator. Type: BAPIEKKOC-CREATED_BY.

PREQ_NAME

Name of purchase requester. Type: BAPIEKPOC-PREQ_NAME.

WITH_PO_HEADERS

Also provide header data. Type: BAPIMMPARA-SELECTION.

DELETED_ITEMS

Including deleted items. Type: BAPIMMPARA-SELECTION.

ITEMS_OPEN_FOR_RECEIPT

Items with still-open goods receipt only. Type: BAPIMMPARA-SELECTION.

PUR_MAT

Material number. Type: BAPIEKPO-PUR_MAT.

MATERIAL_EVG

Long material number. Type: BAPIMGVMATNR.

PUR_MAT_EVG

Long material number. Type: BAPIMGVMATNR.

Tables

PO_HEADERS

Table of PO header data. Type: BAPIEKKOL.

PO_ITEMS

Table of purchase order items. Type: BAPIEKPOC.

RETURN

Return messages. Type: BAPIRETURN.

---

613 BAPI_PO_RELEASE

This FM is used to release or approve the PO. The PO number and release code are required entries in the import parameter. Release status and release indicator are returned. Authorization object M_EINK_FRG is checked.

Example: You're the manager and have received a PO. If you're okay with the PO, you can release it by passing the parameter in this FM.

[↓] Import

PURCHASEORDER

Purchase order number. Type: BAPIMMPARA-PO_NUMBER

PO_REL_CODE

Release code with which you can process the purchase order. Type: BAPIMMPARA-PO_REL_COD.

USE_EXCEPTIONS

You set this indicator, and the exceptions are passed on in addition to the messages in the RETURN parameter. In the standard system, the indicator is set.

If you don't set the indicator, the messages are only passed on in the RETURN parameter. Type: BAPIMMPARA-SELECTION.

NO_COMMIT

You can use the parameter to determine whether or not the COMMIT WORK is to occur directly. Possible values:

- X: No COMMIT_WORK occurs within the FM.
- [Space]: This FM performs the COMMIT_WORK.

Type: BAPIMMPARA-SELECTION.

[↑] Export

REL_STATUS_NEW

New release status is returned in this parameter. Type: BAPIMMPARA-REL_STATUS.

REL_INDICATOR_NEW

New release indicator is returned in this parameter. Type: BAPIMMPARA-PO_REL_IND.

RET_CODE

Return value, return value after ABAP statements. Type: SY-SUBRC.

Table

RETURN

Return messages. Type: BAPIRETURN.

[!] Exceptions

AUTHORITY_CHECK_FAIL

No authorization to release.

DOCUMENT_NOT_FOUND

Purchase order does not exist.

ENQUEUE_FAIL

Purchase order blocked.

PREREQUISITE_FAIL

Release prerequisite not satisfied.

RELEASE_ALREADY_POSTED

Release already effected.

RESPONSIBILITY_FAIL

Responsibility for release missing.

614 BAPI_PO_RESET_RELEASE

This FM cancels or revokes the already effected releases of a PO. The PO number and release code are required entries in the import parameter. The release status and release indicator valid prior to the release are returned. Authorization object M_EINK_FRG is checked.

Example: You're the manager and received a PO. You want to cancel the already released PO. You can pass the parameter in this FM to cancel the already released purchase order.

[↓] Import

PURCHASEORDER

Purchase order number. Type: BAPIMMPARA-PO_NUMBER.

PO_REL_CODE

Release code with which you can process the purchase order. Type: BAPIMMPARA-PO_REL_COD.

USE_EXCEPTIONS

If this indicator is set, the exceptions are passed on in addition to the messages in the RETURN parameter. In the standard system, the indicator is set.

If you don't set the indicator, the messages are only passed on in the RETURN parameter. Type: BAPIMMPARA-SELECTION.

[↑] Export

REL_STATUS_NEW

New release status is returned in this parameter. Type: BAPIMMPARA-REL_STATUS.

REL_INDICATOR_NEW

New release indicator is returned in this parameter. Type: BAPIMMPARA-PO_REL_IND.

Table

RETURN

Return messages. Type: BAPIRETURN.

[!] Exceptions

AUTHORITY_CHECK_FAIL

No authorization to release.

DOCUMENT_NOT_FOUND

Purchase order does not exist.

ENQUEUE_FAIL

Purchase order blocked.

PREREQUISITE_FAIL

Release prerequisite not satisfied.

RELEASE_ALREADY_POSTED

Release already effected.

RESPONSIBILITY_FAIL

Responsibility for release missing.

615 BAPI_PR_CHANGE

This FM changes the purchase requisition. A purchase requisition is the request or instruction to purchase a certain quantity of a material or service so that it can be available at a certain point in time.

Example: You want to purchase four pieces of material MAT1 from your vendor and send the purchase requisition to him. Later, you released that you want six pieces, and you want to change the purchase requisition. You can pass the relevant parameter in this FM and call this FM to change the purchase requisition.

[↓] Import

NUMBER

Purchase requisition number. Type: BAPIMEREQHEADER-PREQ_NO.

PRHEADER

This structure contains the purchase requisition header data. Type: BAPIMEREQHEADER.

PRHEADERX

Change parameter for Enjoy purchase requisition header data. Type: BAPIMEREQHEADERX.

TESTRUN

If this parameter has the value X, the purchase requisition isn't saved to the database; otherwise, it is saved. Type: BAPIFLAG-BAPIFLAG.

[↑] Export

PRHEADEREXP

Purchase requisition header data returned. Type: BAPIMEREQHEADER.

Tables

RETURN

Return parameters with messages from this FM. Type: BAPIRET2.

PRITEM

Toolbar structure for Enjoy purchase requisition: item data. Type: BAPIMEREQITEMIMP.

PRITEMX

Change parameter for Enjoy purchase requisition: item data. Type: BAPIMEREQITEMX.

PRITEMEXP

Transfer structure for Enjoy purchase requisition: item data. Type: BAPIMEREQITEM.

PRITEMSOURCE

Return structure for Enjoy purchase requisition: source of supply. Type: BAPIMEREQSOURCE.

PRACCOUNT

Transfer structure for Enjoy purchase requisition: account assignment. Type: BAPIMEREQACCOUNT.

PRACCOUNTPROITSEGMENT

Profitability segment of purchase requisition. Type: BAPIMEREQACCOUNTPROFITSEG.

PRACCOUNTX

Change toolbar for Enjoy purchase requisition: account assignment. Type: BAPIMEREQACCOUNTX.

PRADDRDELIVERY

PO item: address structure BAPIADDR1 for inbound delivery. Type: BAPIMERQADDRDELIVERY.

PRITEMTEXT

Transfer structure for Enjoy purchase requisition: item text. Type: BAPIMEREQITEMTEXT.

PRHEADERTEXT

Change toolbar for Enjoy purchase requisition: header text. Type: BAPIMEREQHEADTEXT.

EXTENSIONIN

Reference structure for BAPI parameters EXTENSIONIN/EXTENSIONOUT. Type: BAPIPAREX.

EXTENSIONOUT

Reference structure for BAPI parameters EXTENSIONIN/EXTENSIONOUT. Type: BAPIPAREX.

PRVERSION

Version data for purchase requisition item (BAPI). Type: BAPIMEREQDCM.

PRVERSIONX

Change parameter Enjoy purchase requisition: version data. Type: BAPIMEREQDCMX.

ALLVERSIONS

Version management: all version data. Type: BAPIMEDCM_ALLVERSIONS.

PRCOMPONENTS

You can use this table parameter to influence the subcontracting components of a purchase requisition item. Type: BAPIMEREQCOMPONENT

PRCOMPONENTSX

Change structure for PRCOMPONENTS. Type: BAPIMEREQCOMPONENTX.

SERVICEOUTLINE

This parameter contains information about the outline of the service package. Type: BAPI_SRV_OUTLINE.

SERVICEOUTLINEX

Change parameter for Enjoy purchase requisition change service package information. Type: BAPI_SRV_OUTLINEX.

SERVICELINES

This parameter contains information on the individual service lines of the service package. Type: BAPI_SRV_SERVICE_LINE.

SERVICELINESX

Change parameter for Enjoy purchase requisition change service lines. Type: BAPI_SRV_SERVICE_LINEX.

SERVICELIMIT

This parameter contains information about the limit of the service package. Type: BAPI_SRV_LIMIT_DATA.

SERVICELIMITX

Change parameter for Enjoy purchase requisition change service limit. Type: BAPI_SRV_LIMIT_DATAX.

SERVICECONTRACTLIMITS

This parameter contains information on the apportionment of the overall limit among certain contracts. Type: BAPI_SRV_CONTRACT_LIMITS.

SERVICECONTRACTLIMITSX

Change parameter for Enjoy purchase requisition change service limit. Type: BAPI_SRV_CONTRACT_LIMITSX.

SERVICEACCOUNT

This parameter contains information about the account assignment of the service lines and the limit. Type: BAPI_SRV_ACC_DATA.

SERVICEACCOUNTX

Change parameter for Enjoy purchase requisition change service account. Type: BAPI_SRV_ACC_DATAX.

SERVICELONGTEXTS

This parameter contains information about long texts. Type: BAPI_SRV_LONGTEXTS.

SERIALNUMBER

Serial numbers in purchase requisition BAPI. Type: BAPIMEREQSERIALNO.

SERIALNUMBERX

Change parameter for Enjoy purchase requisition serial number. Type: BAPIMEREQSERIALNOX.

---

616 BAPI_PR_CREATE

This FM creates the purchase requisition. A purchase requisition is the request or instruction to purchase a certain quantity of a material or service so that it can be available at a certain point in time.

Example: You want to create the purchase requisition programmatically. You can pass the relevant parameter in this FM and call this FM to create the purchase requisition.

[↓] Import

PRHEADER

This structure contains the purchase requisition header data. Type: BAPIMEREQHEADER.

PRHEADERX

Change parameter for Enjoy purchase requisition header data. Type: BAPIMEREQHEADERX.

TESTRUN

If this parameter has the value X, the purchase requisition isn't saved to the database; otherwise, it is saved. Type: BAPIFLAG-BAPIFLAG.

[↑] Export

NUMBER

Purchase requisition number. Type: BAPIMEREQHEADER-PREQ_NO.

PRHEADEREXP

Purchase requisition header data returned. Type: BAPIMEREQHEADER.

Tables

RETURN

Return parameters with messages from this FM. Type: BAPIRET2.

PRITEM

Toolbar structure for Enjoy purchase requisition: item data. Type: BAPIMEREQITEMIMP.

PRITEMX

Change parameter for Enjoy purchase requisition: item data. Type: BAPIMEREQITEMX.

PRITEMEXP

Transfer structure for Enjoy purchase requisition: item data. Type: BAPIMEREQITEM.

PRITEMSOURCE

Return structure for Enjoy purchase requisition: source of supply. Type: BAPIMEREQSOURCE.

PRACCOUNT

Transfer structure for Enjoy purchase requisition: account assignment. Type: BAPIMEREQACCOUNT.

PRACCOUNTPROITSEGMENT

Profitability segment of purchase requisition. Type: BAPIMEREQACCOUNTPROFITSEG.

PRACCOUNTX

Change toolbar for Enjoy purchase requisition: account assignment. Type: BAPIMEREQACCOUNTX.

PRADDRDELIVERY

PO item: address structure BAPIADDR1 for inbound delivery. Type: BAPIMERQADDRDELIVERY.

PRITEMTEXT

Transfer structure for Enjoy purchase requisition: item text. Type: BAPIMEREQITEMTEXT.

PRHEADERTEXT

Change toolbar for Enjoy purchase requisition: header text. Type: BAPIMEREQHEADTEXT.

EXTENSIONIN

Reference structure for BAPI parameters EXTENSIONIN/EXTENSIONOUT. Type: BAPIPAREX.

EXTENSIONOUT

Reference structure for BAPI parameters EXTENSIONIN/EXTENSIONOUT. Type: BAPIPAREX.

PRVERSION

Version data for purchase requisition item (BAPI). Type: BAPIMEREQDCM.

PRVERSIONX

Change parameter Enjoy purchase requisition: version data. Type: BAPIMEREQDCMX.

ALLVERSIONS

Version management: all version data. Type: BAPIMEDCM_ALLVERSIONS.

PRCOMPONENTS

You can use this table parameter to influence the subcontracting components of a purchase requisition item. Type: BAPIMEREQCOMPONENT.

PRCOMPONENTSX

Change structure for PRCOMPONENTS. Type: BAPIMEREQCOMPONENTX.

SERVICEOUTLINE

This parameter contains information about the outline of the service package. Type: BAPI_SRV_OUTLINE.

SERVICEOUTLINEX

Change parameter for Enjoy purchase requisition change service package information. Type: BAPI_SRV_OUTLINEX.

SERVICELINES

This parameter contains information on the individual service lines of the service package. Type: BAPI_SRV_SERVICE_LINE.

SERVICELINESX

Change parameter for Enjoy purchase requisition change service lines. Type: BAPI_SRV_SERVICE_LINEX.

SERVICELIMIT

This parameter contains information about the limit of the service package. Type: BAPI_SRV_LIMIT_DATA.

SERVICELIMITX

Change parameter for Enjoy purchase requisition change service limit. Type: BAPI_SRV_LIMIT_DATAX.

SERVICECONTRACTLIMITS

This parameter contains information on the apportionment of the overall limit among certain contracts. Type: BAPI_SRV_CONTRACT_LIMITS.

SERVICECONTRACTLIMITSX

Change parameter for Enjoy purchase requisition change service limit. Type: BAPI_SRV_CONTRACT_LIMITSX.

SERVICEACCOUNT

This parameter contains information about the account assignment of the service lines and the limit. Type: BAPI_SRV_ACC_DATA.

SERVICEACCOUNTX

Change parameter for Enjoy purchase requisition change service account. Type: BAPI_SRV_ACC_DATAX.

SERVICELONGTEXTS

This parameter contains information about long texts. Type: BAPI_SRV_LONGTEXTS.

SERIALNUMBER

Serial numbers in purchase requisition BAPIs. Type: BAPIMEREQSERIALNO.

SERIALNUMBERX

Change parameter for Enjoy purchase requisition serial number. Type: BAPIMEREQSERIALNOX.

617 BAPI_PRODORD_CHANGE

This FM changes the production order header data. Authorization object C_AFKO_AWK is checked.

Example: You've created a production order. Now you want to change the planned order quantity from 7 to 9. You can pass the relevant parameters in this FM and call this FM to change the production order.

[↓] Import

NUMBER

Production order number that uniquely identifies the production order. Type: BAPI_ORDER_KEY-ORDER_NUMBER.

ORDERDATA

This structure contains the header information of the production order. The field ORDER_NUMBER must be filled with the number of the order that is to be changed. Type: BAPI_PP_ORDER_CHANGE.

ORDERDATAX

If you want to change any detail of the production order header data, you need to pass that field as X in this parameter. Type: BAPI_PP_ORDER_CHANGEX.

[↑] Export

RETURN

Return parameters with messages. Type: BAPIRET2.

ORDER_TYPE

Order type of the changed production or process order. Type: BAPI_ORDER_COPY-ORDER_TYPE.

ORDER_STATUS

Order status indicator is filled if the order can't be changed. An order can't be changed if, for example, the deletion flag has been set or if the order has been technically completed. Possible values:

- DELETED: Deletion flag is set.
- DELETE_IND: Deletion indicator is set.
- CLOSED: Order is closed.
- COMPLETED: Order is closed technically.
- LOCKED: Order is locked.

Type: BAPI_ORDER_STATUS.

MASTER_DATA_READ

This parameter MASTER_DATA_READ specifies whether master data was read or not when the order was changed. Possible values:

- X: Master data was read.
- [Space]: Master data was not read.

Type: BAPIFLAG-BAPIFLAG.

---

618 BAPI_PRODORD_CREATE

This FM creates the production order. Authorization object C_AFKO_AWK is checked.

Example: You want to create the production order programmatically. You can pass the relevant parameters in this FM and call this FM to change the production order.

[↓] Import

ORDERDATA

This structure contains the header information of the production order. The field ORDER_NUMBER must be filled with the number of the order that is to be changed. Type: BAPI_PP_ORDER_CHANGE.

[↑] Export

RETURN

Return parameters with messages. Type: BAPIRET2.

ORDER_NUMBER

Production order number. Type: BAPI_ORDER_KEY-ORDER_NUMBER.

ORDER_TYPE

Order type of the changed production or process order. Type: BAPI_ORDER_COPY-ORDER_TYPE.

---

619 BAPI_PRODORDCONF_CREATE_HDR

You can use this FM to enter order confirmations for production orders. Goods movements that are posted together with a confirmation can also be passed on. If no goods movements are specified for a confirmation, they are determined according to the standard logic for backflushes and automatic goods receipts relating to a confirmation. Authorization object C_AFRU_AWK for activity 01 (create) is checked for this FM.

Example: If you want to create the production order confirmation, you can pass the relevant parameters in this FM and call this FM.

[↓] Import

POST_WRONG_ENTRIES

This parameter controls which confirmations should be transferred to the error pool. Error pool entries can be corrected and posted through postprocessing of confirmation Transaction CO16N. Type: BAPI_CORU_PARAM-INS_ERR.

TESTRUN

Indicator that can be used to control whether confirmation data is to be provided for posting or whether the confirmation data is merely to be checked in test mode. In test mode, none of the objects is locked against any competing accesses. Value X doesn't save the data and only checks in the test mode. Value [Space] tests and saves the data to the database. Type: BAPI_CORU_PARAM-TESTRUN.

NO_DATA_RESET

Data isn't reset. Type: XFELD.

[↑] Export

RETURN

If confirmations can't be processed due to a major error, you receive information about the error in the return parameter. Type: BAPIRET1.

Tables

ATHDRLEVELS

The confirmation data is passed on in the ATHDRLEVELS table. Each entry in the ATHDRLEVELS table represents a separate confirmation that is checked and processed in the sequence in which it's transferred. Type: BAPI_PP_HDRLEVEL.

GOODSMOVEMENTS

Goods movements that are to be posted in dependence on the confirmations. Automatic goods movements can be posted with a confirmation. Type: BAPI2017_GM_ITEM_CREATE.

LINK_CONF_GOODSMOV

Linkage table for confirmation with goods movement. Items that aren't assigned to a confirmation can't be included in the process when the confirmation is saved. Type: BAPI_LINK_CONF_GOODSMOV.

DETAIL_RETURN

The parameter table DETAILRETURN provides information about the successful posting of each individual confirmation in the ATHDRLEVELS table or about any exceptions that occurred due to incorrect data or current lock situations. Type: BAPI_CORU_RETURN.

CHARACTERISTICS_BATCH

Table of classification characteristics to be entered for a batch that is to be created. Type: BAPI_CHAR_BATCH.

LINK_GM_CHAR_BATCH

The assignment of each individual characteristic valuation to a batch in a goods movement item is carried out via the parameter table LINKGMCHARBATCH. Characteristic valuations that aren't assigned to a goods movement item aren't taken into account. Type: BAPI_LINK_GM_CHAR_BATCH.

---

620 BAPI_PRODORDCONF_CREATE_TT

This FM is used to enter confirmation time tickets for operations in production orders. Goods movements that are posted together with a confirmation can also be passed on. If no goods movements are specified for a confirmation, they are determined according to the standard logic for backflushes and automatic goods receipts relating to a confirmation. Authorization object C_AFRU_AWK for activity 01 (create) is checked for this FM.

Example: You have a production order to produce a car. You have various operations to be performed such as fitting an engine, tires, transmission, chassis, and so on, and you want to enter time ticket confirmations for each of these operations in your production order. You can pass the relevant parameters in this FM and call this FM.

[↓] Import

POST_WRONG_ENTRIES

You can use the POSTWRONGENTRIES parameter to control whether in an exception situation (incorrect data/locked objects), the confirmations to be entered are nevertheless accepted by the SAP system and included in the error pool. Possible values:

- [Space]: Confirmations in which no exception situation occurred are posted. All other confirmations are rejected and not included in the error pool.
- 1: Confirmations in which no exception situation occurred are posted. All confirmations with incorrect data are included in the error pool. On the other hand, confirmations that can't be posted due to a lock situation aren't processed. That is to say, such confirmations aren't included in the error pool.
- 2: Confirmations in which no exception situation occurred are posted. All confirmations in respect of which faulty data or lock situations were detected are included in the error pool.

Type: BAPI_CORU_PARAM-INS_ERR.

TESTRUN

Indicator that can be used to control whether confirmation data is to be provided for posting or whether the confirmation data is merely to be checked in test mode. In test mode, none of the objects is locked against any competing accesses. Value X doesn't save the data and only checks in the test mode. Value [Space] tests and saves the data to the database. Type: BAPI_CORU_PARAM-TESTRUN.

[↑] Export

RETURN

If confirmations can't be processed due to a major error, you receive information about the error in the return parameter. Type: BAPIRET1.

Tables

TIMETICKETS

The data of the confirmation time tickets is passed on in the TIMETICKETS table. Type: BAPI_PP_TIMETICKET.

GOODSMOVEMENTS

Goods movements that are to be posted in dependence on the confirmations. Automatic goods movements can be posted with a confirmation. Type: BAPI2017_GM_ITEM_CREATE.

LINK_CONF_GOODSMOV

Linkage table for confirmation with goods movement. Items that aren't assigned to a confirmation can't be included in the process when the confirmation is saved. Type: BAPI_LINK_CONF_GOODSMOV.

CHARACTERISTICS_WIPBATCH

Table of characteristics of a WIP batch that is to be created. In the CHAR_NAME field, you have to specify the characteristic name, and in the CHAR_VALUE field the associated characteristic value. Type: BAPI_CHAR_WIPBATCH.

LINK_CONF_CHAR_WIPBATCH

The assignment of each individual characteristic valuation for a WIP batch is effected via the parameter table LINKCONFCHARWIPBATCH. Characteristic valuations that aren't assigned to any WIP batch aren't taken into consideration. Type: BAPI_LINK_CONF_CHAR_WIPBATCH.

DETAIL_RETURN

This table provides information for each confirmation to be entered regarding whether the confirmation was entered successfully, whether the entry was prevented due to a lock conflict, and/or which error occurred. Type: BAPI_CORU_RETURN.

CHARACTERISTICS_BATCH

Table of classification characteristics to be entered for a batch that is to be created. Type: BAPI_CHAR_BATCH.

LINK_GM_CHAR_BATCH

The assignment of each individual characteristic valuation to a batch in a goods movement item is carried out via the parameter table LINKGMCHARBATCH. Characteristic valuations that aren't assigned to a goods movement item aren't taken into account. Type: BAPI_LINK_GM_CHAR_BATCH.

621 BAPI_PS_INITIALIZATION

This FM is used to initialize the document tables, buffers, and so on, in the application in the project system module. Processing units are initialized, which are a series of related processing steps. In the first step processing unit is initialized by this FM. Next, individual project system FMs can be used several times. Finally, the processing unit ends when the final precommit FM BAPI_PS_PRECOMMIT (FM 622) is called.

Example: You want to change the status of the project to CLSD. First, you need to call this FM BAPI_PS_INITIALIZATION to initialize the processing unit. You'll use the following FMs to complete the process:

1. BAPI_BUS2001_SET_STATUS (FM 667) to change the status of the project.
2. FM BAPI_PS_PRECOMMIT (FM 622) to end the processing unit.
3. FM BAPI_TRANSACTION_COMMIT (FM 644) to save the processing unit changes to the database.

622 BAPI_PS_PRECOMMIT

This FM should be executed before the processing unit ends and before calling the `COMMIT WORK`. The final checks are carried out here for all objects created, changed, and deleted in the current processing unit, and the temporary IDs or numbers are swapped with the final IDs or numbers. The final network number is displayed in the return table of the precommit. The temporary network number is displayed in brackets behind the final network number in the message for CREATE NETWORK in the ET_RETURN table.

Example: You've changed the status of your project. Now you want to save the changes to ABAP Data Dictionary tables. You need to call this FM to check all of the objects and to get a permanent ID from the temporary ID that is saved in the tables.

Table

ET_RETURN

Return parameter. Type: BAPIRET2.

623 BAPI_QUOTATION_CREATEFROMDATA2

This FM is used to create a customer quotation.

Example: A customer has enquired about the price of the product. You want to give him the quotation for the enquiry. You can pass the relevant parameter in this FM and call this FM to create the quotation.

[↓] Import

SALESDOCUMENTIN

This parameter is for entering the sales document or quotation number. Type: BAPIVBELN-VBELN.

QUOTATION_HEADER_IN

Quotation header data. Type: BAPISDHD1.

QUOTATION_HEADER_INX

Fields of the quotation header data are changed when the respective checkbox of the field is checked. Type: BAPISDHD1X.

SENDER

Sender logical system details are passed in this parameter. Type: BAPI_SENDER.

BINARY_RELATIONSHIPTYPE

This parameter specifies the binary relationship type that is needed when writing object references.

The binary relationship type is defined in the VRBINRELATION view. Possible relationships are the following:

- VORL: Object references at header level.
- VORA: Object references at item level.

Type: BAPIRELTYPE-RELTYPE.

INT_NUMBER_ASSIGNMENT

This parameter activates internal item number assignment. Sometimes, internal item numbering from SAP is needed, for example, for articles that cause a BOM to explode in the sales order. Type: BAPIFLAG-BAPIFLAG.

BEHAVE_WHEN_ERROR

This parameter controls how the program behaves when an error occurs. This parameter makes it possible to create an order even if some of the items can't be created. Possible values:

- [Space]: If an error occurs, processing stops, and the sales document isn't saved.
- P: If an error occurs, the sales document can be saved. Problematic items aren't processed, but they are logged.
- R: Same as for P, but the document isn't saved.

Type: BAPIFLAG-BAPIFLAG.

LOGIC_SWITCH

This parameter controls how the program behaves internally. It de-activates certain types of processing when data is copied externally. Type: BAPISDLS.

TESTRUN

If this parameter has the value X, the document won't be saved; otherwise, it is saved. Type: BAPIFLAG-BAPIFLAG.

CONVERT

This parameter checks whether your customizing has been configured so that you can enter the abbreviation in your own language, and it's then converted into the German key. Value [Space] won't convert. Value X will convert. Type: BAPIFLAG-BAPIFLAG.

[↑] Export

SALESDOCUMENT

This parameter displays the quotation number created by this BAPI. Type: BAPIVBELN-VBELN.

Tables

RETURN

Return messages from this FM call. Type: BAPIRET2.

QUOTATION_ITEMS_IN

This parameter contains item input data for the quotation. Type: BAPISDITM.

QUOTATION_ITEMS_INX

Table to change item input data. Type: BAPISDITMX.

QUOTATION_PARTNERS

This table parameter is used to enter partners, such as sold-to party or ship-to party, both at the header and item level. Type: BAPIPARNR.

QUOTATION_SCHEDULES_IN

This table contains the schedule line data. Type: BAPISCHDL.

QUOTATION_SCHEDULES_INX

Table to change the schedule line data. Type: BAPISCHDLX.

QUOTATION_CONDITIONS_IN

This parameter contains condition data. Type: BAPICOND.

QUOTATION_CONDITIONS_INX

Table to change the condition data. Type: BAPICONDX.

QUOTATION_CFGS_REF

Configuration: reference data. Type: BAPICUCFG.

QUOTATION_CFGS_INST

Configuration: instances. Type: BAPICUINS.

QUOTATION_CFGS_PART_OF

Configuration: part-of specifications. Type: BAPICUPRT.

QUOTATION_CFGS_VALUE

Configuration: characteristic values. Type: BAPICUVAL.

QUOTATION_CFGS_BLOB

Configuration: BLOB internal data (SCE). Type: BAPICUBLB.

QUOTATION_CFGS_VK

Configuration: variant condition key. Type: BAPICUVK.

QUOTATION_CFGS_REFINST

Configuration: reference item/instance. Type: BAPICUREF.

QUOTATION_KEYS

Output table of reference keys. Type: BAPISDKEY.

QUOTATION_TEXT

Texts. Type: BAPISDTEXT.

EXTENSIONIN

You can use this parameter to transfer user-specific enhancements to a BAPI. The customer has to complete the EXTENSION structure, and the system automatically continues processing it. If you haven't defined any user-specific fields, then you can transfer a blank EXTENSIONIN structure. Type: BAPIPAREX.

PARTNERADDRESSES

You can enter document addresses in this entry table. Type: BAPIADDR1.

NFMETALLITMS

BAPI communication structure. Type: /NFM/BAPIDOCITM.

---

624 BAPI_REQUIREMENT_CREATE

This FM creates a requirement coverage request. A requirement coverage request can be a purchase requisition (if the requirement can't be covered on the desired date by existing stocks on hand in the warehouse or stores) or a reservation (if adequate stocks are on hand).

Example: You have a requirement from the customer that isn't fulfilled by the available stock. You need to create a purchase requisition requirement by passing the relevant parameter and calling this FM.

[↑] Export

NUMBER

Purchase requisition or a reservation number. Type: BAPIEBAN-PREQ_NO.

ITEM

This parameter is required for technical reasons only (the item number is defined in the BOR as a key field for the object PURCHASE requisition) and is always passed on with its initial value. Type: BAPIEBAN-PREQ_ITEM.

Tables

REQUIREMENT_ITEMS

Item data of requirement coverage request. Type: BAPIEBAN.

REQUIREMENT_ACCOUNT_ASSIGNMENT

Account assignment data of requirement coverage request. Type: BAPIEBKN.

REQUIREMENT_ITEM_TEXT

Text lines for requirement coverage request. Type: BAPIEBANTX.

RETURN

This table contains the error messages. Type: BAPIERRORS.

[!] Exception

NO_OBJECT_ID

Not possible to assign a number.

---

625 BAPI_REQUIREMENTS_CHANGE

This FM is used to change planned independent requirements (PIRs). The authorization object C_PPBD with the activity V is checked. Planned independent requirement is a SAP technical term for a forecast. Customer independent requirement (CIR) is a SAP technical term for a sales order. Based on the requirement strategy, CIRs can consume PIRs at specific statuses of sales orders. PIRs and CIRs together thus form the demand of an independent nature.

Example: You want to change the PIR that you've created as forecasting of your product because of the change in the forecasting of a material quantity. Pass the relevant parameter in this FM and call this FM to make the changes.

[↓] Import

MATERIAL

Material. Type: BAPISITEMR-MATERIAL.

PLANT

Plant. Type: BAPISITEMR-PLANT.

REQUIREMENTSTYPE

The requirements type determines the planning strategy to be used for a particular requirement. Type: BAPISITEMR-REQU_TYPE.

VERSION

Version number is used for organizational purposes for the creation and further use of the demand program. Type: BAPISITEMR-VERSION.

REQMTSPLANNUMBER

Key that can be given to create and to identify one or more planned independent requirements. Type: BAPISITEMR-REQ_NUMBER.

VERS_ACTIV

This parameter determines if the version is active and relevant for MRP. Type: BAPISITEMR-VERS_ACTIV.

REQUIREMENT_PARAM

This parameter isn't used in this FM. Type: CM60R.

MRP_AREA

An MRP area represents an organizational unit for which you can perform MRP separately. Type: BAPISITEMR-MRP_AREA.

DO_COMMIT

Value X defines that the FM will commit the changes; value [Space] defines that the FM won't commit the changes. Type: BAPISPARAM-DO_COMMIT.

UPDATE_MODE

You use this parameter to determine whether the BAPI is executed in the simulation mode. Value X determines that the planned independent requirements that have been created are posted. Value [Space] defines that no database changes are carried out. Type: BAPISPARAM-UPDATE_MODE.

DELETE_OLD

This parameter is used to determine if the existing schedule lines are deleted. Possible values:

- X: All existing schedule lines are deleted, and the transferred schedule lines are entered.
- [Space]: Existing schedule lines aren't deleted. Newly transferred schedule lines for existing requirement dates overwrite the existing schedule lines. Type: BAPISPARAM-DELETE_OLD.

NO_WITHDR

You can use this parameter to define whether the transferred planned quantity is reduced by the withdrawal quantity of the same period.

- X: The schedule lines are created with the transferred planned quantity.
- [Space]: The transferred planned quantities are reduced by the existing withdrawal quantity.

Type: BAPISPARAM-NO_WITHDR.

MATERIAL_EVG

Long material number. Type: BAPIMGVMATNR.

[↑] Export

REQUIREMENT_ITEM_OUT

This parameter contains the data for the item whose schedule lines have been changed. Type: BAPISITEMR.

Tables

REQUIREMENTS_SCHEDULE_IN

This parameter is used to define scheduling lines that are to be created. Type: BAPISSHDIN.

REQUIREMENTS_CHAR_IN

This parameter is used to create configuration supporting points for planned independent requirements. Type: BAPISCHARR.

RETURN

Return parameter with error messages. Type: BAPIRETURN1.

626 BAPI_REQUIREMENTS_CREATE

This FM is used to create planned independent requirements. The authorization object C_PPBD with activity H is checked.

Example: You want to create the planned independent requirement. You can pass the relevant parameter in this FM and call this FM to create the PIR.

[↓] Import

REQUIREMENTS_ITEM

Item data for which the planned independent requirements are to be created. Type: BAPISITEMR.

REQUIREMENT_PARAM

This parameter isn't used in this FM. Type: CM60R.

DO_COMMIT

Value X defines that the FM will commit the changes; value [Space] defines that the FM won't commit the changes. Type: BAPISPARAM-DO_COMMIT.

UPDATE_MODE

You use this parameter to determine whether the BAPI is executed in the simulation mode. Value X determines that the planned independent requirements that have been created are posted. Value [Space] defines that no database changes are carried out. Type: BAPISPARAM-UPDATE_MODE.

REFER_TYPE

This parameter defines which reference type (e.g., sales plan, material forecast) is preset if you want to use a reference to create the planned independent requirements. Type: BAPISPARAM-REFER_TYPE.

PROFILID

You use this parameter to define the planning profile for which configuration data should be created. Type: BAPISPARAM-PROFILID.

[↑] Export

MATERIAL

Material number. Type: BAPISITEMR-MATERIAL.

PLANT

Plant. Type: BAPISITEMR-PLANT.

REQUIREMENTSTYPE

This parameter determines the planning strategy to be used for a particular requirement. Type: BAPISITEMR-REQU_TYPE.

VERSION

Version. Type: BAPISITEMR-VERSION.

REQMTSPLANNUMBER

This parameter defines the key that can be given to create and to identify one or more planned independent requirements. Type: BAPISITEMR-REQ_NUMBER.

MRP_AREA

MRP area is the organizational unit for which you can perform MRP. Type: BAPISITEMR-MRP_AREA.

MATERIAL_EVG

This parameter can be used to transfer long material numbers that can contain a maximum of 40 characters. To use this parameter, you must activate the switch for the LAMA (long material number) business function in the system. The LAMA business function is available as part of the business function set for DIMP. Type: BAPIMGVMATNR.

Tables

REQUIREMENTS_SCHEDULE_IN

This parameter is used to define scheduling lines that are to be created. Type: BAPISSHDIN.

REQUIREMENTS_CHAR_IN

This parameter is used to create configuration supporting points for planned independent requirements. Type: BAPISCHARR.

RETURN

Return parameter with error messages. Type: BAPIRETURN1.

---

627 BAPI_REQUISITION_CHANGE

This FM is used to change purchase requisitions. Authorization objects M_BANF_BSA (document type in requisition), M_BANF_EKG (purchasing group in requisition), M_BANF_EKO (purchase organization in requisition), and M_BANF_WRK (plant in requisition) are checked.

Example: You want to change the purchase requisition data such as purchase requisition quantity. You can pass the relevant parameters in this FM and call this FM to change the requisition.

[↓] Import

NUMBER

Purchase requisition number. Type: BAPIEBAN-PREQ_NO.

Tables

REQUISITION_ITEMS_OLD

The table contains the current database status of the purchase requisition item (i.e., prior to the change). Type: BAPIEBANV.

REQUISITION_ITEMS_NEW

The table contains the purchase requisition item with the changed values. Type: BAPIEBANV.

REQUISITION_ACCOUNT_OLD

The table contains the current database status of the account assignment data for the purchase requisition item (i.e., prior to the change). Type: BAPIEBKNV.

REQUISITION_ACCOUNT_NEW

The table contains the account assignment data with the changed values. Type: BAPIEBKNV.

REQUISITION_TEXT_OLD

The table contains the current database status of the texts for the purchase requisition item (i.e., prior to the change). Type: BAPIEBANTX.

REQUISITION_TEXT_NEW

The table contains the changed texts for the purchase requisition item. Type: BAPIEBANTX.

RETURN

This table contains the return messages from the method call. Type: BAPIRETURN.

---

628 BAPI_REQUISITION_CREATE

This FM is used to create purchase requisitions. Authorization objects M_BANF_BSA (document type in requisition), M_BANF_EKG (purchasing group in requisition), M_BANF_EKO (purchase organization in requisition), and M_BANF_WRK (plant in requisition) are checked.

Example: You want to purchase 10 quantities of Material M1. You need to create the purchase requisition data for it. Pass the relevant parameters in this FM and call this FM to create the purchase requisition.

[↓] Import

SKIP_ITEMS_WITH_ERROR

If the indicator has been set, faulty items are skipped, and only the error-free items are created. Type: BAPIMMPARA-SELECTION.

AUTOMATIC_SOURCE

Automatic source determination. The default value is X. Type: BAPIMMPARA-SELECTION.

[↑] Export

NUMBER

Purchase requisition number. Type: BAPIEBANC-PREQ_NO.

Tables

REQUISITION_ITEMS

Item data. Type: BAPIEBANC.

REQUISITION_ACCOUNT_ASSIGNMENT

Account assignment data for purchase requisition item. Type: BAPIEBKN.

REQUISITION_ITEM_TEXT

Text for purchase requisition item. Type: BAPIEBANTX.

REQUISITION_LIMITS

Limits for the purchase requisition. Type: BAPIESUHC.

REQUISITION_CONTRACT_LIMITS

Purchase requisition limits with contract reference. Type: BAPIESUCC.

REQUISITION_SERVICES

This table contains services and the service outline of the purchase requisition. Type: BAPIESLLC.

REQUISITION_SRV_ACCASS_VALUES

The table contains the values of the account assignments at the service level and a link to the overall account assignment of the PO item. Type: BAPIESKLC.

RETURN

Return messages from the method call. Type: BAPIRETURN.

REQUISITION_SERVICES_TEXT

Text for the service line. Type: BAPIESLLTX.

REQUISITION_ADDRDELIVERY

PO item: address structure BAPIADDR1 for inbound delivery. Type: BAPIMERQADDRDELIVERY.

EXTENSIONIN

Reference structure for BAPI parameters EXTENSIONIN/EXTENSIONOUT. Type: BAPIPAREX.

629 BAPI_REQUISITION_DELETE

This FM is used to delete or close the purchase requisition. Authorization objects M_BANF_BSA (document type in requisition), M_BANF_EKG (purchasing group in requisition), M_BANF_EKO (purchase organization in requisition), and M_BANF_WRK (plant in requisition) are checked. All requisition items for which the DELETED indicator and/or the CLOSED indicator are set are passed on in the table REQUISITIONITEMSTODELETE. Because this method changes the purchase requisition, change documents are created when you save.

Example: You want to delete the purchase requisition data. You can pass the relevant parameters in this FM and call this FM to delete the purchase requisition.

[↓] Import

NUMBER

Purchase requisition number. Type: BAPIEBANC-PREQ_NO.

Tables

REQUISITION_ITEMS_TO_DELETE

The table contains all purchase requisition items for which the DELETED and/or CLOSED indicator are to be set. Type: BAPIEBAND.

RETURN

Messages are returned in the return parameter. Type: BAPIRETURN.

---

630 BAPI_REQUISITION_RELEASE

This FM is used to release or approve the purchase requisition item by item. Items that satisfy the condition defined in the release strategy are released/approved.

Example: All purchase requisition with the value equal to $4,000 must be released. You can pass the relevant parameters in this FM and call this FM to release the purchase requisition item.

[↓] Import

NUMBER

Purchase requisition. Type: BAPI2009OB-PREQ_NO.

REL_CODE

Two-character identifier with which a person responsible for processing a document can release (approve) an item of a purchase requisition, a complete purchase requisition, or a complete external purchasing document, or cancel a release (i.e., revoke an approval). Type: BAPIMMPARA-REL_CODE.

ITEM

Purchase requisition item. Type: BAPI2009OB-PREQ_ITEM.

USE_EXCEPTIONS

If you set this indicator, the exceptions are passed on in addition to the messages in the parameter RETURN. In the standard system, the indicator is set. Type: BAPIMMPARA-SELECTION.

NO_COMMIT_WORK

If this parameter is set to X, this FM won't call `COMMIT WORK` internally. Type: BAPIFLAG-BAPIFLAG.

[↑] Export

REL_STATUS_NEW

In this parameter, the new release status is returned (i.e., the release status after the release has been effected). Type: BAPIMMPARA-REL_STATUS.

REL_INDICATOR_NEW

In this parameter, the new release indicator is returned (i.e., the release status after the release has been effected). Type: BAPIMMPARA-REL_IND.

Table

RETURN

Return messages. Type: BAPIRETURN.

[!] Exceptions

AUTHORITY_CHECK_FAIL

No authorization to release.

REQUISITION_NOT_FOUND

Purchase requisition does not exist.

ENQUEUE_FAIL

Purchase order blocked.

PREREQUISITE_FAIL

Release prerequisite not satisfied.

RELEASE_ALREADY_POSTED

Release already effected.

RESPONSIBILITY_FAIL

Responsibility for release missing.

631 BAPI_RESERVATION_CHANGE

This FM is used to change the reservation. A reservation can be created for an article or for a number of articles. Each reservation contains information on the use, article number, planned quantity, planned date of requirement/receipt, and the valuation type. A reservation enables your actual stock on hand to be divided into unrestricted-use stock and reserved stock. This also enables you to define the sequence in which customers or other recipients are supplied. A reservation for a goods issue is a request to the warehouse to reserve articles for issue at a later date for a specific purpose. With this FM, you can change the reservation items such as storage location, batch number, quantity in unit of entry, requirement dates of component, item text, goods receipt, and so on.

Example: You've reserved 10 units of a stock good, but you want to change the quantity to 7 so that you can use the other 3 units for something else. First you need to get the details of the reservation you want to change via FM BAPI_RESERVATION_GETDETAILS1 (FM 633). Then use this FM to change the reservation.

[↓] Import

RESERVATION

Reservation number you want to change. Type: BAPI2093_RES_KEY-RESERV_NO.

TESTRUN

If this parameter is set to X, reservation changes aren't saved in the database. Type: BAPI2093_TEST.

ATPCHECK

Available to promise (ATP) is the function to check if the product availability can be confirmed. If this parameter is set to X, the ATP check is executed if you change the reservation. Type: BAPI2093_ATPCHECK.

Tables

RESERVATIONITEMS_CHANGED

This parameter contains the item data of the reservation to be changed. Type: BAPI2093_RES_ITEM_CHANGE.

RESERVATIONITEMS_CHANGEDX

This parameter contains the change structure of the reservation item table. Type: BAPI2093_RES_ITEM_CHANGEX.

RESERVATIONITEMS_NEW

This parameter contains the item data of the reservation items to be added. Type: BAPI2093_RES_ITEM_NEW.

RETURN

Return parameters messages from this FM call. Type: BAPIRET2.

EXTENSIONIN

This parameter is used for customer enhancement. Type: BAPIPAREX.

632 BAPI_RESERVATION_CREATE1

This FM is used to create the reservation for a material. A reservation can be created for an article or for a number of articles. Each reservation contains information on the use, article number, planned quantity, planned date of requirement/receipt, and the valuation type. A reservation enables your actual stock on hand to be divided into unrestricted-use stock and reserved stock. This also enables you to define the sequence in which customers or other recipients are supplied. Reservation for a goods issue is a request to the warehouse to reserve articles for issue at a later date for a specific purpose. With this FM, you can change the reservation items such as storage location, batch number, quantity in unit of entry, requirement dates of component, item text, goods receipt, and so on.

Example: You want to reserve 10 quantity of a stock good. You can pass the relevant parameter in this FM to create the reservation.

[↓] Import

RESERVATIONHEADER

The structure with the header data of the reservation is transferred in this parameter. Type: BAPI2093_RES_HEAD.

TESTRUN

If this parameter is set to X, reservation changes aren't saved in the database. Type: BAPI2093_TEST.

ATPCHECK

Available to promise (ATP) is the function to check if the product availability can be confirmed. If this parameter is set to X, the ATP check is executed if you change the reservation. Type: BAPI2093_ATPCHECK.

CALCHECK

If this parameter is set to X, the system checks during creation of the reservation whether the particular reservation date falls on a working day. If this isn't the case, the system issues a corresponding message via the return table. Type: BAPI2093_CAL-CHECK.

RESERVATION_EXTERNAL

External reservation number. Type: BAPI2093_RES_KEY-RESERV_NO.

[↑] Export

RESERVATION

Reservation number created from this FM is returned. This reservation number uniquely identifies the reservation. If the BAPI is called in Testrun mode, this parameter is initial ([Space]). Type: BAPI2093_RES_KEY-RESERV_NO.

Tables

RESERVATIONITEMS

The table contains the item data for the reservation that is to be created. Type: BAPI2093_RES_ITEM.

PROFITABILITYSEGMENT

Characteristics and characteristic values for the profitability segment. Type: BAPI_PROFITABILITY_SEGMENT.

RETURN

Return parameters messages from this FM call. Type: BAPIRET2.

EXTENSIONIN

This parameter is used for customer enhancement. Type: BAPIPAREX.

---

633 BAPI_RESERVATION_GETDETAIL1

This FM is used to get the reservation details. Authorization objects M_RES_BWA (reservation movement type authorization) and M_MRES_WWA (reservation plant authorization) are checked for activity 03.

Example: You have reserved some quantity of a material MAT, and you want to get the reservation details such as reservation date, user name who created reservation, cost center, sales order number, and so on. Pass the parameter in this FM. See FM BAPI_RESERVATION_CHANGE (FM 631).

[↓] Import

RESERVATION

Reservation number created from this FM is returned. This reservation number uniquely identifies the reservation. If the BAPI is called in Testrun mode, this parameter is initial (Space). Type: BAPI2093_RES_KEY-RESERV_NO.

[↑] Export

RESERVATIONHEADER

The structure with the header data of the reservation is transferred in this parameter. Type: BAPI2093_RES_HEAD.

Tables

RESERVATIONITEMS

The table contains the item data for the reservation that is to be created. Type: BAPI2093_RES_ITEM.

RETURN

Return parameters messages from this FM call. Type: BAPIRET2.

EXTENSIONIN

This parameter is used for customer enhancement. Type: BAPIPAREX.

634 BAPI_ROUTING_CREATE

This FM is used to create a routing with all its subobjects. You can also generate the key date view and the complete history of the routing. Routing is the representation of the flow or sequence that should be followed to build a product such as a car in the shop floor.

Example: You're making a car, so you need a chassis, transmission, engine, tires, and so on. You have to perform some steps or route one after the other to assemble all of these parts together to make the car. For this, you want to create a routing with the sequence, operations, and inspection characteristics for assembling these products together to produce a car. You can pass the relevant parameters and call this FM to create the routing.

[↓] Import

TESTRUN

If this parameter has the value X, routing data isn't saved to the database; otherwise, it is saved. Type: BAPIFLAG.

PROFILE

Profile to be used. When entering a profile, you can define default values for specific attributes of the routing. The system always goes back to these if the relevant value hasn't been transferred. If the profile entry is missing, the following attributes have to be supplied in the task list header (structure TASK):

- PLANT
- TASK_MEASURE_UNIT
- TASK_LIST_USAGE
- TASK_LIST_STATUS

Type: BAPI1012_CONTROL_DATA-PROFILE.

BOMUSAGE

BOM usage. You have to enter the BOM usage when creating a routing if you want to create component assignments as well. The BOM usage controls the selection of the relevant BOM in the component assignment. Type: BAPI1012_CONTROL_DATA-BOM_USAGE.

APPLICATION

BOM application. You have to enter the BOM application when creating a routing if you want to create component assignments as well. The BOM application, along with the BOM usage, controls the selection of the correct alternative BOM in the component assignment. Type: BAPI1012_CONTROL_DATA-APPLICATION.

[↑] Export

GROUP

Key for task list group. In this parameter, the identifier of the generated routing is available for the successful transfer of the copied data. You can access the routing by this parameter. Type: BAPI1012_TSK_C-TASK_LIST_GROUP.

GROUPCOUNTER

Group counter. By the successful transfer of the copied data, the group counter of the generated routing is available in this parameter. You can access the routing by this parameter along with the GROUP parameter. Type: BAPI1012_TSK_C-GROUP_COUNTER.

Tables

TASK

Header data of the routing to be created is transferred in this structure. Type: BAPI1012_TSK_C.

MATERIALTASKALLOCATION

Material task list assignment. The data of the component assignment to be generated are copied in this table. Type: BAPI1012_MTK_C.

SEQUENCE

Task list sequences to be generated are passed in this table. You can only enter alternative or parallel sequences. You're not permitted to enter standard sequences because these are generated automatically when creating the task list header. Type: BAPI1012_SEQ_C.

OPERATION

The data of the operations to be generated is copied into this table. Type: BAPI1012_OPR_C.

SUBOPERATION

Suboperations details to be generated can be passed in this table. Type: BAPI1012_SUB_OPR_C.

REFERENCEOPERATION

The data of the reference operation set reference to be generated is copied into this table. Type: BAPI1012_REF_OPR_C.

WORKCENTERREFERENCE

The data for the standard tasks to be generated is copied into this table. Type: BAPI1012_WC_REF_OPR_C.

COMPONENTALLOCATION

The data of the component assignments to be generated is copied into this table. The assignment takes place by entering the operation and item data. Type: BAPI1012_COM_C.

PRODUCTIONRESOURCE

PRT assignments to be created are copied in this table. Type: BAPI1012_PRT_C.

INSPCHARACTERISTIC

Inspection characteristics data is copied in this table. Type: BAPI1012_CHA_C.

TEXTALLOCATION

Text assignment. You can assign long texts when creating or changing routings. Type: BAPI1012_TXT_HDR_C.

TEXT

Long texts to be processed when creating or changing routings are transferred into this table line by line. Using both attributes LINE_FROM and LINE_TO from the table TEXT ALLOCATION, you determine the long text lines assigned to an object. Type: BAPI1012_TXT_C.

RETURN

Return parameter. All of the messages that occur when creating a routing are present in this table. Type: BAPIRET2.

635 BAPI_SAG_CHANGE

This FM is used to change the details of the scheduling agreement. A scheduling agreement is similar to a contract, which contains the quantity of the article to be ordered from the vendor and the price. A schedule agreement item is entered for each article you want to procure. Authorization objects M_RAHM_BSA (document type in outline agreement), M_RAHM_EKG (purchasing group in outline agreement), M_RAHM_EKO (purchasing organization in outline agreement), and M_RAHM_EKG (plant in outline agreement) for activity 02 are checked.

Example: You want to purchase an item from a vendor, and you've created the schedule agreement. Now, you want to change some details of the items of the scheduling agreement. First you need to get the details of the schedule agreement via FM BAPI_SAG_GETDETAIL (FM 637). Then pass the schedule agreement details and other details you want to change in this FM to change a schedule agreement.

[↓] Import

PURCHASINGDOCUMENT

Purchasing document number. Type: BAPIMEOUTHEADER-NUMBER.

HEADER

Header data of the outline schedule agreement. Type: BAPIMEOUTHEADER.

HEADERX

Header data change parameter of the outline agreement. Type: BAPIMEOUTHEADERX.

VENDOR_ADDRESS

Vendor address that differs from the master record. Type: BAPIMEOUTADDRVENDOR.

HEAD_EXPORT_IMPORT

The parameter contains the data fields for foreign trade processing at the header level of the scheduling agreement instance. Type: BAPIMEOUTEIKP.

HEAD_EXPORT_IMPORTX

The parameter contains the change parameter for the foreign trade data of a scheduling agreement. Type: BAPIMEOUTEIKPX.

TESTRUN

Value X defines the simulation mode, and changes aren't saved to the database. Type: BAPIFLAG-BAPIFLAG.

TECHNICAL_DATA

Technical parameters for the outline agreement. The following fields can be populated:

- MESSAGING_APPL: Selection of application for message output.
- NO_MESSAGING: No message determination.
- NO_MESSAGE_REQ: Without message output.
- NO_AUTHORITY_CHECK: Don't perform authorization check.

Type: BAPIMEOUTTECH.

[↑] Export

EXP_HEADER

Header data of schedule outline agreement returned. Type: BAPIMEOUTHEADER.

EXP_HEAD_EXPORT_IMPORT

The parameter contains the foreign trade data of a scheduling agreement. Type: BAPIMEOUTEIKP.

Tables

RETURN

Return table. Type: BAPIRET2.

ITEM

Item data for scheduling the outline agreement. Type: BAPIMEOUTITEM.

ITEMX

Change parameter for item data outline agreement. Type: BAPIMEOUTITEMX.

ACCOUNT

The parameter contains the account assignment data for scheduling agreement items. Type: BAPIMEOUTACCOUNT.

ACCOUNTPROFITSEGMENT

The parameter contains the data on the basis of which a new profitability segment (CO-PA) is created for the account assignment. Type: BAPIMEOUTPROFITSEGMENT.

ACCOUNTX

The parameter contains the change parameter for the account assignment data for scheduling agreement items. Type: BAPIMEOUTACCOUNTX.

SCHEDULE

The parameter contains the delivery schedule lines of a scheduling agreement. Type: BAPIMEOUTSCHEDULE.

SCHEDULEX

The parameter contains the change parameter for the delivery schedule lines of a scheduling agreement. Type: BAPIMEOUTSCHEDULEX.

SC_COMPONENT

Subcontracting components. the CHANGE_ID field can have the following instances:

- I: Add new component.
- U: Change existing component.
- D: Delete existing component. In the case of an assembly, the entire assembly is deleted. The deletion of assembly components isn't possible.
- R: The existing subcontracting components are to be deleted, and the components belonging to the schedule line from the BAPI interface are to be added.

Type: BAPIMEOUTCOMPONENT.

SC_COMPONENTX

Change parameter for subcontracting components. Type: BAPIMEOUTCOMPONENTX.

SHIPPING

Shipping data scheduling agreement. In the case of a stock transfer item, the parameter contains the shipping data information (EKPV), which can also be changed in the transaction. Type: BAPIMEOUTSHIPP.

SHIPPINGX

Change parameter shipping data scheduling agreement. Type: BAPIMEOUTSHIPPX.

SHIPPING_EXP

Export structure for shipping data. Type: BAPIMEPOSHIPPEXP.

DELIVERY_ADDRESS

Delivery address in the outline agreement. Type: BAPIMEOUTADDRDELIVERY.

ITEM_COND_VALIDITY

Validity period of conditions in agreement. Type: BAPIMEOUTVALIDITY.

ITEM_COND_VALIDITYX

Change parameter of validity period of condition. Type: BAPIMEOUTVALIDITYX.

ITEM_CONDITION

This parameter contains data such as the condition amount and condition currency. Type: BAPIMEOUTCONDITION.

ITEM_CONDITIONX

Change parameter for conditions in outline agreement. Type: BAPIMEOUTCONDITIONX.

ITEM_COND_SCALE_VALUE

This parameter contains the value scales for a condition of an outline agreement item. Type: BAPIMEOUTITEMSCALEVAL.

ITEM_COND_SCALE_QUAN

This parameter contains the quantity scales for a condition of an outline agreement item. Type: BAPIMEOUTITEMSCALEQUAN.

EXPORT_IMPORT

Foreign trade data: outline agreement item. Type: BAPIMEOUTEIPO.

EXPORT_IMPORTX

Change parameter for foreign trade item. Type: BAPIMEOUTEIPOX.

ITEM_TEXT

The parameter contains the long texts for a scheduling agreement item. Type: BAPIMEOUTITEMTEXT.

HEADER_TEXT

This parameter contains the long texts of the scheduling agreement header. Type: BAPIMEOUTTEXT.

HEAD_COND_VALIDITY

This parameter contains the validity periods for conditions for the document header. Type: BAPIMEOUTHEADVALIDITY.

HEAD_COND_VALIDITYX

Change parameter for header condition validity outline agreement. Type: BAPIMEOUTHEADVALIDITYX.

HEAD_CONDITION

This parameter contains data such as condition amount and condition currency for the header. Type: BAPIMEOUTHEADERCOND.

HEAD_CONDITIONX

Change parameter header conditions outline agreement. Type: BAPIMEOUTHEADERCONDX.

HEAD_COND_SCALE_VAL

This parameter contains the value scales for a header condition. Type: BAPIMEOUTSCALEVALUE.

HEAD_COND_SCALE_QUAN

This parameter contains the quantity scales for a header condition. Type: BAPIMEOUTSCALE.

PARTNER

The parameter contains the partner data for a scheduling agreement. You can specify the partners at the following levels:

- Purchasing organization
- Purchasing organization and plant
- Purchasing organization, plant, and vendor subrange

Type: BAPIMEOUTPARTNER.

PARTNERX

Change parameter for partner outline agreement. Type: BAPIMEOUTPARTNERX.

EXTENSIONIN

With the parameters EXTENSIONIN and EXTENSIONOUT, it's possible to enhance the interface of a BAPI without modification to achieve the automatic processing of customer-specific data. Type: BAPIPAREX.

EXTENSIONOUT

With the parameters EXTENSIONIN and EXTENSIONOUT, it's possible to enhance the interface of a BAPI without modification to achieve the automatic processing of customer-specific data. Type: BAPIPAREX.

636 BAPI_SAG_CREATE

This FM is used to create a scheduling agreement. Scheduling agreements are similar to contracts, which contain the quantity of the article that is to be ordered from the vendor and the price. A schedule agreement item is entered for each article you want to procure. Authorization objects M_RAHM_BSA (document type in outline agreement), M_RAHM_EKG (purchasing group in outline agreement), M_RAHM_EKO (purchasing organization in outline agreement), and M_RAHM_EKG (plant in outline agreement) for activity 02 are checked.

Example: If you want to purchase an item from the vendor, you can create the schedule agreement of the items you want to procure, including all of their details. You can pass the relevant parameter in this FM and call this FM to create a schedule agreement.

[↓] Import

HEADER

Header data of the outline schedule agreement. Type: BAPIMEOUTHEADER.

HEADERX

Header data change parameter of the outline agreement. Type: BAPIMEOUTHEADERX.

VENDOR_ADDRESS

Vendor address that differs from the master record. Type: BAPIMEOUTADDRVENDOR.

HEAD_EXPORT_IMPORT

The parameter contains the data fields for foreign trade processing at the header level of the scheduling agreement instance. Type: BAPIMEOUTEIKP.

HEAD_EXPORT_IMPORTX

The parameter contains the change parameter for the foreign trade data of a scheduling agreement. Type: BAPIMEOUTEIKPX.

TESTRUN

Value X defines the simulation mode, and changes aren't saved to database. Type: BAPIFLAG-BAPIFLAG.

TECHNICAL_DATA

Technical parameters for outline agreement. The following fields can be populated:

- MESSAGING_APPL: Selection of application for message output.
- NO_MESSAGING: No message determination.
- NO_MESSAGE_REQ: Without message output.
- NO_AUTHORITY_CHECK: Don't perform authorization check.

Type: BAPIMEOUTTECH.

[↑] Export

PURCHASINGDOCUMENT

Purchasing document number. Type: BAPIMEOUTHEADER-NUMBER.

EXP_HEADER

Header data of schedule outline agreement returned. Type: BAPIMEOUTHEADER.

Tables

RETURN

Return table. Type: BAPIRET2.

ITEM

Item data for scheduling outline agreement. Type: BAPIMEOUTITEM.

ITEMX

Change parameter for item data outline agreement. Type: BAPIMEOUTITEMX.

ACCOUNT

The parameter contains the account assignment data for scheduling agreement items. Type: BAPIMEOUTACCOUNT.

ACCOUNTPROFITSEGMENT

The parameter contains the data on the basis of which a new profitability segment (CO-PA) is created for the account assignment. Type: BAPIMEOUTPROFITSEGMENT.

ACCOUNTX

The parameter contains the change parameter for the account assignment data for scheduling agreement items. Type: BAPIMEOUTACCOUNTX.

SCHEDULE

The parameter contains the delivery schedule lines of a scheduling agreement. Type: BAPIMEOUTSCHEDULE.

SCHEDULEX

The parameter contains the change parameter for the delivery schedule lines of a scheduling agreement. Type: BAPIMEOUTSCHEDULEX.

SC_COMPONENT

Subcontracting components. the CHANGE_ID field can have the following instances:

- I: Add new component.
- U: Change existing component.
- D: Delete existing component. In the case of an assembly, the entire assembly is deleted. The deletion of assembly components isn't possible.
- R: The existing subcontracting components are to be deleted, and the components belonging to the schedule line from the BAPI interface are to be added.

Type: BAPIMEOUTCOMPONENT.

SC_COMPONENTX

Change parameter for subcontracting components. Type: BAPIMEOUTCOMPONENTX.

SHIPPING

Shipping data scheduling agreement. In the case of a stock transfer item, the parameter contains the shipping data information (EKPV), which can also be changed in the transaction. Type: BAPIMEOUTSHIPP.

SHIPPINGX

Change parameter shipping data scheduling agreement. Type: BAPIMEOUTSHIPPX.

SHIPPING_EXP

Export structure for shipping data. Type: BAPIMEPOSHIPPEXP.

DELIVERY_ADDRESS

Delivery address in the outline agreement. Type: BAPIMEOUTADDRDELIVERY.

ITEM_COND_VALIDITY

Validity period of conditions in agreement. Type: BAPIMEOUTVALIDITY.

ITEM_COND_VALIDITYX

Change parameter of validity period of condition. Type: BAPIMEOUTVALIDITYX.

ITEM_CONDITION

This parameter contains data such as the condition amount and condition currency. Type: BAPIMEOUTCONDITION.

ITEM_CONDITIONX

Change parameter for conditions in the outline agreement. Type: BAPIMEOUTCONDITIONX.

ITEM_COND_SCALE_VALUE

This parameter contains the value scales for a condition of an outline agreement item. Type: BAPIMEOUTITEMSCALEVAL.

ITEM_COND_SCALE_QUAN

This parameter contains the quantity scales for a condition of an outline agreement item. Type: BAPIMEOUTITEMSCALEQUAN.

EXPORT_IMPORT

Foreign trade data: outline agreement item. Type: BAPIMEOUTEIPO.

EXPORT_IMPORTX

Change parameter for foreign trade item. Type: BAPIMEOUTEIPOX.

ITEM_TEXT

The parameter contains the long texts for a scheduling agreement item. Type: BAPIMEOUTITEMTEXT.

HEADER_TEXT

This parameter contains the long texts of the scheduling agreement header. Type: BAPIMEOUTTEXT.

HEAD_COND_VALIDITY

This parameter contains the validity periods for conditions for the document header. Type: BAPIMEOUTHEADVALIDITY.

HEAD_COND_VALIDITYX

Change parameter for header condition validity outline agreement. Type: BAPIMEOUTHEADVALIDITYX.

HEAD_CONDITION

This parameter contains data such as condition amount and condition currency for the header. Type: BAPIMEOUTHEADERCOND.

HEAD_CONDITIONX

Change parameter header conditions outline agreement. Type: BAPIMEOUTHEADERCONDX.

HEAD_COND_SCALE_VAL

This parameter contains the value scales for a header condition. Type: BAPIMEOUTSCALEVALUE.

HEAD_COND_SCALE_QUAN

This parameter contains the quantity scales for a header condition. Type: BAPIMEOUTSCALE.

PARTNER

The parameter contains the partner data for a scheduling agreement. You can specify the partners at the following levels:

- Purchasing organization
- Purchasing organization and plant
- Purchasing organization, plant, and vendor subrange.

Type: BAPIMEOUTPARTNER.

PARTNERX

Change parameter for partner outline agreement. Type: BAPIMEOUTPARTNERX.

EXTENSIONIN

With the parameters EXTENSIONIN and EXTENSIONOUT, it's possible to enhance the interface of a BAPI without modification to achieve the automatic processing of customer-specific data. Type: BAPIPAREX.

EXTENSIONOUT

With the parameters EXTENSIONIN and EXTENSIONOUT, it's possible to enhance the interface of a BAPI without modification to achieve the automatic processing of customer-specific data. Type: BAPIPAREX.

---

637 BAPI_SAG_GETDETAIL

This FM is used to view the scheduling agreement details. Authorization objects M_RAHM_BSA (document type in outline agreement), M_RAHM_EKG (purchasing group in outline agreement), M_RAHM_EKO (purchasing organization in outline agreement), and M_RAHM_EKG (plant in outline agreement) for activity 02 are checked.

Example: You've created a scheduling agreement for the items you want to order from the vendor. You want to view the scheduling agreement and its details such as agreement number, company code, agreement type, discount if agreed on, and so on. See FM BAPI_SAG_CHANGE (FM 635).

[↓] Import

PURCHASINGDOCUMENT

Purchasing document number. Type: BAPIMEOUTHEADER-NUMBER.

ITEM_DATA

Item data that you want to read. Type: BAPIFLAG-BAPIFLAG.

ACCOUNT_DATA

Account assignment data. Possible values:

- X: The account assignment data for all items is read out.
- [Space]: The account assignment data isn't read.

Type: BAPIFLAG-BAPIFLAG.

SCHEDULE_DATA

Schedule line data. Possible values:

- X: The scheduling data for all items is read out.
- [Space]: The scheduling data isn't passed on to the invoker.

Type: BAPIFLAG-BAPIFLAG.

SHIPPING_DATA

Shipping data. Possible values:

- X: The shipping data for all items is read out.
- [Space]: The shipping data isn't passed on to the invoker.

Type: BAPIFLAG-BAPIFLAG.

SC_COMPONENT_DATA

Subcontracting components. Possible values:

- X: The subcontracting components for all items are read out.
- [Space]: The subcontracting components aren't passed on to the invoker.

Type: BAPIFLAG-BAPIFLAG.

EXPORT_DATA

Foreign trade data. Possible values:

- X: The foreign trade data for all items are read out.
- [Space]: The foreign trade data aren't passed on to the invoker.

Type: BAPIFLAG-BAPIFLAG.

CONDITION_DATA

Conditions. Possible values:

- X: The condition data for all items are read out.
- [Space]: The condition data aren't passed on to the invoker.

Type: BAPIFLAG-BAPIFLAG.

TEXT_DATA

Long texts. Possible values:

- X: The long text data for all items are read out.
- [Space]: The long text data aren't passed on to the invoker.

Type: BAPIFLAG-BAPIFLAG.

PARTNER_DATA

Partner data. Possible values:

- X: The partner data for all items are read out.
- [Space]: The partner data aren't passed on to the invoker.

Type: BAPIFLAG-BAPIFLAG.

[↑] Export

HEADER

Header data of schedule outline agreement returned. Type: BAPIMEOUTHEADER.

HEAD_EXPORT_IMPORT

The parameter contains the foreign trade data of a scheduling agreement. Type: BAPIMEOUTEIKP.

Tables

ITEM

Item data for scheduling outline agreement. Type: BAPIMEOUTITEM.

ACCOUNT

The parameter contains the account assignment data for scheduling agreement items. Type: BAPIMEOUTACCOUNT.

SCHEDULE

The parameter contains the delivery schedule lines of a scheduling agreement. Type: BAPIMEOUTSCHEDULE.

DELIVERY_ADDRESS

Delivery address in the outline agreement. Type: BAPIMEOUTADDRDELIVERY.

ITEM_COND_VALIDITY

Validity period of conditions in agreement. Type: BAPIMEOUTVALIDITY.

ITEM_CONDITION

This parameter contain data such as the condition amount and condition currency. Type: BAPIMEOUTCONDITION.

ITEM_COND_SCALE_VALUE

This parameter contains the value scales for a condition of an outline agreement item. Type: BAPIMEOUTITEMSCALEVAL.

ITEM_COND_SCALE_QUAN

This parameter contains the quantity scales for a condition of an outline agreement item. Type: BAPIMEOUTITEMSCALEQUAN.

SHIPPING

Shipping data. Type: BAPIMEPOSHIPPEXP.

EXPORT_IMPORT

Foreign trade data: outline agreement item. Type: BAPIMEOUTEIPO.

ITEM_TEXT

The parameter contains the long texts for a scheduling agreement item. Type: BAPIMEOUTITEMTEXT.

HEADER_TEXT

This parameter contains the long texts of the scheduling agreement header. Type: BAPIMEOUTTEXT.

HEAD_COND_VALIDITY

This parameter contains the validity periods for conditions for the document header. Type: BAPIMEOUTHEADVALIDITY.

HEAD_CONDITION

This parameter contains data such as condition amount and condition currency for header. Type: BAPIMEOUTHEADERCOND.

HEAD_COND_SCALE_VAL

This parameter contains the value scales for a header condition. Type: BAPIMEOUTSCALEVALUE.

HEAD_COND_SCALE_QUAN

This parameter contains the quantity scales for a header condition. Type: BAPIMEOUTSCALE.

PARTNER

The parameter contains the partner data for a scheduling agreement. You can specify the partners at the following levels:

- Purchasing organization
- Purchasing organization and plant
- Purchasing organization, plant, and vendor subrange

Type: BAPIMEOUTPARTNER.

EXTENSIONOUT

With the parameters EXTENSIONIN and EXTENSIONOUT, it's possible to enhance the interface of a BAPI without modification to achieve the automatic processing of customer-specific data. Type: BAPIPAREX.

RETURN

Return parameter(s). Type: BAPIRET2.

638 BAPI_SALESORDER_CHANGE

This FM is used to change or delete a sales order. You can change header, item, schedule line, and configuration data of the sales order. You should follow these points while changing the sales order:

- Only specify fields that should be changed.
- Select these fields by entering an "X" in the checkboxes.
- Enter "U" in the UPDATEFLAG field (other possible values are D for delete and I for add).
- Always specify key fields when changing the data, including in the checkboxes.
- If you want to change the configuration, you need to complete it again fully.
- Maintain quantities and dates in the schedule line data.

Example: You've created a sales order based on the purchase order of the customer for selling 4 quantities of Item1, and you want to change the sales order and its item details because your customer has changed his order from quantity 4 to 6. You can pass the parameters you want to change in this FM and execute this FM to change the sales order.

[↓] Import

SALESDOCUMENT

Sales order number you want to change. Type: BAPIVBELN-VBELN.

ORDER_HEADER_IN

Order header data. Type: BAPISDH1.

ORDER_HEADER_INX

Sales order header data change structure. Type: BAPISDH1X.

SIMULATION

Switch to simulation mode for writing BAPI. Value X doesn't update the data. Value Space updates the data (i.e., posts the data). Type: BAPIFLAG-BAPIFLAG.

BEHAVE_WHEN_ERROR

This parameter controls how the program behaves when an error occurs. Possible values:

- Space: If an error occurs, processing stops, and the sales document isn't saved.
- P: If an error occurs, the sales document can be saved. Problematic items aren't processed, but they are logged.
- R: Similar to P, but the document isn't saved.

Type: BAPIFLAG-BAPIFLAG.

INT_NUMBER_ASSIGNMENT

This parameter activates internal item number assignment. Sometimes, internal item numbering from SAP is needed, for example, for articles that cause a BOM to explode in the sales order. Type: BAPIFLAG-BAPIFLAG.

LOGIC_SWITCH

Logic switch for internal control parameter. Type: BAPISDLS.

NO_STATUS_BUF_INIT

If this parameter is set to X, no refresh of status buffer occurs. Type: BAPIFLAG-BAPIFLAG.

Tables

RETURN

This table contains messages from this FM call. Type: BAPIRET2.

ORDER_ITEM_IN

Sales order item data. Type: BAPISDITM.

ORDER_ITEM_INX

Sales order items change structure table. Type: BAPISDITMX.

PARTNERS

Partner data. Type: BAPIPARNR.

PARTNERCHANGES

Partner data change structure. Type: BAPIPARNRC.

PARTNERADDRESSES

Reference structure for partner address. Type: BAPIADDR1.

ORDER_CFGS_REF

Configuration reference data. Type: BAPICUCFG.

ORDER_CFGS_INST

Configuration instances. Type: BAPICUINS.

ORDER_CFGS_PART_OF

Configuration part-of specifications. Type: BAPICUPRT.

ORDER_CFGS_VALUE

Configuration: characteristic values. Type: BAPICUVAL.

ORDER_CFGS_BLOB

Internal configuration data. Type: BAPICUBLB.

ORDER_CFGS_VK

Configuration: variant condition key. Type: BAPICUVK.

ORDER_CFGS_REFINST

Configuration: reference item/instance. Type: BAPICUREF.

SCHEDULE_LINES

This parameter contains the data for schedule lines in sales documents. Type: BAPISCHDL.

SCHEDULE_LINESX

Check table for schedule lines data. Type: BAPISCHDLX.

ORDER_TEXT

Sales order text table. Type: BAPISDTEXT.

ORDER_KEYS

Output table of reference keys. Type: BAPISDKEY.

CONDITIONS_IN

This parameter contains condition data. If neither the CURRENCY or the CURRENCY_ISO fields are completed, the entry is issued as a percentage value. This could lead to an incomplete entry if the condition type isn't in percentages. Type: BAPICOND.

CONDITIONS_INX

Change structure for condition data. Type: BAPICONDX.

EXTENSIONIN

You can use this parameter to transfer user-specific enhancements to a BAPI. The customer has to complete the EXTENSION structure, and the system automatically continues processing it. If you haven't defined any user-specific fields, then you can transfer a blank EXTENSIONIN structure. Type: BAPIPAREX.

NFMETALLITMS

BAPI communication structure. Type: /NFM/BAPIDOCITM.

639 BAPI_SALESORDER_CREATEFROMDAT2

This FM is used to create a sales order. It's required that you enter sales order header data and partner data as input parameters.

Example: You want to create the sales order programmatically. You can pass the relevant parameter in this FM and call this FM to create the sales order data.

[↓] Import

SALESDOCUMENTIN

Sales order number you want to change. Type: BAPIVBELN-VBELN.

ORDER_HEADER_IN

Order header data. Type: BAPISDH1.

ORDER_HEADER_INX

Sales order header data change structure. Type: BAPISDH1X.

SENDER

This parameter defines the sender system of the sales order. Type: BAPI_SENDER.

BINARY_RELATIONSHIPTYPE

This parameter specifies the binary relationship type that is needed when writing object references. The binary relationship type is defined in the VRBINRELATION view. The following types are available in SD:

- VORL object references at the header level
- VORA object references at the item level

Type: BAPIRELTYPE-RELTYPE.

INT_NUMBER_ASSIGNMENT

This parameter activates internal item number assignment. Sometimes, internal item numbering from SAP is needed, for example, for articles that cause a BOM to explode in the sales order. Type: BAPIFLAG-BAPIFLAG.

BEHAVE_WHEN_ERROR

This parameter controls how the program behaves when an error occurs. Possible values:

- [Space]: If an error occurs, processing stops, and the sales document isn't saved.
- P: If an error occurs, the sales document can be saved. Problematic items aren't processed, but they are logged.
- R: Similar to P, but the document isn't saved.

Type: BAPIFLAG-BAPIFLAG.

LOGIC_SWITCH

Logic switch for internal control parameter. Type: BAPISDLS.

TESTRUN

Switch to the simulation session for writing BAPI. Value X doesn't update the data to database. Value [Space] updates the data (i.e., posts the data). Type: BAPIFLAG-BAPIFLAG.

CONVERT

This parameter checks whether your customizing has been configured so that you can enter the abbreviation in your own language and it's then converted into the German key. Value [Space] doesn't convert. Value X converts. Type: BAPIFLAG-BAPIFLAG.

[↑] Export

SALESDOCUMENT

Sales document number generated from this FM. Type: BAPIVBELN-VBELN.

Tables

RETURN

This table contains messages from this FM call. Type: BAPIRET2.

ORDER_ITEM_IN

Sales order item data. Type: BAPISDITM.

ORDER_ITEM_INX

Sales order items change structure table. Type: BAPISDITMX.

ORDER_PARTNERS

This table parameter is used to enter partners such as sold-to party, or ship-to party, both at the header and item level. Type: BAPIPARNR.

ORDER_SCHEDULES_IN

Schedule line data. Type: BAPISCHDL.

ORDER_SCHEDULES_INX

Change table for schedule line data. Type: BAPISCHDLX.

ORDER_CONDITIONS_IN

Conditions data. Type: BAPICOND.

ORDER_CONDITIONS_INX

Change table for condition data. Type: BAPICONDX.

ORDER_CFGS_REF

Configuration reference data. Type: BAPICUCFG.

ORDER_CFGS_INST

Configuration instances. Type: BAPICUINS.

ORDER_CFGS_PART_OF

Configuration part-of specifications. Type: BAPICUPRT.

ORDER_CFGS_VALUE

Configuration: characteristic values. Type: BAPICUVAL.

ORDER_CFGS_BLOB

Internal configuration data. Type: BAPICUBLB.

ORDER_CFGS_VK

Configuration: variant condition key. Type: BAPICUVK.

ORDER_CFGS_REFINST

Configuration: reference item/instance. Type: BAPICUREF.

ORDER_CCARD

Credit card data. Type: BAPICCARD.

ORDER_TEXT

Sales order text table. Type: BAPISDTEXT.

ORDER_KEYS

Output table of reference keys. Type: BAPISDKEY.

EXTENSIONIN

You can use this parameter to transfer user-specific enhancements to a BAPI. The customer has to complete the EXTENSION structure, and the system automatically continues processing it. If you haven't defined any user-specific fields, then you can transfer a blank EXTENSIONIN structure. Type: BAPIPAREX.

PARTNERADDRESSES

Document address table. Type: BAPIADDR1.

NFMETALLITMS

BAPI communication structure. Type: /NFM/BAPIDOCITM.

640 BAPI_SALESORDER_GETLIST

This FM is used to display the customer's sales order. You must enter the relevant customer number in the SALESDOCUMENT parameter and the requested sales organization in the SALES_ORGANIZATION parameter.

Example: You have a sales order parameter field and customer parameter field in the selection screen of the report program. You want to get the list of sales order for a customer. You can pass the relevant parameter in this FM and call this FM to get the details.

[↓] Import

CUSTOMER_NUMBER

Customer number. Type: BAPI1007-CUSTOMER.

SALES_ORGANIZATION

Sales organization. Organizational unit responsible for the sales and distribution of certain products and services. The responsibility of the sales organization can also include the legal liability for the product and the customer's right to compensation. Type: BAPIORDERS-SALES_ORG.

MATERIAL

Material number. Type: BAPIORDERS-MATERIAL.

DOCUMENT_DATE

Starting date for the time period for which you want to create a list of sales documents. Type: BAPIORDERS-DOC_DATE.

DOCUMENT_DATE_TO

End date of period, for which you want to create a list of sales documents. Type: BAPIORDERS-DOC_DATE.

PURCHASE_ORDER

Customer purchase order number. Type: BAPIORDERS-PURCH_NO.

TRANSACTION_GROUP

The transaction group controls which sales document types you can control with which system transactions during sales processing. Possible values:

- 0: Sales order.
- 1: Customer inquiry.
- 2: Quotation.
- 3: Scheduling agreement.
- 4: Contract.
- 5: Item proposal.
- 6: Delivery.
- 7: Billing document.

Type: BAPIFLAG-BAPIFLAG.

PURCHASE_ORDER_NUMBER

Customer purchase order number. Type: BAPIORDERS-PURCH_NO_C.

MATERIAL_EVG

Long material number. Type: BAPIMGVMATNR.

[↑] Export

RETURN

Error text. Type: BAPIRETURN.

Table

SALES_ORDERS

Customer sales order is returned in this table. Type: BAPIORDERS.

---

641 BAPI_SALESORDER_SIMULATE

This FM is used to simulate the sales order. This FM doesn't update the sales order; instead, it simulates the sales order.

Example: You want to create the sales order in the simulation mode to review it, but you don't want to save the sales order. You can pass the relevant parameter in this FM and call this FM to simulate the sales order data.

[↓] Import

ORDER_HEADER_IN

Order header. Type: BAPISDHEAD.

CONVERT_PARVW_AUART

This parameter checks whether your customizing has been configured so that you can enter the abbreviation in your own language and it's then converted into the German key. Value [Space] doesn't convert. Value X converts. Type: BAPIFLAG-BAPIFLAG.

[↑] Export

SALESDOCUMENT

Sales document number generated from this FM. Type: BAPIVBELN-VBELN.

SOLD_TO_PARTY

This parameter contains data for the sold-to party. Type: BAPISOLDTO.

SHIP_TO_PARTY

This parameter contains data for the ship-to party. Type: BAPISHIPTO.

BILLING_PARTY

This parameter contains data for the payer. Type: BAPIPAYER.

RETURN

Error message. Type: BAPIRETURN.

Tables

ORDER_ITEM_IN

Sales order item data. Type: BAPISDITM.

ORDER_PARTNERS

This table parameter is used to enter partners such as sold-to party, or ship-to party, both at the header and item level. Type: BAPIPARNR.

ORDER_SCHEDULE_IN

Schedule line data. Type: BAPISCHDL.

ORDER_ITEMS_OUT

Item data output. Type: BAPIITEMEX.

ORDER_CFGS_REF

Configuration reference data.
Type: BAPICUCFG.

ORDER_CFGS_INST

Configuration instances. Type: BAPICUINS.

ORDER_CFGS_PART_OF

Configuration part-of specifications.
Type: BAPICUPRT.

ORDER_CFGS_VALUE

Configuration: characteristic values.
Type: BAPICUVAL.

ORDER_CFGS_BLOB

Internal configuration data.
Type: BAPICUBLB.

ORDER_CCARD

Credit card data input table.
Type: BAPICCARD.

ORDER_CCARD_EX

Credit card data output. Type: BAPICCARD_EX.

ORDER_SCHEDULE_EX

Schedule line data output.
Type: BAPISDHEDU.

ORDER_CONDITION_EX

Conditions output. Type: BAPICOND.

ORDER_INCOMPLETE

Incompleteness log. Type: BAPIINCOMP.

MESSAGETABLE

This table contains the variable messages corresponding to the BAPIRET2 structure. Type: BAPIRET2.

EXTENSIONIN

You can use this parameter to transfer user-specific enhancements to a BAPI. The customer has to complete the EXTENSION structure, and the system automatically continues processing it. If you haven't defined any user-specific fields, then you can transfer a blank EXTENSIONIN structure. Type: BAPIPAREX.

PARTNERADDRESSES

Document address table. Type: BAPIADDR1.

NFMETALLITMS

BAPI communication structure.
Type: /NFM/BAPIDOCITM.

NFMETALLITMS_EX

BAPI communication structure. Type: /NFM/BAPIDOCITM.

---

642 BAPI_SCHEDULE_MAINTAIN

This FM creates or changes the scheduling agreement delivery schedule lines. The items in the sales order are further divided into one or more schedule lines. These schedule lines differ from each other with respect to date and quantity.

Example: To create schedule agreement or schedule lines for the sales document, you can pass the relevant parameter in this FM and call this FM.

[↓] Import

PURCHASINGDOCUMENT

Purchasing document number of the scheduling agreement. Type: BAPIMEOUTHEADER-NUMBER.

TESTRUN

Simulation mode. If this parameter is set to X, data isn't saved to the database; otherwise, it is saved. Type: BAPIFLAG-BAPIFLAG.

TECHNICAL_DATA

Collection of technical parameters to control general functions such as message determination and the authorization check at document level. Type: BAPIMEOUTTECH.

[↑] Export

EXP_HEADER

Header information data of the scheduling agreement are returned in this parameter. Type: BAPIMEOUTHEADER.

Tables

SCHEDULE

Delivery schedule line data outline agreement table. Type: BAPIMEOUTSCHEDULE.

SCHEDULEX

Delivery schedule line change table. Type: BAPIMEOUTSCHEDULEX.

SC_COMPONENT

Component data of outline agreement. In case of subcontracting items, this parameter contains the assigned subcontracting components. Type: BAPIMEOUTCOMPONENT.

SC_COMPONENTX

Change table for component data. Type: BAPIMEOUTCOMPONENTX.

EXTENSIONIN

With the parameters EXTENSIONIN and EXTENSIONOUT, it's possible to enhance the interface of a BAPI without modification to achieve the automatic processing of customer-specific data. Type: BAPIPAREX.

EXTENSIONOUT

With the parameters EXTENSIONIN and EXTENSIONOUT, it's possible to enhance the interface of a BAPI without modification to achieve the automatic processing of customer-specific data. Type: BAPIPAREX.

RETURN

This parameter contains the messages returned from this FM call. Type: BAPIRET2.

643 BAPI_SOURCEDETERMIN_GETSOS

This FM is used to determine the source of supply for the material or material group. Source of supply is the master data object that is used to model procurement of a product. Sources of supply can be classified as those for in-house production, which contain all BOM and routing data that is relevant for production, and those for external procurement.

Example: You want to determine the source of supply of your material. You can pass the relevant parameter in this FM and call this FM to determine the source of supply.

[↓] Import

MATERIAL

Material number. Type: BAPIEBAN-MATERIAL.

MAT_GRP

Material group is used to group together several materials or services that have some common attribute. Type: BAPIEBAN-MAT_GRP.

PLANT

Plant. Type: BAPIEBAN-PLANT.

DELIV_DATE

Delivery date is the date on which the item is to be delivered. Type: BAPIEBAN-DELIV_DATE.

QUANTITY

Quantity that is to be ordered. Type: BAPIEBAN-QUANTITY.

UNIT

Unit of measure in which the requested material is to be ordered. Type: BAPIEBAN-UNIT.

ITEM_CAT

Item category. Type: BAPIEBAN-ITEM_CAT.

ACCTASSCAT

Account assignment category determines which account assignment data (such as cost center, account number, etc.) is necessary for the item. Type: BAPIEBAN-ACCTASSCAT.

PURCH_ORG

Purchasing organization procures materials and services, negotiates conditions of purchase with vendors, and is responsible for such transactions. Type: BAPIEBAN-PURCH_ORG.

FIXED_VEND

Fixed vendor. Type: BAPIEBAN-FIXED_VEND.

PCKG_NO

Package number specifies an internal package number grouping together all services belonging to an item of a purchasing document. Type: BAPIEBAN-PCKG_NO.

UNIT_ISO

Specifies the ISO code for the unit of measure in which the requested material is to be ordered. Type: BAPIEBAN-PREQ_UNIT_ISO.

MATERIAL_EVG

This parameter can be used to transfer long material numbers that can contain a maximum of 40 characters. To use this parameter, you must activate the switch for the LAMA (long material number) business function in the system. The LAMA business function is available as part of the business function set for DIMP. Type: BAPIMGVMATNR.

Tables

SOURCES_OF_SUPPLY

This table contains all source of supply. Type: BAPISOSY.

RETURN

Return messages received from this FM call. Type: BAPIRETURN.

---

644 BAPI_TRANSACTION_COMMIT

This FM executes the `COMMIT WORK` command. When you call any BAPI that has changed the data in the SAP system, you need to call this FM to commit those changes to the database.

Example: You've called any BAPI that has created or changed the data such as BAPI_SALESORDER_CHANGE (FM 638). To save the changes to the database, you must call this FM.

[↓] Import

WAIT

The default value of this parameter is [Space]. If the parameter contains the value [Space] or it doesn't contain a value at all, then a simple `COMMIT WORK` is executed.

If the parameter WAIT contains a value other than [Space], a `COMMIT WORK AND WAIT` command is executed. Type: BAPITA-WAIT.

[↑] Export

RETURN

Return messages from this FM call are returned in this FM. If the `COMMIT WORK` command isn't successfully executed, the system crashes. If the parameter WAIT is set to a value, the error message "Posting could not be carried out" may be returned. Type: BAPIRET2.

---

645 BAPI_TRANSACTION_ROLLBACK

This FM executes the `ROLLBACK WORK` command. When you call any BAPI that has changed data in the SAP system, it's possible that in between the saving, some data didn't save successfully—you therefore want to roll back all of the changes made. You need to call this FM to roll back those changes. The prerequisite for using rollback is that the changes haven't been already transferred to the database with `COMMIT WORK`.

Example: You've called any BAPI that has created or changed the data such as BAPI_SALESORDER_CHANGE (FM 638). However, changes aren't successfully written to the database table, and you need to roll back the changes. To roll back the changes you must call this FM.

[↑] Export

RETURN

No messages are returned if an error occurs. If the ROLLBACK WORK command isn't successfully executed, the system crashes. Type: BAPIRET2.

---

646 BAPI_USER_ACTGROUPS_ASSIGN

This FM is used to change a user role assignment. The user role assignments are transferred in the table ACTIVTYGROUPS where all of the existing assignments are overwritten by the assignment in the table. Existing assignment are also deleted if they are no longer present in the ACTIVITYGROUPS table.

Example: User USR1 has been assigned with the roles ROLE1, ROLE2, and ROLE3. You want to add ROLE4. You can pass ROLE1,

ROLE2, ROLE3, and the new ROLE4 in the ACTIVITYGROUPS table in this FM and call this FM. This inserts ROLE4 to the user.

[↓] Import

USERNAME

User name. Type: BAPIBNAME-BAPIBNAME.

Tables

ACTIVITYGROUPS

This table contains activity group assignment for the user. This table contain the following fields:

- ACTIVITYGROUPS-AGR_NAME: Name of the activity group.
- ACTIVITYGROUPS-FROM_DAT: Beginning of the validity of the assignment of the user to the activity group.
- ACTIVITYGROUPS-TO_DAT: End of the validity of the assignment of the user to the activity group.
- ACTIVITYGROUPS-AGR_TEXT: Activity group description.
- ACTIVITYGROUPS-ORG_FLAG: Flag indicating whether the activity group was assigned indirectly using HR Organization Management.

Type: BAPIAGR.

RETURN

Return structure with messages from this FM call. Type: BAPIRET2.

---

647 BAPI_USER_ACTGROUPS_DELETE

You can use this FM to delete a user role assignment. This FM deletes all of the activity group assignments for the USERNAME.

Example: You've created a temporary user for testing purpose and have assigned various roles to it. After the testing is complete, you want to delete all of the roles assigned to the user name. You can pass the user name in this FM to delete the roles.

[↓] Import

USERNAME

User name. Type: BAPIBNAME-BAPIBNAME.

Table

RETURN

Return structure with messages from this FM call. Type: BAPIRET2.

---

648 BAPI_USER_CHANGE

This FM changes the attributes of the user. The existing attributes are overwritten by the attributes transferred to the interface with the change parameter as set to X.

Example: You want to change the user's user group from DEVELOPER to TESTER. You can pass the relevant parameter in this FM and call this FM to do so.

[↓] Import

USERNAME

User name. Type: BAPIBNAME-BAPIBNAME.

LOGONDATA

Logon data for the user, including user valid from, user valid to, user type, account ID, user group in the user master maintenance, personal time zone, and so on. Type: BAPILOGOND.

LOGONDATAX

Select change-relevant fields for user logon data. Type: BAPILOGONX.

DEFAULTS

User default data such as start menu, spools output device, date format, logon language, decimal format, and so on. Type: BAPIDEFAUL.

DEFAULTSX

Select change-relevant fields for the user defaults. Type: BAPIDEFAX.

ADDRESS

Address data of the user. Type: BAPIADDR3.

ADDRESSX

Select change-relevant fields for user address data. Type: BAPIADDR3X.

PARAMETERX

Select change-relevant fields for the parameter structure, that is, get/set parameter ID, parameter value, and description. Type: BAPIPARAMX.

COMPANY

ID is assigned to the company address to identify it in other systems. Type: BAPIUSCOMP.

COMPANYX

Select change-relevant fields for the company address. Type: BAPIUSCOMX.

SNC

Secure network communication data. Type: BAPISNCU.

SNCX

Change structure for secure network communication data. Type: BAPISNCUX.

BACK_DISTRIBUTION

Redistribution into the central system. Type: BAPIFLAG.

PASSWORD

Password for logon. Type: BAPIPWD.

PASSWORDX

Select change-relevant fields for password. Type: BAPIPWDX.

ADDCOMX

Change structure for communication data. Type: BAPIADCOMX.

REF_USER

Reference user name. Type: BAPIREFUS.

REF_USERX

Reference user change structure. Type: BAPIREFUSX.

ALIAS

User name alias. Type: BAPIALIAS.

ALIASX

User name alias change flag. Type: BAPIALIASX.

GROUPSX

User group change flag. Type: BAPIGROUPX.

UCLASS

User classification in accordance with the license agreement between the customer and SAP. Type: BAPIUCLASS.

UCLASSX

Change indicators for license-related user classification. Type: BAPIUCLASSX.

EXTIDSX

Change indicators for external IDs. Type: BAPIUSEXTIDX.

PRODUCTIVE_PWD

If this parameter is set to X, it sets the productive password. Type: BAPIFLAG-BAPIFLAG.

Tables

PARAMETER

Table with user parameters. Type: BAPIPARAM.

RETURN

Return structure with messages returned from this FM call. Type: BAPIRET2.

ADDTEL

Table for telephone numbers. Type: BAPIADTEL.

ADDFAX

Table for fax numbers. Type: BAPIADFAX.

ADDTTX

Table for teletex numbers. Type: BAPIADTTX.

ADDTLX

Table for teletex numbers for business address services. Type: BAPIADTLX.

ADDSMTP

Table for email address. Type: BAPIADSMTP.

ADDRML

Table for RML address. Type: BAPIADRML

ADDX400

Table for X400 address. Type: BAPIADX400.

ADDRFC

Table for RFC address. Type: BAPIADRFC.

ADDPRT

Table for PRT address. Type: BAPIADPRT.

ADDSSF

Table for SSF address. Type: BAPIADSSF.

ADDURI

Table for URI address. Type: BAPIADURI.

ADDPAG

Table for pager number. Type: BAPIADPAG.

ADDCOMREM

Table for communication type comments. Type: BAPICOMREM.

GROUPS

Transfer structure for a list of user groups. Type: BAPIGROUPS

PARAMETER1

Replaces parameter. Type: BAPIPARAM1.

UCLASSSYS

System-specific license-related user classification. Type: BAPIUCLASSSYS.

EXTIDHEAD

Header data for external ID of a user. Type: BAPIUSEXTIDHEAD.

EXTIDPART

Part of a long field for the external ID of a user. Type: BAPIUSEXTIDPART.

---

649 BAPI_USER_CLONE

This FM changes or creates a user. This method is needed to distribute the users via ALE. This method should only be used to call other methods of object USER and should not be called directly. This FM can only be used in the Central User Administration (CUA) system.

Example: User has been locked because of entering the wrong password several times. You can unlock the user by passing the user name in parameter USERNAME and UNLOCK in the method parameter in this FM.

[↓] Import

USERNAME

User name. Type: BAPIBNAME-BAPIBNAME.

METHOD

Method to be called. Type: BAPIMETHOD.

LOGONDATA

Logon data for the user, including user valid from, user valid to, user type, account ID, user group in the user master maintenance, personal time zone, and so on. Type: BAPILOGOND.

PASSWORD

Password. Type: BAPIPWD.

DEFAULTS

User default data such as start menu, spools output device, date format, logon language, decimal format, and so on. Type: BAPIDEFAUL.

ADDRESS

Address data of the user. Type: BAPIADDR3.

COMPANY

ID is assigned to the company address to identify it in other systems. Type: BAPIUSCOMP.

SNC

Secure network communication data. Type: BAPISNCU.

REF_USER

Reference user name. Type: BAPIREFUS.

ALIAS

User name alias. Type: BAPIALIAS.

UCLASS

User classification in accordance with the license agreement between the customer and SAP. Type: BAPIUCLASS.

UCLASSX

Change indicators for license-related user classification. Type: BAPIUCLASSX.

EXTIDSX

Change indicators for external IDs. Type: BAPIUSEXTIDX.

USATTRX

Change indicators for the additional user attributes. Type: BAPIUSATTRX.

ADD_FIELDS

Additional USERCLONE IDoc fields. Type: BAPIADDFLD.

ADD_FIELDSX

Change indicator for ADD_FIELDS. Type: BAPIADDFLDX.

Tables

PARAMETER

Table with user parameters. Type: BAPIPARAM.

PROFILES

Profiles. Type: BAPIPROF.

ACTIVITYGROUPS

Activity groups. Type: BAPIAGR.

RETURN

Return structure with messages returned from this FM call. Type: BAPIRET2.

ADDTEL

Table for telephone numbers. Type: BAPIADTEL.

ADDFAX

Table for fax numbers. Type: BAPIADFAX.

ADDTTX

Table for teletex numbers. Type: BAPIADTTX.

ADDTLX

Table for teletex numbers for business address services. Type: BAPIADTLX.

ADDSMTP

Table for email address. Type: BAPIADSMTP.

ADDRML

Table for RML address. Type: BAPIADRML.

ADDX400

Table for X400 address. Type: BAPIADX400.

ADDRFC

Table for RFC address. Type: BAPIADRFC.

ADDPRT

Table for PRT address. Type: BAPIADPRT.

ADDSSF

Table for SSF address. Type: BAPIADSSF.

ADDURI

Table for URI address. Type: BAPIADURI.

ADDPAG

Table for pager number. Type: BAPIADPAG.

ADDCOMREM

Table for communication type comments. Type: BAPICOMREM.

GROUPS

Transfer structure for a list of user groups. Type: BAPIGROUPS

APPLREF

References to application objects for user. Type: BAPIAPREF.

PERSDATA

Personalization distribution exchange format. Type: BAPIPERS.

PARAMETER1

Replaces parameter. Type: BAPIPARAM1.

EXTIDHEAD

Header data for external ID of a user. Type: BAPIUSEXTIDHEAD.

EXTIDPART

Part of a long field for the external ID of a user. Type: BAPIUSEXTIDPART.

USATTR

Additional user attributes can be stored in this parameter. Type: BAPIUSATTRTAB.

650 BAPI_USER_CREATE1

This FM is used to create a user in all systems.

Example: A new employee has joined the organization; you need to create his details in the user master. You can pass the relevant parameter in this FM to create a new user.

[↓] Import

USERNAME

User name. Type: BAPIBNAME-BAPIBNAME.

NAME_IN

User name. Type: BAPIBNAME.

LOGONDATA

Logon data for the user, including user valid from, user valid to, user type, account ID, user group in the user master maintenance, personal time zone, and so on. Type: BAPILOGOND.

PASSWORD

Password. Type: BAPIPWD.

DEFAULTS

User default data such as start menu, spools output device, date format, logon language, decimal format, and so on. Type: BAPIDEFAUL.

ADDRESS

Address data of the user. Type: BAPIADDR3.

COMPANY

ID is assigned to company address to identify it in other systems. Type: BAPIUSCOMP.

SNC

Secure network communication data. Type: BAPISNCU.

REF_USER

Reference user name. Type: BAPIREFUS.

ALIAS

User name alias. Type: BAPIALIAS.

EX_ADDRESS

Don't use this parameter. Type: BAPIADDR3-PERS_NO.

UCLASS

User classification in accordance with the license agreement between the customer and SAP. Type: BAPIUCLASS.

FORCE_SYSTEM_ASSIGNMENT

This indicator is only relevant in a CUA landscape.

If it's set, a system assignment for the current system is forced, even if it's the central system.

If it isn't set, a system assignment is only made if the module is called in the child system. Type: BAPIFLAG-BAPIFLAG.

SELF_REGISTER

If this indicator is set to X, it isn't necessary to change the newly created user's password at the next logon. This indicator should only be used when creating users in self-registration scenarios. Type: BAPIFLAG-BAPIFLAG.

TECH_USER

Technical user. Type: BAPITECHUSER.

LOCK_LOCALLY

User is created with a local lock so he can't log on immediately. Type: BAPIUSLOCKX-BAPIUSLOCK.

Tables

PARAMETER

Table with user parameters. Type: BAPIPARAM.

RETURN

Return structure with messages returned from this FM call. Type: BAPIRET2.

ADDTEL

Table for telephone numbers. Type: BAPIADTEL.

ADDFAX

Table for fax numbers. Type: BAPIADFAX.

ADDTTX

Table for teletex numbers. Type: BAPIADTTX.

ADDTLX

Table for teletex numbers for business address services. Type: BAPIADTLX.

ADDSMTP

Table for email address. Type: BAPIADSMTP.

ADDRML

Table for RML address. Type: BAPIADRML.

ADDX400

Table for X400 address. Type: BAPIADX400.

ADDRFC

Table for RFC address. Type: BAPIADRFC.

ADDPRT

Table for PRT address. Type: BAPIADPRT.

ADDSSF

Table for SSF address. Type: BAPIADSSF.

ADDURI

Table for URI address. Type: BAPIADURI.

ADDPAG

Table for pager number. Type: BAPIADPAG.

ADDCOMREM

Table for communication type comments. Type: BAPICOMREM.

GROUPS

Transfer structure for a list of user groups. Type: BAPIGROUPS.

PARAMETER1

Replaces parameter. Type: BAPIPARAM1.

EXTIDHEAD

Header data for external ID of a user. Type: BAPIUSEXTIDHEAD.

EXTIDPART

Part of a long field for the external ID of a user. Type: BAPIUSEXTIDPART.

651 BAPI_USER_DELETE

This FM deletes a user from the system. All of the user assignments to system, profiles, and roles are also deleted.

Example: An employee has left the organization. You have to delete his details from SAP system. You can pass the relevant parameter in this FM and call this FM to delete the user.

[↓] Import

USERNAME

User name. Type: BAPIBNAME-BAPIBNAME.

Tables

RETURN

Return structure with messages returned from this FM call. Type: BAPIRET2.

652 BAPI_USER_EXISTENCE_CHECK

This FM checks if a user exists. The message indicating whether or not the user exists is returned in the return table.

Example: You've given a user name field in the selection screen, and you want to validate whether the user entered in this parameter exists or not. You can pass the relevant parameter in this FM and execute this FM to check the existence of the user.

[↓] Import

USERNAME

User name. Type: BAPIBNAME-BAPIBNAME.

[↑] Export

RETURN

Return structure with messages returned from this FM call. Type: BAPIRET2.

---

653 BAPI_USER_GET_DETAIL

This FM is used to get the details of the user. This FM also returns the profiles and activity groups assigned to the user.

Example: You want to change the profile detail of the user. First, you need to know the details of the profile attached to the user. You can pass the relevant parameter in this FM and call this FM to get the details of the user.

[↓] Import

USERNAME

User name. Type: BAPIBNAME-BAPIBNAME.

CACHE_RESULTS

Temporarily buffer results in work process. Type: FLAG_X.

[↑] Export

LOGONDATA

Logon data for the user, including user valid from, user valid to, user type, account ID, user group in the user master maintenance, personal time zone, and so on. Type: BAPILOGOND.

DEFAULTS

User default data such as start menu, spools output device, date format, logon language, decimal format, and so on. Type: BAPIDEFAUL.

ADDRESS

Address data of the user. Type: BAPIADDR3.

COMPANY

ID is assigned to the company address to identify it in other systems. Type: BAPIUSCOMP.

SNC

Secure network communication data. Type: BAPISNCU.

REF_USER

Reference user name. Type: BAPIREFUS.

ALIAS

User name alias. Type: BAPIALIAS.

UCLASS

User classification in accordance with the license agreement between the customer and SAP. Type: BAPIUCLASS.

LASTMODIFIED

Date and time of last modification of the user master data. Type: BAPIMODDAT.

ISLOCKED

User lock contains the following fields:

- WRNG_LOGON: Password logon locked by incorrect logon attempts.
- LOCAL_LOCK: Logon generally locked for the local system.
- GLOB_LOCK: Logon generally locked for all systems (globally).
- NO_USER_PW: Password logon deactivated for this user.

All the fields can have one of the two values: L (locked) or U (unlocked). Type: BAPISLOCKD.

Tables

PARAMETER

Table with user parameters. Type: BAPIPARAM.

PROFILES

Profiles. Type: BAPIPROF.

ACTIVITYGROUPS

Activity groups. Type: BAPIAGR.

RETURN

Return structure with messages returned from this FM call. Type: BAPIRET2.

ADDTEL

Table for telephone numbers. Type: BAPIADTEL.

ADDFAX

Table for fax numbers. Type: BAPIADFAX.

ADDTTX

Table for teletex numbers. Type: BAPIADTTX.

ADDTLX

Table for teletex numbers for business address services. Type: BAPIADTLX.

ADDSMTP

Table for email address. Type: BAPIADSMTP.

ADDRML

Table for RML address. Type: BAPIADRML.

ADDX400

Table for X400 address. Type: BAPIADX400.

ADDRFC

Table for RFC address. Type: BAPIADRFC.

ADDPRT

Table for PRT address. Type: BAPIADPRT.

ADDSSF

Table for SSF address. Type: BAPIADSSF.

ADDURI

Table for URI address. Type: BAPIADURI.

ADDPAG

Table for pager number. Type: BAPIADPAG.

ADDCOMREM

Table for communication type comments. Type: BAPICOMREM.

PARAMETER1

Replaces parameter. Type: BAPIPARAM1.

GROUPS

Transfer structure for a list of user groups. Type: BAPIGROUPS.

UCLASSSYS

System-specific license-related user classification. Type: BAPIUCLASSSYS.

EXTIDHEAD

Header data for external ID of a user. Type: BAPIUSEXTIDHEAD.

EXTIDPART

Part of a long field for the external ID of a user. Type: BAPIUSEXTIDPART.

SYSTEMS

BAPI structure for CUA target systems. Type: BAPIRCVSYS.

654 BAPI_USER_GETLIST

This FM is used to return the list of users based on the selection criteria entered. You can pass the selection criteria either in table SELECTION_RANGE or in table SELECTION_EXP. If the SELECTION_RANGE table is filled, the SELECTION_EXP table isn't used.

Example: You want to get the detail of all of the employees whose names start with "A." You can pass the parameter in this FM to get the list of all those employees.

[↓] Import

MAX_ROWS

Maximum number of lines of hits. Type: BAPIUSMISC-BAPIMAXROW.

WITH_USERNAME

If this parameter is specified, first name, family name, and the complete name are also returned in the USERLIST table. Type: BAPIUSMISC-WITH_NAME.

[↑] Export

ROWS

Number of users selected. Type: BAPIUSMISC-BAPIROWS.

Tables

SELECTION_RANGE

This parameter is a generic ranges table, with which you can restrict the selection of users using various attributes. The most important user attributes delivered by the FM BAPI_USER_GET_DETAIL (FM 653) are supported.

In the PARAMETER field, specify a parameter name of the FM BAPI_USER_GET_DETAIL (FM 653). The field FIELD must be filled with the component name of the reference type. Type: BAPIUSSRGE.

SELECTION_EXP

You can use this parameter for a generic search with free selections. You can restrict the selection by users using different attributes. The most important user attributes delivered by the FM BAPI_USER_GET_DETAIL (FM 653) are supported.

In the field PARAMETER, specify a parameter name for the FM BAPI_USER_GET_DETAIL (FM 653). The field FIELD must be filled with the component name of the reference type. Type: BAPIUSSEXP.

USERLIST

The table contains all users found for the selection. If the parameter WITH_USERNAME was set, the first name, family name, and the complete name are also returned. Type: BAPIUSNAME.

RETURN

Return parameter with messages from this FM call. Type: BAPIRET2.

---

655 BAPI_USER_LOCK

This FM is used to lock the user from logging on to the system. You can unlock the user using FM BAPI_USER_UNLOCK (FM 656).

Example: You want to lock the user; for example, you don't want a user to have logon access to the production or development system. You can pass the relevant parameter in this FM to lock the user.

[↓] Import

USERNAME

User name. Type: BAPIBNAME-BAPIBNAME.

Table

RETURN

Return structure with messages from this FM call. Type: BAPIRET2.

---

656 BAPI_USER_UNLOCK

This FM is used to unlock the user and allows the user to log on to the system. You can lock the user from FM BAPI_USER_LOCK.

Example: User USR01 is locked. You want to unlock this user so he can log on to the system. You can pass the relevant parameter in this FM to unlock the user.

[↓] Import

USERNAME

User name. Type: BAPIBNAME-BAPIBNAME.

Table

RETURN

Return structure with messages from this FM call. Type: BAPIRET2.

---

657 BAPI_USER_PROFILES_ASSIGN

This FM is used to change the profile assignment of the user. The profile assignments for user USERNAME are passed in a table called PROFILES. All existing assignments are overwritten with the assignments in the table. This also means that existing assignments can be deleted if they are no longer contained in the table. Only the field PROFILES-PROFILE must be filled.

Example: User USR1 has been assigned with the profiles P1, P2, and P3. You want to add profile P4. You can pass P1, P2, P3, and the new profile P4 in the PROFILES table in this FM and call this FM. This will insert the profile P4 to the user.

[↓] Import

USERNAME

User name. Type: BAPIBNAME-BAPIBNAME.

Tables

PROFILES

Table containing all of the profile assignments for a user. This table has the following fields:

- PROFILES-BAPIPROF: Profile name.
- PROFILES-BAPIPTEXT: Profile text.
- PROFILES-BAPITYPE: Profile type (composite, single, generated).
- PROFILES-BAPIAKTPS: Active or maintenance version.

Type: BAPIPROF.

RETURN

Return structure with messages from this FM call. Type: BAPIRET2.

658 BAPI_USER_PROFILES_DELETE

This FM is used to delete all of the profile assignments of the user.

Example: You've created a temporary user for testing purposes and have assigned various profiles to it. After the testing is complete, you want to delete all of the profiles assigned to the username. You can pass the username in this FM to delete the profile.

[↓] Import

USERNAME

User name. Type: BAPIBNAME-BAPIBNAME.

Table

RETURN

Return structure with messages from this FM call. Type: BAPIRET2.

659 BAPI_USR01DOHR_GETEMPLOYEE

This FM can be used to get an employee's ID number from his user name.

Example: You want to get the employee ID of the user. Based on the ID, you want to look into the transports assigned to that user from Transaction SE01. You can pass the relevant parameter in this FM to get the employee number.

[↓] Import

ID

User name. Type: BAPIUSR01-UNAME.

BEGINDATE

Begin date from which user is valid. Type: BAPIPAKEY-VALIDBEGIN.

ENDDATE

End date unto which user is valid. Type: BAPIPAKEY-VALIDEND.

Tables

RETURN

Return structure with messages from this FM call. Type: BAPIRET2.

EMPLOYEENUMBER

Employee personal number is returned in this parameter. Type: BAPIUSR01-EMPLOYEENO.

---

660 BAPI_VENDOR_CREATE

This FM is used to create or change the vendordetails in your system online. This FM calls Transaction XK01, and you need to enter the details in that transaction screen. Creating vendors from different systems without dialog screens isn't possible from this FM.

Example: To create the new vendor details in your system, call this FM.

[↑] Export

VENDORNO

Newly created vendor number. Type: BAPIVENDOR_01-VENDOR_NO.

VENDOR

Newly created vendor number. Type: BAPIVENDOR_01-VENDOR_NO.

RETURN

Return parameter with messages from this FM. Type: BAPIRET1.

---

661 BAPI_VENDOR_DELETE

This FM is used to mark the vendor for deletion.

Example: Your vendor V1 has closed his business, so you want to delete the entry from the system. You can pass the relevant parameter in this FM to delete the vendor.

[↓] Import

VENDORNO

Newly created vendor number. Type: BAPIVENDOR_01-VENDOR_NO.

[↑] Export

RETURN

Return parameter with messages from this FM. Type: BAPIRET1.

---

662 BAPI_VENDOR_GETDETAIL

This FM is used to get the detailed data of the vendor. General data, bank data, and company code data for the vendor are retrieved.

Example: You want to get the details of the vendor so that you can use that detail in your program. You can pass the relevant parameter in this FM and call this FM to get the vendor details.

[↓] Import

VENDORNO

Vendor. Type: BAPIVENDOR_01-VENDOR_NO.

COMPANYCODE

Company code. Type: BAPIVENDOR_01-COMP_CODE.

[↑] Export

GENERALDETAIL

This table contains general attributes of the vendor such as name, address components, and language. Type: BAPIVENDOR_04.

COMPANYDETAIL

This parameter contains company-code data. Type: BAPIVENDOR_05.

RETURN

Return code with messages returned from this FM call. Type: BAPIRET1.

Table

BANKDETAIL

Bank data of the vendor such as vendor account number, country code, bank key, partner bank type, and so on. Type: BAPIVENDOR_06.

---

663 BAPI_WHSE_STOCK_GET_DETAIL

This FM returns the detailed quant data in the warehouse.

Example: You want to know the detailed stock data for a quant from the external system. You can pass the relevant parameter in this FM and call this FM to get the detailed quant data.

[↓] Import

WHSENUMBER

Warehouse number/warehouse complex. Type: BAPI3016C-WHSENUMBER.

QUANT

Unique code within a warehouse number used to identify a material stock of a particular category in a storage bin. Type: BAPI3016C-QUANT.

Tables

STOCKDATA

In this table, the system returns the quant details that have been read. Type: BAPIQUANT.

EXTENSIONOUT

Parameter for transmission of customer data. Type: BAPIPAREX.

RETURN

Return parameter that contains messages from this FM call. Type: BAPIRET2.

---

664 BAPI_XBP_JOB_CLOSE

This FM is used to finish the background job processing. After the method has been executed, the job is transferred to the background processing system with the status Job without Start Date.

Example: When you're done with the job processing, you can call this FM to close the background job.

[↓] Import

JOBNAME

Job name. This parameter, along with the job number, uniquely identifies the background job. Type: BAPIXMJOB-JOBNAME.

JOBCOUNT

Job ID. This parameter, along with the job name, uniquely identifies the job. Type: BAPIXMJOB-JOBCOUNT.

EXTERNAL_USER_NAME

Name of the SAP external user who called this FM. Type: BAPIXMLOGR-EXTUSER.

TARGET_SERVER

Instance for execution of a job. Type: BAPIXMJOB-EXECSERVER.

RECIPIENT_OBJ

Structure for the object ID. Type: SWOTOBJID.

RECIPIENT

Structure for the spool list recipients. Type: BAPIXMRECIP.

[↑] Export

RETURN

This table contains the return messages from this FM call. Type: BAPIRET2.

---

665 BAPI_XBP_JOB_OPEN

This FM creates a job for background processing from an external system.

Example: You want to start the job in the SAP system from an external system. You can pass the relevant parameter in this FM and call this FM to start the job. When you're done with the background job processing, you can call the FM BAPI_XBP_JOB_CLOSE (FM 664) to close the job.

[↓] Import

JOBNAME

Job name. This parameter, along with the job number, uniquely identifies the background job. Type: BAPIXMJOB-JOBNAME.

EXTERNAL_USER_NAME

Name of the SAP external user who called this FM. Type: BAPIXMLOGR-EXTUSER.

JOBCLASS

Job classification. Type: BAPIXMJOB-JOBCLASS.

[↑] Export

JOBCOUNT

Job ID. This parameter, along with the job name, uniquely identifies the job. Type: BAPIXMJOB-JOBCOUNT.

RETURN

This table contains the return messages from this FM call. Type: BAPIRET2.

666 BAPI_XBP_JOB_STATUS_GET

This FM is used to get the status of the job of the background processing system.

Example: You want to get the status of the background system from the external program. You can pass the relevant parameter in this FM and call this FM to get the job status.

[↓] Import

JOBNAME

Job name. This parameter, along with the job number, uniquely identifies the background job. Type: BAPIXMJOB-JOBNAME.

JOBCOUNT

Job ID. This parameter, along with the job name, uniquely identifies the job. Type: BAPIXMJOB-JOBCOUNT.

EXTERNAL_USER_NAME

Name of the SAP external user who called this FM. Type: BAPIXMLOGR-EXTUSER.

[↑] Export

STATUS

This parameter contains the status of the specified job. The following statuses are possible:

- R: Active, is being executed.
- Y: Ready, the job is released and meets the start requirements. However, the job is waiting for a free work process.
- P: Scheduled but not yet released.
- S: The job is released and is waiting to meet the start requirements. It can then be executed.
- A: Terminated.
- F: Complete, all job steps were successfully executed.

Type: BAPIXMJOB-STATUS.

RETURN

This table contains the return messages from this FM call. Type: BAPIRET2.

HAS_CHILD

This parameter is set to X if the job has child jobs. Type: BTCH0000-CHAR1.

---

667 BAPI_BUS2001_SET_STATUS

This FM sets or resets the status for project definitions. One system and user status can be set and one system status and user status can be revoked every time for the project definition. System status REL (Released), TECO (Technically Completed), CLSD (Closed), MDLK (Master Data Locked), and DLFL (Deletion Flag) are possible. Processing of sequence statuses is first the system status is revoked, and then the user status is revoked. Next, the system status is set, and at last the user status is set. This BAPI can be called repeatedly until the logical unit of work is closed by FM BAPI_PS_PRECOMMIT (FM 622).

Example: You want to change the status of the project from REL to CLSD. First, you need to call this FM BAPI_PS_INITIALIZATION (FM 621) to initialize the processing unit. Next, you call FM BAPI_BUS2001_SET_STATUS to change the status of the project. Finally, you call FM BAPI_PS_PRECOMMIT (FM 622) and BAPI_TRANSACTION_COMMIT (FM 644) to save the changes to the database.

[↓] Import

PROJECT_DEFINITION

Project definition number Type: BAPIPR-PROJECT_DEFINITION.

UNDO_SYSTEM_STATUS

System status you want to revoke for the project definition. Type: BAPI_SYSTEM_STATUS-SYSTEM_STATUS.

UNDO_USER_STATUS

User status you want to revoke for the project definition. User statuses can be set and revoked in accordance with the status profile assigned to the project definition. Type: BAPI_USER_STATUS-USER_STATUS.

SET_SYSTEM_STATUS

System status you want to set for the project definition. Type: BAPI_SYSTEM_STATUS-SYSTEM_STATUS.

SET_USER_STATUS

User status you want to set for the project definition. User statuses can be set and revoked in accordance with the status profile assigned to the project definition. Type: BAPI_USER_STATUS-USER_STATUS.

[↑] Export

RETURN

Return values with error messages. Type: BAPIRETURN1.

Table

E_RESULT

If an errors occur when setting/revoking system/user statuses in the project definition, the relevant object and error message are output in this table. Type: BAPI_STATUS_RESULT.

Part XV
Miscellaneous

This part contains the function modules (FMs) that are important and commonly used, but can't be categorized into any of the previous parts. Many domain-specific FMs that aren't BAPIs are also covered in this part. FMs related to change documents, BOM explosions, reading a material, and statuses are covered as well.

668 CALL_BROWSER

This FM is used to call a web browser with the URL that you've specified. You can use this FM on the frontend of all platforms supported by SAP.

Example: You've created a report that has an external web page's URL as a hyperlink. On clicking the hyperlink, you want to open the web page in the new browser. You can pass the relevant parameter and call this FM to open the browser.

[↓] Import

URL

URL of browser call. Type: C.

WINDOW_NAME

Name of browser target window. Type: C.

NEW_WINDOW

This parameter opens a new window for the browser. Type: SY-DATAR.

BROWSER_TYPE

Obsolete: Don't use. Type: TOLE-APP.

CONTEXTSTRING

Obsolete: Don't use. Type: SDOK_IFACE-CONTEXTSTR.

[!] Exceptions

FRONTEND_NOT_SUPPORTED

Frontend not supported.

FRONTEND_ERROR

Error occurred in SAP GUI.

PROG_NOT_FOUND

Program not found or not in executable form.

NO_BATCH

Frontend function cannot be executed in background.

UNSPECIFIED_ERROR

Unspecified exception.

669 CHANGEDOCUMENT_CLOSE

This FM is used to close change document creation. This FM also writes the change document header and completes the processing of the change document creation for an object.

Change documents are created for business objects that are to be changed frequently. Change documents are recommended for critical objects that are suspicious for audits such as for changing sales orders or production orders. Change document data is saved in table CDHDR (Change Document Header Data) and CDPOS (Change Document Item Data). You'll also use the following FMs in this process:

- CHANGEDOCUMENT_OPEN (FM 673)
- CHANGEDOCUMENT_SINGLE_CASE (FM 678)

Example: When you're done with change document processing, you'll need to save the changes and close the change document. You can pass the relevant parameter in this FM to close the change document.

[↓] Import

DATE_OF_CHANGE

Changed on date. Type: CDHDR-UDATE.

OBJECTCLASS

Name of the object class for which the change documents were created. The object class must be the same as that specified in CHANGEDOCUMENT_OPEN. Type: CDHDR-OBJECTCLAS.

OBJECTID

ID or value of the object characteristic. The object characteristic must be the same as that in CHANGEDOCUMENT_OPEN. Type: CDHDR-OBJECTID.

TCODE

Transaction in which the change was made. Type: CDHDR-TCODE.

TIME_OF_CHANGE

Changed time. Type: CDHDR-UTIME.

USERNAME

Changed by user name. Type: CDHDR-USERNAME.

OBJECT_CHANGE_INDICATOR

This flag indicates whether the commercial object whose changes are logged here was inserted, changed, or deleted. Possible values:

- I: Object was inserted.
- U: Object was changed. This is the default value.
- D: Object was deleted.

Type: CDHDR-CHANGE_IND.

PLANNED_OR_REAL_CHANGES

This parameter isn't used. Type: CDHDR-CHANGE_IND.

NO_CHANGE_POINTERS

If this parameter is set to X, documents won't be written to the database. Type: CDHDR-CHANGE_IND.

[↑] Export

CHANGENUMBER

Change document number. Type: CDHDR-CHANGENR.

[!] Exceptions

HEADER_INSERT_FAILED

SQL error during INSERT of header.

NO_POSITION_INSERTED

No items were entered.

OBJECT_INVALID

OPEN was called with another object.

OPEN_MISSING

No OPEN was carried out.

POSITION_INSERT_FAILED

G/L account number.

670 CHANGEDOCUMENT_DELETE

This FM is used to delete the change documents for all clients or one clients.

Example: You've created the change document for material changes, but you no longer need to log these changes. To delete the change document, pass the relevant parameters and call this FM.

[↓] Import

CLIENT

Client in which deletion is to take place. Possible values:

- Client name: Change document for that client will be deleted.
- Space: Change document for all clients will be deleted.

Type: CDHDR-MANDANT.

COMMIT_COUNTER

The parameter is only effective when a COMMIT WORK should be performed in the FM; that is, parameter WITH_COMMIT is set.

OBJECTCLASS

Object class. Type: CDHDR-OBJECTCLAS.

OBJECTID

Object ID. Type: CDHDR-OBJECTID.

UP_TO_DATE

All change documents until the specified date are deleted. Type: CDHDR-UDATE.

WITH_COMMIT

The flag determines whether a COMMIT WORK should be called in the FM. This can be necessary when a lot is to be deleted. In this case, the delete operations are divided up to avoid a rollback segment overflow. The division is by clusters, so that no inconsistencies occur in the case of cancellation. Possible values:

- Space: No COMMIT WORK.
- X: With COMMIT WORK.

DISABLE_AUTHORITY_CHECK

This parameter disables the authority check for deleting the change documents.

CHANGENR

If only one change document number is to be deleted, it can be specified in this parameter. Type: CDHDR-CHANGENR.

[↑] Export

NUMBER_OF_DELETED_HEADERS

Number of deleted change document headers. Type: SY-DBCNT.

NUMBER_OF_DELETED_POSITIONS

Number of deleted change document items. Type: SY-DBCNT.

NUMBER_OF_DELETED_UIDS

Number of deleted change document ITEM_GUIDs. Type: SY-DBCNT.

NUMBER_OF_DELETED_STRINGS

Number of deleted change document item strings. Type: SY-DBCNT.

[!] Exceptions

NO_AUTHORITY

No authorization to delete change documents.

NO_CHANGES_FOUND

No change documents found.

---

671 CHANGEDOCUMENT_DISPLAY

This FM is used to display the change document.

Example: You've created the change document for material changes. To display the change document log to the end user, pass the relevant parameters and call this FM.

[↓] Import

I_APPLICATIONID

Report name. Type: REPID.

IS_LAYOUT

Layout details for the ALV grid display. This FM calls FM REUSE_ALV_GRID_DISPLAY for the ALV display. Type: SLIS_LAYOUT_ALV.

IS_VARIANT

Layout variant details for the ALV grid details. Type: DISVARIANT.

FLG_AUTOCONDENSE

If this parameter is set to X, it condenses the display value. Type: C.

I_CB_PROGRAM

Name of ABAP/4 program. Type: SY-REPID.

I_OBJECTCLAS

Object class name. Type: CDHDR-OBJECTCLAS.

IS_SEL_HIDE

This parameter isn't supported. Type: SLIS_SEL_HIDE_ALV.

I_SCREEN_START_LINE

Coordinate of start line of the dialog box.

I_SCREEN_START_COLUMN

Coordinate of start column of the dialog box.

I_SCREEN_END_LINE

Coordinate of end line of the dialog box.

I_SCREEN_END_COLUMN

Coordinate of end column of the dialog box.

I_CALLBACK_PF_STATUS_SET

PF status for the ALV display. Type: SLIS_FORMNAME.

IT_EVENTS

Events for the ALV display. Type: SLIS_T_EVENT.

IT_CDRED_STR

Change documents and structure for display of STRING fields. Type: CDRED_STR_TAB.

Table

I_CDRED

Change document display table. Type: CDRED.

---

672 CHANGEDOCUMENT_MULTIPLE_CASE

This FM is used to create change document items.

Example: You've created a change document and want to create multiple object characteristics for the change documents programmatically. After you've prepared the tables with FM CHANGEDOCUMENT_PREPARE_TABLES (FM 674), call this FM to create the multiple change document items.

[↓] Import

CHANGE_INDICATOR

This parameter determines the processing logic of the FM. Possible values:

- D: Delete. Change document items present in table TABLE_OLD will be marked for deletion.
- I: Insert. Change document present I the table TABLE_NEW will be created.

- U: UPDATE. The keys of TABLE_OLD and TABLE_NEW will be compared.

DOCU_DELETE

The parameter determines how the document items are to be created for deleted records:

- X: A document item is created for every field in the deleted record.
- [Space]: A document item is created only with the key of the deleted record.

DOCU_INSERT

This parameter determines how document items are created when records are inserted.

- X: Document item is created for every field of the inserted record.
- [Space]: Document is created with the key of the inserted record.

REFTABLENAME

Name of the reference table structure for currency or unit fields. Type: CDPOS-TABNAME.

TABLENAME

Name of the table structure of object class. Type: CDPOS-TABNAME.

Tables

TABLE_NEW

Table contains the changed data.

TABLE_OLD

Table contains the unchanged data.

---

673 CHANGEDOCUMENT_OPEN

This FM is used to open change document creation. This FM initializes the internal fields and object characteristics. The following are other FMs you'll use in this process:

- CHANGEDOCUMENT_SINGLE_CASE (FM 678)
- CHANGEDOCUMENT_DISPLAY (FM 671)
- CHANGEDOCUMENT_CLOSE (FM 669)
- CHANGEDOCUMENT_DELETE (FM 670)

Example: You want to log the changes of the critical object such as material changes. You need to first open the change document creation. You can pass the relevant parameters in this FM to open the change document processing. Next, you need to create the single change document item.

[↓] Import

OBJECTCLASS

Object class name for which change documents are to be created. Type: CDHDR-OBJECTCLAS.

OBJECTID

Object value ID. Type: CDHDR-OBJECTID.

PLANNED_CHANGE_NUMBER

Plan change number. Type: CDHDR-PLANCHNGNR.

PLANNED_OR_REAL_CHANGES

This parameter specifies whether the changes that are logged are actual or planned changes. Possible values:

- R: Actual or real changes
- P: Planned changes
- [Space]: If a plan number exists, then used planned changes; if a plan number doesn't exist, then use actual changes. Type: CDHDR-CHANGE_IND.

[!] Exception

SEQUENCE_INVALID

No CLOSE was called for the last object.

---

674 CHANGEDOCUMENT_PREPARE_TABLES

This FM is used to edit a table for a change document. This FM is used to check if the two transferred tables in the table parameters; that is, TABLE_NEW and TABLE_OLD, are identical. Also, this FM decides if the internal tables are to be processed for the multiple cases.

Example: You're editing a material, and you've created a change document to log your changes. You need to find out if there are any changes made by the end user. You'll have two tables: one with old data and one with new data. After you prepare these tables, you can use FM CHANGEDOCUMENT_MULTIPLE_CASE (FM 672).

[↓] Import

CHECK_INDICATOR

This parameter checks if the tables are to be changed. Value X determines that tables are only to be compared, and value [Space] determines that the tables are compressed and formatted for MULTIPLE_CASE.

TABLENAME

Name of the table structure of the internal tables. Type: CDPOS-TABNAME.

CHECK_ONLY_LOGFLAGS

This control parameter checks change document-relevant fields only. Type: CDCHECK_LOGFLAGS.

[↑] Export

RESULT

This parameter returns whether the tables TABLE_NEW and TABLE_OLD are identical. Value F determines TABLE_OLD is different from TABLE_NEW, and value T determines that both tables are identical.

Tables

TABLE_NEW

This table contains the changed data. This table structure must have the structure of the table specified in parameter TABLENAME along with the additional processing flag. During compression, identical lines from TABLE_NEW and TABLE_OLD are deleted, and the processing flag is set. This FM sets the processing flag according to the following logic and with the following values:

- I: Insert. Lines that exist with the table key in TABLE_NEW but not in TABLE_OLD.
- U: Update. Lines that exist with the table key in TABLE_NEW and in TABLE_OLD, the contents of the TABLENAME structure fields in TABLE_NEW aren't the same as in TABLE_OLD.
- Delete the line in TABLE_NEW; delete the line in TABLE_OLD If the lines that exist with the table key in TABLE_NEW and in TABLE_OLD, the contents of the TABLENAME structure fields in TABLE_NEW are the same as in TABLE_OLD.

TABLE_OLD

Table contains the changed data. This table structure must have the structure of the table specified in parameter TABLENAME along with the additional processing flag. During compression, identical lines from the TABLE_NEW and TABLE_OLD are deleted, and the processing flag are set. This FM sets the processing flag according to the below logic:

- D: Delete. Line exists with the table key in TABLE_OLD but not in TABLE_NEW.
- [Space]: Line exists with the table key in TABLE_OLD and in TABLE_NEW, and the contents of the structure TABLENAME fields are not the same in TABLE_OLD and TABLE_NEW.
- Delete the line in TABLE_OLD; delete the line in TABLE_NEW if Line exists with the table key in TABLE_OLD and in TABLE_NEW, and the contents of the structure TABLENAME fields are the same in TABLE_OLD and TABLE_NEW.

675 CHANGEDOCUMENT_READ

This FM reads change document headers and change document items and processes the old and new values according to their category; that is, update, insert, or delete. In some cases, the units and currencies are added on to the value fields when a unit or currency has value.

Example: You've created a change document for a material. To read the change document for the same object and display all of the changes made on that object, pass the relevant parameter and execute this FM.

[↓] Import

ARCHIVE_HANDLE

This parameter controls whether change documents are to be read from the database or the archive.

If 0 (preset) is passed for this parameter, the database is read; otherwise, an archive is read.

If an archive is to be read, the handle is used to specify from which archive you want to read.

The handle must be generated by the FM ARCHIVE_OPEN_FOR_READ; otherwise, the exception WRONG_ARCHIVE_HANDLE is generated. Type: SY-TABIX.

CHANGENUMBER

Change document number. Type: CDPOS-CHANGENR.

DATE_OF_CHANGE

From change date for the search. If no change date is specified, all of the change document numbers from the current system date are determined. Type: CDHDR-UDATE.

OBJECTCLASS

Object class. Type: CDHDR-OBJECTCLAS.

OBJECTID

Object ID. Type: CDHDR-OBJECTID.

TABLEKEY

Object class table key. The table key is the key of a changed table entry.

If no table key is specified, all change document positions with the specified change document number are determined.

If a table key is specified, only the change document positions with the specified change document number and table key are determined. Type: CDPOS-TABKEY.

TABLENAME

Object class table name. If no table name is specified, all change document positions with the specified change document number are determined.

If a table name is specified, only the change document positions with the specified change document number and table name are determined. Type: CDPOS-TABNAME.

TIME_OF_CHANGE

Change time for search. Type: CDHDR-UTIME.

USERNAME

Last changed by. If this parameter isn't specified, all change documents are determined. Type: CDHDR-USERNAME.

LOCAL_TIME

If set (<> [Space]) the local date and time are used for the parameters DATE_OF_CHANGE and TIME_OF_CHANGE.

TABLEKEY254

Table key for CDPOS in character 254 format. This table key determines the key of the changed row in table CDPOS. Type: CDPOS_UID-TABKEY.

KEYGUID

UUID in character form. The GUID of an entry in the table CDPOS_UID can be passed in this field if the TABKEY stored in CHAR shape is longer than 70 characters. This GUID is then read in the table CDPOS_UID. Type: CDPOS_UID-KEYGUID.

DATE_UNTIL

Change date up to which you want to search. Type: CDHDR-UDATE.

TIME_UNTIL

Change time up to which you want to search. Type: CDHDR-UTIME.

KEYGUID_STR

UUID in character format. Type: CDPOS_STR-KEYGUID.

READ_CHANGEDOCU

This parameter defines how the data is read from the archive.

If the parameter is set to X, the data is read from the database and from the archive. The archive is accessed only for those change documents that have the archiving object as CHANGEDOCU.

Change documents archived using the archiving class CHANGEDOCU can only be accessed with the parameter ARCHIVE_HANDLE. Type: C.

TIME_ZONE

Time zone used as the basis for calculating local time. Type: TTZZ-TZONE.

[↑] Export

ET_CDRED_STR

The table CDRED_STR is returned. Type: CDRED_STR_TAB.

Table

EDITPOS

Table with edited change document items are returned in this parameter. Type: CDRED.

[!] Exceptions

NO_POSITION_FOUND

No item found.

WRONG_ACCESS_TO_ARCHIVE

Incorrect access to archive.

TIME_ZONE_CONVERSION_ERROR

Error converting local to system time.

---

676 CHANGEDOCUMENT_READ_HEADERS

This FM reads the change document header either from the database or from the archive.

Example: You want to read the change document header data for material changes and display those changes so that you can analyze what changes have been made, who made those changes, and when those changes are made on that material object. To do so, you can pass the relevant parameter in this FM and execute it.

[↓] Import

ARCHIVE_HANDLE

This parameter controls whether change documents are to be read from the database or the archive.

If 0 (preset) is passed for this parameter, the database is read; otherwise, an archive is read.

If an archive is to be read, the handle is used to specify from which archive you want to read.

The handle must be generated by the FM ARCHIVE_OPEN_FOR_READ; otherwise, the exception WRONG_ARCHIVE_HANDLE is generated. Type: SY-TABIX.

DATE_OF_CHANGE

From change date for the search. If no change date is specified, all of the change document numbers from the current system date are determined. Type: CDHDR-UDATE.

OBJECTCLASS

Object class. Type: CDHDR-OBJECTCLAS.

OBJECTID

Object ID. Type: CDHDR-OBJECTID.

TIME_OF_CHANGE

Change time for the search. Type: CDHDR-UTIME.

USERNAME

Last changed by. If this parameter isn't specified, all change documents are determined. Type: CDHDR-USERNAME.

LOCAL_TIME

If set (<> [Space]), the local date and time are used for the parameters DATE_OF_CHANGE and TIME_OF_CHANGE.

TIME_ZONE

If LOCAL_TIME = X, the time zone in which change documents were created is used. Type: TTZZ-TZONE.

DATE_UNTIL

Change date up to which you want to search. Type: CDHDR-UDATE.

TIME_UNTIL

Change time up to which you want to search. Type: CDHDR-UTIME.

NOPLUS_ASWILDCARD_INOBJID

If an application is certain that a PLUS in the OBJECTID parameter isn't to be interpreted as a WILDCARD, then it sets the parameter NOPLUS_ASWILDCARD_INOBJID = X (not [Space]). The FM then cancels the switch of the selection option from EQ to CP. The read action is then performed with the actual PLUS character as part of the specified OBJECTID. Any applications that have this problem must check the situation before CHANGEDOCUMENT_READ_HEADERS is called and, if required, attempt to influence the read performance with the new parameter. Possible values:

- [Space]: The PLUS character in the object ID is interpreted as a WILDCARD.

- X: The PLUS character in the object ID is interpreted as a regular character and as part of the key.

Type: C.

READ_CHANGEDOCU

This parameter defines how the data is read from the archive.

If the parameter is set to X, the data is read from the database and from the archive. The archive is accessed only for those change documents that have the archiving object as CHANGEDOCU.

Change documents archived using the archiving class CHANGEDOCU can only be accessed with the parameter ARCHIVE_HANDLE. Type: C.

Table

I_CDHDR

Table with determined header information. Type: CDHDR.

[!] Exceptions

NO_POSITION_FOUND

No item found.

WRONG_ACCESS_TO_ARCHIVE

Incorrect access to archive.

TIME_ZONE_CONVERSION_ERROR

Error converting local to system time.

677 CHANGEDOCUMENT_READ_POSITIONS

The FM reads change document items and processes both the old and the new values according to their category. In some cases, the units and currencies are added on to the value fields. Additionally, the corresponding change document header data is also returned in the export parameter.

Example: You've made the material changes and you want to read and display the item level changes. Pass the relevant parameter in this FM and call it to get the item data of the material changes.

[↓] Import

ARCHIVE_HANDLE

This parameter controls whether change documents are to be read from the database or the archive.

If 0 (preset) is passed for this parameter, the database is read; otherwise, an archive is read.

If an archive is to be read, the handle is used to specify from which archive you want to read.

The handle must be generated by the FM ARCHIVE_OPEN_FOR_READ; otherwise, the exception WRONG_ARCHIVE_HANDLE is generated. Type: SY-TABIX.

CHANGENUMBER

Change document number. Type: CDPOS-CHANGENR.

TABLEKEY

Object class table key. The table key is the key of a changed table entry.

If no table key is specified, all change document positions with the specified change document number are determined.

If a table key is specified, only the change document positions with the specified change document number and table key are determined. Type: CDPOS-TABKEY.

TABLENAME

Object class table name. If no table name is specified, all change document positions with the specified change document number are determined.

If a table name is specified, only the change document positions with the specified change document number and table name are determined. Type: CDPOS-TABNAME.

TABLEKEY254

Table key for CDPOS in character 254 format. This table key determines the key of the changed row in table CDPOS. Type: CDPOS_UID-TABKEY.

KEYGUID

UUID in character form. The GUID of an entry in the table CDPOS_UID can be passed in this field if the TABKEY stored in CHAR shape is longer than 70 characters. This GUID is then read in the table CDPOS_UID. Type: CDPOS_UID-KEYGUID.

KEYGUID_STR

UUID in character format. Type: CDPOS_STR-KEYGUID.

[↑] Export

HEADER

Change document header (structure CDHDR). Type: CDHDR.

ET_CDRED_STR

The table CDRED_STR is returned. Type: CDRED_STR_TAB.

Tables

EDITPOS

Change document formatting table is returned in this table. Type: CDRED.

EDITPOS_WITH_HEADER

Change document display structure is returned in this table. Type: CDRED.

[!] Exceptions

NO_POSITION_FOUND

No item found.

WRONG_ACCESS_TO_ARCHIVE

Incorrect access to archive.

678 CHANGEDOCUMENT_SINGLE_CASE

This FM is used to create a single change document item. The work area for creation of the item must have the same structure of the DDIC table specified in parameter TABLENAME. Only those fields that are relevant to the change document in DDIC are considered for creation.

Example: In the process for CHANGEDOCUMENT_OPEN (FM 673), you need to create the single change document item to which you can pass the relevant parameter in this FM to create the single item.

[↓] Import

CHANGE_INDICATOR

The change flag determines the processing logic of the FM and must have one of the following values:

- D: DELETE. A change document item is to be created to flag the record in WORKAREA_OLD as deleted.
- I: INSERT. A change document item is to be created to flag the record in WORKAREA_NEW as newly created.
- U: UPDATE. WORKAREA_OLD and WORKAREA_NEW are compared, and a change document item is created for each changed field. The keys of the two work areas must be identical.

DOCU_DELETE

The parameter determines how the document positions are to be created after delete:

- X: Document position is created for each field of the deleted record.
- [Space]: Document positions are only created with the key of the deleted record.

REFAREA_NEW

Work area of the new structure with reference information for quantity and currency fields.

REFAREA_OLD

Work area of the old structure with reference information for quantity and currency fields.

REFTABLENAME

Name of the reference table structure. Type: CDPOS-TABNAME.

TABLENAME

Name of the table structure of the object class. Type: CDPOS-TABNAME.

WORKAREA_NEW

The work area must have the structure of the table specified under TABLENAME and contain the data changed by the user.

WORKAREA_OLD

The work area must have the structure specified under TABLENAME and contain the unchanged original data.

[!] Exceptions

NAMETAB_ERROR

Error when calling NAMETAB_GET.

OPEN_MISSING

No OPEN was carried out.

POSITION_INSERT_FAILED

SQL error occurred during insert item.

---

679 CLOI_PUT_SIGN_IN_FRONT

This FM is used to put a sign in front of a numeric field.

Example: You've received the field as 123–, and you want the negative sign in front of the numeric field instead. You can pass the relevant parameters and call this FM to get the value as –123.

[Δ] Changing

VALUE

Value you want to change. Type: CLIKE.

---

680 CS_BOM_EXPL_EQU_V2

This FM is used to explode the BOM of an equipment item.

Example: You want to explode an equipment item to see its BOM hierarchy (all of the individual parts/materials). Call this FM to explode the BOM.

[↓] Import

ALTVO

Alternative priority. Type: CSDATA-XFELD.

AUFSW

This parameter determines and enters the order level and channel. Type: CSDATA-XFELD.

AUSKZ

This parameter checks for scrap equipment. Type: CSDATA-XFELD.

BAGRP

Assembly restriction. Type: MARA-MATNR.

BEIKZ

Material provision indicator. Type: STPO-BEIKZ.

BESSL

Material provision selection indicator. Type: CSDATA-XFELD.

BGIXO

Load assembly information for exploded assemblies only. Type: CSDATA-XFELD.

BREMS

Limited explosion. Type: CSDATA-XFELD.

CAPID

Application ID of BOM. Type: TC04-CAPID.

COSPR

This parameter is for use with the job-specific material preread function. Type: CSDATA-XFELD.

DATUV

Validity date. Type: STKO-DATUV.

DELNL

Delete items not kept in stock from list. Type: CSDATA-XFELD.

EHNDL

This parameter is used for the probability handle. Type: CSDATA-XFELD.

EMENG

Required quantity. Type: STKO-BMENG.

EQNRV

Equipment number. Type: EQUI-EQUNR.

ERSKZ

Spare part indicator. Type: STPO-ERSKZ.

ERSSL

Spare part selection indicator. Type: CSDATA-XFELD.

FBSTP

Limited multilevel parameter. Explosion externally procures the item. Type: CSDATA-XFELD.

MBWLS

Read values for material. Type: CSDATA-XFELD.

MDMPS

Limited multilevel explosion. This parameter will at least explode phantom assemblies. Type: CSDATA-XFELD.

MEHRS

Multilevel explosion. Type: CSDATA-XFELD.

MKMAT

Limited multilevel; explode KMAT. Type: CSDATA-XFELD.

MMORY

Memory version. Value 1 is for setting memory on, and Value 2 is for off. Type: CSDATA-XFELD.

POSTP

Item category. Type: STPO-POSTP.

RVREL

Relevant to sales. Type: STPO-RVREL.

SANFR

Production. Type: STPO-SANFE.

SANIN

Maintenance. Type: STPO-SANIN.

SANKA

Costing. Type: STPO-SANKA.

SANKO

Engineering/design. Type: STPO-SANKO.

SANVS

Shipping. Type: STPO-SANVS.

SCHGT

Bulk material. Type: STPO-SCHGT.

STKKZ

PM assembly. Type: STPO-STKKZ.

STLAL

Alternative BOM. Type: STKO-STLAL.

STLAN

BOM usage. Type: STZU-STLAN.

WERKS

Plant. Type: MARC-WERKS.

[↑] Export

TOPEQUI

Start object of equipment is returned. Type: CSTEQUI.

Tables

STB

Collective item data table. This table contains the children of the BOM. Type: STPOX.

MATCAT

This table contains the material catalog subassemblies. Type: CSCMAT.

[!] Exceptions

ALT_NOT_FOUND

Alternative not found.

CALL_INVALID

Incorrect parameter combination.

EQUIPMENT_NOT_FOUND

Equipment not found.

MISSING_AUTHORIZATION

No authorization (BOM).

NO_BOM_FOUND

No BOM for this material.

NO_PLANT_DATA

Material not maintained in the plant.

NO_SUITABLE_BOM_FOUND

BOM does not exist.

681 CS_BOM_EXPL_MAT_V2

This FM is used to explode the BOM for a material.

Example: You have a base machine model of a car, such as the Mercedes S550. You want to get the details of the factory top materials and components so that you can add those details in the activities required to create a subassembly in order to ultimately produce the model of the car. You can pass the relevant parameter and call this FM to explode the material's BOM.

[↓] Import

FTREL

Limited multilevel. Stop explosion at items not relevant to production. Type: CSDATA-XFELD.

ALEKZ

Features: ALE environment. Type: CSDATA-XFELD.

ALTVO

Alternative priority. Type: CSDATA-XFELD.

AUFSW

Determine and enter order level and channel. Type: CSDATA-XFELD.

AUMGB

Calculate scrap quantity. Type: CSDATA-XFELD.

AUMNG

Scrap quantity. Type: STKO-BMENG.

AUSKZ

Scrap. Type: CSDATA-XFELD.

AMIND

Scrap quantity indicator. Type: CSDATA-XFELD.

BAGRP

Assembly restriction. Type: MARA-MATNR.

BEIKZ

Material provision indicator. Type: STPO-BEIKZ.

BESSL

Material provision selection indicator. Type: CSDATA-XFELD.

BGIXO

Load assembly information for exploded assemblies only. Type: CSDATA-XFELD.

BREMS

Limited explosion. Type: CSDATA-XFELD.

CAPID

Application ID. Type: TC04-CAPID.

CHLST

Activate the function's homogeneity list. Type: CSDATA-XFELD.

COSPR

Internal: (CO) order-specific MATPREREAD. Type: CSDATA-XFELD.

CUOBJ

Object dependencies. Type: MARC-CUOBJ.

CUOVS

CUOBJ version. Type: IBVERSION.

CUOLS

Read characteristic value for position. Type: CSDATA-XFELD.

DATUV

Validity date. Type: STKO-DATUV.

DELNL

Delete items not kept in stock from list. Type: CSDATA-XFELD.

DRLDT

Determine (document) revision level. Type: CSDATA-XFELD.

EHNDL

Usage probability handling value: value 1 is for modifying item quantity, and value 2 is for a required quantity. Type: CSDATA-XFELD.

EMENG

Required quantity. Type: STKO-BMENG.

ERSKZ

Spare part indicator. Type: STPO-ERSKZ.

ERSSL

Spare part selection indicator. Type: CSDATA-XFELD.

FBSTP

Limited multilevel. Stop explosion at externally procured item. Type: CSDATA-XFELD.

KNFBA

Set this parameter as X to explode configurable material Type: CSDATA-XFELD.

KSBVO

Calculation specific special procurement code. Type: CSDATA-XFELD.

MBWLS

Read material valuation. Type: CSDATA-XFELD.

MKTLS

Read material description Type: CSDATA-XFELD.

MDMPS

Limited multilevel. Explode phantom assemblies at least. Type: CSDATA-XFELD.

MEHRS

Multilevel explosion. Type: CSDATA-XFELD.

MKMAT

Limited multilevel. Explode KMAT. Type: CSDATA-XFELD.

MMAPS

Limited multilevel. Explode at least M assembly (M order). Type: CSDATA-XFELD.

SALWW

Limited multilevel. Stop for general Plant change. Type: CSDATA-XFELD.

SPLWW

Limited multi level. Stop for routing changes Type: CSDATA-XFELD.

MMORY

Memory use (1 = on; 0 = off; [Space] = no reaction). Type: CSDATA-XFELD.

MTNRV

Material. Type: MARA-MATNR.

NLINK

Identify and submit edge structure. Type: CSDATA-XFELD.

POSTP

Item category. Type: STPO-POSTP.

RNDKZ

Round off: Value [Space]: always, 1: never, 2: only levels > 1. Type: CSDATA-XFELD.

RVREL

Relevant to sales. Type: STPO-RVREL.

SANFR

Production. Type: STPO-SANFE.

SANIN

Maintenance. Type: STPO-SANIN.

SANKA

Costing. Type: STPO-SANKA.

SANKO

Engineering/design. Type: STPO-SANKO.

SANVS

Shipping. Type: STPO-SANVS.

SCHGT

Bulk material. Type: STPO-SCHGT.

STKKZ

PM assembly. Type: STPO-STKKZ.

STLAL

Alternative BOM. Type: STKO-STLAL.

STLAN

BOM usage. Type: STZU-STLAN.

STPST

Level up to which material is to be exploded. Type: STPOX-STUFE.

SVWVO

Precedence of BOM entered. Type: CSDATA-XFELD.

WERKS

Plant. Type: MARC-WERKS.

NORVL

Revision level isn't determined. Type: CSDATA-XFELD.

MDNOT

Restriction on MDMPS: don't explode phantom M. Type: CSDATA-XFELD.

PANOT

Restriction on MDMPS: no parallel discontinue. Type: CSDATA-XFELD.

QVERW

Application key (use quote). Type: TMQ2-USEQU.

VERID

Production version. Type: MKAL-VERID.

VRSVO

Precedence of entered production version. Type: CSDATA-XFELD.

[↑] Export

TOPMAT

Data for start material. Type: CSTMAT.

DSTST

Structure destroyed by filter. Type: CSDATA-XFELD.

Tables

STB

Collective item data table. This table contains the children of the BOM. Type: STPOX.

MATCAT

This table contains the material catalog subassemblies. Type: CSCMAT.

[!] Exceptions

ALT_NOT_FOUND

Alternative not found.

CALL_INVALID

Incorrect parameter combination.

MATERIAL_NOT_FOUND

Material not found.

MISSING_AUTHORIZATION

No authorization (BOM).

NO_BOM_FOUND

No BOM for this material.

NO_PLANT_DATA

Material not maintained in the plant.

NO_SUITABLE_BOM_FOUND

BOM does not exist.

CONVERSION_ERROR

Conversion error of units.

---

682 CS_WHERE_USED_MAT

This FM is used to find the where-used list for the BOM.

Example: You want to know in which factory top materials a specific component is used so that you can generate the bottom-to-top hierarchy of the BOM. You can pass the relevant parameters in this FM and call this FM to get the where-used list of materials.

[↓] Import

DATUB

Date valid to. Type: RC29L-DATUB.

DATUV

Date valid from. Type: RC29L-DATUV.

MATNR

Material. Type: RC29L-MATNR.

POSTP

Item category. Type: RC29L-POSTP.

RETCODE_ONLY

If this parameter is set to X, only the return code is set to success or failure, and no data is returned. Type: CSDATA-XFELD.

STLAN

BOM usage. Type: RC29L-STLAN.

WERKS

Plant. Type: RC29L-WERKS.

MCLMT

Maximum BOM number at a mass change. Type: STZU-STUEZ.

MNSTL

BOM number from which the search is to be made. Type: STZU-STLNR.

MXSTL

BOM number up to which the search is to be made. Type: STZU-STLNR.

STLTP

Type of BOM. Type: RC29A-ATABR.

NEWSI

New session initialization. Type: CSDATA-XFELD.

[↑] Export

TOPMAT

Data on the initial material. Type: MC29S.

Tables

WULTB

Where-used list table of items. Type: STPOV.

EQUICAT

Assembly table: equipment. Type: CSCEQUI.

KNDCAT

Assembly table: sales orders. Type: CSCKND.

MATCAT

Assembly table: materials. Type: CSCMAT.

STDCAT

Assembly table: standard objects. Type: CSCSTD.

TPLCAT

Assembly table: functional locations. Type: CSCTPL.

PRJCAT

Project module table. Type: CSCPRJ.

[!] Exceptions

CALL_INVALID

FM call incomplete or invalid.

MATERIAL_NOT_FOUND

Material not available.

NO_WHERE_USED_REC_FOUND

No where-used record available.

NO_WHERE_USED_REC_SELECTED

No where-used record selected.

NO_WHERE_USED_REC_VALID

No valid where-used record available.

683 CSAP_MAT_BOM_READ

This FM can be used to display simple material BOMs. You can also display BOM groups such as all variant of a variant BOM as displayed in Transaction CS03. With this FM, you can't display long text, subitems, or the classification date of BOM items for batches.

You also can only display alternatives or variant 01.

Example: You want to read the BOM details of an engine so that you'll know which parts or components (and in which quantities) comprise the engine. To display the BOM details in your program, pass the relevant parameter and call this FM.

[↓] Import

MATERIAL

Material. Type: CSAP_MBOM-MATNR.

PLANT

Plant. Type: CSAP_MBOM-WERKS.

BOM_USAGE

BOM usage. Type: CSAP_MBOM-STLAN.

ALTERNATIVE

Alternative BOM. Type: CSAP_MBOM-STLAL.

VALID_FROM

Valid-from date. Type: CSAP_MBOM-DATUV.

VALID_TO

Valid-to date. Type: CSAP_MBOM-DATUB.

CHANGE_NO

Change number. Type: CSAP_MBOM-AENNR.

REVISION_LEVEL

Revision level. Type: CSAP_MBOM-REVLV.

FL_DOC_LINKS

If this parameter is set to X, it reads document assignments. Type: CSDATA-XFELD.

FL_DMU_TMX

If this parameter is set to X, it reads the DMU transformation. Type: CSDATA-XFELD.

[↑] Export

FL_WARNING

Log contains warning messages. Type: CAPIFLAG-FLWARNING.

Tables

T_STPO

BOM items. Type: STPO_API02.

T_STKO

BOM headers. Type: STKO_API02.

T_DEP_DATA

Object dependencies: basic data. Type: CSDEP_DAT.

T_DEP_DESCR

Object dependencies: description. Type: CSDEP_DESC.

T_DEP_ORDER

Object dependencies: sort sequence. Type: CSDEP_ORD.

T_DEP_SOURCE

Object dependencies: source code. Type: CSDEP_SORC.

T_DEP_DOC

Object dependencies: documentation. Type: CSDEP_DOC.

T_DOC_LINK

Document assignments. Type: CSDOC_LINK.

T_DMU_TMX

DMU-transformation. Type: CSDMU_TMX.

T_LTX_LINE

Bill of materials: long texts. Type: CSLTX_LINE.

T_STPU

API structure for internal APIs: STPU and identifier. Type: STPU_API01.

---

684 CSEP_MAT_BOM_SELECT

This FM is used to select a BOM for a material or for a material, plant, or a BOM usage.

Example: You want to know if a BOM exists in other plants and with different BOM usages. Execute this FM to select the BOM you're querying, and the FM provides you with the relevant details.

[↓] Import

MATERIAL

Material. Type: CSAP_MBOM-MATNR.

PLANT

Plant. If the value in PLANT is *, which is the default setting, all of the BOMs for the given material in all of the plants are returned. If the PLANT value is specified, only BOMs for that plant are selected. If PLANT is [Space], only group materials are selected. Type: CSAP_MBOM-WERKS.

BOM_USAGE

BOM usage. Type: CSAP_MBOM-STLAN.

FL_MATERIAL_CHECK

If this parameter is set to X, the system checks whether the specified material exists. Type: CSDATA-XFELD.

FL_FOREIGN_KEY_CHECK

If this parameter is set to X, the system checks whether BOM usage and PLANT exists. Type: CSDATA-XFELD.

Tables

T_BOM_HEADER

Header information of the material returned. Type: BOM_HEADER_API01.

T_BOM_HEADER_REV

Header information depending on the revision status is returned. Type: BOM_HEADER_API01.

[!] Exception

ERROR

Terminate processing.

---

685 CSEP_MAT_BOM_SELECT_CHANGE

This FM returns the list of the BOMs that are valid for the engineering change number (ECN).

Example: You want to display the list of BOMs that you've changed with a particular change number. You can pass the relevant parameters and call this FM to display the list of relevant BOMs.

[↓] Import

CHANGE_NO

Engineering change number. Type: AENNR.

Tables

T_BOM_HEADER

Header information of the material returned. Type: BOM_HEADER_API01.

T_BOM_HEADER_REV

Header information depending on revision status is returned. Type: BOM_HEADER_API01.

[!] Exception

ERROR

Terminate processing.

---

686 CSEP_MAT_BOM_SELECT_WHERE_USED

This FM is used to find the where-used list of the material in the BOMs.

Example: You want to know all of the parents in which a certain material is used. You can pass the material number in the importing parameter and call this FM, which returns the list of all of the parents of the material.

[↓] Import

MATERIAL

Material. Type: MATNR.

Tables

T_BOM_HEADER

Header information of the material returned. Type: BOM_HEADER_API01.

T_BOM_HEADER_REV

Header information depending on revision status is returned. Type: BOM_HEADER_API01.

[!] Exception

ERROR

Terminate processing.

---

687 DEQUEUE_ESFUNCTION

This FM is used to release the lock that was applied from FM ENQUEUE_ESFUNCTION (FM 688).

Example: You've locked the FM processing from FM ENQUEUE_ESFUNCTION. When you're finished editing the FM, you need to unlock the FM so that other users can start their changes. Pass the relevant parameter and call this FM to unlock the FM. See FM ENQUEUE_ESFUNCTION.

[↓] Import

MODE_TFDIR

Lock mode for table TFDIR. Lock mode can be E (write lock), S (read lock), or X (exclusive, noncumulative). The default value is E. Type: ENQMODE.

FUNCNAME

First enqueue lock argument. Type: TFDIR-FUNCNAME.

X_FUNCNAME

This parameter defines the lock behavior when the initial value is passed in the lock field. If the initial value is assigned to <field> and X_<field>, then a generic lock is initialized with respect to <field>. If <field> is assigned the initial value, and X_<field> is defined as X, the lock is set with exactly the initial value of <field>. The default value is [Space].

_SCOPE

This defines the lock behavior when the update task is called. Possible values:

- _Scope = 1: The lock or lock releases aren't passed to the update program. The lock is removed when the transaction is ended.
- _Scope = 2: The lock or lock release is passed to the update program. The update program is responsible for removing the lock. The interactive program with which the lock was requested no longer has an

influence on the lock behavior. This is the standard setting for the ENQUEUE FM.

- _SCOPE = 3: The lock or lock release is also passed to the update program. The lock must be removed in both the interactive program and in the update program. This is the standard setting for the DEQUEUE FM.

The default value for this parameter is 3.

_SYNCHRON

If X is passed, the DEQUEUE function waits until the entry has been removed from the lock table. Otherwise, it's deleted asynchronously; that is, if the lock table of the system is read directly after the lock is removed, the entry in the lock table may still exist. The default value is [Space].

_COLLECT

This parameter controls whether the lock request or lock release should be performed directly or whether it should first be written to the local lock container. Possible values:

- [Space]: The lock request or lock release is sent directly to the lock server.
- X: The lock request or lock release is placed in the local lock container. The lock requests and lock releases collected in this lock container can then be sent to the lock server at a later time as a group by calling the FM FLUSH_ENQUEUE.

The default value is [Space]. Type: DDENQCOLL.

---

688 ENQUEUE_ESFUNCTION

This FM requests the lock for ESFUNCTION, which is the lock object for FMs and locks the table TFDIR. The key parameter for this lock object is the FM name (field FUNCNAME).

Example: To lock the FM so that no one else can edit it, you can pass the relevant parameter and call this FM to lock the FM. To unlock the locked FM, you can pass the parameter in FM DEQUEUE_ESFUNCTION (FM 687).

[↓] Import

MODE_TFDIR

Lock mode for table TFDIR. Lock mode can be the following:

- E: Write lock.
- S: Read lock
- X: Exclusive, noncumulative.

The default value is E. Type: ENQMODE.

FUNCNAME

First enqueue lock argument. Type: TFDIR-FUNCNAME.

X_FUNCNAME

This parameter defines the lock behavior when the initial value is passed in the lock field. If the initial value is assigned to <field> and X_<field>, then a generic lock is initialized with respect to <field>. If <field> is assigned the initial value, and X_<field> is defined as X, the lock is set with exactly the initial value of <field>. The default value is [Space].

_SCOPE

This defines the lock behavior when the update task is called. Possible values:

- _Scope = 1: The lock or lock releases aren't passed to the update program. The lock is removed when the transaction is ended.
- _Scope = 2: The lock or lock release is passed to the update program. The update program is responsible for removing the lock. The interactive program with which the lock was requested no longer has an influence on the lock behavior. This is the standard setting for the ENQUEUE FM.
- _SCOPE = 3: The lock or lock release is also passed to the update program. The lock must be removed in both the interactive program and in the update program. This is the standard setting for the DEQUEUE FM.

The default value for this parameter is 3.

_WAIT

This parameter determines the lock behavior when there is a lock conflict. Possible values:

- [Space]: If a lock attempt fails because there is a competing lock, the exception FOREIGN_LOCK is triggered.
- X: If a lock attempt fails because there is a competing lock, the lock attempt is repeated after waiting for a certain time.

  The exception FOREIGN_LOCK is triggered only if a certain time limit has elapsed since the first lock attempt. The waiting time and the time limit are defined by profile parameters.

The default value is [Space].

_COLLECT

This parameter controls whether the lock request or lock release should be performed directly or whether it should first be written to the local lock container. Possible values:

- [Space]: The lock request or lock release is sent directly to the lock server.
- X: The lock request or lock release is placed in the local lock container. The lock requests and lock releases collected in this lock container can then be sent to the lock server at a later time as a group by calling the FM FLUSH_ENQUEUE.

The default value is [Space]. Type: DDENQCOLL.

[!] Exceptions

FOREIGN_LOCK

If a lock attempt fails because there is a competing lock, the exception FOREIGN_LOCK is triggered. See parameter _WAIT for more details.

SYSTEM_FAILURE

Internal error from enqueue server.

689 ENQUE_SLEEP

This FM is used to sleep or delay the processing of a program for the specified number of seconds.

Example: You're saved some material data to various material-related ABAP Data Dictionary tables. You want to read the same data that you have saved. Since the data is not committed to database tables, you don't have the same data that you have saved just now. Ideally, while saving the data to the database, you should use the command `COMMIT WORK` and `WAIT`, but you can also use this FM if you want your program to wait for a specified number of seconds.

[↓] Import

SECONDS

Number of seconds you want the system to wait.

[!] Exception

SYSTEM_FAILURE

System error in lock operation.

690 FUNCTION_EXISTS

This FM is used to check whether the specified FM exists, and finds the function group in which it exists.

Example: You've created a program that dynamically comes up with a FM name based on your temporary table field value. Based on the value generated, you need to first check if the FM exists. If the FM exists, you need to execute that FM. To check the existence of the FM, you can pass the relevant parameters and call this FM.

[↓] Import

FUNCNAME

Name of FM. Type: RS38L-NAME.

[↑] Export

GROUP

Name of function group returned. Type: RS38L-AREA.

INCLUDE

Name of include. Type: RS38L-INCLUDE.

NAMESPACE

Namespace of the FM. Type: RS38L-NAMESPACE.

STR_AREA

Name of function group without namespace. Type: RS38L-STR_AREA.

[!] Exception

FUNCTION_NOT_EXIST

This exception is raised if the FM does not exist.

---

691 FUNCTION_IMPORT_INTERFACE

This FM provides the interface for all of the details such as export, import, table, changing parameters, and enhancement parameters of the FM.

Example: You have a database table where you've stored the FM name. You need to find the details of this FM and call it dynamically in the report program. Pass the relevant parameter such as the FM name to get the interface and other parameters in this FM and call this FM to get the interface details.

[↓] Import

FUNCNAME

Name of FM. Type: RS38L-NAME.

INACTIVE_VERSION

Reads inactive version according to worklist. Type: CHAR1.

WITH_ENHANCEMENTS

If this parameter has the value X, then enhancement parameters are also returned. Type: CHAR1.

IGNORE_SWITCHES

If this parameter has the value X, switches are ignored. Type: CHAR1.

[↑] Export

GLOBAL_FLAG

This parameter indicates whether it's a global or local interface. Type: RS38L-GLOBAL.

REMOTE_CALL

FM can be called with a remote function call. Type: RS38L-REMOTE.

UPDATE_TASK

FM running in the update task. Type: RS38L-UTASK.

EXCEPTION_CLASSES

Additional attributes for FMs. Type: ENLFDIR-EXTEN3.

Tables

EXCEPTION_LIST

FM exception parameters are returned in this table. Type: RSEXC.

EXPORT_PARAMETER

FM export parameters are returned in this table. Type: RSEXP.

IMPORT_PARAMETER

FM import parameters are returned in this table. Type: RSIMP.

CHANGING_PARAMETER

FM changing parameters are returned in this table. Type: RSCHA.

TABLES_PARAMETER

FM tables parameters are returned in this table. Type: RSTBL.

P_DOCU

FM documentation interface table. Type: RSFDO.

ENHA_EXP_PARAMETER

FM enhancement export parameters are returned in this table. Type: RSEXP.

ENHA_IMP_PARAMETER

FM Enhancement Import parameters are returned in this table. Type: RSIMP.

ENHA_CHA_PARAMETER

FM enhancement changing parameters are returned in this table. Type: RSCHA.

ENHA_TBL_PARAMETER

FM enhancement tables parameters are returned in this table. Type: RSTBL.

ENHA_DOCU

FM enhancement documentation interface table. Type: RSFDO.

[!] Exceptions

ERROR_MESSAGE

Some error messages occur.

FUNCTION_NOT_FOUND

FM does not exist.

INVALID_NAME

Invalid FM.

692 FUNCTION_INCLUDE_CONCATENATE

This FM is used to display the details of the FM includes based on the program name, function group, namespace, or complete area given in the changing parameters.

Example: You want to get the source code that you've written in the FM. First you need to get the include name of the FM. You can pass the relevant parameter in this FM to get the include name. Next you need to use the ABAP command `READ REPORT` to read the source code of that include.

[↓] Import

INCLUDE_NUMBER

Number of the FM include. Type: TFDIR-INCLUDE.

[↑] Export

TOP

Name of the TOP include. Type: RS38L-INCLUDE.

UXX

Name of the UXX include. Type: RS38L-INCLUDE.

INCLUDE

Name of a function group include. Type: RS38L-INCLUDE.

[Δ] Changing

PROGRAM

Name of main program of a function group. Type: RS38L-INCLUDE.

GROUP

Function group name without namespace. Type: RS38L-AREA.

NAMESPACE

Namespace. Type: RS38L-NAMESPACE.

COMPLETE_AREA

Entire function group name. Type: RS38L-AREA.

[!] Exceptions

NOT_ENOUGH_INPUT

Input entered is not enough to determine the include details.

NO_FUNCTION_POOL

Function pool does not exist.

DELIMITER_WRONG_POSITION

Wrong position was entered.

693 FUNCTION_INCLUDE_INFO

This FM returns the includes for a FM name or for a function group. For the function group, generic entries such as % and + are possible.

Example: You want to display all of the FMs that are present in the function group as well as the details of the function group. You can pass the function group name in the GROUP parameter for which you want to get the FM details.

[↑] Export

FUNCTAB

Details of FMs are returned in this parameter. Type: SUNI_FUNCTAB.

NAMESPACE

Namespace of the returned FM. Type: RS38L-NAMESPACE.

PNAME

Program name of the FM. Type: TFDIR-PNAME.

[Δ] Changing

FUNCNAME

FM name. Type: RS38L-NAME.

GROUP

Function group name. Type: RS38L-AREA.

INCLUDE

Function include name. Type: RS38L-INCLUDE.

[!] Exceptions

FUNCTION_NOT_EXISTS

Function name entered in changing parameter does not exist.

INCLUDE_NOT_EXISTS

Include entered in changing parameter does not exist.

GROUP_NOT_EXISTS

Function group entered in changing parameter does not exist.

NO_SELECTIONS

No selections found.

NO_FUNCTION_INCLUDE

No function include returned.

694 FUNCTION_INCLUDE_SPLIT

This FM returns the information of the FM and function group. If the PROGRAM name or COMPLETE_AREA parameter is specified, then the namespace and function group are returned. If the INCLUDE parameter is specified, the namespace, include (without namespace), include number (for FMs), three-character suffix, and FM name are returned.

Example: You want to find all of the text elements used in a FM. You have a selection screen parameter of FM name, and you need to get the function group name. You should add SAPL as a prefix to the function group name and get the FM program name by passing the relevant parameters in this FM. When you get the FM program name, use the ABAP statement `READ REPORT` to read the details of the report and find the text elements. You can pass the relevant parameters and call this FM to do so.

[↓] Import

PROGRAM

Name of a program with namespace. Type: RS38L-INCLUDE.

SUPPRESS_SELECT

If this parameter is set to X, entry from TFDIR isn't checked. Type: C.

COMPLETE_AREA

Full name of the group with a namespace. Type: RS38L-AREA.

[↑] Export

NAMESPACE

Namespace. Type: RS38L-NAMESPACE.

FUNCTION_NOT_EXISTS

The specified FM doesn't exist. Type: C.

GROUP

Name of function group. Type: RS38L-AREA.

FUNCNAME

FM name. Type: RS38L-NAME.

INCLUDE_NUMBER

Number of the FM include according to TFDIR. Type: TFDIR-INCLUDE.

NO_FUNCTION_INCLUDE

This parameter is set to X if the include specified isn't an include of a function group. Type: C.

NO_FUNCTION_MODULE

This parameter is set to X if the include specified isn't a FM include. Type: C.

SUFFIX

Three-character suffix of the include according to the naming standard, such as TOP and UXX. Type: RS38L-SUFFIX.

RESERVED_NAME

This parameter is set to X if the name of the include is reserved for the function builder; that is, if the include name starts with SAPL, which is reserved for the function builder name. Type: C.

TOO_MANY_DELIMITERS

Too many namespace delimiters in the name. Type: C.

RESERVED_FOR_EXITS

Name is reserved for exit modules. Type: C.

HIDDEN_NAME

This parameter is for internal use. Type: C.

[Δ] Changing

INCLUDE

Name of include you want to get the details of. Type: RS38L-INCLUDE.

[!] Exceptions

INCLUDE_NOT_EXISTS

Include does not exist.

GROUP_NOT_EXISTS

Function group does not exist.

NO_SELECTIONS

Nothing was selected.

NO_FUNCTION_INCLUDE

The include you specified is not a FM.

NO_FUNCTION_POOL

Not a function group.

DELIMITER_WRONG_POSITION

Namespace delimiters are in the wrong position.

NO_CUSTOMER_FUNCTION_GROUP

Not a customer exit group.

NO_CUSTOMER_FUNCTION_INCLUDE

Not a customer exit include.

RESERVED_NAME_CUSTOMER

Reserved for customer exits.

NAMESPACE_TOO_LONG

Namespace is more than 10 characters.

AREA_LENGTH_ERROR

This exception is raised if the include name is less than four characters.

695 G_DECIMAL_PLACES_GET

This FM determines where the decimal place for the currency should be placed.

Example: You're getting the currency from the database table. Based on the decimal places for the currency, you want to display it to the user like if you get $1000 from database table and the decimal places for currency is at 2 digits, you will display $10.00 to the user. To get the decimal places for the currency, you need to pass the parameter and call this FM.

[↓] Import

CURRENCY

Currency key.

[↑] Export

DECIMAL_PLACES

Number of the places after the decimal point. Type: TCURX-CURRDEC.

---

696 GET_COMPONENT_LIST

This FM returns the list of all components of the structure that is used in the specified program.

Example: To get the list of components for the structure of a program based on which you want to make a dynamic select query, pass the relevant parameters and call this FM.

[↓] Import

PROGRAM

Report name from which you want to get the structure detail. Type: SY-REPID.

FIELDNAME

Structure for which you want to get the component details.

Table

COMPONENTS

List of components of the structure. Type: RSTRUCINFO.

---

697 GET_FIELDTAB

This FM returns all of the fields and information of the database table.

Example: You have a table name that you're going to use as the input parameter of a screen. Based on the entered table name, you want to get the table field name, create a dynamic select query, and build that data to send to other systems. You can pass the relevant parameters in this FM to get the details of the fields of the table.

[↓] Import

LANGU

Language. Type: DFIES-LANGU.

ONLY

This parameter returns only the header information of the table entered if the value is H or complete information along with the field details if the value is [Space].

TABNAME

Table name whose information you want to get. Type: DFIES-TABNAME.

WITHTEXT

If this parameter is set to X, details are returned with text.

[↑] Export

HEADER

Table header data. Type: X030L.

RC

SY SUBRC return value. Type: SY-SUBRC.

Table

FIELDTAB

Output table with field information of the table. Type: DFIES.

[!] Exceptions

INTERNAL_ERROR

Internal error.

NO_TEXTS_FOUND

All texts are not available.

TABLE_HAS_NO_FIELDS

Table has no fields.

TABLE_NOT_ACTIV

Table does not exist actively.

698 GET_GLOBAL_SYMBOLS

This FM returns a list of all global data of an ABAP program; that is, internal tables, work area, global variable, screen parameters, and so on.

Example: Your development team isn't following the coding standards, and you've come up with a new standard, such as you need to declare all of the global variables with g_<varname> and internal tables with i_<internal_tablename>. To create a report program where you give the program name on the selection screen to find all of the global data and variables used in that program, pass the relevant parameters and call this FM.

[↓] Import

PROGRAM

Program name of which you want the global data. Type: SY-REPID.

NAME_PATTERN

Pattern of the output. The default value * returns all of the global data of the program. Type: C.

Table

FIELDLIST

List of returned global data of table. Type: RFIELDLIST.

---

699 GET_INCLUDETAB

This FM returns all of the includes of the specified program.

Example: You need to create a program where you give the program name in the selection screen parameter to get and analyze all of the include programs used in that program. You can pass the program name and call this FM to get all of the includes present.

[↓] Import

PROGRAM

Program name for which you want the global data. Type: SY-REPID.

Table

INCLTAB

This table contains the list of includes present in the program.

---

700 K_WERKS_OF_BUKRS_FIND

This FM returns all of the plant and valuation area details for the company code entered in the selection criteria.

Example: To get a list of all of the plants that are assigned to the specified company code, pass the relevant parameters and call this FM for the same purpose.

[↓] Import

BUKRS

Company code. Type: T001K-BUKRS.

Tables

ITAB_001K

This table returns the valuation area (table T001K) details. Type: T001K.

ITAB_001W

This table returns the plant (table T001W) details. Type: T001W.

701 MATERIAL_BTCI_SELECTION_NEW

This FM creates the batch input data for the SELECTION SCREEN view of the material master maintenance Transaction MM01.

Example: You want to upload some materials from the legacy system to the SAP system, so you create batch input data for the material master. You can pass the relevant parameters and call this FM to get the batch input data for all of the views of the material master.

[↓] Import

MATERIAL

Material number for which you want to get the view details. Type: MARA-MATNR.

MATERIALART

Material type. This parameter should be specified only when you're creating a new material according to the specification. For existing materials, this is determined from the material master. Type: MARA-MTART.

SELECTION

Maintenance status. Type: T130M-PSTAT.

TCODE

Transaction code of the material master from where the material type and status should be fetched for the existing material. Type: T130M-TCODE.

[↑] Export

SELSTATUS

Union of the maintenance status of the selected material is returned in this parameter. Type: T130M-PSTAT.

SELSTATUS_IN

This parameter returns the union of the maintenance status of the specified views without considering if the screen is active in view selection. Type: T130M-PSTAT.

Tables

BTCI_D0070

This table contains all batch input data for the selection screen in material master maintenance, including the entry for the start screen, according to the given input parameters. The content of this table is to take into the batch input table of the calling program, after filling the entry-level image view 0060 and then filling the organizational level image view 0080. Type: BDCDATA.

[!] Exceptions

MATERIAL_NOT_FOUND

Material specified does not exist.

MATERIAL_NUMBER_MISSING

Material number is not entered.

MATERIAL_TYPE_MISSING

Material type is not entered.

MATERIAL_TYPE_NOT_FOUND

Material type entered is not found.

NO_ACTIVE_DYNPRO_SELECTED

No active screen is selected.

NO_AUTHORITY

No authorization to run the FM.

---

702 MD_MRP_LIST_API

This FM determines the MRP list for the material or MRP area. This FM has a similar result that you get from Transaction MD06 (MRP List).

Example: To determine the MRP run and MRP area for the specified plant, as well as what details to analyze and display in your customized report output, pass the relevant parameter and call this FM.

[↓] Import

PLSCN

Planning scenario of long-term planning. Type: PLSC-PLSCN.

MATNR

Material number. Type: MARC-MATNR.

WERKS

Plant. Type: MARC-WERKS.

BERID

MRP area. Type: MDLV-BERID.

AFIBZ

Name of display filter. Type: T463-AFIBZ.

SINFG

Single-source indicator. Type: CHAR1.

LIFNR

Vendor key. Type: T463-LIFNR.

AFHOR

Period after today's date. Type: T463-ESPAN.

DTNUM

Number of MRP table. Type: MDKP-DTNUM.

INPER

Internal period indicator. Type: MDST-INPER.

DISPLAY_LIST_MDPSX

If this parameter is set to X, internal table MDPSX is displayed. Type: CHAR1.

DISPLAY_LIST_MDEZX

If this parameter is set to X, internal table MDEZX is displayed. Type: CHAR1.

DISPLAY_LIST_MDSUX

If this parameter is set to X, internal table MDSUX is displayed. Type: CHAR1.

[↑] Export

E_MT61D

Material master MRP data is returned in this parameter. Type: MT61D.

E_MDKP

Header data for MRP document is returned in this parameter. Type: MDKP.

Tables

MDPSX

Item in MRP document data is returned in this table. Type: MDPS.

MDEZX

Individual lines of the MRP element data are returned in this table. Type: MDEZ.

MDSUX

Total lines for MRP element data are returned in this table. Type: MDSU.

[!] Exceptions

MRP_LIST_NOT_FOUND

MRP list is not found.

MATERIAL_PLANT_NOT_FOUND

Material is not valid for the plant.

ERROR

Some other error occurred.

---

703 MD_STOCK_REQUIREMENTS_LIST_API

This FM determines the stock requirements list for the material or plant. This FM has the similar result that you get from Transaction MD04 (Stock Requirement List).

Example: To determine the stock of the material required for the plant and analyze and display the details in a report output, pass the relevant parameter and call this FM.

[↓] Import

PLSCN

Planning scenario of long-term planning. Type: PLSC-PLSCN.

MATNR

Material number. Type: MARC-MATNR.

WERKS

Plant. Type: MARC-WERKS.

BERID

MRP area. Type: MDLV-BERID.

ERGBZ

Selection rule. Type: T462-ERGBZ.

AFIBZ

Name of display filter. Type: T463-AFIBZ.

INPER

Internal period indicator. Type: MDST-INPER.

DISPLAY_LIST_MDPSX

If this parameter is set to X, internal table MDPSX is displayed. Type: CHAR1.

DISPLAY_LIST_MDEZX

If this parameter is set to X, internal table MDEZX is displayed. Type: CHAR1.

DISPLAY_LIST_MDSUX

If this parameter is set to X, internal table MDSUX is displayed. Type: CHAR1.

NOBUF

If this parameter is set to X, then data isn't read from memory. Type: CHAR1.

PLAUF

Type: of planning run. Type: CM61X-PLAUF.

[↑] Export

E_MT61D

Material master MRP data is returned in this parameter. Type: MT61D.

E_MDKP

Header data for the MRP document is returned in this parameter. Type: MDKP.

E_CM61M

Control indicator in the MRP level. Type: CM61M.

E_MDSTA

Statistics fields for stocks. Type: MDSTA.

E_ERGBZ

Selection rule. Type: SFILT-ERGBZ.

Tables

MDPSX

Item in MRP document data is returned in this table. Type: MDPS.

MDEZX

Individual lines of the MRP elements data is returned in this table. Type: MDEZ.

MDSUX

Total lines for MRP elements data is returned in this table. Type: MDSU.

[!] Exceptions

MATERIAL_PLANT_NOT_FOUND

Material is not valid for the plant.

PLANT_NOT_FOUND

Plant is not found.

---

704 NUMERIC_CHECK

This FM is used to edit and check numeric fields. The return type is NUMC if the input value is numeric, or it is CHAR if the input value contains characters as well.

Example: You have a parameter in the selection screen. To validate whether the value entered is numeric or not, pass the value in the import parameter and call this FM.

[↓] Import

STRING_IN

Input value.

[↑] Export

STRING_OUT

Output value.

HTYPE

Data type NUMC or CHAR returned. Type: DD01V-DATATYPE.

705 QF05_RANDOM

This FM generates random numbers between 0 and 1. The random numbers are generated by the linear congruence method.

Example: You want to generate the random number in your program; for example, you may want to reset the password to have an auto-generated password with numbers. You can pass the parameter and call this FM to generate the random number. The below formula is linear congruence generator formula for generating random numbers.

*X(n) = A * X(n-1) + C mod M*
RAN_NUMBER = X(n) / M

The concrete parameters are

*M = 2**32 (modulus)*
A = 67301 (multiplier)
C = 1 (increment)

The initial value X(n-1) can be determined in various ways. If you need a sequence of random numbers, you should only specify the initial value in the first call, if at all.

[↓] Import

RAN_SEED

You can pass the start value for the random number generation. Type: QF00-RAN_SEED.

[↑] Export

RAN_NUMBER

Random number generated is returned in this parameter. Type: QF00-RAN_NUMBER.

RAN_SEED

Start value from which random number is generated. Type: QF00-RAN_SEED.

706 QF05_RANDOM_INTEGER

This FM generates the random number, which lies in the range specified in the parameters RAN_INT_MAX and RAN_INT_MIN.

Example: You want to generate the random number in your program; for example, you may want to reset the password to have an auto-generated password with numbers. You can pass the parameter and call this FM to generate the random number.

[↓] Import

RAN_INT_MAX

Greatest required value. Type: QF00-RAN_INT.

RAN_INT_MIN

Smallest required value. Type: QF00-RAN_INT.

[↑] Export

RAN_INT

Random number. Type: QF00-RAN_INT.

---

707 RS_GET_ALL_INCLUDES

This FM returns all of the includes of the program.

Example: You're creating a program where you give the report name in a selection screen parameter and want to get all of the includes of the report program. You can pass the program name in this FM and call this FM to get all of the includes present.

[↓] Import

PROGRAM

Program name whose include you want to determine. Type: SY-REPID.

WITH_INACTIVE_INCLS

If this parameter is set to X, inactive includes are also returned.

WITH_RESERVED_INCLUDES

This parameter is for internal use. Type: SY-INPUT.

Table

INCLUDETAB

This table contains the list of includes present in the program.

---

708 RV_ORDER_FLOW_INFORMATION

This FM reads the sales document flow of sales documents after the delivery and billing.

Example: If you want to get the flow of the sales order after the delivery and billing, you can pass the relevant parameters and execute this FM.

[↓] Import

AUFBEREITUNG

Type of flow editing. The default value is 2.

BELEGTYP

Sales document category. Type: VBUK-VBTYP.

COMWA

Outgoing document for flow processing. This is a mandatory parameter. Type: VBCO6.

NACHFOLGER

Successor display wanted.

N_STUFEN

Number of levels of the successors. Type: VBFA-STUFE.

VORGAENGER

Predecessor wanted.

V_STUFEN

Number of levels of the predecessors. Type: VBFA-STUFE.

[↑] Export

BELEGTYP_BACK

Sales document category returned. Type: VBUK-VBTYP.

Table

VBFA_TAB

Document flow information is returned in this table. Type: VBFA.

[!] Exceptions

NO_VBFA

Sales document flow (VBFA) data is not found.

NO_VBUK_FOUND

Sales document header status and administrative data (VBUK) are not found.

709 RZL_SLEEP

This FM is used to sleep or delay the processing of the program for the specified number of seconds.

Example: You're saving the data to the material database for various material-related tables and reading the same table at the same time, which causes inaccurate data. Ideally, while saving the data to the database, you should use the command `COMMIT WORK` and `WAIT`, but you can also use this FM if you want your program to wait for a specified number of seconds.

[↓] Import

SECONDS

Number of seconds you want the system to wait.

[!] Exception

ARGUMENT_ERROR

Some error occurred.

---

710 SAPGUI_SET_FUNCTIONCODE

This FM sets and simulates the function code or OK code in a command field. This enables you to run screen sequences without user input.

Example: A screen contains a single count field. When the field is changed, the default function code = should be called, which triggers PBO and increments the count by 1. You can pass the relevant parameters and call this FM in the PBO to set the function code.

[↓] Import

FUNCTIONCODE

Function code. This function code is processed as a command field command by SAP GUI. It processes the function codes exactly as you enter them. You can enter "/h" for switching on the debugger or any other command field function code. Function code [Space] isn't processed as ENTER but is ignored. Instead, you must use = as the function code to process ENTER. Type: C.

[!] Exception

FUNCTION_NOT_SUPPORTED

Function code entered isn't supported on this frontend platform.

---

711 SCI_INSPECT_LIST

This FM is used to start the code inspector check for the object specified in the parameter P_OBJECTS.

Example: You want to create a code review tool for your organization where, based on the program name given on the screen, the system displays the code inspector analysis with additional logics of code review. Call this FM and input the object details you want to analyze to display the code inspector result.

[↓] Import

P_USER

Code inspector: person responsible. Type: SCI_USER.

P_OBJECTS

Code inspector: table type for objects. Type: SCIT_OBJS.

P_NAME

Code inspector: element name you want to analyze. Type: SCI_NAME.

P_SHOW_RESULTS

If this parameter is X, analysis results are displayed. Type: FLAG.

[↑] Export

P_RESULTS

Code inspector results structure. Type: SCIT_REST.

P_RESULTS_HD

Header entries of code inspector result table. Type: SCIT_RESH.

P_VARIANT

Code inspector test variant structure. Type: SCI_TSTVAR.

[!] Exceptions

NO_OBJECT

Object is not found.

OBJS_ALREADY_EXISTS

Object set already exists.

NO_DEFAULT_CHECKVARIANT

No default check variant is found.

TOO_MANY_OBJECTS

Too many objects are found.

COULD_NOT_READ_VARIANT

Could not read check variant.

712 STATUS_BUFFER_REFRESH

This FM deletes the internal status management buffer. A status is an indicator which informs the user that a particular result or status has been reached. It also influences the number of business transaction that the user can perform. If you run a business transaction, it sets or deletes one or more statuses for an object.

Example: You're done with the processing of the business transaction and want to refresh the status buffer and release the memory of the buffer.

[↓] Import

I_FREE

If this parameter is set to X, the status buffer is deleted, and the table is released. Type: SY-BATCH.

---

713 STATUS_CHANGE_EXTERN

This FM is used to set or delete the user status. The status change is first executed in the internal buffer, and the changes are written to the database only with the `COMMIT_WORK` statement.

Example: You want to change the user status of the sales order from unstructured to structured when you allocate the sales order for payment. First, you need to get the user status of the sales order. You can pass the relevant parameters in FM STATUS_TEXT_EDIT (FM 724) for the same purpose. After you get the sales order status, which is unstructured in this case, you can change the sales order status to structured by passing the relevant parameters and executing this FM.

[↓] Import

CHECK_ONLY

If this parameter is set to X, no changes are made. If the parameter is set to =, only the validity of the change is checked. Type: XFELD.

CLIENT

Client. Type: SY-MANDT.

OBJNR

Object number of status object. This object number must exist in the status object information table JSTO. Type: JSTO-OBJNR.

USER_STATUS

User status to be set in internal format. Status can be between E0000 and E9999. More detail regarding which status has what meaning can be found in table TJ30. Type: JEST-STAT.

SET_INACT

If this parameter is set to X, status is inactive or deactivated. Type: XFELD.

SET_CHGKZ

If this parameter is set to X, the creation of the change document is activated for the object number. Type: JSTO-CHGKZ.

NO_CHECK

Checkbox field. Type: XFELD.

[↑] Export

STONR

Status number after the status change. Type: TJ30-STONR.

[!] Exceptions

OBJECT_NOT_FOUND

Status object does not exist.

STATUS_INCONSISTENT

Status situation is inconsistent.

STATUS_NOT_ALLOWED

User status not allowed.

---

714 STATUS_CHANGE_INTERN

This FM is used to set or delete the system status. The status change is first executed in the internal buffer, and the changes are written to the database only with the `COMMIT_WORK` statement.

Example: You want to change the system status of the sales order. First, you need to get the system status of the sales order. You can pass the relevant parameters in STATUS_TEXT_EDIT (FM 724) for the same purpose. After you get the sales order status, you can change the sales order status by passing the relevant parameters in this FM and executing it.

[↓] Import

CHECK_ONLY

If this parameter is set to X, no changes are made; if the parameter is =, only the validity of the change is checked. Type: XFELD.

CLIENT

Client. Type: SY-MANDT.

OBJNR

Object number of the status object. This object number must exist in the status object information table JSTO. Type: JSTO-OBJNR.

ZEILE

Predefined line number for the message handler. Type: MESG-ZEILE.

SET_CHGKZ

If this parameter is set to X, the creation of the change document is activated for the object number. Type: JSTO-CHGKZ.

[↑] Export

ERROR_OCCURRED

A general error has occurred.

OBJECT_NOT_FOUND

When the message handler is active, this flag indicates that the object number given could not be found in the table JSTO.

STATUS_INCONSISTENT

When the message handler is active, this flag indicates that an attempt was made to set and delete a status at the same time. Each status may only appear once in the table STAT.

STATUS_NOT_ALLOWED

When the message handler is active, this flag indicates that the status could not be set.

Tables

STATUS

Table of single statuses to be set or deleted. The STAT field contains the internal format of the status. Status is set if the STAT-INACT flag is initial; otherwise, it's deleted. Status can be between I0000 and I9999. Type: JSTAT.

[!] Exceptions

OBJECT_NOT_FOUND

Status object does not exist.

STATUS_INCONSISTENT

Status situation is inconsistent.

STATUS_NOT_ALLOWED

User status not allowed.

---

715 STATUS_CHECK

This FM checks whether the status (user and system based on the import parameter STATUS) is set or not. If the status isn't set, the exception STATUS_NOT_ACTIVE is returned.

Example: You're displaying the sales order and all its active status in a report program, and you want to check whether the status of the sales order is active and display it in the report. You can pass the relevant parameters and call this FM to check this.

[↓] Import

BYPASS_BUFFER

If this parameter is set to X, the status object bypasses the buffer and is read from the database. Type: XFELD.

CLIENT

Client. Type: JEST-MANDT.

OBJNR

Object number. The object number must exist in table JSTO. Type: JEST-OBJNR.

STATUS

System status (Ixxxx) or user status (Exxxx). Type: JEST-STAT.

[!] Exceptions

OBJECT_NOT_FOUND

Status object does not exist.

STATUS_NOT_ACTIVE

Status is not active.

716 STATUS_DELETE_FROM_BUFFER

This FM deletes status objects and status single records from the internal status buffer. This means that when rereading the current status, it is read from the database.

Example: To delete the buffer for some order so that you can read the status from the database, you can pass the relevant parameters and call this FM.

[↓] Import

CLIENT

Client. Type: SY-MANDT.

OBJNR

Object number that is to be deleted from the buffer. Type: JSTO-OBJNR.

OBJNR_GENERIC

Generic object number. All of the object numbers are deleted from the buffer that begin with the content specified in this parameter. Type: JSTO-OBJNR.

[!] Exception

OBJNR_TAB

Type: JSTO_PRE.

717 STATUS_MAINTAIN

This FM is used to maintain or display object statuses. If the FM is called from the list with status objects, the option for the next, first, and so on objects in the list can be activated. Status profile changes aren't allowed if the user status was already active.

Example: You've created a report showing all of the changes you've made in the sales order. You also want to display the status of the sales order in one output column. You can pass the relevant parameters and call this FM to maintain the status.

[↓] Import

CLIENT

Client. Type: SY-MANDT.

DISPLAY_ONLY

If this parameter is set, objects can't be changed and are in display status only. Type: XFELD.

FLG_NO_STSMA_CHANGE

If this parameter is set to X, no changes can be made to the status profile. Otherwise, changes can be made if the user status of the old status profile was not active. Type: XFELD.

FLG_OBJECT_LIST

A list of status objects is to be processed. If this flag is set to X, additional functions list first object, previous object, next object, and last object are displayed in the pushbutton. When these functions are confirmed, you exit status maintenance, and a return code is transferred to the parameter EXIT_CODE and the FLG_* flags parameters. The caller can then recall the FM for the appropriate object. Type: XFELD.

OBJNR

Object number. Type: JSTO-OBJNR.

SUBSCREEN_EXT

Screen number for screen displaying object-specific additional information. Possible values can be from 0000 to 9999. Type: SY-DYNNR.

SUBSCREEN_PROG_EXT

Module pool name for subscreen displaying object-specific additional information. Type: SY-REPID.

SET_CHGKZ

If this parameter is set to X, the creation of a change document is activated for the object number. Type: JSTO-CHGKZ.

FLG_NO_SYSTEMSTATUS

If this parameter is set to X, it suppresses the display of system status. Type: XFELD.

FLG_NO_USERSTATUS

If this parameter is set to X, it suppresses the display of user status. Type: XFELD.

STSMA_ALWAYS_CHANGABLE

Schema can be changed if user status is already archive. Type: XFELD.

[↑] Export

FLG_FIRST_OBJ

Status maintenance was exited with the function key FIRST OBJECT. Call the FM then with the last object in your list. This parameter has the value X if the first object pushbutton is selected. Type: XFELD.

FLG_LAST_OBJ

Status maintenance was exited with the function key LAST OBJECT. Call the FM then with the last object in your list. This parameter has the value X if the last object pushbutton is selected. Type: XFELD.

FLG_NEXT_OBJ

Status maintenance was exited with the function key NEXT OBJECT. Call the FM then with the last object in your list. This parameter has the value X if the next object pushbutton is selected. Type: XFELD.

FLG_PREVIOUS_OBJ

Status maintenance was exited with the function key PREVIOUS OBJECT. Call the FM then with the last object in your list. This parameter has the value X if the previous object pushbutton is selected. Type: XFELD.

STONR

Status number after ending the status maintenance. Possible values are from 00 to 99. Type: TJ30-STONR.

EXIT_CODE

OK code for exiting the FM. Possible values:

- EEND: Exit with Exit button.
- BACK: Exit with Back button.
- ABBR: Exit with Cancel button.
- FOBJ: Exit with First Object button.
- LOBJ: Exit with Last Object button.
- POBJ: Exit with Previous Object button.
- NOBJ: Exit with Next Object button.

Type: RSEUL_KEYS-CODE.

[!] Exception

OBJECT_NOT_FOUND

Object status not found in table JSTO.

---

718 STATUS_NUMBER_CONVERSION

This FM returns the 4-character, language-dependent abbreviation and the 30-character, language-dependent long text for a status in internal format. This FM is a counterpart of FM STATUS_TEXT_CONVERSION (FM 723).

Example: You want to get the language-dependent abbreviation of the internal status and display it on the screen so you can see how a user will view the text for status I0001. You can pass the relevant parameter and execute this FM to get the status CRTD (Created).

[↓] Import

CLIENT

Client. Type: SY-MANDT.

LANGUAGE

Language in which you require status conversion. Type: SY-LANGU.

OBJNR

The object number is only needed to determine the status profile. If parameter STSMA is filled or if only the text for the status profile is required, this parameter isn't required to be filled. Type: JSTO-OBJNR.

STATUS_NUMBER

Status in internal format. Possible values are from I0000 to I9999 for system status and E0000 to E9999 for user status. Type: JEST-STAT.

STSMA

Status profile for user status. This parameter is filled if the user status is given and the object number isn't available. This parameter value is checked from table TJ20. Type: JSTO-STSMA.

[↑] Export

LANGUAGE_FOUND

Language in which parameter text TXT04 and TXT30 is found. First the attempt is made to deliver the texts in the required language, according to parameter LANGUAGE. If the texts don't exist in this language, they are delivered in the maintenance language of the status profile (user status) or in German (system status). Type: SY-LANGU.

TXT04

This 4-character, language-dependent status ID is displayed in the status line. Type: TJ02T-TXT04.

TXT30

This 30-character, language-dependent text is displayed in this table. Type: TJ02T-TXT30.

[!] Exceptions

INSUFFICIENT_INPUT

Neither the STSMA parameter nor the OBJNR parameter is filled. An attempt was made to fetch a text for a user status, but neither the object number nor the status profile was given.

OBJECT_NOT_FOUND

The object number does not exist in table JSTO.

STATUS_NOT_FOUND

The status does not exist. A system status must be entered in table TJ02 (maintain using Transaction BS22); a user status must be entered in table TJ30 (maintain using Transaction BS02).

STSMA_NOT_FOUND

Status profile not found.

719 STATUS_OBJECT_CREATE

This FM generates the status object and takes the object number from the object number management. If no object number is given, the FM takes a temporary object number that must be replaced later.

Example: You want to create an internal order for status profile STATSCHEM. You can pass the relevant parameter in this FM and call this FM to create the status object. If the object number isn't given, you should call FM STATUS_OBJECT_SWITCH_NUMBER (FM 720) to generate the temporary object number and save it to the database.

[↓] Import

CHGKZ

If this parameter is set to X, the creation of a change document is activated for the object number. Type: JSTO-CHGKZ.

CLIENT

Client. Type: SY-MANDT.

OBJNR

Object number. Type: JSTO-OBJNR.

OBTYP

Object category for which the status object is to be created. Object type should be present in table TJ03. Type: JSTO-OBTYP.

STSMA

Status profile for the user status. This parameter is filled if the user status is given and the object number isn't available. This parameter value is checked from table TJ20. Type: JSTO-STSMA.

IONRA_IMP

This parameter isn't used now.

I_OBJECTKEY

Object key for object number management.

ONLY_E_STATUS

If this parameter is checked, it checks only the user status. Type: XFELD.

[↑] Export

OBJNR

Object number returned from the object number management (FMs OBJECT_NUMBER_GET_xx). Entry of the object number is optional. If one isn't entered, a temporary object number (TMxxxxxxx) is given by the system.

These temporary objects can't be written to the database, so before the `COMMIT WORK`, the temporary object numbers must be replaced with real ones using the FM STATUS_OBJECT_SWITCH_NUMBER. Type: JSTO-OBJNR.

STONR

Status number. If the user status with status number is present, it's displayed; otherwise, 00 is returned. Type: TJ30-STONR.

[!] Exceptions

OBTYP_INVALID

Invalid object type.

STATUS_OBJECT_ALREADY_EXISTS

Status object already exists.

STSMA_INVALID

Invalid status profile.

STSMA_OBTYP_INVALID

Status profile not created for object type.

---

720 STATUS_OBJECT_SWITCH_NUMBER

This FM replaces temporary object numbers with real object numbers in the status buffer. This FM should be called before `COMMIT_WORK` to generate the permanent object from the temporary object and saving them to the database.

Example: You've created an internal order for status profile STATSCHEM by using FM STATUS_OBJECT_CREATE (FM 719). You've received a temporary object number from FM STATUS_OBJECT_CREATE. You should call this FM to generate the permanent object number from the temporary object number and save it to the database.

[↓] Import

CLIENT

Client. Type: SY-MANDT.

OBJNR_NEW

New object number. This object number is assigned by the object number management (FMs OBJECT_NUMBER_GET_xx). Type: JSTO-OBJNR.

OBJNR_OLD

Old (temporary) object number. The object number must begin with "TM". It's delivered by the FM STATUS_OBJECT_CREATE if this is called without entering an object number. Type: JSTO-OBJNR.

[!] Exceptions

OBJECT_NOT_FOUND

Old object number does not exist.

OBJNR_NEW_INVALID

New object number is temporary.

OBJNR_OLD_INVALID

Old object number is not temporary.

STATUS_OBJECT_ALREADY_EXISTS

New status object already exists.

---

721 STATUS_PRE_READ

This FM imports the status of several objects into the internal array buffer. With this FM, you can read the single status (i.e., JEST file) and status management records (i.e., JSTO file) of several status objects.

Example: You want to read the status management record for the status and display them in a report. You can pass the relevant parameters in this FM and call it to load the status of all of the objects at once into the buffer, and then you can read an individual status from FM STATUS_READ (FM 722). This will help improve performance of your program because now the required statuses are already stored in the buffer, so you won't hit the database for a single status.

[↓] Import

CLIENT

Client. Type: SY-MANDT.

GET_CHANGE_DOCUMENTS

Load change documents in buffer also. Type: SY-DATAR.

[↑] Export

NUMBER_OF_FETCHES

This counter indicates how many status objects were read from the database in the internal status buffer. Objects already in the buffer aren't read a second time. Object numbers that don't exist are ignored. Type: SY-TABIX.

Table

JSTO_PRE_TAB

Object number table from which objects are read. Type: JSTO_PRE.

---

722 STATUS_READ

The FM reads a status object in the status management internal buffer and delivers all active single statuses of the status object.

Example: You want to read the status from the internal buffer so that you can change the status from one user status to another and display it in the report program output. You can pass the relevant parameters in this FM and call it to read the status. After you receive the relevant status, you can pass it to FM STATUS_CHANGE_EXTERN (FM 714) to change the status.

[↓] Import

CLIENT

Client. Type: SY-MANDT.

OBJNR

Object number. Type: JSTO-OBJNR.

ONLY_ACTIVE

If this parameter is set as X, only the active status is returned. Type: JEST-INACT.

[↑] Export

OBTYP

Object category. Type: JSTO-OBTYP.

STSMA

Status profile assigned to the object. Type: JSTO-STSMA.

STONR

Status number of the currently active user status. If the status isn't active, value 00 is returned. Type: TJ30-STONR.

Table

STATUS

Table of individual statuses for the object. STATUS-STAT contains the status in internal format (Ixxxx, Exxxx). If the STATUS-INACT field is [Space], the status is active; if it's X, the status is inactive. Type: JSTAT.

[!] Exception

OBJECT_NOT_FOUND

Status object not found.

---

723 STATUS_TEXT_CONVERSION

This FM converts the four-character, language-dependent abbreviation into an internal format (e.g., Exxx or Ixxx). This FM is a counterpart of FM STATUS_NUMBER_CONVERSION (FM 718).

Example: You want to get the internal status based on the language-dependent abbreviation and save it into the database; for example, you might want to save the internal format for status CRTD (Created). You can pass the relevant parameter in this FM and execute this FM to get the status I0001, which can be saved in the database.

[↓] Import

CLIENT

Client. Type: SY-MANDT.

LANGUAGE

Language of TXT04. Type: SY-LANGU.

MODE

Mode: I (system status), E (user status).

OBJNR

Object number, only evaluated if the status is a user status (mode = E), and the status profile isn't given. Type: JSTO-OBJNR.

STSMA

Status profile for user status (mode = E). Type: JSTO-STSMA.

TXT04

The four-character, language-dependent status abbreviation. Type: TJ02T-TXT04.

[↑] Export

LANGUAGE_FOUND

Language in which the 4-character, language-dependent status abbreviation was found. Type: SY-LANGU.

STATUS_NUMBER

Internal status number (Ixxxx or Exxxx) found for the status abbreviation. Type: JEST-STAT.

[!] Exceptions

INSUFFICIENT_INPUT

When searching for the user status, either the object number or the status profile must be given, so that the search can be made in the correct status profile. This is necessary because neither the internal status numbers nor the language-dependent status abbreviation can be uniquely identified via the status profile (MODE = E).

NOT_FOUND

The status with the abbreviation entered does not exist.

OBJECT_NOT_FOUND

The status object does not exist (MODE = E).

WRONG_MODE

The parameter MODE does not have value as I or E.

---

724 STATUS_TEXT_EDIT

This FM converts the 4-character language-dependent status to a 40-character line and displays it. This FM is used both for system status and user status.

Example: You want to change the user status of the sales order from unstructured to structured when you allocate the sales order for payment. First you need to get the user status of the sales order. You can pass the relevant parameters in this FM for the same purpose.

[↓] Import

CLIENT

Client (only for exceptional cases). Type: SY-MANDT.

FLG_USER_STAT

If this parameter is set to X, the user status line is also edited; otherwise, only the system status is edited. Type: XFELD.

OBJNR

Status object number. Type: JEST-OBJNR.

ONLY_ACTIVE

If this flag isn't set, all of the statuses that were sometime active but currently inactive are taken into consideration. Type: XFELD.

SPRAS

Language in which statuses are edited. Type: SY-LANGU.

BYPASS_BUFFER

If this parameter is set to X, the status line reflects the status of the object in the database. The status buffer isn't changed by this call. Type: XFELD.

[↑] Export

ANW_STAT_EXISTING

This flag is set to X if at least one user status is active or has been active. Type: XFELD.

E_STSMA

This parameter returns the status profile for the status object. If no status profile is assigned, this parameter has value [Space]. Type: JSTO-STSMA.

LINE

The 40-character text field with the prepared-for system status line. Type: BSVX-STTXT.

USER_LINE

The 40-character text field with the prepared-for user status line. Type: BSVX-STTXT.

STONR

Status number of the currently active user status. If the status isn't active, value 00 is returned. Type: TJ30-STONR.

[!] Exception

OBJECT_NOT_FOUND

Status object does not exist.

725 SWC_ELEMENT_GET

This FM is used to read an ABAP field from a container. A container is used with reference to business objects and workflows. The container contains the import and export parameters associated with any method of the business object at runtime. Whenever you're calling any method of business object, you need to populate the container for import parameters. After the method gets executed, it returns the values (export parameters) in the container. The business object container is of structure SWCONT.

Example: You've created a workflow for material creation, and you want to read the PLANT field from the container element. Pass the relevant parameters and call this FM to read the ABAP field.

[↓] Import

ELEMENT

Container element.

[↑] Export

FIELD

ABAP/4 field.

Table

CONTAINER

Container instance. Type: SWCONT.

[!] Exceptions

IS_NULL

Null value.

NOT_FOUND

Element not found.

TYPE_CONFLICT

Type or length conflict.

726 SWC_ELEMENT_SET

This FM is used to write an ABAP field in a container.

Example: You want to call the business object programmatically and display the material that you've created. You need to pass the PLANT field in the business object importing parameter in the business object container. You can do so by passing the parameter in this FM and writing it to the business object container.

[↓] Import

ELEMENT

Container element.

FIELD

ABAP/4 field.

Tables

CONTAINER

Container instance. Type: SWCONT.

[!] Exception

TYPE_CONFLICT

Type or length conflict.

727 SWE_EVENT_CREATE

This FM triggers an event for the business object. This event must be defined in the corresponding object type. The triggering of an event documents a change to the status of an object. The change to the status should be executed on the database before the corresponding event is triggered. If an event is triggered before the database change, then the event receivers that are registered there are also started. However, if an event receiver uses start conditions (e.g., in the form of a check FM) that access the object attributes, then this can lead to incorrect results because outdated or obsolete values need to be accessed.

Example: To trigger the CREATE event of a material business object from your program because you're creating a material, pass the relevant parameters in this FM and call it.

[↓] Import

OBJTYPE

Type of triggering object. Type: SWETYPECOU-OBJTYPE.

OBJKEY

ID (key) of triggering object. Type: SWEINSTCOU-OBJKEY.

EVENT

Name of event. Type: SWETYPECOU-EVENT.

CREATOR

Event creator. Type: SWHACTOR.

TAKE_WORKITEM_REQUESTER

If this parameter is set to X, the last work item is used as the requester. Type: SWEFLAGS-WIREQFLAG.

START_WITH_DELAY

Flag for delayed start of a RFC. Type: SWEFLAGS-DELAYFLAG.

START_RECFB_SYNCHRON

Flag for a synchronous call to the receiver FM. Type: SWEFLAGS-SYNCFLAG.

NO_COMMIT_FOR_QUEUE

Entry in event queue without the explicit COMMIT. Type: SWEFLAGS-COMMITFLAG.

DEBUG_FLAG

Flag for debugging in the receiver FM. Type: SWEFLAGS-DEBUGFLAG.

NO_LOGGING

Flag for no logging. Type: SWEFLAGS-LOGFLAG.

IDENT

ID of event (only if previously in queue). Type: SWEDUMEVID-EVTID.

[↑] Export

EVENT_ID

ID of triggered event. Type: SWEDUMEVID-EVTID.

RECEIVER_COUNT

ID of an event instance. Type: SWEDUMEVID-EVTID.

Table

EVENT_CONTAINER

Data container of event. Type: SWCONT.

[!] Exception

OBJTYPE_NOT_FOUND

Triggering object type not found.

---

728 SWC_TABLE_GET

This FM is used to read an internal table from the container.

Example: You've created a workflow where you want to read the internal table with work centers from the business object container element. You can pass the relevant parameters and call this FM to read the internal table.

[↓] Import

ELEMENT

Container element.

Tables

CONTAINER

Container instance. Type: SWCONT.

TABLE

Internal table read from container.

[!] Exceptions

IS_NULL

Null value.

NOT_FOUND

Element not found.

TYPE_CONFLICT

Type or length conflict.

729 SWC_TABLE_SET

This FM is used to write an internal table in the container.

Example: You want to call the business object programmatically, so you need to pass the business object table parameter in the business object container. You can do so by passing the table parameter in this FM and writing it to the business object container.

[↓] Import

ELEMENT

Container element.

Tables

CONTAINER

Container instance. Type: SWCONT.

TABLE

Internal table to be passed to container.

[!] Exception

TYPE_CONFLICT

Type or length conflict.

730 SWO_CREATE

This FM is used to instantiate the business object from the program. The other FMs you'll use in this process are listed here:

- SWO_INVOKE (FM 731)
- SWO_FREE (FM 732)

Example: You want to call the business object method from the program. First, you need to call this FM to instantiate the business object.

You'll get the runtime information of the business object. See the listed FMs for more information.

[↓] Import

OBJTYPE

Business object type. Type: SWOTOBJID-OBJTYPE.

OBJKEY

Business object key. Type: SWOTOBJID-OBJKEY.

OBJNAME

Business object name. Type: TOJTB-EDITELEM.

LOGICAL_SYSTEM

Logical system name. Type: SWOTOBJID-LOGSYS.

[↑] Export

OBJECT

Runtime object name is returned in this parameter. Type: SWOTRTIME-OBJECT.

RETURN

Return value. Type: SWOTRETURN.

[!] Exception

NO_REMOTE_OBJECTS

No remote object found.

---

731 SWO_INVOKE

This FM calls or starts the business object method.

Example: To call the business object method from the program, pass the relevant parameter in this FM.

[↓] Import

ACCESS

Access to an attribute. Value G is to get the attribute, and value P is to set the attribute. Value C is for invoking the method. The default value is C. Type: SWOTP-ACCESS.

OBJECT

Runtime object. This is a mandatory parameter. Type: SWOTRTIME-OBJECT.

VERB

Object type component. Type: SWOTINVOKE-VERB.

PERSISTENT

Persistent object IDs (type = o). Type: SWOTP-PERSISTENT.

REQUESTER

Requester object ID. Type: SWOTOBJID.

SYNCHRON

Synchronous mode. Type: SWOTINVOKE-SYNCHRON.

UNSORTED_CONTAINER

Container hasn't been sorted, so the FM must do the sorting. Type: SWOTINVOKE-SYNCHRON.

NO_ARFC

Don't execute ARFC. Type: SEU_BOOL.

[↑] Export

RETURN

Return value. Type: SWOTRETURN.

VERB

Object type component. Type: SWOTINVOKE-VERB.

MODE_ID

ID of new mode with an asynchronous call. Type: SY-INDEX.

Table

CONTAINER

Business object container. Type: SWCONT.

---

732 SWO_FREE

This FM is used to release the business object method.

Example: You've already invoked the business object method and are finished using it. To release the business object method, pass the relevant parameter in this FM.

[↓] Import

OBJECT

Runtime object. Type: SWOTRTIME-OBJECT.

[↑] Export

RETURN

Return value. Type: SWOTRETURN.

The Author

Tanmaya Gupta is a senior SAP consultant and an SAP ABAP developer working in a well-known organization in India. He has more than five years of dynamic and challenging SAP experience. Currently, he is working in ABAP development in the area of SAP manufacturing execution and production planning. He has experience with basic and advanced topics of ABAP including reports, dialog programming, forms, function modules, ABAP Data Dictionary, queries, LSMW, IDocs, OOPs, Web Dynpro, enhancements, and mobility. He has been involved in many development, implementation, and support projects.

Tanmaya holds a bachelor of technology degree in computer science and engineering and is currently pursuing his master of technology in information technology. He is the author of *ABAP Data Dictionary* (SAP PRESS, 2011), which he considers one of his biggest achievements.

Tanmaya currently lives in Pune, India, and loves to involve himself in activities that help him to diversify his experience. He likes swimming and playing chess.

Index

A

C

D

G

H

I

J

K

L

M

P

Q

R

T

V

W

X

- Provides 100 little-known, time-saving tips and tricks
- Features step-by-step instructions and guiding screenshots
- Helps you use the ABAP Workbench more effectively

Abdulbasit Gülsen

ABAP Workbench

100 Things You Should Know About...

Don't hesitate in learning how to perform ABAP Workbench tasks more easily and efficiently than ever before! Whether you're a beginner or advanced user, this book provides tips and tricks that give you different and valuable ways of working with the tool. Based on your specific needs, you'll easily navigate the 100 tips and workarounds to help you increase productivity, save time, and improve the overall ease-of-use of working with the ABAP Workbench.

341 pp., 2012, 49,95 Euro / US$ 49.95
ISBN 978-1-59229-427-5
www.sap-press.com

Galileo Press